The Pilgrim Art

THE CALIFORNIA WORLD HISTORY LIBRARY

Edited by Edmund Burke III, Kenneth Pomeranz, and Patricia Seed

1. *The Unending Frontier: Environmental History of the Early Modern World,* by John F. Richards
2. *Maps of Time: An Introduction to Big History,* by David Christian
3. *The Graves of Tarim: Genealogy and Mobility across the Indian Ocean,* by Engseng Ho
4. *Imperial Connections: India in the Indian Ocean Arena, 1860–1920,* by Thomas R. Metcalf
5. *Many Middle Passages: Forced Migration and the Making of the Modern World,* edited by Emma Christopher, Cassandra Pybus, and Marcus Rediker
6. *Domesticating the World: African Consumerism and the Genealogies of Globalization,* by Jeremy Prestholdt
7. *Servants of the Dynasty: Palace Women in World History,* edited by Anne Walthall
8. *Island World: A History of Hawai'i and the United States,* by Gary Y. Okihiro
9. *The Environment and World History,* edited by Edmund Burke III and Kenneth Pomeranz
10. *Pineapple Culture: A History of the Tropical and Temperate Zones,* by Gary Y. Okihiro
11. *The Pilgrim Art: Cultures of Porcelain in World History,* by Robert Finlay
12. *The Quest for the Lost Nation: Writing History in Germany and Japan in the American Century,* by Sebastian Conrad; translated by Alan Nothnagle
13. *The Eastern Mediterranean and the Making of Global Radicalism, 1860–1914,* by Ilham Khuri-Makdisi
14. *The Other West: Latin America from Invasion to Globalization,* Marcello Carmagnani

The Pilgrim Art

Cultures of Porcelain in World History

Robert Finlay

UNIVERSITY OF CALIFORNIA PRESS

Berkeley Los Angeles London

University of California Press, one of the most distinguished university presses in the United States, enriches lives around the world by advancing scholarship in the humanities, social sciences, and natural sciences. Its activities are supported by the UC Press Foundation and by philanthropic contributions from individuals and institutions. For more information, visit www.ucpress.edu.

University of California Press
Berkeley and Los Angeles, California

University of California Press, Ltd.
London, England

Library of Congress Cataloging-in-Publication Data
Finlay, Robert, 1940–.
The pilgrim art : Cultures of porcelain in world history / Robert Finlay.
p. cm.
Includes bibliographical references and index.
ISBN 978-0-520-24468-9 (cloth : alk. paper)
1. Porcelain industry—China—Social aspects—History. 2. Porcelain, Chinese—Social aspects—History. 3. Porcelain—Social aspects—History. 4. Art and society. I. Title.
HD9610.8.C62F56 2010
338.4'766650951—dc22 2009040698

Manufactured in the United States of America

23 22 21 20 19
10 9 8 7 6 5 4 3 2

To Caitlin

CONTENTS

ILLUSTRATIONS

MAPS

PLATES

following page 172

NOTE ON TERMINOLOGY

CERAMICS

The distinctions among *earthenware, stoneware,* and *porcelain,* in terms of appearance, firing range, and material composition, are explained in the opening section of chapter 3. Earthenware covered with a tin-based glaze is known as *maiolica* in Spain and Italy, *faience* in France, and *delftware* in Holland and England. Although there are only negligible differences among these wares, the terms are employed here for variety of language and ease of exposition. *Chinaware* always refers to porcelain exported from China.

GEOGRAPHY

Southwest Asia is used instead of *Middle East; western Asia* encompasses both Southwest Asia and Europe from the period after the collapse of the Roman Empire in the fifth century. Maritime Southeast Asia, the chain of islands stretching from the Philippines to Sumatra, is also called the *Archipelago. Eurasia* is characterized as the *ecumene*—from an ancient Greek term for the "inhabited" or known quarters of the world—and should be understood as including the northern and eastern coasts of Africa.

PRINCIPAL CHINESE DYNASTIES

Shang, ca. 1700–1027 B.C.E.
Zhou, 1027–221 B.C.E.
Qin, 221–207 B.C.E.

Han, 206 B.C.E.–220 C.E.
Sui, 581–618
Tang 618–906
Song, 960–1279
Northern Song, 960–1127
Southern Song, 1127–1279
Yuan (or Mongol), 1279–1368
Ming, 1368–1644
Qing (or Manchu), 1644–1911

ACKNOWLEDGMENTS

I should like to thank Mimi Gardner Gates, director of the Seattle Art Museum, for inviting me to participate in the conference "Porcelain Stories from China to Europe" in March 2000. As she and Julie Emerson, curator of decorative arts, explained, an essay I wrote on the global influence of Chinese porcelain played a role in inspiring the exhibition. The conference represents my only exposure to the congenial world of porcelain scholars and aficionados, so I am especially grateful for the invitation. Moreover, the museum kindly permitted me to reproduce images from the exhibition catalogue for the present book—the sort of round-robin exchange that nicely reflects the circuit of cultural influence typical of porcelain across the centuries. A glance at the references makes clear how much this book depends on the work of countless students of porcelain. As an outlander to the subject, I brazenly used their publications for my own purposes while never forgetting how much I am indebted to the labors of those more expert authorities. I wish to thank George Huppert, editor of the *Journal of the Historical Society (TJHS)* for permission to incorporate parts of my article, "The Voyages of Zheng He: Ideology, State Power, and Maritime Trade in Ming China," *TJHS* 8/3 (2008): 327–47, into chapter 7 of this book. I am also grateful to Jerry Bentley, editor of the *Journal of World History (JWH),* for permission to use an article, with the same title as the present work, published in *JWH* 9/2 (1998): 141–87. Rose Kerr, formerly Keeper of the Far Eastern Department at the Victoria and Albert Museum, London, and an expert on ceramic technology, provided helpful criticism of that article when we met at Cambridge University in 1998. I am also obliged to the Dean's Office of Fulbright College, University of Arkansas, for a grant to assist in the collection of images from the Seattle Art Museum, the Cleveland Museum of Art, and the Peabody Essex Museum, Salem, Massachusetts. Jiang Jin, once my graduate student and now a faculty member at East China Normal University, Shanghai, provided invaluable assistance in helping me deal with materials in the Chinese language. William H. McNeill greatly encouraged me after reading the manuscript, recognizing that although the discussion at times seemed to stray from the subject of porcelain, it turned out to be carefully plotted meandering. In addition, his prodigious labors as a world historian have been my model of excellence since he served long ago as my dissertation director at the University of Chicago.

My daughter Conwill sustained me with her companionship and good cheer during the writing of this book. My family at all times has shown heartening interest in my research, despite the curious byways down which it sometimes led me. Constance has been an invariable source of support and affection, and, as always, she remains my ideal reader. I owe a great deal to our daughter Adrianne, who applied her exceptional writing skills to editing the text and eliminating many stylistic failings. At every stage of my work, her sister, Caitlin, has helped track down out-of-the-way references and publications, despite the demands on her time in Olin Library at Cornell University, the marvelous and hospitable institution where much of my research was carried out. Dedicating this book to Caitlin is the most pleasing aspect of having finished it at last.

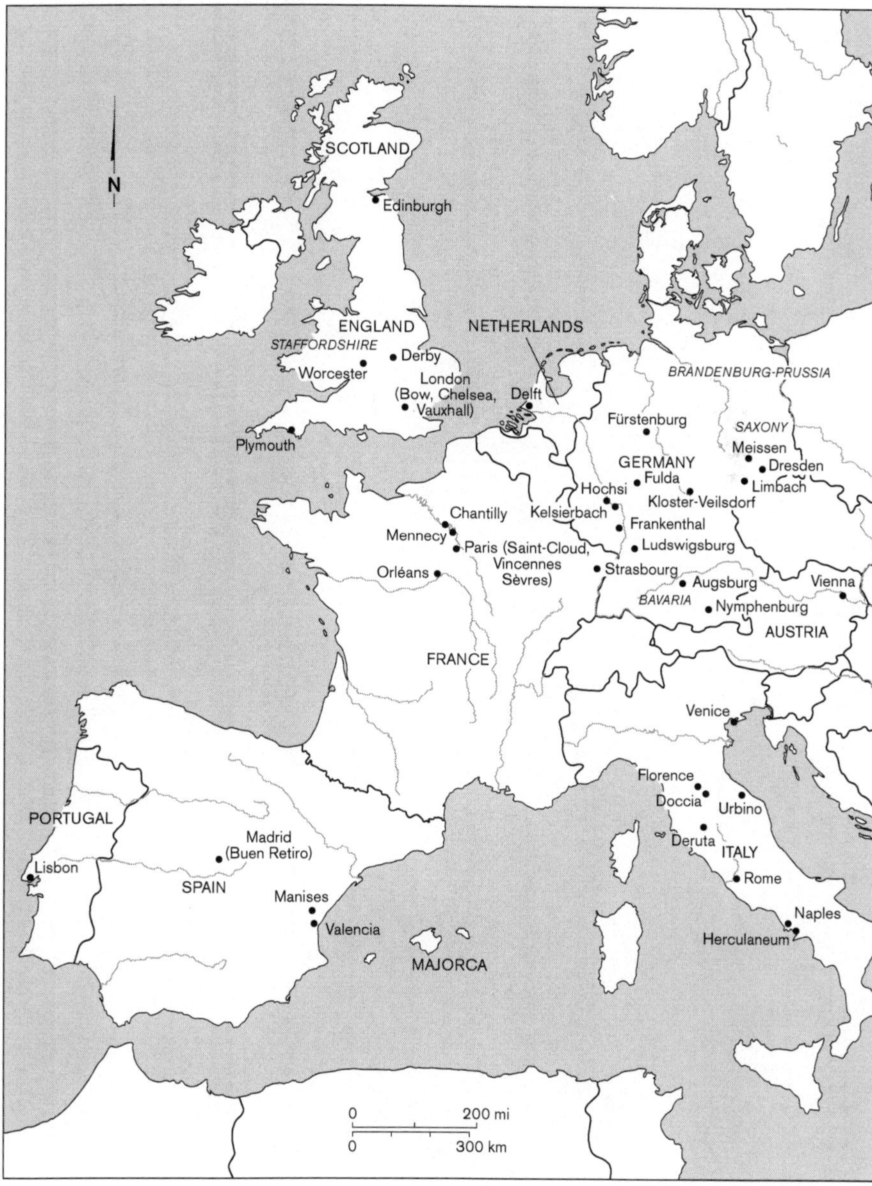
N
SCOTLAND
Edinburgh
ENGLAND
STAFFORDSHIRE
Derby
Worcester
London
(Bow, Chelsea,
Vauxhall)
Plymouth
NETHERLANDS
Delft
BRANDENBURG-PRUSSIA
Fürstenburg
SAXONY
Meissen
Dresden
GERMANY
Fulda
Limbach
Hochsi
Kloster-Veilsdorf
Kelsierbach
Frankenthal
Ludswigsburg
Strasbourg
Augsburg
BAVARIA
Nymphenburg
Vienna
AUSTRIA
Chantilly
Mennecy
Paris (Saint-Cloud,
Vincennes
Sèvres)
Orléans
FRANCE
Venice
Florence
Doccia
Urbino
Deruta
ITALY
Rome
Naples
Herculaneum
PORTUGAL
Lisbon
Madrid
(Buen Retiro)
SPAIN
Manises
Valencia
MAJORCA
0
200 mi
0
300 km

MAP 1. Europe

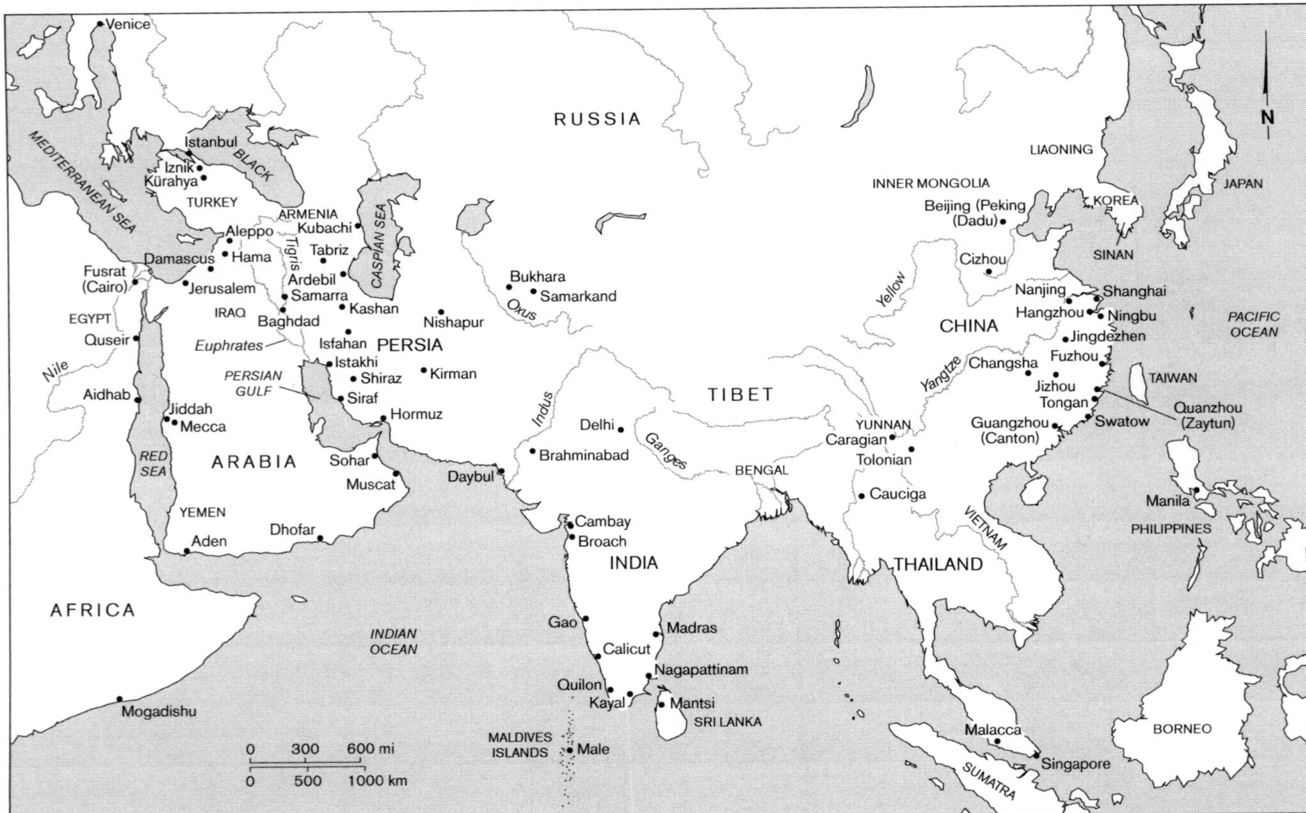
N
Venice
MEDITERRANEAN SEA
Istanbul
BLACK
Iznik
Kürahya
TURKEY
ARMENIA
Kubachi
CASPIAN SEA
Aleppo
Hama
Tigris
Tabriz
Damascus
Fusrat
(Cairo)
Jerusalem
Ardebil
Samarra
Kashan
EGYPT
IRAQ
Baghdad
Nishapur
Quseir
Euphrates
Isfahan
PERSIA
Istakhi
Shiraz
Kirman
Nile
PERSIAN
GULF
Siraf
Aidhab
Jiddah
Mecca
Hormuz
RED
SEA
ARABIA
Sohar
Muscat
Daybul
YEMEN
Dhofar
Aden
AFRICA
INDIAN
OCEAN
Mogadishu
0
300
600 mi
0
500
1000 km
RUSSIA
Bukhara
Samarkand
Oxus
Indus
Delhi
Brahminabad
Ganges
TIBET
BENGAL
Cambay
Broach
INDIA
Gao
Madras
Calicut
Nagapattinam
Quilon
Kayal
Mantsi
SRI LANKA
MALDIVES
ISLANDS
Male
LIAONING
INNER MONGOLIA
Beijing (Peking
(Dadu)
KOREA
JAPAN
SINAN
Cizhou
Yellow
Nanjing
Shanghai
Hangzhou
Ningbu
CHINA
Jingdezhen
PACIFIC
OCEAN
Fuzhou
Yangtze
Changsha
Jizhou
Tongan
TAIWAN
Quanzhou
(Zaytun)
Swatow
YUNNAN
Caragian
Guangzhou
(Canton)
Tolonian
Cauciga
Manila
PHILIPPINES
VIETNAM
THAILAND
Malacca
Singapore
SUMATRA
BORNEO

MAP 2. Asia

Introduction

In 1598 Philip II of Spain was buried in the Escorial palace north of Madrid in a coffin made from the keel of the *Cinco Chagas de Cristo,* a vessel that had served as the flagship of five viceroys of Goa in India, the center of the Portuguese maritime empire in Asia. Sailing for the Portuguese crown for over a quarter of a century, the teak-built carrack had made about nine round-trip voyages between Goa and Lisbon, twice as many as the usual transport. The two legs of the *carreira da Índia,* "roadway to India," added up to 37,000 kilometers, a journey that took at least eighteen months and levied a frightful toll in men and vessels. Although Portuguese seamen piously declared that "God takes them out and God brings them back," the number shipwrecked or lost on the return voyage, when captains invariably overloaded their vessels with Asian merchandise, was disproportionately great.[1] Perhaps Philip, who believed that Providence guided his realm, considered that the fortunate *Cinco Chagas,* named for the "Five Wounds" of the Crucifixion, had benefited from the same dispensation. The great carrack also evoked a global vista that appealed to the king, for mariners celebrated it as a remarkable link between East and West, connecting the far sides of the world just as the lordship of Philip himself had done in life. The monarch, who paid exacting attention to mortuary details, evidently regarded his carrack coffin in the claustrophobic, subterranean vault of the Escorial as an emblem of his wide-ranging dominion.

The *Cinco Chagas* had been moored in Lisbon harbor for some years before Philip II died, serving a degrading retirement as a demasted storage hulk. The monarch could appropriate its keel for his tomb because twenty years earlier he had seized Portugal after King Sebastian I (r. 1557–78), the last of the Avis dynasty, and seven thousand of his nobles were slaughtered at the battle of Alcázar-Quibir in

Morocco. Uniting Portuguese and Spanish territories in his own person, with possessions in Europe, the Americas, Africa, India, and Southeast Asia, Philip thus came to rule the first global empire. One of the triumphal arches lining his ceremonial entry into Lisbon in 1581 carried a legend proclaiming him "lord of everything in the East and West."[2] For contemporaries who shared his pious outlook, Philip's power and wealth seemed to bring within reach the ancient Christian dream of universal imperium, mankind united under one crown and one faith; hope soared among them that heretics and infidels finally would be crushed. The king's Mexican and Peruvian mines produced tons of silver that subsidized Spanish power throughout Europe, including war against Protestant rebels in the Netherlands and Ottoman Turks in the Mediterranean. Some of Philip's military commanders urged that he follow up the conquest of Portugal by invading Elizabethan England.

The commercial networks of the Iberian kingdoms meant that Philip controlled the world's most lucrative maritime trade, including that in pepper and spice from India to Europe, in silk and silver between China and Japan, and in slaves and gold between Africa and the New World. Walter Raleigh (1552–1618) recognized the economic significance of maritime supremacy: "Hee that commaunds the sea, commaunds the trade, and hee that is Lord of the trade of the worlde is Lord of the wealth of the worlde."[3] Philip II's American silver streamed around the globe, quickening economic activity in India, Southeast Asia, and China. To the despair of the Dutch and the English, his ships dominated the Indian Ocean and Atlantic. They also held a monopoly on voyaging across the Pacific, though Francis Drake (ca. 1540–96) made a lone incursion there in 1579 in the *Golden Hind* during his famous circumnavigation of the world. A raid he planned on Manila, the Spanish headquarters in the Philippine Islands, never came off, but near Panama he captured the Spanish *Cacafuego*, which carried bales of silk, twenty-six tons of gold, and fifteen hundred porcelains. He traded much of the latter to Miwok Indians, near what is now San Francisco Bay, and later he presented several impressive items to Queen Elizabeth I (r. 1588–1603) after grandly sailing up the Thames, his rigging festooned with colorful Chinese silks.[4]

His world-girdling triumph no doubt fortified Drake, for he shared Philip II's providential perspective on human affairs. "Our enemies are many," he said, "but Our Protector commandeth the world."[5] In general, however, the deity seemed to be favoring Spain. In the Mediterranean in 1571, with the aid of Venetian and papal galleys, Philip inflicted a sensational defeat on the Ottoman navy at Lepanto, off the coast of Greece. Some of the Muslim veterans of that encounter ended up in the Philippines a few years later, ready to fight the Spanish on the battlefield of Southeast Asia. Ruler of the first global empire, Philip found himself engaged in the first world war.[6]

For some of the king's servants, the union of Spain and Portugal suggested a breathtaking project. In the decade after Sebastian fell in battle, administrators in

Manila repeatedly urged Philip to launch the conquest of China from there, expanding his wealth and power to an incomparable extent. A memorial of 1586 from the governing council of Manila argued that as a consequence of holding the Philippines, "there is offered to his Majesty the greatest occasion and the grandest beginning that ever in the world was offered to a monarch."[7] Yet, though a medal with Philip's portrait flaunted the legend *Non Sufficit Orbis* (The World Is Not Enough) after the takeover of Portugal, the monarch in fact decided that he had a sufficiency.[8] Bedeviled by rebellion in the Netherlands and cobbling together an armada to invade Protestant England in 1588, he did not think it prudent to march into China. In fact, he sided with the view that the Philippines provided a command center for spiritual rather than military conquest: Jesuit priests and mendicant friars working for the conversion of China and Japan would receive crucial support from the strategically located possession, a pendant of Christianity suspended off the Asian mainland.[9]

Philip also saw the Philippines as commercially valuable, even though the islands lacked profitable native products. A stronghold there, only two weeks' sea journey from the Chinese coast, would circumvent the Portuguese monopoly on trade with China. Miguel López de Legazpi (1502–72), the conqueror of the islands, predicted in 1569 that "we shall gain the commerce with China, whence come silks, porcelains, benzoin, musk, and other articles."[10] Soon after, the first of many Spanish galleons carrying Chinese merchandise left Manila for Acapulco (Mexico), its cargo paid for by silver from the New World. As ruler of Portugal after 1580, however, Philip also could command agents in Macao, the Portuguese trading post near Canton (Guangzhou) in southern China. A modest advantage of this extension of power was that the king could order porcelains at their source for the first time.

THE PILGRIM FLASK OF PHILIP II

The greatest art patron of the century, with a collection of fifteen hundred paintings as well as numerous manuscripts, prints, tapestries, clocks, jewelry, and exotic natural specimens, Philip had long admired and purchased Chinese porcelain. In the 1570s he directed the earthenware potters of Talavera de la Reina, a town ninety-five kilometers southwest of Madrid, to turn out blue-and-white tiles for the Escorial in imitation of the dominant color scheme of Chinese ceramics. He believed that blue and white suited the sober architecture of his palace better than the flamboyant Italo-Flemish style of polychrome decoration. In Portugal for his coronation in 1581, he stayed in the royal Santos Palace in Lisbon, where blue-and-white porcelains sumptuously adorned the ceiling of a domed room. After the union with Portugal, a flood of blue-and-white Lisbon earthenware entered Spanish cities, where buyers referred to it generically as *mariposas* (butterflies) because of a common decorative motif copied from Chinese porcelain. Philip fostered good relations with allies and clients in Europe by sending gifts, often of porcelain. His cousin Archduke Ferdinand II of Austria (1529–96) also received some Aztec feather-work, looted by conquistadors decades earlier, which he put on display in a *kunstkammer* (art cabinet) with his porcelains. By the time of his death, Philip owned the largest collection of chinaware in Europe. According to an inventory of 1598, it amounted to three thousand porcelains, most of them tableware, including serving platters, carafes, sauceboats, and wide-mouthed jars.[11]

Philip possessed a number of idiosyncratic pieces, including a jar with a crowned double-headed eagle clasping a heart pierced by arrows, a motif identified with the Order of St. Augustine, a missionary society with many friars in the Philippines. But the most revealing porcelains were several known as "pilgrim flasks," almost certainly commissioned after the union with Portugal. Like the coffin in the Escorial, the blue-and-white flask has global referents. It was fired in the kilns of the city of Jingdezhen in southeastern China, the most important manufacturer of porcelain since the fourteenth century. Two hundred years later, Westerners began commissioning works there, though a Chinese broker in Canton probably relayed the order for the king's flask to the potters. While obviously Chinese in its clay materials and craftsmanship, Philip's flask otherwise reflects an extensive circuit of influence. In form, color, and decoration, it represents the end product of centuries of entwined traditions, industry, and artistry, the flowing together of wide-ranging cultural currents. (See figure 1.)

Shaped much like a canteen, with a tall, tapering neck and flattened spherical body, the pilgrim flask is decorated on one side with a portrait of a seated scholar and a servant boy in a rocky landscape, a theme combining elements from ancient Persia and Tang China.[12] The arms of Spain, *Castile and León quarterly,* appear on the other side, most likely copied from a Spanish coin provided by Philip's agents in Macao or Manila. The slender, sloping sides of the piece are adorned with lotus-flower patterns deriving from early Buddhist India; Chinese images of insects and rocks line the neck. The white of the flask's body may be traced to Chinese ceramic imitations of Persian silverwork in the seventh century and, somewhat later, to simulations in pottery of the lustrous sheen of pale jade from Central Asia. The blue shades on the flask derive from cobalt pigment blended from ore mined in Persia and China, while the overall format of the ornament represents a convergence of Chinese and Islamic approaches to design and spatial organization developed since the fourteenth century. In short, taken altogether, the pilgrim flask of Philip II is a product of Eurasian cross-cultural contact, a representative climax to centuries of long-distance interaction and mingled traditions.

European enthusiasm for Chinese porcelain represents a late episode in the millennium-long history of the ceramic. Peoples in the jungles of Borneo and the Philippines had been employing it for many centuries; potters in Korea, Japan, Vietnam, Egypt, and Iraq had been emulating it for a thousand years. But few pieces reached the West before the sixteenth century, and when they did, their owners treated them as cherished possessions, mounting them in precious metal and inscribing them with heraldic crests. A land of drab earthenware during the Middle Ages, Europe regarded porcelain with wonder and envy. When Vasco da Gama (ca. 1460–1524) sailed from Portugal in 1497 for his epoch-making voyage around Africa to India, he had instructions from King Manuel I (r. 1495–1521) to bring back the two things the West most desired—spices and porcelain. Two years later, having lost more than half of his 170 men to disease and starvation, da Gama presented the monarch with sacks of black pepper, cinnamon, and cloves, as well as a dozen pieces of chinaware, the first of 300 million to be shipped to Europe over the next three centuries. After Portuguese captains reached China in 1517, Manuel ordered a number of porcelains: the earliest known blue-and-white vessel with European decoration is a ewer of 1520 decorated with an armillary sphere, a symbol of discovery and the king's personal emblem. King John III (r. 1521–57) owned plates bearing the Portuguese royal arms and a circular medallion formed by a crown of thorns and the insignia of the Jesuits. Sebastian, Manuel's ill-fated grandson, had a dish with his heraldic crest encircled by four Buddhist-style lions chasing a ball.[13]

In commissioning his pilgrim flasks, then, Philip II acted as much like a conventional Portuguese sovereign as an avid art collector. Philip III (r. 1598–1621), who shared his father's enthusiasm for porcelain, went to Portugal in 1619 for his belated coronation as king of that nation—a visit with unhappy consequences since he died in a few years from an illness contracted there. Making his ceremonial entrance into Lisbon, he paraded under the customary triumphal arches. The one erected by the guild of potters depicted Portuguese carracks unloading Chinese porcelain in the city's harbor and other ships taking aboard Portuguese imitations of the Chinese ware for export to European countries. An inscription boasted that "ours also go to different regions of the world." Holding up a blue-and-white earthenware vase labeled "porcelains" *(porcelanas),* an allegorical figure declared:

> Here most gracious Majesty
> We offer you the pilgrim art
> Made in the Lusitanian Kingdom
> Which China sold us at such high prices![14]

CHINESE PORCELAIN AND CROSS-CULTURAL EXCHANGE

This book explores the cultural role of Chinese porcelain in world history. In replicating chinaware, the Lisbon potters unwittingly trailed far behind craftsmen in many other cultures. For over a thousand years, porcelain was both the most universally admired and the most widely imitated product in the world. From the time of its creation in the seventh century, it played a central role in cultural exchange in Eurasia: it was a prime material vehicle for the assimilation and transmission of artistic symbols, themes, and designs across vast distances. Porcelain had a commanding impact by virtue of influencing all ceramic traditions it encountered, from Japan and Java to Egypt and England. In some cases, it displaced local traditions altogether, thereby reaching deeply into indigenous cultural life.

Porcelain provides evidence for artistic, commercial, and technological interaction between China and other regions of the world from the origins of the pottery to the beginning of the modern age at the close of the eighteenth century. In the fourteenth century, porcelain inspired a commercial enterprise unprecedented in range and volume in the premodern world—raw material (cobalt ore) shipped from Persia to China to make huge quantities of blue-and-white chinaware for Muslims in India, Egypt, Iraq, and Persia. From the sixteenth century, by way of the Philippines and Acapulco, Spanish galleons delivered blue-and-white to Mexico City and Lima (Peru) while aristocrats in Europe took delivery of custom-made tableware from Canton. Given its volume and circulation by the eighteenth century, porcelain yields the earliest and most extensive physical evidence for sustained cultural encounter on a worldwide scale, perhaps even for emergence of the first genuinely global culture.

Porcelain acted as a sensitive barometer of human affairs, more than any other commodity. It registered the impact of artistic conventions, international trade, industrial development, political turmoil, elite expenditure, ceremonial rites, and cultural contact. It therefore is central to a wide range of topics: commercial exchange, domestic economy, consumption patterns, interior design, architecture, ornamental motifs, fashion styles, dining etiquette, foodways, transportation networks, political propaganda, manufacturing technology, product innovation, scientific research, gender relations, religious beliefs, and social values.

Naturally, other articles of trade also touch on some of these topics. In recent years, aspects of world history have been examined from the point of view of salt, tea, chocolate, coffee, silver, tobacco, opium, sugar, flowers, wine, codfish, corn, rhubarb, coal, clay, potatoes, spices, firearms, glass, and silk. All the products examined in these works are shaped by human intervention, and indeed most are impossible without it. Contemporary anthropology teaches that none are culturally neutral; they all carry a valence, a certain meaning and context for everyone who uses the given item.

Some things that people consume, however, carry a more powerful charge than others. As products of nature, even though processed and altered by humans, coal, corn, and cod can say only so much. As crafted objects, entirely stuff of human invention, silk fabric, glass, silverwares, and ceramics perform at higher levels of cultural abstraction and metaphor, closer to sculpture and painting than to salt and sugar. They function as acts of imagination, expressions of tradition, assertions of communal identity, demonstrations of social cohesion, vehicles of prestige management, objectifications of self, and embodiments of social value. Nor is there a formal order to these various aspects, for the same object may represent a range of meanings as well as change those meanings over time. Human artifacts mediate between nature and culture, appetite and reflection, disorder and intentionality, exploitation and innovation, the ephemeral and the enduring, the raw and the cooked. With their iconographic message about the vanity of worldly possessions forgotten by all but art historians, seventeenth-century Dutch still lifes continue to enthrall modern viewers in part because the works powerfully exemplify the same tensions, with natural things—lobsters, tulips, nautilus shells, peeled lemons, oysters, ruddy hams, a brace of game—displayed alongside manufactured objects, such as pewter plates, Venetian crystal goblets, French silverware, crimson Turkish carpets, Japanese lacquerware, folds of damask, and blue-and-white porcelain bowls. (See figure 2.)

Ceramic vessels, however, have an exceptional status even within the category of made things. Pots are shaped by the pressure of hands on wet clay, making ceramics the most physically intimate of the arts, virtually impossible to abstract as a form from the earthy matter employed and the functional artifact created. Although classed as a decorative art, pottery has attributes that make it distinct from jewelry, gems, carpets, wallpaper, fabric, marble, bookbinding, cloisonné, silverware, and furniture. Since the eighteenth century, pottery (usually porcelain) has been put on a pedestal to be admired, yet it resists that lofty isolation by virtue of its utility in cooking, eating, storing, and, until recently, excreting. Bowls, jars, and vases in museums have a forlorn, derelict quality: vacant and untouchable, they are severed from the everyday uses for which potters fashioned them.

Significantly, all cultures project an anthropomorphic image onto standing ceramics, seeing them symbolically as three-dimensional analogues for the human body. And since women use pottery as a receptacle within domestic settings, the bodily reflection generally is feminine, predictably with erotic overtones: shaped in curves, the pot is characterized as possessing a foot (or bottom), belly, arms, shoulders, neck, mouth, and lip.[15] Soon after reaching the Philippines, Legazpi reported that he had captured two Chinese junks that carried "some fine porcelain jars, which they call *sinoratas*."[16] This figurative dimension sets ceramics apart from all other commodities. Indeed, some cultures, especially in eastern and southern Africa, where women are the chief potters, identify metaphorical links between pots, kilns, and females, because all are involved in irreversible, heat-mediated transformations: pots warming food, kilns firing pots, and the womb nurturing the embryo.[17] Porcelain even extends the sexual correspondence with the female inasmuch as its satiny surface has evoked comparison, in all cultures, with the complexion of a beautiful woman.

THE CULTURAL SIGNIFICANCE OF CHINESE PORCELAIN

As an examination of cross-cultural interaction in world history, this book considers how various societies integrated Chinese porcelain into their art, religion, politics, and economy. It also looks at how the commodity reflects important events in world history. But even though porcelain has its intrinsic fascinations, making it a source of endless appeal and affection to collectors and connoisseurs, it was not itself central to historical action. A couple of late-sixteenth-century invasions of Korea by Japan are called the Potters' Wars, but no war was ever fought by potters or over pottery, and although King Frederick William I of Prussia (r. 1713–40) marshaled what contemporaries termed the Porcelain Regiment, the ceramic never played a role in war except as incidental, if prized, plunder. Massive imports of chinaware to the West after 1500 resulted in development of a European version of porcelain by the early eighteenth century; but the Chinese vessels just hurried along an innovation that unquestionably would have taken place anyway within a few generations.

Excluding the likely effect of porcelain on reducing death from disease and thereby stimulating population increase, little in history would have been different if the ceramic had never been created. China did without it until the seventh century, Europe until the sixteenth, and common earthenware serves most purposes in many countries today. Usually referring to porcelain generically (and eponymously) as *china,* people in industrial societies overwhelmingly prefer it for their tableware, and scientists annually turn out a mountain of highly technical publications on its nature and uses. Porcelain is employed in numerous important ways, including in guided missiles, space shuttles, jet turbines, internal combustion engines, laser technology, body armor, dental surgery, and the bathroom—it remains an unsurpassed material for the sink, tub, and toilet—but these considerations go beyond the chronological scope of this book.[18]

In addition, while porcelain was a product in international commerce from the seventh century, it was by no means the most outstanding in quantity and influence. Merchants traded textiles, especially silk and cotton, in greater amounts, a consideration that made fabrics the chief material carrier of cultural messages in the form of designs, motifs, and colors. According to Matteo Ricci (1552–1610), the great pioneer of the Jesuit mission to China, "There is no other staple of commerce with which the Portuguese prefer to lade their ships than Chinese silk, which they carry to Japan and India, where it finds a ready market."[19] He also notes that Spanish merchants in the Philippines transshipped huge quantities of silk to the Americas and other parts of the world.

At all times, chinaware trailed behind spices in trade between East and West, and in the eighteenth century, Chinese tea also exceeded it in importance. Porcelain, however, played an exceptional role in cultural exchange between China and the distant reaches of Eurasia, one that other commodities intrinsically could not perform. Spices not only came from various parts of Asia, they naturally were intended for immediate use and consumption, and although pepper, nutmeg, cloves, and cinnamon were seen as possessing pharmacological and cultural significance, these meanings were imposed by their consumers, not intrinsic to the merchandise.

Silk was regarded as indispensable for elite apparel and for employment in religious ritual in Roman and Byzantine Christianity, but China lost its silk monopoly by the sixth century, when other countries obtained the technology of sericulture. Silk exported from China also was often plain and in the form of yarn, and while Chinese silks with embroidered designs conveyed cultural messages about China, craftsmen in Southwest Asia and Europe frequently unwove the fabric and recycled the thread. Moreover, like all textiles, silk deteriorated rapidly if not kept in tombs, shrines, or reliquaries. Glass vessels also had a short life span inasmuch as they were easily broken. And because glass is made from a mixture of sand (which provides silica), the vessels could be melted at a relatively low temperature and refashioned in other shapes and colors. Artisans made silverwares in all places the white metal could be acquired, and merchants traded them internationally; but the high value of silver meant the vessels regularly went into the melting pot to be recycled for ready cash or into more fashionable utensils.[20] Whereas precious plate and glass perishes, porcelain lives on. Easy to break yet hard to destroy, it retains its color and decoration with perfection, even after centuries at the bottom of the sea.[21] It invariably conveyed cultural meaning in its shapes and decoration, though often this was confusedly, if creatively, apprehended by foreign patrons. The exceptional longevity of porcelain (and pottery in general) results in the curious circumstance that much of the history of metalwork forms can be reconstructed only by looking at their relatively inexpensive ceramic replications.

In contrast to other commodities, porcelain remained a Chinese monopoly until just three centuries ago. Koreans made it from the ninth century and Japanese from the early seventeenth; yet this was done under Chinese tutelage, wholly dependent on more than two millennia of Chinese craft expertise and technology. The Chinese (or Sino-centered) monopoly truly was broken only in the early eighteenth century when, stimulated by imports of Chinese and Japanese porcelain, researchers at Meissen in Germany created a version of the ceramic, an achievement soon duplicated by Sèvres in France and by many other European pottery manufactories.

Another distinctive characteristic of Chinese porcelain was that it was exported in finished form and, unlike glass, could not be recycled, though the Dutch sometimes added decoration to pieces in the seventeenth century. In the absence of recycling, artisans in Japan, Iraq, Turkey, Holland, England, and France repaired broken porcelains with wire and metal clamps. An eighteenth-century Parisian man-about-town recorded that craftsmen from Normandy made a living as sellers of rabbit skins and "menders of broken china-ware."[22] A London advertising handbill from around 1770 broadcast the services of "Edmund Morris China-Rivetter, at the China Jarr, in Grays Inn. . . . Mends all sorts of China Wares with a Peculiar Art . . . so as a Rivetted Piece of China will do as much Service as when New."[23]

Various peoples found porcelain so compelling and even magical that fragments of shattered vessels were pulverized for medicine, framed for decorative hangings, handed out in religious ritual, used as auspicious gambling counters, and plastered into the walls of towers, shrines, churches, and mosques. Now and again, it also had its uses in war. Chinese soldiers in the thirteenth century loaded bamboo flamethrowers with porcelain shards and metal chips. An enemy of William Kidd (ca. 1645–1701), the British sea captain who turned to piracy, had his men stuff broken chinaware dishes into a cannon and fire them to shred the sails of his opponents.[24] In *Annus Mirabilis,* John Dryden (1631–1700) makes sport with this notion in describing the English victory over the Dutch in the 1665 battle of Lowestoft off the Suffolk coast, an encounter in which the Dutch lost seventeen ships, including a few that had been employed in trade with the Moluccas (the Spice Islands) and China:

> Amidst whole heaps of spices lights a Ball,
> And now their Odours arm'd
> against them flie:
> Some preciously by shatter'd Porc'lain fall,
> And some by Aromatick splinters
> die.[25]

Porcelain also stands out from other commodities inasmuch as it had a uniquely ecumenical impact. Spices and silk went on a one-way journey, from east to west, at the end of which people consumed the spices and the silk frayed, faded, and finally vanished. Porcelain, however, not only endured but also played a central role in reciprocal cultural influence. Chinese artistic motifs and designs taken from porcelain were embraced by distant societies, after which, reshuffled, reinterpreted, and frequently misconstrued, they were sent back from where they came as decoration on merchandise such as cotton cloth, carpets, and silverwares. In like fashion, Chinese potters often adapted alien designs for their products, which merchants then exported to the very foreign realms where the designs had originated generations before. Thus a Chinese-influenced version of exotic decoration would be imitated by craftsmen half a world away, not suspecting that they were heirs of the cultural tradition that had first inspired that which they currently were emulating. In connection with other media, chiefly textiles, metal vessels, and architectural adornment, porcelain had a central part in a sometimes dizzying loop of cultural association and amalgamation.

The cultural impact of porcelain provides an illuminating but unexplored theme in the writing of world history. Inasmuch as broken pottery usually ends up buried in stratified deposits in the ground, it has been an indispensable source of information for archaeologists since the late seventeenth century. Historical study of material culture, commodities, and consumption, however, began only in recent decades.[26] In particular, it is not surprising that historians have disregarded porcelain. Although there is an enormous literature on the subject, it appears in publications rarely consulted by them, such as exhibition catalogues, auction house magazines, antique monthlies, museum booklets, specialized monographs, art journals, pottery periodicals, and archaeology bulletins. In the study of history, porcelain has been given no more respect than museum visitors habitually accord it as they bustle past cabinets of platters and pots to contemplate well-known (and comparatively comprehensible) paintings and statues.

Despite the significance of porcelain in long-distance commerce for centuries, economic historians have shown scant interest in it. For their part, most writers on the subject, typically aficionados, connoisseurs, collectors, and museum curators, concentrate on the aesthetic qualities of the vessels rather than on their economic ramifications. Thus while there has been no attempt to quantify and chart the massive Chinese export trade in ceramics, there is a surfeit of antiquarian research devoted to identifying eighteenth-century British armorial bearings on chinaware. Of course, experts on porcelain commonly approach their subject from more significant perspectives: design motifs, pottery forms, specific vessels, kiln complexes, archaeological finds, and notable collections. But though valuable and interesting in its own right, such work seldom addresses larger historical considerations, economic ones in particular.

Exceptional and long-standing barriers segregate the study of art from that of economics. In the most basic ways—themes investigated, sources examined, training required, practices followed, questions asked—art historians and economic historians obviously work quite differently. Porcelain, however, is most revealing when treated as a cultural cynosure, a nexus where art and commerce converge, drawn together by an artifact that in some measure incarnates and articulates the beliefs, customs, and mentalities of those who make, purchase, and esteem it. Positioned at the intersection of everyday life, commerce, and art, porcelain vessels were often simultaneously functional wares, profitable merchandise, and treasured possessions. Linked to social behavior, long-distance trade, and elite taste, porcelain affords a distinctive standpoint from which to view world history, casting light on many topics other than itself.

Porcelain is employed here as a sort of organizing principle, a way to examine the tangled interactions that make up human history. This book adopts the perspective expressed by Wallace Stevens in *Anecdote of a Jar*, in which (as in a Dutch still life) the crafted object imposes form and order on the unruly profusion of natural things:

> I placed a jar in Tennessee,
> And round it was, upon a hill.
> It made the slovenly wilderness
> Surround that hill.
>
> The wilderness rose up to it,
> And sprawled around, no longer wild.
> The jar was round upon the ground
> And tall and of a port in air.

It took dominion everywhere.
The jar was gray and bare.
It did not give of bird or bush,
Like nothing else in Tennessee.[27]

THE FALL OF CHINA AND THE RISE OF THE WEST

In a manner of speaking, this study is concerned more with the wilderness than with the jar: that is, it focuses on what porcelain reveals about cultures around the world, not on the commodity itself. It emphatically does not provide a history of Chinese porcelain, still less a history of ceramics. Accordingly, technical material regarding clays, glazes, and kilns is kept to a minimum. The presentation follows the admirable example set by Cipriano Piccolpasso (ca. 1523–79), who explains in *The Three Books of the Potter's Art,* an introduction to Italian pottery manufacture, that his work steers clear of much specialized detail "so as not to perplex other men's thoughts with what is not needful."[28]

Although the subject matter spans more than a millennium of history, most attention in this book is devoted to the early modern period, 1500 to 1800. All chapters draw material from those years, which have been more thoroughly investigated by students of porcelain than have earlier centuries. Furthermore, a study of the cultural influence of Chinese porcelain in world history necessarily highlights developments in early modern times. From the origins of civilization around 4000 B.C.E. to the voyages of Christopher Columbus (1451–1506) and da Gama, the most wide-ranging, long-term cultural interaction took place across the super-continent formed by the isthmus linking Eurasia and Africa. That landmass effectively constituted an ecumene—from Greek *oikoumenē,* the "inhabited quarter" of the world—a series of civilizations or extensive regional societies in communication, however shaky and sporadic, between the Atlantic and Pacific.[29]

Some travelers, most famously Marco Polo (1254–1324) and Ibn Battuta (1304–69), crossed much of that expanse. Polo could make his journey from Venice to China because Mongol conquerors provided security on the so-called Silk Road, a network of trails and oases linking West and East Asia. Ibn Battuta journeyed from Morocco to Canton, wending his way along a trading network (or diaspora) of Muslim merchants on the sea lanes connecting Southwest Asia and the Indian Ocean to the coast of southern China. Counting his later trip to West Africa, he traveled some 120,000 kilometers in twenty-nine years, passing through much of the ecumene.

Discovery of the New World and the Cape Route to India, however, gave rise to an ecumene on a global scale, with peoples everywhere caught up in ever-increasing commercial, technological, and intellectual exchange. A result of that ecumenical transition was the availability of Asian products in Europe and the Americas. For the first time, porcelain became a truly worldwide commodity. What may be called the "globalization" of material culture began in the era of Columbus and da Gama, with chinaware motifs, colors, and shapes its earliest and most wide-ranging manifestation. The ewer of Manuel I and the pilgrim flask of Philip II are fitting illustrations of this.

Reflecting the acknowledged superiority of Chinese manufactured goods, globalization of material culture initially came to pass under Chinese auspices. Columbus sailed the Ocean Sea with a letter from the Spanish crown recommending him to the grand khan of China and the kings of India, and in his copy of *The Travels of Marco Polo* (1298), Columbus earmarked the passage in which the Venetian promised that "an incalculable amount of trade" awaited Westerners in the land of Cathay.[30] When the second Portuguese fleet under Pedro Álvares Cabral (ca. 1467–1520) returned from the Indian Ocean to Lisbon in 1501, Manuel I reported to a fellow monarch that "porcelains, musk, amber, and aloes wood" could be acquired from a land to the east of India known as "Malchina" (from Sanskrit *Mahacina*, "Great China").[31]

China had the world's most advanced economy throughout much of history, providing goods to an enormous domestic market, as well as to Korea, Japan, Southeast Asia, and countries on the Indian Ocean.[32] Europeans gained direct access to Asian markets after 1500, and along with Indian cottons and Asian spices, it was Chinese merchandise—tea, silk fabric, lacquer furniture, hand-painted wall hangings, and porcelains—that Westerners most desired. From the seventeenth century, desperate to halt the flow of silver to Asia to pay for the goods, European rulers promoted efforts to replicate chinaware and other manufactured items.

From 1800, however, globalization of material culture proceeded swiftly under Western rather than Chinese sponsorship. The industrial revolution had been gaining steam for several decades before the turn of the century, with pottery manufacturers among the most important pacesetters in devising new manufacturing techniques and sketching the outlines of factory organization. The first global consequence of the industrial revolution appeared with the precipitous collapse of Chinese porcelain in international markets in the late eighteenth century as a result of competition from British ceramics, mainly those produced by Josiah Wedgwood (1730–95), the famous pottery baron.

Yet while the fall of Chinese porcelain clearly is noteworthy in the contexts of the history of ceramics, an expansive chronicle of material culture, and the preliminary global impact of the industrial revolution, its greatest value for an examination of world history lies in it mirroring a transformation of the greatest magnitude. That change is clarified by placing it within a lengthy perspective. By around the year 1000, long-distance commercial exchange had integrated the ecumene into what contemporary historians call a *world system*, a series of interacting, overlapping economies. A complex trading network of silver and gold bullion, as well as spices, precious stones, metals, textiles, and pottery, connected much of Eurasia. More than any other region, China functioned as the fulcrum of the world system, the motor that made it run.[33]

The overseas reach of the huge Chinese economy intensified ancient patterns of trade and cultivation throughout Southeast Asia and the countries of the Indian Ocean. The distant ripples of its effect could be felt even in Europe as the market for Asian products expanded in the Mediterranean and north of the Alps. Because of its role as a powerhouse from the turn of the first millennium, China had some economic justification for its ancient self-designation in cultural terms as the "Middle Kingdom" *(zhongguo)*, the axis of the world, with lesser states on its periphery and regrettably barbarian peoples, such as Africans and Europeans, remote beneficiaries of its blessings.

China lost that dominance and self-assurance soon after 1800, however, as the center of gravity of the world system shifted toward the countries of northwestern Europe. In *The Wealth of Nations* (1776), Adam Smith (1723–90) famously stated a crucial precondition for that realignment: "The discovery of America, and that of a passage to the East Indies by the Cape of Good Hope, are the two greatest and most important events recorded in the history of mankind."[34] The widespread possessions of Philip II represented the most remarkable early manifestation of Europe's new, elevated position around the world. Warning of the menace Philip's empire posed to England, William Camden (1551–1623) said it extended so far that the king "might truly say, *Sol mihi semper lucet:* the sunne always shineth upon me."[35] But the threat receded as the Spanish-Portuguese empire fractured at the seams in the seventeenth century. Still, as Baron de Montesquieu (1689–1755) suggested in *The Spirit of the Laws* (1748), the Iberian collapse did not mark a retreat of the West from the global arena: "Europe carries on the commerce and navigation of the other three parts of the world, just as France, England, and Holland carry on nearly all the navigation and commerce of Europe."[36]

Building on their economic dominance, those nations went on to establish their own overseas dominions, upon which the sun indeed did not set until the late twentieth century. From having been for centuries no more than a marginal participant among the associated territories of Eurasia, the West began seizing center stage in the early modern period, pioneering global maritime routes, setting up overseas trading posts, planting European-style societies in the Americas, colonizing most of Asia, fashioning new political and economic institutions, and ultimately emerging as the driving force of modernity.[37]

This revolutionary shift of the world system from East to West was paralleled in the international career of Chinese porcelain, the fortunes of which moved in concert with those of China itself. European zeal for importing porcelain after da Gama's voyage suggests the awe and envy with which Westerners regarded China from the time they read Marco Polo's account of the Middle Kingdom. European dedication to replicating porcelain from the seventeenth century points to Western determination to escape economic dependence on China and mount a challenge to its industrial might. Finally, the commercial success Europeans achieved with their own pottery in ousting chinaware from international markets at the end of the eighteenth century foreshadowed Western international dominance in the modern world. In the widest perspective, then, the fall of Chinese porcelain by 1800 closely tracks the epochal decline of China in world affairs and the corresponding rise of the West to primacy within the global ecumene.

Of course, at least on some important counts, that supremacy is a thing of the past. Since the end of World War II, Western nations have been forced to give up their colonial possessions, and they have lost the power to dictate military and political affairs across the board. There are also signs that China will contest the West for economic dominance in the modern-day, rapidly shifting world system, thereby perhaps resurrecting the standing it held through much of history. Whether East or West (or neither) emerges triumphant, however, porcelain will remain above the battle: if it no longer attracts the wonder and high regard bestowed on it for ages, that is a result of the ceramic having become irrevocably a possession of the world, manufactured and used virtually everywhere.

Yet while China has long since lost its monopoly, Jingdezhen, the town that once produced nearly all the world's porcelain, continues to turn out more than 300 million pieces annually, most of it blandly interchangeable with that manufactured in Italy, Denmark, Chile, and Malaysia. Still, mindful of its spectacular past and its willing contemporary customers, Jingdezhen makes a handsome profit by reproducing exactly some of the dazzling porcelains that captivated the world for so many centuries.

1

The Porcelain City

Jingdezhen in the Eighteenth Century

In the opening years of the eighteenth century, François-Xavier Dentrecolles established a church in Jingdezhen, the great porcelain center on the Chang River in the province of Jiangxi, southeastern China. A recruit for the French mission of the Jesuits, he was thirty-five years old when he arrived in Canton in 1698 on board the *Amphitrite,* a ship purchased by the Compagnie des Indes orientales (French East India Company), a state-sponsored syndicate, from Louis XIV (r. 1643–1715).[1]

Dentrecolles was not the most eminent or controversial of the approximately fifty Jesuits who served with him over the next four decades, but he had a passion for the curious and unusual, along with a gift for sifting and marshaling information. After working in Jingdezhen for more than two decades, he presided over the French missionary residence in Beijing until 1732, during which he translated and commented on Chinese accounts of medicine, currency, and government administration. He also sent reports home on the raising of silkworms, the crafting of artificial flowers in silk and paper, the manufacture of synthetic pearls, methods of smallpox inoculation, and the cultivation of tea, ginseng, and bamboo. This represented the sort of engagement with indigenous culture that the Society of Jesus expected of its learned priests. A fellow Jesuit declared in a funeral eulogy for Dentrecolles in Beijing in 1741 that everyone had "a high opinion of his wisdom."[2] His assignment to Jingdezhen suggests that Dentrecolles's superiors recognized his talent for inquiry and analysis from the start. Significant information was expected from the man posted there.

In 1712 and 1722 Dentrecolles wrote lengthy letters on the manufacture of porcelain to Louis-François Orry, treasurer of Jesuit missions to China and India. They soon appeared in the *Lettres édifiantes et curieuses de Chine par des missionnaires jésuites* (1702–76), a production in thirty-four volumes that provided the first wide-ranging source for European knowledge about China. The material resurfaced in the influential *Description de l'Empire de la Chine* (1735) by Jean-Baptiste du Halde (1674–1743), who once served as chaplain to Louis XIV. Du Halde's work shaped the admiring view of China promoted by Voltaire (1694–1778) and other philosophes. In the epitome of Enlightenment thought, the *Encyclopedia* (1751–72), Denis Diderot (1713–84) introduces the article on porcelain by confessing that he can do no better than provide Dentrecolles's account. Du Halde's work was translated into English in 1738, and some years later the young Josiah Wedgwood, already dreaming of transforming the world of pottery, copied parts of Dentrecolles's text into his Commonplace Book. In his widely consulted *Universal Dictionary of Trade and Commerce* (1757–74), Malachy Postelthwayt (ca. 1707–67) reproduces the Jesuit's observations, albeit without crediting his source.

Dentrecolles won a measure of reputation and influence because his letters on Jingdezhen comprised the first accurate and comprehensive account of the manufacture of Chinese porcelain ever sent to the West. They promised to reveal secrets Europeans had been seeking for centuries.

"THE TOWN OF YEAR-ROUND THUNDER AND LIGHTNING"

Readers learned more from the letters of Dentrecolles than technical detail about clay, glazes, and kilns. He evoked the bustling life of the porcelain city, presenting a view of provincial China and its urban workers that is unique in Jesuit relations. He estimates that Jingdezhen held 18,000 families or 100,000 persons, a figure roughly equal to that in government records. Visitors, however, commonly believed that as many as a million souls lived in the city, no doubt because hectic activity and crowded shops conveyed the impression of teeming multitudes.

In Jingdezhen, Dentrecolles records, "one seems to be in the midst of a carnival." Porters trying to make passage in the streets raised cries on all sides. Merchants from every quarter of the empire thronged the alleys and warehouses, mingling with a handful of traders from Japan, Southwest Asia, and Europe. An inscription on a temple boasted that "the town is producing imperial porcelain for the entire country, couriers are coming and going day and night, officials are arriving from everywhere, merchants doing their business incessantly."[3] A generation before Dentrecolles arrived, a Dutch visitor described the scene:

> Upon the 25 of April, we came to a Village famous for Shipping, called Vcienjen, where lay great store of vessels of several sorts and sizes, which were come thither from all

> parts of China, to lade with China Earthenware, whereof great store is sold in this Village. Quite through the middle of this rich Village runs a broad Street, full of shops on both sides, where all manner of Commodities are sold; but the chiefest Trade is in Purceline, or China dishes, which is to be had there in great abundance.[4]

The shopkeepers on Porcelain Street (as Dentrecolles calls it) paid rents and had to purchase government licenses, but a flea market on a small island in the river welcomed anyone who had odds and ends to sell. Some dealers, known as "island basket carriers," collected wares with blotches and hairline cracks from the kilns and sold them piecemeal at the market, concealing defects with plaster, wheat gluten, and mulberry juice. Customers euphemistically called these sorry items "goods that have crossed the river."[5]

Dentrecolles marvels at men striding through the narrow streets carrying planks topped with porcelains on their shoulders, never losing their balance amid swarming crowds. In fact, pedestrians gave them a wide berth as anyone who bumped into them and shattered the porcelains had to pay for the damage—a costly lesson that profits came first in Jingdezhen. Dentrecolles describes chains of workers hauling clay to "vast storehouses surrounded by walls, where one sees, in row upon row, a great number of jars of earth." He expresses astonishment that laborers unloading the sweltering kilns "put salt in their tea in order to drink a lot without becoming ill." Artisans going to work made hasty sacrifice at shrines dedicated to Tung, tutelary god of fire and porcelain. Hucksters peddled adulterated clay and phony glaze to would-be potters, landless villagers hoping to make their fortunes in the booming city. Furnaces operated around the clock, and a large floating population anchored in boats along the riverbank, providing lodging, delivering foodstuffs, and loading porcelains. A visiting imperial magistrate complained, "The noise of tens of thousands of pestles thundering on the ground and the heavens alight with the glare from the fires kept me awake all night."[6] In his description of the approach to Jingdezhen at night, Dentrecolles evokes a moonlit vision of the city as swept by conflagration, silhouetted by billows of smoke and flame, the surrounding mountains forming the walls of a single gigantic furnace, its countless "fire-eyes" (vent holes) tended by shadowy laborers.

Jingdezhen is on the eastern bank of the Chang River, which flows from the mountains to the north that separate northeastern Jiangxi province from neighboring Anhui. The city stands at the point where the river exits rocky gorges and loses its swiftness, broadening into a shallow, curving basin five kilometers long. Dozens of streams flowing into the valley powered undershot waterwheels and iron trip-hammers that crushed rock to be used for making pottery. Hong Yanzu (1267–1329), an official stationed in Jiangxi province, portrayed the scene in a poem: "The bones of the mountain in the end turn to powder, / On the outskirts, many pestles pound the earth, / On the river, half the boats transport mud."[7] Mills produced the best material in the drizzling spring, when the force of water on the pestles was robust and regular, making the pulverized earth fine and dense.

Dentrecolles explains that in official imperial nomenclature, Jingdezhen did not rank as a city because it had no encircling wall, "perhaps because one could not then enlarge and extend it as one wishes." Its designation in the early Song period as a market town *(zhen)* meant that it relied on trade and therefore could expand as commercial needs dictated, perpetually rebuilding and enlarging after floods and blazes. According to Dentrecolles, a fire once wiped out eight hundred porcelain shops, yet the owners made such handsome profit from rents that they immediately employed scores of masons and carpenters for reconstruction.

Hemmed in by mountains, Jingdezhen looked across the river to tombs built by merchants and shopkeepers on low hills made up largely of porcelain shards dumped there over the centuries. Dentrecolles records that bodies of the poor were thrown into a place known as "the pit to infinity, the grave for all the world," at the foot of a hill flaunting the sepulchers of the rich. It distressed him to contemplate the generations of unfortunates flung into that abyss, the flesh on their bodies consumed by quicklime. Every winter, Buddhist monks gathered and burned the bones to make room for yet more, an especially grueling task during frequent periods of plague. Streaming past the graveyard and town, the Chang leaves the valley toward the southwest, where it once more plunges into gorges on its journey to Lake Boyang, the gateway by which the porcelains of Jingdezhen reached the wider world.

Renowned in China as "the Town of Year-Round Thunder and Lightning," Jingdezhen was the largest industrial complex in the world when Dentrecolles arrived.[8] Its inhabitants depended for their livelihood on the three thousand kilns scattered through the city and cluttering the surrounding slopes; craftsmen also worked in numerous kilns nearby, especially in the village of Hutian, four kilometers southeast of Jingdezhen. As a Qing official remarked, "The soil can be poor and local customs unhealthy, and when the people did not have the means to provide for themselves, they molded the soil into vessels for eating and drinking to provide for themselves."[9] A sixteenth-century observer noted that in northern Jiangxi province, "the wealthy become merchants and the clever people become artisans, for there is not enough food to feed so many people where the mountains are dense and the fields cramped."[10] Tang Ying (1682–1756), an imperial scholar-official associated with Jingdezhen for almost three decades from his first appointment there in 1728, wrote that "the fire stands in the same relation to them as fine weather and rain to others, and they depend on porcelain as others do on millet and corn." Or as a poet put it centuries earlier: "Ten thousand chimneys smoke to fill ten thousand mouths."[11]

Dentrecolles aspired both to win converts among the artisans and to discover their secrets of porcelain, a commodity desired and imitated everywhere, not least in the France of Louis XIV. In the workshops of the potters, the Jesuit "preached Him who made the first man out of clay and from whose hands we depart to become vessels of splendor or of shame." Despite the costly sea green antiques and gilded vases for sale on Porcelain Street, he most treasured a gift from a parishioner, a crude plate found in the rubbish of a shop and decorated with the Virgin and St. John flanking the Cross, a relic he valued "more highly than the finest porcelain made a thousand years ago." One of his converts told him that similar ceramic souvenirs had been smuggled into Japan in cases of ordinary wares until the "enemies of religion" halted the traffic shortly before Christianity came to Jingdezhen.

"SENDING PORCELAIN TO ALL PARTS OF THE WORLD"

Dentrecolles notes that some of his parishioners kneaded clay for a living: "But this work is very grueling, and those Christians who are employed in it have difficulty attending church; they receive permission only if they can get a substitute, for when this labor is stopped, all the other workers are held up." Kneading clay was just one of the many coordinated steps needed for the manufacture of porcelain. Jingdezhen used methods of mass production centuries before the advent of machine power and the assembly line.[12] Since, as Dentrecolles proclaims, "Jingdezhen alone has the honor of sending porcelain to all parts of the world," such techniques were essential. The connection would not have surprised Adam Smith. Familiar with Chinese porcelain from massed displays in the drawing rooms of Edinburgh and Paris, he articulated the economic principle that governs the production of well-traveled commodities. In the famous third chapter of *The Wealth of Nations,* he explains that division of labor in production increases as the market for merchandise expands.[13] Coordinated effort, specialized skills, and standardized replication of wares were the only way for Jingdezhen to fill short-term orders for huge amounts of porcelain from seagoing merchants in Canton and other ports.

Francesco Carletti (ca. 1573–1636), a Florentine merchant, expressed astonishment at the porcelain he saw in Macao around 1600: "The quantity of it is so great that whole fleets, let alone single ships, could be laden with it."[14] Even before the Portuguese arrived in China in the early sixteenth century, they routinely transported as many as 60,000 porcelains from India in a single carrack; cargoes of 200,000 became common after they established direct trade with China.[15] Ships of the Vereenigte Oost-Indische Compagnie (United East India Company), or VOC, of the United Provinces of the Netherlands carried more than 600,000 ceramics from China every year between 1600 and 1700, 20 percent of which went to Europe. The Dutch also kept some 900,000 porcelains in stock at a transfer depot at Anping Gang on the coast of Taiwan (Formosa), just as the English East India Company (EIC) kept large stores in its London warehouse. A VOC vessel shipped 150,000 ceramics in 1700, and an English one took away forty tons (or some 500,000 pieces) ten years later. In 1721 four ships of the EIC loaded 210,000 pieces each. The sales catalogue of a cargo reveals that a Swedish ship brought back precisely 499,061 porcelains in 1732. Another Swedish vessel, the *Götheborg*, transported 700,000 Chinese ceramics in 1745, as well as silk, tea, rattan, mother-of-pearl, and spices; but it famously sank within sight of its home port of Gothenburg after a round-trip journey of over two years and 40,000 kilometers. In the 1777–78 sailing season, the VOC, the EIC, and other European East Indies companies exported a total of 697 tons (more than 8,700,000 pieces) of porcelain from Canton on twenty-two vessels.

It all added up to at least 300 million pieces of chinaware arriving on European docks in the three centuries after the Portuguese reached China. Huge amounts also were shipped throughout East Asia and to Southwest Asia, bringing the export of porcelains during those centuries to an average of some three million pieces every year. Most came from Jingdezhen, although hundreds of kilns on the coast of the southern provinces of Guangdong and Fujian also produced substantial quantities of less highly regarded chinaware for Korea, Japan, and Southeast Asia. A late Ming writer belittled those kilns for making "porcelain buddhas and delicate figurines, things of no great practical value."[16] Dentrecolles points out that even foreigners did not mistake the ceramics for Jingdezhen porcelains, for coastal wares are "snow-white, without luster, and never decorated with colors." (See figure 3.)

Even with millions of pieces being exported, however, porcelain was never the main Chinese export. In the early eighteenth century, it generally ran a distant third behind silk and tea destined for western Asia, with fans, lacquerware furniture, quicksilver, vermilion (or cinnabar), sugar, dye, crude zinc, camphor, dried rhubarb (a medicinal drug), copper, and gold as supplementary exports. In 1698 the EIC's Court of Directors in London, headquartered in Leadenhall Street, instructed the captain of the *Fleet Frigate* to acquire "the very best sorts of China Goods," including fabrics of silk, damask, and velvet ("as much differing as possible from English Patterns"), and tea of "the very best sort" and to fill all space otherwise available with chinaware of the "greatest variety of Colours and Paints."[17] Yielding steady profits of 80 to 100 percent, porcelain represented 5 percent of the value of all VOC shipments and 2 percent of the value of all Asian exports of the EIC. In 1752 the VOC's *Geldermalsen* sank on its voyage from Canton to Batavia (Jakarta) carrying 162,000 porcelains, including 27,531 dinner utensils, 63,623 teacups and saucers, 578 teapots, 19,535 coffee cups and saucers, 821 beer tankards, and 606 vomit pots—all of which amounted to 5 percent of the cargo's value. But along with 125 gold bars, it was the only part of the shipment eventually to reach market: excavated from the bottom of the South China Sea in the 1980s, the porcelains, still in excellent condition, were auctioned in London for £10 million.[18]

Mass production also was essential for large commissions from domestic entrepreneurs and from the imperial court in what is now Beijing. The latter sometimes called for table services and ritual vessels in a wide variety of hues and shapes.

A devotee of porcelain, the Xuande emperor (r. 1426–35), ordered more than 400,000 pieces in some years. The Wanli emperor (r. 1573–1620) called for only 100,000 per year, a burden still great enough to cause disgruntlement among the potters. In the early eighteenth century, on top of other orders, imperial officials annually sent a consignment of 50,000 bowls, dishes, and plates to the capital. In addition, the emperor often commissioned similar quantities as diplomatic gifts to kings and chiefs as part of the tributary trade with overseas polities that China regarded as clients. The founder of the Ming dynasty, the Hongwu emperor (r. 1368–98), sent an envoy in 1375 to the Ryukyu Kingdom (modern Okinawa), southwest of Japan, with 70,000 porcelains for local potentates.[19]

Some three hundred kilns won designation as suppliers for the emperor, with private producers also drafted into service when Beijing's demands outran the capacity of the imperial furnaces. Eunuch agents of the Son of Heaven (Tianzi) supervised the kiln complex, which comprised over 50 master craftsmen and some 350 workers serving two dozen departments. The people of northern Jiangxi paid high taxes to fund imperial costs for kiln construction, raw materials, labor, and shipping. Moreover, the standards for imperial (or "official") articles were lofty, even oppressive at times. Porcelains regarded by the eunuchs as unsuitable for the emperor supposedly were smashed and buried so that nonimperial hands would not sully them; but the overseers actually sold huge amounts clandestinely on the home market, despite severe penalties if found out. Some connoisseurs, however, preferred the output of private kilns because it generally displayed greater invention and imagination. A Qing authority explained that potters working on official wares are "unwilling to take risks," whereas those serving private kilns "scribble freely trusting to their hands. Experienced brushes are 'given their heads.' . . . [T]hey alone reach heights to which others cannot attain."[20]

As Dentrecolles reports, some items demanded by the court were so fine and delicate that potters had "to place them on cotton wool since they could not be handled without breaking." Others proved too intricate or unwieldy to be molded and fired, such as thick-sided tanks, perhaps to be used as bathtubs, which certain potters labored on for three years without success. The imperial kilns, however, represented no more than a fraction of those in Jingdezhen, and they were the only ones subject entirely to government direction and, for all practical purposes, held captive to a single customer. While artisans in official and private kilns, along with all other handicraft workers, had to join guilds that supervised conditions of employment and living quarters, most potters working in private operations effectively escaped government supervision and control. Instead, they had to satisfy consumers by means of their own expertise and industry.

Far-flung and varied markets fostered an artisan mentality in Jingdezhen that was exceptionally open to innovation. Virtuosity and flexibility were as essential for the prosperity of the porcelain city as standardization and mass production. Such adaptability was unusual, for in peasant societies potters are notoriously conservative craftsmen: they are close to their materials, committed to repetitive tasks, and constrained by local mores, and they service isolated markets. Unlike farmers, whose livelihood is at the mercy of the weather, potters depend on their skill, on techniques developed by trial and error. Because poor potting and a bad firing could wipe out months of labor and destroy a household, potters usually are devoted to turning out the same sort of wares by time-honored methods.[21]

In contrast, Jingdezhen's orientation to distant markets encouraged creative, resourceful enterprise. Change came from the outside world, forcing potters to look beyond their mountain fastness. The spread of Buddhism in China in the Sui and Tang periods meant new ceremonial paraphernalia were required; hence Jingdezhen and other pottery centers produced ceramic versions of reliquaries, alms bowls, oil lamps, and stem-cups. A popular form in China (and later in Southeast Asia) proved to be the *kendi,* a small Indian pouring jar (or jarlet) used for ritual ablution, with a bulbous body, no handle, and a spout set at an angle to the shoulder.[22] Jingdezhen potters also produced a range of unique products for the studies of Chinese literati *(rujia),* such as inkstones, water droppers, brush pots, wrist rests, paperweights, and chess sets. At least one kiln specialized in musical instruments, turning out flutes, flageolets, and miniature carillons with nine chimes. According to Dentrecolles, artisans displayed notable skill in their crafting of idiosyncratic items (or "toys"), such as tortoises that float on water and a cat "whose head held a lamp, the light of which gleamed through its two eyes." "They assured me," Dentrecolles reports, "that in the night the rats were terrified by it." Obeying commands from a son of the emperor, potters made a massive lantern in a single piece that could light up a whole room; but they botched an attempt to produce a fourteen-pipe organ that he also demanded. (See figure 3.)

Beyond exotic articles for the court and common wares for the domestic market, Jingdezhen produced works catering to the tastes and needs of consumers around the world: Edo (Tokyo), Manila, Batavia, Delhi, Isfahan, Cairo, Venice, Amsterdam, and Paris. In fact, Dentrecolles arrived in the porcelain city shortly before the VOC and other Western joint-stock companies opened offices in Canton to facilitate relaying commissions to kiln owners. After 1700 European orders for wig stands, picture frames, close-stool pans, shaving basins, colanders, hyacinth vases, bulb pots, walking-stick handles, mustard jars, saltcellars, fork handles, sauceboats, chafing dishes, cheese cradles, and pudding molds stretched further the proficiency of the artisans. English traders ordered newly fashionable monteiths (glass chillers) in the late seventeenth century—bowls with semicircular cuts in the rim for propping wine glasses resting on ice in the center—providing wooden models alien to the Chinese. The Dutch sent glass cruets, vessels with double spouts and a vertical partition inside to separate oil and vinegar, to be copied. For less specialized items, potters substituted familiar objects: when Dutch merchants requested spittoons in

1700, the order was filled by adapting the shape of an octagonal vase used for displaying a single lotus flower.

Dentrecolles claims that in filling a European commission twenty years later, the workmen "made some designs which were supposed to be impossible": urns fourteen centimeters tall, topped with pyramidal caps, each made in several pieces, yet molded together so adroitly that joinings could not be detected. "I was told," he relates, "that eighty urns were made, but that only eight of them were successful, all the others being ruined." The appearance of Westerners in China led to production of figurines of them, invariably unflattering caricatures rather than realistic portraits. According to Lan Pu's *Potteries of Jingdezhen,* a late-eighteenth-century compilation of views on porcelain, traders in such eccentric pieces came mainly from Guangdong province, where Europeans tended to cluster: "They sell them to the foreign devils to fill their markets. The shapes are usually very strange." An early-eighteenth-century, gnomelike figure of a Dutchman, commissioned by VOC merchants and christened "Mr. Nobody" (after a character in an English play), doubled as a drinking vessel. A porcelain cup from the same time bears a depiction of Dutch merchants, big-nosed and oddly garbed, shopping for such souvenirs at a Canton pottery stall.[23]

Beyond serving a wide and diverse market, the artisans of Jingdezhen were impelled toward novelty by their production of skeuomorphs—objects that imitate the form or shape of one material in another.[24] As the apocryphal book the Wisdom of Solomon (15:9) declares, potters "must compete with workers in gold and silver and imitate workers in bronze." Everywhere they worked, potters provided an attractive, down-market substitute for vessels of precious metal and semiprecious gemstone. Furthermore, the pliability of clay allowed an adept potter to mold it to mimic commonplace materials, such as wood, horn, and leather, as well as turn out fanciful sculptures in baked clay, such as crayfish, lotus flowers, and crabapple blossoms. From the late Shang period (ca. 1000 B.C.E.), pottery imitated ritual bronze vessels used as ceremonial utensils and funerary goods. During the Tang, potters in Jingdezhen and elsewhere in China time and again developed new skills and designs by replicating vessels from models in jade and silver. From the Song, they also supplied the markets of Southwest Asia with ceramic versions of artifacts, such as brass hand warmers, rock-crystal (quartz) ewers, ivory chessmen, and rosewood prayer screens.

Not surprisingly, Jingdezhen also profited from making imitations of ancient ceramics, especially those of the Song period. A talented potter and antiquarian, Tang Ying learned the art of making "close copies of famous wares of the past," elegant ceramics in a sea green hue that he would present to his patrons at the imperial court.[25] Without actually naming Tang, Dentrecolles describes how "the mandarin who has honored me with his friendship" would put porcelains into a fatty soup, after which they would be fired a second time and then stowed in a foul sewer for at least a month. They emerged from this noxious treatment looking several centuries old, and because they were thickly potted, "they do not ring when struck and make no humming noise when held close to the ear."

By the time Dentrecolles arrived in Jingdezhen, potters had broadened their repertoire to turn out porcelain facsimiles of Dutch pewter beer mugs, Venetian crystal vases, and French silver loving cups. Zhu Yan, a former governor of Jiangxi province and the author of *A Description of Pottery* (1774), the first thorough survey of Chinese ceramics, states that "among all the works of art in carved gold, embossed silver, chiseled stone, lacquer, mother-of-pearl, bamboo and wood, gourd and shell, there is not one that is not now produced in porcelain, a perfect copy of the original piece." He also notes that potters decorated porcelains with gleaming enamel colors that mimicked dyes on fashionable brocaded silks, complete with designs of sporting fish, sacred storks, and sky-soaring dragons.[26]

The very plasticity of clay, then, combined with a wide variety of consumers, demanded that Jingdezhen potters tirelessly expand their technical expertise. Yet, although they showed versatility in adapting alien shapes and decoration, their reliance on distant markets rendered them vulnerable to shifting fashions and foreign economies. Dentrecolles observes that "for one workman who becomes rich, there are a hundred others who are ruined but who don't stop striving for their fortune because of their ambition to earn enough to open a merchant shop." Hard luck inescapably arose from the bright possibilities held out by the porcelain city.

Jingdezhen functioned with ungainly efficiency: with predictable impoverishment of potters, with rivalry among kilns and entrepreneurs, with considerable waste and worker dissatisfaction, without direct contact with its most important customers, without central direction over several thousand furnaces—yet with effective and flexible division of labor as a whole. It achieved domination of the global market in ceramics not only by virtue of the superiority of its product but also by the scale and organization of its production. It represented the climax of handicraft industry, the grandest achievement of wholesale, concentrated manufacture before the age of steam-driven machines. More than a hallucinatory vision of the city at night, Dentrecolles's evocation of Jingdezhen as a single gigantic furnace reflected the reality of its daily production.

"THESE VESSELS PASS THROUGH SO MANY HANDS"

To keep all the kilns of the city supplied, workers shoveled clay through a series of suspension ponds, then others skimmed off the creamy surface residue, from which organic impurities had been eliminated.[27] Dentrecolles notes that "one hair or one grain of sand could ruin all the work," that is, cause the porcelain to crack or warp. After this laborious procedure, refining and kneading the material were broken down into additional stages, including beating the clay with wooden spatulas by day and night. One-tenth of the vessels produced were "pressed ware," made by packing the clay paste into clay molds. According to Tang, only a small number of "really clever hands" could make the molds, ensuring that the identical pattern and form of a vessel could be replicated by ordinary craftsmen thousands of times. Mold makers enjoyed some prestige among the potters and lodged in separate quarters. A ready supply of an assortment of molds, Dentrecolles reports, meant that a merchant increased his profit by supplying his customers "much quicker as well as cheaper than another who would have to make the molds."

Twenty artisans worked in sequence on a single piece of porcelain before it was put into a kiln for the first time. Workers blew glaze through bamboo tubes and gauze onto some large vessels as many as seventeen times. By Dentrecolles's count, at least seventy craftsmen worked on polishing, decorating, and glazing the fired porcelain before it was returned to the oven for a second firing. "It is surprising," he remarks, "to see with what speed these vessels pass through so many hands." A portrait of a chrysanthemum on a vase had its petals outlined by one decorator, its stalk by the next, and additional embellishments by others. As Dentrecolles explains, "One workman draws only the first color line on the rim of the porcelain; another traces flowers, which a third one paints; this man is painting water and mountains, that one either birds or other animals." This anticipates nicely Adam Smith's celebrated description of the division of labor in the production of pins:

> One man draws out the wire, another straights it, a third cuts it, a fourth points it, a fifth grinds it at the top for receiving the head; to make the head requires two or three distinct operations . . . and the important business of making a pin in this manner divided into about 18 distinct operations.[28]

Tang claims that to maintain uniformity as well as excellence in painted decoration, pottery painters who sketched designs and those who filled in the colors were forbidden to develop other skills so that "their minds are not distracted."

Zhu Yan provides an inventory from 1529 of over fifty designs on porcelain vessels—including dragons piercing scrolls of Indian lotus flowers, flying storks amid sprays of blossoms, and phoenixes rising into clouds of propitious omen—and he assures his readers that "in a short summary like this it is impossible to give a complete list of all the different designs."[29] For generations, illiterate decorators had copied Chinese and Arabic calligraphy stroke by stroke, an experience that stood them in good stead in the eighteenth century when they began ornamenting table services with the equally incomprehensible heraldic crests of European nobility. In the eighteenth century, more than half the directors, captains, and supercargoes of the EIC purchased armorial dinner and tea services, part of the five thousand English armorial sets commissioned in Canton, including some for regiments and societies. In London, various craft and guild institutions—including the Companies of the Fishmongers, Butchers, Bakers, Poulterers, Bricklayers, and Merchant

Taylors—purchased porcelain tureens decorated with their coats of arms.[30] Jingdezhen produced a tea set decorated with the impaled arms of John Drummond and Lady Charlotte Beauclerk (granddaughter of Charles II and Nell Gwyn), a wedding gift for the highborn couple. (See figures 5 and 6.)

Working from written instructions and sketches, the Chinese decorators naturally made mistakes, such as superimposing one coat of arms on another, facing crests in the wrong direction, muddling the coloring, appending feathers to a wolf, mistaking dolphins for birds, and transforming fearsome bear claws into clumps of grass. An artisan copied a coat of arms from a bookplate sent as a model for a set of armorial porcelains by enclosing it in a neat rectangular border, precisely as it appeared on the handy bookplate. A tureen in a Swedish service had an odd grayish cloud painted next to the coat of arms, evidently the result of a water stain on the sketch during the voyage to Canton.

Armorial porcelains cost ten times as much as standard tableware. A complete service sent to England in the early eighteenth century, counting shipping and payment of customs duty, came to £100 (or roughly $17,200 in today's terms). It therefore must have been distressing for a landed gentleman to commission a service, which took about three years from order to delivery, only to have it finally arrive with his directives inscribed next to the requisite decoration on every piece: *Our coat of arms* or *This color is red.* Persian potters compounded the confusion when they copied Chinese renditions of European armorial bearings onto their own earthenwares: Latin maxims descended into gibberish, and heraldic arms loomed over landscapes of gigantic lotus blossoms.[31]

Decoration from Louis XIV's court, such as festoons of tendrils and floral swags, began to appear on Jingdezhen pottery in the early 1700s. By the late eighteenth century, Jingdezhen potters treated motifs on plates and platters for the European market as interchangeable parts (peony, bird, willow tree, fence, jagged rock, pavilion) that they selected to depict a simple garden vignette on a teacup or a full-fledged panorama on a soup tureen. A dinner service commissioned by a French courtier achieved an elegant note, however: decorated with border cartouches of a carp in a stream, which in China symbolized a student struggling forward in the Confucian civil examination system, the motif was gracefully adapted as a punning homage to Mme de Pompadour (1721–64), née Jeanne-Antoinette Poisson, mistress of Louis XV (r. 1715–74). Around the same time, the Dutch used similar dishes to serve fish, surely thinking potters painted the motif with that in mind. A generation later, the same design performed additional pedestrian duty as the main decoration on a punch bowl made for the Schuykill Fishing Company of Pennsylvania.[32]

Porcelain painters fulfilling Western commissions had to decipher a host of baffling representations from Roman literature, the Bible, and European current events. These included Neptune with his trident, Venus rising from the sea, Adam and Eve in the Garden of Eden, the Ascension of Christ, sword-wielding Scots in kilts, urban riots in Holland, and even a porcelain shop in Amsterdam. In the 1720s Dutch merchants, always delighted to needle the English, ordered plates embellished with satires on the financial scandal of the South Sea Bubble: harlequins frolicked around legends proclaiming "Away foolish shareholders!" and "By God, lost all my shares!"[33] On a porcelain cup depicting the Crucifixion, potters mistook the dice thrown by Roman soldiers at the foot of the Cross for tiny roses and put a garland of flowers on Jesus instead of a crown of thorns. Sometimes potters were required to copy illustrations of classical love stories ranging from the mildly erotic to the unreservedly pornographic. A Jesuit noted that Chinese artists and craftsmen expressed astonishment at European prints and engravings that served as models: "In China they laugh at figures that get lost under the frame, princes bareheaded and naked on a charger, princesses with their breasts uncovered and dressed in ermine in the face of a garden that clearly speaks of summer, or Christian virgins dressed up like actresses."[34]

As a result of the special care required, Western scenes on pottery were expensive. Dutch merchants in Canton advised the VOC directors that all "European painting or figures cost twice as much as Chinese." According to Dentrecolles, porcelain decorators depicted human figures feebly, though he cautions that "certain landscapes and illuminated city prospects brought from Europe to China hardly allow us to ridicule the Chinese for the manner in which they portray themselves in their paintings." Muslim religious strictures, however, strictly limited the range of depictions on porcelains. An official of the EIC instructed his porcelain buyer in Canton that "one General Rule must always be observed [regarding porcelains destined for Southwest Asia], and that is, never to pack a piece of Ware that hath the figure of Humane Species, or any Animal whatsoever."[35]

In general, Westerners emphasized the color and shape of wares more than the drawings on them. Dentrecolles scorned the Chinese portrayal of people and animals for the same reason that João Rodrigues (1561–1633), a Jesuit who served in both Japan and China, criticized the Japanese: "They can hardly be compared with our painters as regards the proportions of the body and in respect to the body itself. For they lack a true knowledge of shading figures, for it is this which makes them stand out and gives them strength and beauty."[36] Giovanni Gheradini (1654–ca. 1704), a Modenese painter who came to China with Dentrecolles in 1698, dismissed Chinese art as soon as he arrived:

> The Chinese have as little knowledge of architecture and painting as I of Greek or Hebrew. Yet they are charmed by fine drawing, by a lively and well-managed landscape, by a natural perspective, but as for knowing how to set about such things, that is not their affair. They understand far better how to weigh silver and to prepare rice.[37]

Exotic allure, however, generally trumped supposedly inferior composition. In 1637 the directors of the VOC sent instructions to its agents in Batavia:

> In the sale of the last received porcelains we found that the kinds painted with Dutch figures were not as much esteemed and in demand as the ones which are decorated in the Chinese manner and in the custom of their country, so that you should write to Tayouan [Taiwan] and command them that in the future they should always and until further express order send no other porcelains than those painted in the Chinese way.[38]

Beyond decoration of porcelains, mass production also was evident in the operation of kilns. Manufacturers specialized in certain items, such as storage jars and fishbowls, wine cups and lanterns. Some kilns produced replicas of porcelains from the Song; others copied bronze vessels from the Shang period or jade cups from the Han. At the end of the Ming, two kilns produced only imitations of ancient pieces. Some concentrated on large "dragon jars," painted with motifs of dragons surrounded by clouds, pearls, lotus petals, and flowers of Paradise. One kiln specialized in dishes for Japan, where fishnets were an auspicious symbol, bearing delicate designs of a blue net tied with a bow. Another operation concentrated on "toad ware," bowls modeled as toads, a symbol of wealth and good fortune in business, squatting amid orchid clusters and bamboo leaves.[39]

Crews worked in shifts around the clock, for the firing of a kiln required continual supervision. Experts directed the complicated loading of wares and the stoking of furnaces. Like many specialists, kiln fillers lived in separate housing and had a supervisor and regular hours of employment. By the early Qing period, a tradition existed tracing groups of kiln fillers to various outlying villages of Jingdezhen. Kiln stokers (or bakers) broke down into "the hot fire men, the slow fire men, and circulating fire men," since a variety of wares called for a range of temperatures and baking times.[40] Furnace tenders sprinkled water to keep fire channels clear, peering in the "fire-eyes" of the kiln to direct the blaze where needed. A kiln burning pine faggots (for superior ware) called for different methods than a kiln using brushwood (for coarse ware). Constant firing of kilns naturally dictated constant repair. The Wei family monopolized that job since the Yuan period: they "handed the procedure down from teacher to pupil," using an inimitable, secret mortar with the consistency of molasses.[41]

Dentrecolles thought it remarkable that everything in the kiln is "fluid and flowing": a copper coin on top of a pile of porcelains would melt through all of them to the floor of the oven. As *Potteries of Jingdezhen* explains, firing the furnace called for an exquisite balance of considerations:

> Unless the fire is hot and strong, the pieces will not get cooked evenly. Unless the fire is small and low, the moisture will not dry by degrees, with the result that the colour after baking will not be sleek and glossy. Unless the fire circulates freely, the middle and the rear, the left and the right, cannot get thoroughly baked, and raw patches are bound to occur.[42]

The kiln master used various methods to determine when the firing reached perfection, such as looking in the vent holes to ascertain when the gravel on the bed of the furnace started to glitter or when an individual vessel could be seen emerging from the enveloping flame. In a 1637 treatise on technology, Song Yingxing writes that in a red-hot kiln "the porcelain pieces are soft like cotton wool," and the kiln master would take one out with a pair of iron tongs to determine if it was time to smother the fire.[43] The largest kilns held 100,000 pieces, and firing could take as long as a week. Fireclay cases, or saggars, made in a village five kilometers downriver from Jingdezhen, protected each piece. Using saggars meant that wares could be stacked on top of one another in the kiln, thereby increasing the number that could be fired at a time. Good saggars could be used for up to ten firings, but poorly made ones disintegrated after only a few. According to Dentrecolles, employing saggars ensured that "the complexion of the porcelain is not breathed on by the ardor of the fire."

With or without saggars, the force of fire could result in beautiful, bizarre, or repugnant effects. Citing a Jingdezhen chronicle, Dentrecolles claims that some fine porcelains in ages past were "so much sought after that the furnace was hardly opened before the merchants squabbled over the first pick." Vessels came from the kilns with marks on the glaze in the shape of butterflies, unicorns, and leopards, unforeseen forms and colors that were "most lovely spontaneous creations of the fire, the causes of which it was impossible to explain." A seventeenth-century treatise on pottery observed that fired vessels sometimes changed color from yellow to purple "due to some magic in the fire which is beyond our understanding."[44] Some ancient wares were considered so dazzling that broken ones would be mended with a diamond-tipped needle and brass wire, the sutures too faint to be seen. Sometimes a fragment would be framed and mounted "like a precious stone, it was so rare and highly prized." Zhu quotes a Song connoisseur as exclaiming, "How lucky it would be even for a rich man to get a single bowl in color like a solid jewel, with its rays flashing light on every side!"[45]

Such marvels were always in peril, however. Pieces shattered in the flames of every kiln, and many warped or fractured. In the eighteenth century, potters routinely doubled the quantity ordered in a large commission because half the raw-clay pieces would turn out "knock-kneed, flattened, or otherwise injured and spoilt."[46] Colors frequently went awry, and repellent shades—variously called "pig liver," "camel lung," "rat skin," "nose mucus," and "dribbling spittle"—made the firing a waste of time and money.[47] When tens of thousands of vessels bloated and collapsed into a rock-solid heap, the kiln owner faced ruin from losing months of labor and his capital investment.

As Dentrecolles points out, colossal amounts of ceramic debris piled up around Jingdezhen from more than a thousand years of production; but since "in China, everything is put to use," the city built on its mundane misfortune. Kiln rubbish served as filler for the brick walls of buildings and, mixed with stone and gravel, as masonry material. Dumped on the banks of the river below the town and pounded on by traffic for years, it eventually became the foundation for new markets and streets. Swept away in floods, porcelain shards paved the riverbed with a chromatic shimmer, affording the Jesuit "a delightful sight" as he strolled along the Chang on a sunlit day.

Clearly, Dentrecolles had an eye for the splendors of smoky, polluted Jingdezhen. He expresses wonder at the magic worked by blazing kilns on clay and pigment: "The fire makes [the murky shade of the painting under the glaze] appear in all its beauty, almost in the same way as the natural warmth of the sun makes the most beautiful butterflies, with all their tints, come out of their eggs." A potter showed him a vessel that seemed akin to milky quartz, the only one of hundreds that miraculously survived the firing. Some porcelains displayed vivid portraits of Chinese and Tatar ladies: "The drapery, the complexion, and the features were all exquisite, so that from a distance one might have thought they were pieces of enamel." An ivory-white ware bore designs traced so deftly "it appears as if a thin vapor spread over its surface." A glowing porcelain had a glaze so threaded with veins "one might think it was shattered into a thousand fragments without falling to pieces, so that it resembles a piece of mosaic work."

Similar enthusiasm for the wares of Jingdezhen is expressed by Tang, praised by contemporaries because "His Excellency's heart was in his pottery."[48] To some extent, the same could be said of Dentrecolles: intent on discovering the secrets of porcelain manufacture, he became enchanted by the artifact itself. Yet he also highlighted the excellence of porcelain as a way to win support for his all-important pastoral vocation: "Perhaps some pious individual, who admires the beautiful works that Jingdezhen furnishes to all Europe, would be zealous enough to consecrate a small portion of his wealth to the conversion of the workers who make them." He emphasizes that he has a sizable flock, increasing every year. In 1722, he tells the reader, he baptized nearly fifty adults, so his congregation is in urgent need of additional catechists and a larger temple.

"CHINESE WORKERS WHO MAKE PORCELAIN"

Many kinds of laborers and craftsmen provided materials and skills for pottery production, and besides kneaders of clay, the Christian converts perhaps included clay miners, stone crushers, wood choppers, basket weavers, rope makers, carpenters, ironmongers, barrel makers, brick molders, kiln builders and repairers, saggar makers, clay throwers and stampers, mold makers, wheel spinners, glaze mixers and appliers, kiln loaders, fire stokers, brokers, packers ("mat-men"), porters, and boatmen. "Pile assessors" made bids on odd lots of porcelains with blotches and cracks, goods they smartened up and sold at the flea market on the island in the Chang.

One-armed men stomped clay with their feet, and debtors swathed in sodden rags repaid loans by unloading the blistering kilns.

In workshops thatched with rice-straw, the elderly and blind sat on low stools grinding pigments with unglazed porcelain pestles. Youngsters rubbed decorative gilding on wares with garlic bulbs to fasten the material during firing; they crouched at the feet of potters, spinning their wheels by means of two bamboo sticks. Novice painters brushed fired pots, still warm from the oven, with the boiled juice of old tea leaves to produce dappled patterns in imitation of antique porcelains. Painters and enamelers, groups that included many women, worked in clean surroundings and received relatively good pay; hence they regarded themselves as superior to drudges such as clay humpers and kneaders.

Still, few pottery workers earned more than a pittance, in part because an abundance of laborers kept wages low. As Dentrecolles observes, "Jingdezhen is the refuge for innumerable poor families who can't support themselves in nearby villages." Workers received payment in *jiazhi,* circular copper coins pierced by a square hole. The shape derives from Chinese cosmology—an allusion to the sphere of Heaven encompassing the four-sided Earth—and goes back to the Zhou period. *Jiazhi* ultimately stems from *karshápana,* the Tamil (Indian) word for a small base-metal coin. Malays in Sumatra called their tin coin *kasha,* which the Portuguese rendered as *caixa* and the Dutch as *casjes,* the source for the English *cash.* A string of 1,000 copper-cash was worth one tael (an ounce) of silver in the early eighteenth century.[49]

A run-of-the-mill potter, making commonplace items such as bowls and saucers, received three cash for every twenty-six porcelain pieces (termed a "plank"), with the expectation by the workshop manager that he would turn out one hundred pieces every day. If a diligent potter produced six planks daily, he would earn 6.5 taels a year, roughly the same income as an agricultural laborer. Painters made 9 taels a year, and the most skilled potters received 12. In contrast, an ordinary bowl cost one-tenth of a tael (or 100 copper coins) and good-quality items cost 2 taels apiece (or 2,000 coins)—that is, one-third of a common potter's annual income. With the assistance of Jesuits in Macao around 1600, Carletti purchased an assortment of 650 good plates and bowls for 20 taels (or 300 cash apiece) and five exceptionally fine blue-and-white vases, made from "that which they call the flower of the earth," for 14 taels.[50]

Harsh treatment and miserly pay sometimes drove workers to sabotage, strikes, and rebellion. In the early fifteenth century, four thousand craftsmen tried to flee Jingdezhen, but soldiers dragged them back to their work sheds. Following a devastating flood in 1540, hunger riots erupted in the city and brought work to a standstill. Revolts also took place in 1574, 1597, and 1602. In 1604 independent potters, ordered by an imperial magistrate to provide more wares for Beijing than usual, burned down kilns in protest. Squabbles between gangs of workers from different districts of Jingdezhen at times spiraled into attacks on kiln owners; demands for payment on time escalated into work stoppages and uprisings.[51] Dentrecolles relates that valuable antique porcelains buried for safekeeping during these violent outbreaks occasionally came to light in the ruins of buildings or when cleaning out old, disused wells.

Fighting in the Rebellion of the Three Feudatories (1673–81), a major challenge to the new Qing dynasty, swept through Jingdezhen and caused destruction of the furnaces. Half the kiln owners lost their property, and a Dutch merchant recorded that there was "great mortality among the porcelain makers."[52] In 1681, trying to provide a more secure place for porcelain manufacture, the Kangxi emperor (r. 1661–1722)—who, Dentrecolles says, "never misses a thing"—unsuccessfully tried to transplant some potters and their materials to Beijing. Around the same time, hoping to trade directly with European merchants, a number of craftsmen moved their production south to the coast of Fujian province; but, as Dentrecolles reports, "they were not successful in their manufacture" so far from home.

The Qianlong emperor (r. 1736–95) condemned exploitation by kiln owners, instructing them to abstain from harsh behavior that provoked incidents. Imperial overseers, however, who had less concern for the welfare of their charges than for their own standing with distant superiors, often compounded the tribulations of workers. Qianlong concluded once, "The porcelains Tang Ying sent to the court this time are still of the old fashion. Why did he not follow the new models dispatched to him and produce new wares? All the costs and expenses of their production and transportation shall not be reimbursed from the court; instead they shall be borne by Tang himself."[53] If the superintendent followed common practice, however, he would have passed on his losses to the potters under his command.

According to Dentrecolles, whatever the range of wages in Jingdezhen, all "Chinese workers who make porcelain," including painters and enamelers, endured poor and wretched lives, subject to beatings by their masters if they made mistakes or failed to complete assignments. As expressed by a Song poet, hardship and inequality were the perennial fate of potters:

> Pots cover every inch of space before the door
> But there's not a single tile on the roof.
> Whereas the mansions of those who wouldn't soil their fingers with clay
> Bear tiles overlapping tightly like the scales of a fish.[54]

When wealthy entrepreneurs bought up shops and homes in the center of Jingdezhen in 1730, a local official complained, "Most merchants rent out dwellings in the pursuit of their despicable profession. They invariably throw out the old and the sick who are unable to work."[55] When there was a cutback in production, workers went back to their home villages to eke out an existence or stayed in the city and sold freshwater snails and fish in the streets.

Inspired by Western commissions, imperial agents entreated Dentrecolles to obtain novel designs from his compatriots that might gratify the emperor. Instead, the pastor responded to the pleas of his flock:

> The Christians implore me strongly not to furnish such models, for the mandarins are not so understanding as our merchants when the workers tell them that something is impractical. And the bastinado is often applied liberally before the mandarin gives up on a design which promises great advantages.

Although the faith preached by Dentrecolles no doubt provided a measure of consolation to his converts, for some it replicated the misery of their labor. He laments that in their worship they adopt "holy castigations, which they inflict upon their flesh: I have sometimes been obliged to send them away from the church to make them take a little rest."[56] They had an agitated longing for relics and rosaries; they put the holy water their pastor gave them into sealed porcelain jars to preserve its miraculous properties. Still, however zealous his converts, Dentrecolles probably was chagrined that they did not pay exclusive worship to their Christian savior. Indeed, he found himself sermonizing amid a Babylon of strange gods.

"MIRACLE-FANCIERS": FAITH AND FURNACE TRANSFORMATION

European missionaries in China discovered that both the common people and Confucian literati casually assimilated diverse religious notions and spurned the dogmatism of salvation religion. Religious cults in China did not emphasize doctrinaire belief, theological rigor, or transcendent truth. Rather, they focused on sanctioned tradition, ritual performance, and civic morality.[57] As a consequence, Chinese listeners displayed incredulity or indignation when Christian missionaries preached their one, true religion. According to Matteo Ricci:

> The number of idols in evidence throughout the kingdom of China is simply incredible. Not only are they on exhibition in the temples, where a single temple might contain thousands of them, but in nearly every private dwelling. . . . In public squares, in villages, on boats, and through the public buildings, this common abomination is the first thing to strike the attention of a spectator. Yet it is quite certain that comparatively few of these people have any faith in this unnatural and hideous fiction of idol worship. The only thing they are persuaded of in this respect is, that if their external devotion to idols brings them no good, at least it can do them no harm.[58]

Ricci detected the same spiritual insouciance among the Confucian elite, for they made "the very distracting error of believing that the more different ways there are of talking about religious questions, the more beneficial it will be for the common good."[59] He had to acknowledge, however, that "the great confusion of religious sects" in China had fostered tolerance of spirituality that worked to the advantage of the Jesuit mission.[60] It was just such forbearance that impelled a scholar-official to pay courteous reverence to the Christian god in a shrine erected by Dentrecolles near Jingdezhen. Chinese open-mindedness about religion sometimes meant that even the missionaries themselves were conscripted into folk worship. After his death in 1657, the Jesuit Etienne Faber became revered as a local deity in Shanxi province, with effigies of him placed in village pagodas. Shanghai clockmakers made Ricci their patron idol because of his construction of chiming timepieces, called "self-sounding bells" *(zimingzhong)*, for the imperial court.[61]

In fact, Jesuit missionaries left themselves open to such misunderstandings as a consequence of their policy of accommodation to Chinese practices and beliefs. They permitted Chinese Christians to use the crucifix in rituals at ancestral shrines for the dead, and, following Ricci, they maintained that Shangdi (Lord-on-High), the ancient Chinese term for immanent and universal order, was indistinguishable from Tianzhu (Lord of Heaven), their own coinage for Christ Crucified. In striving to reconcile Chinese culture with Christianity, Ricci ignited the Rites Controversy, a long, acrimonious debate over Jesuit incorporation of homage to Confucius (ca. 551–479 B.C.E.) and the Chinese sages of antiquity in Christian ritual.[62] Ricci believed that Christianity had been brought to China in the first century of the common era, around the same time as Buddhism, as a consequence of the supposed preaching of St. Thomas the Apostle in India. The Jesuit quotes a breviary of the Indian church: "Through St. Thomas the splendor of a life-giving faith flourished through all of India. Through St. Thomas the Kingdom of Heaven took wings and sped its flight to the Chinese."[63] Unfortunately, he goes on, in the course of time the Christian message became corrupted through error or the malice of opponents of the gospel. He believed that the Jesuit Mission was destined to restore the purity of Christianity in China, in part by demonstrating its compatibility with the original teachings of Confucius. These were highly contentious claims, and European critics accused the China Jesuits of discounting Christian saints and instead reciting *Sancti Confuci ora pro nobis* in their devotions.[64]

With confusion and ambiguity typical of Chinese conversion to Christianity at all levels of society, it is likely that Jingdezhen's "fresh faithful" (as Dentrecolles called his parishioners) treasured their jars of holy water while continuing to pay tribute to Tung in one of his many shrines. Dentrecolles explains that "as each profession has its own particular idol, and as divinity is bestowed as easily here as the rank of count or marquis is given in certain European countries, it is hardly surprising that there is a god of porcelain." Christians told him that veneration of Tung originated from the failure of potters to complete a commission from the emperor in the late Ming period for enormous fishbowls decorated with the five-clawed imperial dragon:

> It is said that once an emperor wanted them to make him porcelain from a model he provided. They told him many times that it was impossible, but all these remonstrances only served to excite his desire. During their lives, these emperors are regarded as divinities to be feared throughout China, and they think nothing should oppose their

wishes. His officers therefore redoubled their demands, and applied all kinds of harshness to the workers. These wretches spent all their money and tried their hardest, but they received only beatings in return. In a moment of despair, one threw himself into the burning furnace and was instantly consumed. The porcelains in that lot, so they say, came out perfectly beautiful and to the liking of the emperor, who then asked for nothing finer. From that time, the unfortunate man has been regarded as a hero, and he became in consequence the idol who watches over the workers in porcelain. I do not know whether his elevation has tempted other Chinese to take the same route in hope of a similar honor.

In his *Illustrated Explanation of the Miracles of the God of the Furnace* (1744), Tang Ying records a legend that one supervisor of the kilns occasionally saw Tung pointing out the finest pieces to him amid the flames. According to Song Yingxing, the porcelain idol once spoke to a potter in a dream, revealing the secret of firing a certain red-colored ware desired by the emperor: "News of this event immediately became widespread and [the extraordinary force that produced the vessel] was known as transmutation or chanciness in the kilns."[65]

Creation of the guardian spirit of the potters represents an instance of what Zhu calls "furnace transformations," the miracles and magic that take place in the kiln. "Porcelain," he writes, "is created out of the element *earth*, and combines in itself also the essential powers of the elements *water* and *fire*."[66] According to a Ming connoisseur, "The potter's clay moulds can be made by man's hands; but the magic changes that take place after they are fired in the foundry cannot be predicted." Another pottery expert explained, "The expression *kiln change* means magical transformations effected by fire. It is not merely a matter of the glaze changing colour. There are genuine cases of vessels taking on strange and wonderful shapes."[67]

"Miracle-fanciers," as Song calls them, identified two sorts of kiln transmutation: the rarest took place when "the magical nature of fire" altered the shape of the molded clay or produced an object with supernatural properties; more frequent wonders came to light when the porcelain turned a surprising color during firing or revealed the likenesses of animals not painted on it. Zhu reports the belief that clay vessels sometimes turned into jade in the kiln, an uncanny event that caused terror among officials and impelled them secretly to shatter the pieces with a hammer. Supervisors also destroyed pots when they unpredictably emerged as red as vermilion from the furnace, a perceived result of supernatural forces being stirred up by the planet Mars.[68]

As a consequence of magical change affecting the vessel in the kiln, cut flowers in a waterless vase "blossomed and bore fruit just as though they had coiling roots in the ground." A porcelain jar outside an official's residence resounded with melodies of organs and flutes; another exhaled wind and clouds throughout the day. A jar to which water and pebbles were added every day for years always remained not quite full. An ordinary porcelain bowl, purchased casually on the street, produced a kaleidoscope of images in its basin when water froze in it—a spray of peach blossoms, a peony branch crowned with flowers, wild geese soaring over a winter landscape. In the late Ming, the court ordered potters to make a sizable windscreen; but during the firing, one slab of clay supposedly turned into a bed and another metamorphosed into a ship three meters long, complete with porcelain sails and porcelain rigging. Terrified by the monstrous mutation, local officials "hammered it to pieces, not daring to forward it to the palace."[69]

Such eerie phenomena sprang from furnace transformation. Zhu suggests they "could not be produced by mortal hands" but must result from the oven having tapped into and disrupted supernatural forces.[70] A stele set up at a pottery center in the Song period declared, "Looking into the kiln, with its strong fire, one often sees insects, which must be gods in disguise, moving in shimmering water."[71] The perception was that the potter's kiln works magic, turning mundane material, the clay of the earth, into a substance that partakes of the sacred. Reproducing cosmic and alchemic processes, the kiln possesses the power to transubstantiate matter through employment of fire while the gods themselves act as celestial potters in shaping life from common clay.

Such notions were universal, flourishing wherever pottery was made. Around 26,000 years ago, in the Late Paleolithic (or Old Stone Age), mammoth hunters in the then-tundra region of Moravia (in the present-day Czech Republic) learned how to fire clay in bonfires, thereby creating the first ceramic (from Greek *keramos,* "potter's clay").[72] Decoration on some of the earliest ceramics resembles the crisscross and spiral patterns typical of basket weave, suggesting that clay-smeared textile containers used for cooking over open fires inspired creation of the first pots. Pottery was central to ritual feasting in the Late Paleolithic, and rows of marks inscribed on clay tokens suggest they were used for keeping track of animals and debts. The move toward large-scale herding of sheep, goats, and cattle stimulated further development of pottery inasmuch as people needed containers to store milk, cheese, and yogurt. As revealed by chemical analysis of shards from 6000 B.C.E., the earliest pottery made in England contained those dairy products. By 8000 B.C.E. fired pottery was widespread in western Asia and had been discovered independently in Japan. The oldest known pottery in China, dating to around 12,000 B.C.E., comes from several locations, including the Yangzi River delta, where it preceded the adoption of formal rice agriculture. In the Americas, the first pottery appeared around 6000 B.C.E. among hunter-gatherers in the tropical rainforest of the lower Amazon basin. It did not appear in areas such as Georgia and South Carolina until 2500 B.C.E. and in the midwestern United States until about two thousand years later.

Wherever it emerged, pottery technology most likely derived from making bread and porridge, which also involved grinding, adding water, kneading, shaping, and baking. Fired clay, one of the earliest synthetics, as well as the first material that humans entirely created by employing heat, represented a milestone in human development, a basis for the Neolithic revolution (from ca. 10,000 B.C.E.). Using ceramic vessels for cooking proved significant for processing and preserving foods, for brewing and distilling liquors, for rendering harmless some natural toxins in plant products, and for enhancing the nutritive value of plant and animal food. Knowledge gained from producing baked clay proved crucial in the development of metalworking, a hallmark of civilization, and taxation records kept by Mesopotamian temples on baked-clay cuneiform tablets represent the first step within settled communities toward systems of enumeration, writing, and formal education.

Long before domestication of plants and animals, artisans created ceramic figurines of deities and animal spirits, testimony to the primordial link between baked clay and faith in the supernatural. Two clay sculptures of bison were placed in a cave in southern France 15,000 years ago, and human figurines were molded in the same region 8,000 years later. Strange decoration on burial pottery from Neolithic Greece and Minoan Crete hints that potters made specialized wares for religious ceremonies. In northern China during the Neolithic, funerary artifacts included ceramic statues of humans with eyes of inlaid turquoise. Early rice-growing cultures along the Yangzi River produced earthenwares with incised designs that suggest shamans used the pottery for magical and occult practices.

Furthermore, the ability of clay to mimic metals, minerals, and other costly substances associated with holy beings gave ceramics a central role in spiritual life, as when turquoise, the semiprecious stone identified with Hathor, the Egyptian mother-goddess and bovine deity, was replicated in blue-green pottery as a votive offering. Ceramic decoration even may have originated in many societies as apotropaic protection, talismanic defense against the terrifying, awe-inspiring powers unleashed in furnace transformation.[73]

"GOD HAD KNEADED SOME CLAY": THE DIVINITIES OF POTTERY

Ancient Chinese myths portray demiurges molding human beings from loess, the yellow earth of the northern highlands.[74] Tang Ying records the belief that the five basic elements identified by Confucian and Daoist scholars—earth, fire, water, wood, and metal—obey enigmatic decrees established by "the Great Potter." In Japanese legend, the earliest emperor, the first human ruler descended from the gods, works as a potter in a realm set between Heaven and Earth. In the seventeenth century, Japanese Christians translated this conception into an oral tradition that envisions the Creator God (Deus) fusing clay with his own rib to fashion the first man. Egyptian myths speak of a divine potter who takes the guise of Ptah of Memphis, creator of the gods, or Khnemu, the ram-headed god of fertility. In ancient Mesopotamia, Enkil, the Sumerian god of creation, molded deities and sacred kings from a wad of clay, and "to return to one's clay" signified death for human beings.[75] In *The Epic of Gilgamesh* (ca. 2500 B.C.E.), Aruru, the mother of creation, labors as a potter to form Endiku, Gilgamesh's double, his alter ego:

> She moistened her hands,
> she pinched off some clay,
> she threw it into the wilderness,
> kneaded it, shaped it to her idea.[76]

Across West Africa, from Mali to Cameroon, potters (usually female) make vessels that are seen to govern supernatural interactions, including protection against disease, witchcraft, and psychological affliction.[77] "Spirit pots" made in northeastern Nigeria by the Ga'anda are representations of a community guardian who promotes healing and rich harvests. Iya Mapo, a spirit-goddess of the Yoruba of southern Nigeria, "the mother of mothers, quiet old mother of silent earth," shapes human beings from river mud. At death, they divide once more into water and earth, with the virtuous ascending to the heavens while the wicked fall into the "world of broken pots."[78] Among the Akan of southern Ghana, the deity Ayesu will appear for ritual performances only if an earthenware water pot is intact, for a cracked one signifies that the god has lost all spiritual power. In the same region, pots made by the Aja and Fon embody Legba, a deity associated with danger and deception.

Kane, the primary deity of ancient Hawaiian islanders, the god of procreation and ancestor of all kings, breathes life into an effigy formed from clay, naming the man created *Keli'ikuhonua,* "Red Earth." The Dayaks of Borneo believe that Kadjanka, the moon god, taught humans to mold jars from clay, the same substance from which other deities fashioned the sun and planets. In Hindu rituals of coronation in Khmer kingdoms of seventh-century Cambodia, vessels of clay (and metal) were seen as carriers of divine power, used for pouring consecrated water over pottery sculptures of gods. The Hindu Mother Goddess is associated with clay, the primordial plastic material, amorphous and undifferentiated, that forms the template of all created things. Potters in India pay special devotion to her manifestation as the smallpox goddess *(shitala mata)* and to Ganesha, the playful and propitious elephant-headed deity, son of Siva and Parvati.

Indian potters traditionally are looked upon in their communities as given power by the gods to transmute polluted earth into vessels for ritual and everyday use. The one million potters in contemporary India trace their descent from Prajapati, Lord of Creativity, and regardless of their low social status, they wear the sacred thread, a privilege otherwise granted only to the priestly caste of Brahmins. Viswakarman, another manifestation of Prajapati, represents divine energy, and as architect of the universe, he created Vishnu, the protector and preserver of worlds, from the disc of the sun, a celestial body with which the potter's wheel is identified.[79]

Notions of hallowed potters and potting divinities also are found in the myths and legends of the Kelabit of Borneo, Vietnamese of the Hong (Red) River Valley, Konso of Ethiopia, Fang of Gabon, Bakongo of the Congo, Zapotec of Mexico, Jívaro of the Andes, and Serrano Indians of California.[80] Pueblo Indians of the American Southwest regarded pots as having souls, and when a vessel cracked during firing, the sound signaled the release of a living being. In the *Popol Vuh* (Book of the Community), a mythological narrative and genealogy of the rulers of the Quiché Maya of the highlands of Guatemala, the names of creation gods are the same as those of potters who fashion life from clay.[81] The Incas of Peru believed that their supreme deity, Viracocha the Maker, shaped the world and mankind from the clay of Lake Titicaca, their sacred water. In the eighteenth century, Natchez Indians of Louisiana told a French traveler that "God had kneaded some clay, such as that which potters use, and had made it into a little man; and . . . finding it well formed, he blew up his work, and forthwith that little man had life, grew, acted, walked, and found himself a man perfectly well shaped."[82] This creation story probably derived from the preaching of Jesuit missionaries about how Yahweh, the Lord God of the Hebrew Scriptures, "formed man [Heb. *'adam*] from the clay of the earth [Heb. *adamah*]" (Gen. 2:7).

A version of this passage made its way into the Qur'an (15:26): "Behold, We have created man from potter's clay, moulded out of slime." Five other like statements ensured that the notion of a divine potter became a standard theme in Islam. In a baroque elaboration of the Qur'anic texts, Ibn-Jarir al-Tabari (ca. 838–923), a prominent Muslim historian, devotes pages of commentary to the creation of Adam, describing how the angel Gabriel brought clays of various colors to the deity, who shaped them with his own hands into an effigy of the first human until it was ready for the divine inspiration of spirit, "like potter's clay untouched by fire."[83] Farid al-Din 'Attar (d. 1221), the Persian mystic poet, wrote that inasmuch as God creates humans, he also holds them to strict account: "He's a potter who first makes pots with great skill and then smashes them Himself."[84]

Yahweh warned the prophet Jeremiah that "just like clay in the hands of the potter, so are you in My hands, O House of Israel!" (Jer. 18:8), a text on which Origen of Alexandria (ca. 185–ca. 254), a Christian theologian, based the homily that "God, the potter of our bodies, the Creator of our constitution," directs all human affairs. Origen also expounded on St. Paul's assertion that God created man by wielding the sort of sovereign command that a potter has over clay (Rom. 19:21).[85] The apostle to the Gentiles declared that while the light of Christ shines in the hearts of his creation, "we have this treasure in clay vessels, to show that the transcendent power belongs to God, not to us" (Cor. 4:7). Italian potters of the sixteenth century acknowledged the same dependence on divinity. Making the sign of the cross before firing their kilns "in the name of Christ Jesus," they regarded their work as resonant with spiritual forces. Their beliefs about the latter, however, also seem to derive from a context that has little to do with the Christian religion: the potters held that "if the firing happens to take place at the waning of the moon, the fire lacks brightness in the same manner as the moon its splendor."[86]

When Dentrecolles "preached Him who made the first man out of clay," he drew on the rich tradition of ceramic metaphors in his own culture that had much in common with the daily experience and spiritual perceptions of the potters of Jingdezhen. He clearly believed that if his Chinese converts did not exclusively worship the Christian god, then the new religion eventually would triumph over their pagan superstition by a measure of salutary association with the idols and spirits to whom they still rendered devotion.

In fact, Dentrecolles knew about one such accommodation, an instance in which the image of a Chinese deity entered the Christian communion of saints. Guanyin, the Buddhist goddess of compassion, helped those in need.[87] Among many other kinds of supplicants, prostitutes regarded her as their patron. One of the deity's thirty-three forms was *Baiyi* ("white garment"), and representations of her seated on a white lotus were common on family altars. As part of the cult of "the water moon," she protected seafarers, who built shrines to her on the shorelines of East Asia. A maternal figure promising salvation and fertility—one of her titles was "sender of sons" *(songzi Guanyin)*—by the sixteenth century, she had become identified in southern China with the Virgin Mary. Potters portrayed both figures with a rosary, and Christians, whether converts or Westerners, referred to porcelains of Guanyin as *Sancta Maria*. When a Dominican friar visited a Canton temple in 1556 and saw a statuette of Guanyin holding a baby boy, he paid reverence it to as "the image of Our Lady, made by the ancient Christians," believers who came to China (he assumed) with St. Thomas, the legendary missionary.[88]

Artisans carved statues of Guanyin/Mary in elephant ivory in China and the Philippines, with some making their way to Mexico City by the early seventeenth century. Decades later, a porcelain representation from Fujian, probably copied from ivory, sold for £1 in London, a sum equivalent to a skilled artisan's weekly income (or to three taels of silver in China). Mary II of England (r. 1689–94) purchased one of the Fujian pieces for her celebrated porcelain collection, and the Meissen manufactory began producing its own copies of Guanyin as soon at it created a recipe for porcelain. (See figure 3.)

Partisans of Guanyin in Beijing venerated a porcelain statue of her because it purportedly bore an exact likeness, a miraculous result of the presence of the goddess in the furnace during the firing of the piece. In the Ming period, potters decorated a platter with an image of Guanyin surrounded by supernatural figures and worshipers, a rare motif that a wealthy patron probably commissioned. Dentrecolles notes that figurines of Guanyin with a baby on her lap were made in Jingdezhen and sold in its shops. Some came into the hands of Japanese Christians, who referred to the figure as Maria-Kannon (Mary-Guanyin) and prayed before her effigy in their clandestine worship. The statues may have comforted Dentrecolles's parishioners in their exotic faith by a fruitful confusion between the Chinese and Christian Madonnas. As the Jesuit confesses, "Ministers of the Gospel, especially in China, must join the discretion of the serpent to the simplicity of the dove."[89]

When the town managers of Jingdezhen ordered construction of a new temple to the god of porcelain, Dentrecolles, by virtue of his friendship with an unnamed superintendent of the kilns, won exemption for his converts from forced labor on it. But he coveted the house of worship of the Queen of Heaven (Tianhou), "a palace which surpasses for magnificence all the other temples" in the porcelain city. He told his congregation that he looked forward to the day when "this temple in fact will become a basilica dedicated to the true Queen of Heaven," a figure who shared some attributes with the Chinese goddess. He records that a porcelain merchant who had amassed a fortune in silver from trade with the Philippines and Spanish America financed the Queen's temple to fulfill a vow he made when the goddess appeared at the height of a tempest and rescued him from shipwreck.

The Queen was the patron saint of fishermen and sailors—as was the Virgin Mary, who included among her titles *Stella Maris* (Star of the Sea). Like Guanyin, the Queen attracted homage from men engaged in saltwater commerce; under the cult name Mazu, worship of her was centered on Meizhou Island on the Fujian coast. Shrines and sanctuaries to her stood in Chinese ports, as well as along rivers and the Grand Canal; she was believed to dwell on Putuoshan Island on the Zhejiang littoral. A coastal official declared that "all those who welcome the morning tide and see the evening tide hold her in their hearts."[90] Dependent on remote markets for their livelihood and never setting eyes on the sea, the landlocked residents of the porcelain city shared that piety.

FROM JINGDEZHEN TO THE SEA

Worshipers of the Queen of Heaven in Jingdezhen prayed she would guide their porcelains to the sea. The grandeur of her temple testified to the significance of water transportation for the city's prosperity. Dentrecolles emphasizes that the inhabitants depended on the Chang River for their sustenance and livelihood "because everything that is consumed there has to come from somewhere else." He calculates that the city needed 10,000 loads of rice and 1,000 pigs every day, all of which arrived by boat and barge. Material for the kilns also came by water. An endless line of boats filled with clay came one hundred kilometers downriver from Qimen in Anhui province, north of Jiangxi, since deposits around Jingdezhen largely had been played out by the early eighteenth century. Fir and pine, light and resinous woods that produced the best flames in the kilns, also came from Anhui. An average kiln consumed 11,000 kilograms of wood in a firing, while a large one used some 64 tons a day. Centuries of pottery production had deforested the nearby hills, so potters relied on timber floated down the Chang. Limestone and fern, materials used for making glaze, came by boat from south of the city.

Merchants from guilds in the commercial center of Huizhou (or Xin'an) in southern Anhui sold wood, rice, and cotton in Jingdezhen, some 180 kilometers away by river. According to a Ming chronicler, a few traders were worth a million strings of cash, and a number had fortunes of 250,000 taels. Pawnbrokers and store managers, the Huizhou entrepreneurs first came to prominence in national markets as licensed salt traders during the Song period. In the Ming, they invested heavily in porcelain production, in part by advancing loans to needy potters and taking repayment by purchasing their wares at reduced prices. As promoters of a thriving printing industry, they also provided ceramic decorators with woodblock prints as a source of new images and motifs that would appeal to scholars and their social equivalents among the educated, landowning gentry *(shidafu)*.[91]

Naturally, Huizhou merchants used their commercial networks to distribute the wares of Jingdezhen. Almost all porcelain was sold before it left the city. Buyers belonging to traveling merchants' associations negotiated with one of the fifty-odd brokerage firms in Jingdezhen that controlled sales, packing, and transport. Taking advantage of such services dramatically reduced transaction costs and increased the efficiency of the market. According to an account of procedures from the Qing period, brokers bargained with sellers, checked supply lists, and prepared the tax-declaration certificates required by the government; they also took delivery of merchandise and arranged for its shipping.[92]

Given the volume of Jingdezhen's output, as well as the weight and fragility of porcelain, water transportation was vital. As Adam Smith underscored, "By means of water-carriage a more extensive market is opened to every sort of industry than what land-carriage alone can afford it, so it is upon the sea-coast, and along the banks of navigable rivers, that industry of every kind naturally begins to subdivide and improve itself." The porcelain city employed techniques of mass production as a consequence of far-flung markets being reached (as Smith described it) by "an inland navigation much more extensive than that of the Nile or the Ganges, or perhaps than both of them put together." Travel by barge and boat, Ricci declared, stood out as one of China's wonders: "This country is so thoroughly covered by an intersecting network of rivers and canals that it is possible to travel almost anywhere by water."[93]

Tang Ying describes how porcelains were shipped abroad. Workers tied up coarse wares in matting, making a bundle of thirty to sixty pieces wound with bamboo strips; they wrapped high-quality vessels in paper and rice straw, and then packed them in barrels holding six hundred pieces and weighing 225 kilograms. The bundles and barrels went onto long, lightly built boats anchored two and three deep at the riverbank. The craft traveled eighty kilometers southwest to the town of Raozhu, close to where the Chang flows into Lake Boyang. During rainy periods, the passage was rough in the gorges near Jingdezhen: tons of porcelain shards made the river shallow, creating perilous rapids in places.

Dentrecolles knew the trip well inasmuch as he made his home in Raozhou and commuted upriver on a regular basis. There was no inn for travelers in Jingdezhen, and magistrates there required strangers to stay the night on a boat or in the home of a resident who would go surety for their conduct. Furthermore, Raozhou had lower living costs than did Jingdezhen. Porcelain buyers often stayed in Raozhou, sparing themselves a tiring trip upriver. Whatever the fame of Jingdezhen, its pottery commonly was called Rao ware in China and, according to Dentrecolles, it was characterized in the Song period as "jade from Rao" and as "precious jewels of Raozhou." A VOC merchant visited the town in 1656 and recorded:

> Of all the vessels sailing to the South as well as to the North [from Raozhou], some supply and some supplement their cargo-capacity with porcelain, and mostly coppen [that is, cups and bowls] . . . and we found the mandarins accompanying us not sleepy in this, for they laid in as much of this pottery as their vessels could drag away or swallow up . . . to peddle to their profit at Nanchin [Nanjing] and elsewhere; we saw a fine occasion to contract for some rare porcelain, but we were lacking appropriate samples, in demand with us, which was a pity.[94]

At Raozhou, stevedores transferred the porcelains into deep-draught junks for the voyage across Lake Boyang, the largest body of freshwater in China and the scene of innumerable shipwrecks. Mariners appointed to rescue ships endangered by storms had the reputation of being "most forward to work the Merchants Destruction, in order to enrich themselves with the Spoil, especially if they think they can do it without being discover'd."[95] Escaping that hazard, many boats made the four-day sail to Jiujiang on the northern coast of the lake, a port that featured a large porcelain market. From there, cargoes of pottery were transshipped farther northeast on the Yangzi River to Nanjing. According to Fernão Mendes Pinto (ca. 1510–83), a Portuguese writer and traveler, rich men of that important city possessed "endless quantities of very fine porcelain pieces, which are like precious jewels to them."[96] Finally, by way of barges on the Grand Canal, the ceramics made it to Beijing, over a thousand traveling kilometers from Jingdezhen.

Sailing southwest on Lake Boyang, other ships leaving Raozhou headed for Nanchang, capital of Jiangxi province, where they went up the Gan River for portage of their cargoes south over what was popularly known as the "Mountain Pass of the Plum Trees."[97] Ricci describes it as "the most celebrated mountain pass in the whole kingdom," the scene of a never-ending procession of porters, pack mules, and palanquins.[98] In the nine-hundred-kilometer journey from Jingdezhen to Canton, the day's trek over the mountain was the only resort to land transportation. Countless bundles and barrels of porcelain were trundled in wheelbarrows and carried by bearers on the twenty-kilometer route, a task eased by paved walkways and steps carved into the slopes during the early Tang period.

Silver passed in the other direction, for a substantial percentage of the white metal entering southern Chinese ports went north over the mountain to pay for porcelain and other consumer goods, such as silk textile and lacquerware. China adopted a silver standard in the 1430s, so silver fetched high prices there while regions possessing a lot of silver—Japan and Spain (from its Mexican and Peruvian mines)—had an insatiable demand for Chinese merchandise. In 1567 the customs office on the pass collected slightly over one million taels based on a levy of one-thirtieth of declared value of goods; the total climbed higher when silver from Spanish America began streaming in about ten years later.[99] A Portuguese merchant in the early seventeenth century remarked that "silver wanders throughout all the world in its peregrinations before flocking to China, where it remains, as if it were its natural center."[100] With the entry of Westerners to Asian trade, silver became the principal medium of exchange linking the extremities of Eurasia as well as those of the global ecumene. As Montesquieu wrote, silver as a commodity is "the basis for the greatest commerce of the universe."[101] Along with silver ingots, merchants trudging north over the pass carried imports of tortoise shell, ivory, coral, black pepper, and incense woods. In the nineteenth century, crates of Indian opium, shipped to China by the British, went by the same route.

Near the pinnacle of the mountain, Ricci says, travelers passed through "a tremendous gate built into the precipitous rocks" that marked the border between the provinces of Jiangxi and Guangdong.[102] After reaching the southern foot of the mountain, the porcelains went onto boats once again to be shipped south down the Gan to the docks of Canton. In good weather, the trip went quickly. In 1693 Joachim Bouvet (1656–1730), a Jesuit stationed at the court in Beijing, made it from Nanchang to Canton in only twenty days. He was starting a four-year journey to France, marked by long stays in Thailand and India. When he finally returned to China in 1698, he brought Dentrecolles with him.

2

The Secrets of Porcelain

China and the West in the Eighteenth Century

In 1685 Joachim Bouvet and five fellow mathematicians constituted the first French Jesuit mission to China. Bouvet received the prestigious assignment to tutor the Kangxi emperor in geometry and philosophy, a task he believed would further the cause of Christianity. For the Jesuits, the most learned of the clerical orders, conversion and the search for knowledge went hand in hand. The *Constitution* drawn up by Ignatius of Loyola (1491–1556), founder of the Jesuits, called for the systematic gathering, transmission, and publication of information of all kinds. In the first phase of its history, between its establishment in 1534 and its general suppression in 1782, the Society of Jesus published some 5,600 titles on science, including works on medicine, geography, agriculture, and natural history.

From 1735 to 1795 Jesuits in Beijing translated more than four hundred works from Chinese. Dispatching voluminous reports to their superiors back home, Ricci, Bouvet, and Dentrecolles formed part of the first global information network.[1] Their motives were not primarily altruistic or scientific, however: they trusted that spreading Western knowledge to China would clandestinely scatter the seeds of Christianity. As Ricci explained, "Whoever may think that ethics, physics and mathematics are not important in the work of the Church, is unacquainted with the taste of the Chinese, who are slow to take a salutary spiritual potion, unless it is seasoned with an intellectual flavoring."[2] At the same time, Ricci and his companions believed that sending Chinese knowledge back home would help the West compete with the most productive economy in the world.

Jesuits brought Euclidean geometry, Copernican astronomy, Renaissance perspective drawing, and Western musical theory to China. Georg Joseph Kamel (1661–1706), a Jesuit from Moravia who served in the Philippines, collected dried plants from China and sent them to Europe, where the Swedish botanist Carolus Linnaeus (1707–78) used them in constructing his authoritative system of taxonomic classification and binomial nomenclature. Jesuits in China sent models of Chinese sugar milling technology to Peru, and they introduced rhubarb to Europe as a sovereign remedy for stomach disorders. In the 1630s Jesuits serving in Peru carried parcels of the dried bark of the cinchona tree, the source of quinine, to Rome, where it worked wonders on fever victims. A London journal of 1658 referred to it as "the excellent powder known by the name of Jesuits Powder, which cureth all manner of Agues."[3] By way of Spanish galleons sailing from Acapulco to Manila, Jesuits in Peru also sent cinchona bark to China, where in 1691 Bouvet won the goodwill of the Kangxi emperor by treating him for malaria.

The dramatic success of "Jesuits Powder" helped Bouvet persuade Kangxi the next year to extend toleration to Christianity, making it legally possible to establish churches in the provinces. For the first time, all China was open to Christian missionaries, and at least in the eyes of Bouvet and his colleagues, the Middle Kingdom itself seemed ripe for conversion. Soon after, Bouvet journeyed to France to recruit Jesuit experts in science, mathematics, and technology for service in China, for the emperor aspired to establish an equivalent of Louis XIV's *Académie Royale.* As a result of Bouvet's lobbying efforts, Dentrecolles and nine other Jesuits sailed on the *Amphitrite* from La Rochelle in March 1698 and arrived in Canton in November. Bouvet rushed the newcomers to an audience with Kangxi, after which they were assigned to their posts. In all likelihood, Bouvet was instrumental in sending Dentrecolles to Jingdezhen, for there was intense interest at the French court (and everywhere else in the West) in discovering how porcelain was made. Like his Chinese counterpart, Louis XIV sought to profit from knowledge possessed by the most powerful kingdom on the distant side of the world.

"THE LAND OF PROMISE": CHINA AND THE WEST

Dentrecolles begins his letter of 1712 to Louis-François Orry, treasurer of the Jesuit mission in China and India, by declaring that curiosity alone never would have impelled his laborious research into porcelain. This is a hint that he, like his Jesuit confrères working in the enamel and painting workshops of Kangxi's palace, resented being sidetracked from his pastoral obligations. "Nevertheless, I believe," he adds immediately, "that a detailed description of all that is concerned with this sort of work should be of some use in Europe." His superiors in Paris told him that his letter was published in 1717 in the Jesuit compilation of letters from China, *Lettres édifiantes.* They sent him a copy of the volume, for Detrecolles clearly had it before him when he wrote his second letter five years later.

They informed Dentrecolles, however, that his information had not been sufficient for French potters to learn how to make porcelain. He therefore returned to his task, though with diminished enthusiasm. Whereas his first report conveyed a fascination with porcelain and the particulars of life in Jingdezhen, his letter of 1722 (published in 1726) is dry and matter-of-fact, rarely venturing beyond an inventory of glaze components and coloring effects. He expresses the diffident hope that "the knowledge I am about to pass on to you will be of some help to the makers of earthenware *[faience]*, if they are unable to attain the perfection of Chinese porcelain." He puts forward many proposals toward that end: white Malta earth may substitute for a certain Chinese clay, crab-apple wood may be used for making glaze instead of Chinese fern, charcoal made from willow or elder might replace that from bamboo, European soapstone could be employed in place of a chalky substance known only in China. He assures his readers that "those works that are impossible to do in China may easily be done in Europe if one could find the same kind of materials there."

Dentrecolles's superiors plainly sent him to Jingdezhen on a mission of industrial espionage, no doubt with the pious hope he would save some souls along the way. His letters represent one of the earliest and most calculated cases of an effort to implement mercantilist economic strategies of technology transfer, import substitution, and product innovation. His work also inspired later generations to pursue the same aims. After reading Dentrecolles's reports in du Halde's *Description de l'Empire de la Chine,* Sir Joseph Banks (1743–1820), president of the Royal Society of London for forty-two years, implored Josiah Wedgwood to send a skilled potter to Jingdezhen to discover the secrets of making porcelain. Banks also suggested to a British envoy on his way to China that "a few learned men among their [porcelain and tea] workmen could acquire immense knowledge which would be worth untold wealth."[4]

The background of Orry and Bouvet gives further evidence of the nature and context of Dentrecolles's undertaking. Orry belonged to a powerful dynasty that held high positions in French government as secretaries of the treasury *(intendants des finances).* In the late seventeenth century, Orry noblemen also served as controllers of the French East India Company, the syndicate that purchased the *Amphitrite.* Eager to develop a formula for porcelain in France, Orry instructed Dentrecolles to send him samples of clay. A later member of the dynasty, Jean-Henry-Louis Orry, comte de Fulvy (1703–51), the younger brother of Louis XIV's minister of finance, had exceptionally close links to Chinese ceramics. He held office as an *intendant,* superintended trade with Asia, and collected Chinese porcelain, some pieces of which carried his coat of arms. In 1747 he helped found the Vincennes porcelain manufactory, the predecessor of the royal enterprise of Sèvres. An aristocrat close to Dentrecolles also may have been interested in collecting porcelain and learning about its manufacture: Marquis Jean-Claude de Broissia gave financial support to the Jingdezhen church, apparently in memory of his Jesuit nephew, who went to China on the *Amphitrite,* stayed in Raozhou around the same time as Dentrecolles, and died in 1704.[5]

During his stay in France in 1697–98, Bouvet attended the court at Versailles wearing the robes of a Chinese mandarin, a fashion familiar to French nobles, who on occasion masqueraded in the costume at balls and festivals. He consulted with officials of high rank, such as the Orry, and he collaborated with Louis-François Orry, who helped negotiate the sale of the *Amphitrite* to the French East India Company. Since France was at war and the government lacked funds to finance trade ventures, Bouvet convinced the Indies company to extend its ambitions to China, and he raised capital for the venture. It proved profitable: shareholders in the voyages of the *Amphitrite* realized a dividend of 50 percent on their investment.[6]

One of the most gifted and controversial of the China Jesuits, Bouvet played a leading role in the Rites Controversy. He had close associations with the group known as the Figurists, who maintained that ancient Chinese religion was linked to Christianity by virtue of the spread of the sons of Noah around the world. As Bouvet predicted to a correspondent, "One day we will arrive at an analysis which will reduce [Chinese ideographic script] to Egyptian hieroglyphs and demonstrate that both are the writing used among the learned before the Flood."[7] He even believed that the diagrams in the classic *Book of Changes* (Yijing), an ancient Chinese treatise on divination, could be decoded to reveal the truths of his faith. Such extraordinary notions, including the idea that France and China shared a cultural and religious identity, motivated his appeals for royal support for the Jesuit mission.

As an envoy of Kangxi to Louis XIV, Bouvet aspired to bring the West and China closer together, ameliorating the differences between them under the benign auspices of commercial exchange, European science, and Christian evangelism. He received a cordial hearing from Louis XIV, who, as sovereign of the most powerful realm in Europe, believed he must overshadow other monarchs in dealing with the greatest ruler in Asia. The king had sent the first French Jesuits to China just thirteen years earlier, and he considered that more were needed to counter the dominant presence of Portuguese clerics at the imperial court. Bouvet presented Louis XIV with gifts from Kangxi of forty-nine Chinese volumes and a set of engravings. One of the latter most likely provided the royal Beauvais tapestry works with the design for *The Audience of the Emperor,* which depicts the Son of Heaven enthroned in the midst of exotic plants, pagodas, and blue-and-white porcelain bowls.

Bouvet, who regarded Kangxi as a potential Constantine for China, a ruler who would convert his empire to Christianity, wrote *Portrait historique de empéreur de la Chine* (1697), replete with fulsome comparisons between the Chinese and French monarchs, and dedicated it to Louis XIV. Bouvet unctuously informed the Sun King:

> The Jesuits . . . were not a little surprised to meet at the utmost corner of the Earth with what they had never seen before but in *France,* that is to say, a Prince, who, like Yourself, has improved his sublime Genius by the Greatness of Soul, which alone renders him worthy of the greatest Empire of the Universe. . . . In short, a Prince . . . who would without question be accounted the most Glorious Monarch upon Earth, if his Reign had not been coincident with that of Your Majesty.[8]

Bouvet sent a copy of the biography to Gottfried Wilhelm Leibniz (1646–1716), the renowned philosopher and court librarian of the Elector of Hanover. Already engaged in a stimulating correspondence with Jesuits in Beijing, Leibniz responded by translating the work into Latin and publishing it in the second edition of his *Novissima Sinica* (Latest News of China) in 1699.

Leibniz's embrace of Bouvet's work stemmed from the breathtaking potential both men saw in the encounter between the West and China. In *Novissima Sinica,* Leibniz declares that the location of the civilizations at the two extremes of the Eurasian ecumene is a providential sign that eventually they would join in a vital cultural fusion. Identifying China as an "anti-Europe" or an "oriental Europe," he believed that if such dissimilar cultures could come to understand and learn from each other, then the path toward universal peace and harmony surely would be open to all mankind.[9] In general, he thought that Western and Chinese civilizations met on equal terms, each with its respective strengths, "so that now they win, now we."[10] Each had something to offer the other: the West had the lead in science, mathematics, and military technology; China excelled in "the precepts of civil life," that is, in law, ethics, and natural religion.

This evenhanded approach, however, turned out to have significant implications for future perspectives on China since it opened the door for a later focus on a supposed disparity in the respective cultural endowments. In his *Siècle de Louis XIV* (begun in the 1730s and published in 1751), Voltaire suggested that China had achieved perfection in morals and law two millennia earlier but that this very triumph, with its concomitant "respect for the ancient masters," had blocked progress in the sciences, "which is the work of time and boldness of genius."[11] Educated by the Jesuits, Voltaire had high regard for China—he kept a portrait of Confucius on his desk, probably taken from du Halde's *Description*—but he had planted the seeds of an unchanging Asian monolith in European thought. For Leibniz, the "anti-Europe" of China represented a model of virtue; by the later eighteenth century, it denoted a stagnant society that must be broken open by the dynamic, free-trading West.

Echoing Ricci, Leibniz proclaims in *Novissima Sinica* that the Chinese effectively follow the teachings of Christ without the grace of knowing Christianity. He recommends that instead of Jesuits being sent to preach in the Middle Kingdom, Confucian missionaries should come to Europe to instruct its scandalous Christians in morality. (Echoing Leibniz, Voltaire praises China for not having the "rage to make proselytes," a spiritual affliction peculiar to the Christian West.)[12] With all his veneration of China, however, Leibniz remained practical and realistic. He thought that inasmuch as China's several religions were incompatible with Christianity, imperial military force would be used to impose the Christian faith, an eventuality he deplored. He also worried that the balance of civilizations would be tipped toward China by virtue of the zealous activity of those whom Robert Burton (1577–1640), in *The Anatomy of Melancholy* (1621), called "land-leaping Jesuits," the very individuals bringing China and the West so much closer together.[13]

In Rome in 1689, Leibniz urged Claudio Filippo Grimaldi (1638–1712), a Jesuit with whom he had corresponded about Chinese manufacture of paper and porcelain, "not to worry so much about getting things European to the Chinese, but rather about getting remarkable Chinese inventions to us; otherwise little profit will be derived from the China mission."[14] He wrote to a German noble in 1690 that he feared "we are carrying all the secrets of our sciences to [the Chinese] which someday will be employed against Christians."[15] A year later, he warned an envoy of Louis XIV that "when the Chinese will have learned our sciences through these [Jesuit] priests, Europe will no longer have an edge on them and that is where our superiority will end."[16]

Leibniz feared that the Chinese would "close their doors" after they learned what they wanted from Westerners.[17] Not only that, the Chinese might use their expertise in manufacturing to co-opt the European export trade. In his large collection of books on China, the German philosopher possessed the *Tratados históricos* (1676) of Domingo Navarette (1618–86), a Spanish Dominican who lived in the Middle Kingdom from 1659 to 1664. Navarrete encouraged Leibniz's exalted view of China, for he declared that it was the "Land of Promise" mentioned in the Bible and that the earthly Paradise probably was situated there. But Leibniz also read the friar's warning that "the Chinese are very ingenious at imitation, they have imitated to perfection whatsoever they have seen brought out of Europe. In the province of Canton, they have counterfeited several things so exactly, that they sell them Inland for Goods brought out from Europe."[18]

Bouvet sent Leibniz a list of things being shipped from Canton on the return voyage of the *Amphitrite,* including ginseng, musical instruments, lacquer furniture, and "dragon porcelains" intended for the use of the emperor; "but," he writes, "we hope seriously in the future to extract from the Chinese other things much more useful to Europe, so as not to become inferior to them by so freely placing our best knowledge at their disposal, thus benefiting from the excellent advice you have given us."[19] Of course, the Sino-Western balance tilted most conspicuously toward China in production of high-quality ceramics. As Dentrecolles remarks, "The Chinese have not crossed the seas to seek for glass from Europe," even though they admire it, but Westerners, driven by "cupidity and greed," have an insatiable appetite for porcelain.

"AN ABYSS FOR GOLD AND SILVER": ASIAN TRADE AND THE WEST

Recognizing that China was superior in manufacture of pottery, observers such as Leibniz and Jean-Baptiste Colbert (1618–83), Louis XIV's finance minister, believed that European states suffered economically from that dominance. In 1664 Colbert founded the French East India Company, and two years later he ordered construction of the port of Lorient on the Bay of Biscay as a center for commerce with Asia. The minister intended the Compagnie and Lorient to compete with the VOC and Amsterdam, developing trade with the East that would reduce France's loss of precious metal.[20] The astonishing commercial success of the Dutch Republic had taught European ministers that long-distance trade demanded a state that would promote economic and institutional policies aimed at benefiting the nation, including maintaining a merchant marine, establishing systems of patent law, and sponsoring research to develop new technologies. Above all, they must guard against the threat posed by foreign manufactures.

Advocates of mercantilist policies gauged national wealth in terms of possession of bullion and a favorable balance of exchange. The reigning economic opinion was that a country faced injury if its specie went to pay for another country's products. Regrettably, Europeans discovered that China had no desire or need for its manufactured products. An agent in Canton wrote to the directors of the English East India Company in 1669, "We cannot tell what to advise your Honours to send to these parts, the Natives being fond of nothing but silver and lead; and probably if the rest of your Goods were thrown over board at Sea, your Cargoes home would not be much less."[21] European spending on foreign manufactures therefore spurred calls for halting the outflow of precious metal and for legislation against imports such as porcelain, Indian cottons (especially muslins and printed calicoes), Chinese wallpaper, and lacquer furniture. The economic and legal tools of mercantilism were restrictive tariffs, state-approved monopolies, and government subsidies for manufactures.

This outlook dominated economic thinking and official action through most of the eighteenth century. In the *Wealth of Nations,* though, Smith asserted that "the attention of government never was so unnecessarily employed, as when directed to watch over the preservation or increase of the quantity of money in any country."[22] He argued that the annual produce of land and labor (or what is now termed gross domestic product) remains the same no matter how much capital flows abroad. He regarded a nation obsessed with retaining gold and silver as succumbing to a kind of magical thinking, just as if a country hoarded an "incredible augmentation of pots and pans" because the utensils are made of iron. He emphasized the beneficial circulation of gold and silver to distant trading countries and the use of precious metals to facilitate exchange among the numerous nations that together make up "the great mercantile republik" of the world.[23]

Smith's innovative perspective arose from his consideration of the enormous commercial growth that had expanded to encompass much of the globe by the late eighteenth century. Around two hundred years earlier, however, the terms of international exchange seemed more a menace than a benefit to Western observers. Thomas Mun (1571–1641), a director of the EIC and the author of *A Discourse on Trade, from England unto the East Indies* (1621), asserted that the shortage of silver bullion "has been, and is, a general disease of all nations[,] . . . but it seems that the malady is grown mortal here with us, and therefore cries out for remedy."[24] Sir Thomas Roe (1581–1644), English ambassador to the court of the Mughal empire (1526–1720) of India, lamented what he regarded as the drain of his kingdom's lifeblood: "Seeing our state cannot beare the exportation of mony, except some new trade can be discouered from the East to serue this Kingdome, it must fall to ground by the weaknes of itts owne leggs."[25] A late-seventeenth-century English writer complained that imports from Asia "hinder the consumption of our own Manufactures, and more especially when those Imports are chiefly purchased of our own Bullion or Treasure."[26] The French physician François Bernier (1620–88), who spent years at the Mughal court, described India as "an abyss for gold and silver," a phrase mercantilist tracts of the eighteenth century wore to a cliché in bewailing the West's handicap in commerce with all of Asia.[27] In his *Essay on Tea* (1756), Jonas Hanway (1712–86) complained that "India and China are such gainers on their trade with Europe, that they draw away, by sensible degrees, all the gold and silver which are not consumed, or retained in utensils, in this quarter of the globe."[28]

Linnaeus, Smith's senior by sixteen years, vehemently rejected free-trade arguments. Like Roe, Bernier, and Hanway, he regarded Asia as a gigantic parasite sucking away European treasure. Driven by utopian dreams of import substitution, he collected plants from around the world so that frigid Lapland could be transformed eventually into a Baltic East Indies, producing its own sugar, opium, cotton, cinnamon, tea, and silk. He wrote that he "considered nothing more important than to close that gate [to China] through which all silver of Europe disappears," especially since it was being shipped away for no more than "dry leaves of bushes and thin threads, spun by caterpillars." He urged his correspondents in Asia to bring back from China "a Tea bush in a pot" and "an untilled piece of the original porcelain soil." His scorn for China no doubt intensified after one porcelain dinner service he ordered shattered in transport and another arrived intact yet decorated in a disagreeable shade of red.

In the view of many Western observers like Linnaeus, it seemed shockingly profligate to hurl precious metal into the abyss for commodities such as porcelain, cotton, and spice. A mid-seventeenth-century English commentator groused about paying "good silver and gold for Chinese shards, rags, and herbs."[29] Henry Fielding (1707–54), the English novelist, regarded spending on porcelain as indicative of the corruption of far-flung empire—"the gold of one Indies run away with the mud of another."[30] In a book on Parisian life, Louis-Sébastian Mercier (1740–1814) noted the folly of paying high prices for the Chinese ceramic or its European imitations: "What a wretched luxury porcelains are! A cat, with one tap of its paw, can do more damage than the devastation of twenty arpents of land."[31] In 1755 a French writer on lacquerware protested that "immense sums leave Europe each year, only to be absorbed by the vast regions of Asia . . . [with Westerners confronting countless dangers] in order to retrieve for their fellow citizens nothing but some glazed wood that the simplest use and smallest accident could destroy."[32] French experts saw the hemorrhage of bullion as a special problem for their country: in the eighteenth century, all but a small percentage of France's Asian imports, primarily from India and China, were paid for in American silver, cargoes of which French sea captains picked up in southern Spanish ports on their outbound voyages.[33]

If economic harm suffered from the drain of specie actually was more theoretical than real (as Smith maintained), the threat posed by foreign imports seemed acute and immediate to businessmen and artisans. Protests against Asian merchandise multiplied when the East Indies companies shifted their purchases to manufactured goods and cut back on importing spice. Having gained a monopoly in the spice trade by the late seventeenth century, the VOC saw the value of its position go downhill rapidly. The decline presaged the eclipse of Dutch commercial power in India and the rise there of the EIC, the foundation for British takeover of the subcontinent in the mid-nineteenth century.

Spices made up 57 percent of VOC revenue around 1670 but only 37 percent by 1700. The European market was glutted: the price of pepper tumbled, and the VOC resorted to cutting down clove trees in the Moluccas to control supply and sustain high profits.[34] As a director of the company noted, "We find that yearly there are certainly produced twice as many cloves as are consumed in the whole world."[35] In the late seventeenth century, the VOC made dividend payments to its shareholders in packets of cloves rather than cash, a practice that caused disgruntlement while also proving insufficient to dispose of the steadily mounting surplus. In 1735 the VOC resorted to the expedient of torching more than 500,000 kilograms of nutmeg stored in an Amsterdam warehouse, a measure that obscured the sun and cloaked the city in sweet-scented smoke. Smith regarded that particular act of destruction as evidence that state monopolies in trade were self-defeating in the long term, inevitably fruitless and unprofitable.[36]

Ever since the entry of the Portuguese into the Indian Ocean in the early sixteenth century, imports of Asian spices had stimulated a burgeoning consumer market in the West. Because spices were native to the tropics and could not be cultivated in temperate climates, Europeans felt pleased to gain access to them, however apprehensive some puritans were about the impact of the seductive plant products on contemporary morals. But the Western market for spices proved inelastic as households and cooks only needed so much pepper and cinnamon. Asian manufactured products were a different matter altogether, for they created rippling effects with profound consequences for the European economy and society. Exotic manufactures provoked resentment among businessmen and craftsmen, encouraged product innovation and new technologies, and little by little changed perceptions of class status, public behavior, and living standards.

By no later than the end of the eighteenth century, a defining feature of Western modernity had become apparent: in their drive for profit, entrepreneurs and traders were constructing something new, a consuming society enthralled with novel products and pleasures, a moneyed public with a bottomless appetite for costly, stylish trappings as measures of social standing and self-esteem. Extravagant consumption customarily had distinguished the higher ranks of society from the hoi polloi; use of silver plate, sartorial finery, ceremonial feasts, palatial homes, and liveried servants signified the trappings of power and social preeminence. But with the rage for Asian merchandise, fueled by massive imports and steadily declining prices, consumption for the first time became conspicuous for lower social ranks as well.

One of the most dramatic and enduring effects of trading silver for Asian manufactured commodities was to set the wheel of fashion spinning. An Englishman noted that "from the greatest gallants to the meanest cook maids, nothing was thought so fit to adorn their Persons as the fabrick from India! Nor for the ornament of Chambers like India Skreens, Cabinets, Beds, or Hangings, nor for Closets like China and lacquered ware."[37] In the 1630s the EIC imported 150,000 Indian cotton piece goods; fifty years later, it shipped in almost three million annually. The cottons proved a popular alternative to woolen fabrics, which were much less colorful and convenient; they prompted a turn toward lightweight fabrics for clothing as well as for home furnishings—painted curtains, counterpanes, upholstery, and wall hangings. As a matter of policy, the EIC imported Asian textiles to gratify the demands of fashion. This was spelled out in 1681 in instructions its directors sent to their agents in India:

> Note this for a Constant and generall Rule that in all flowred Silkes you Change ye fashion and flower as much as you can every yeare, for English Ladies and they say ye french and other Europeans will give twice as much for a new thing not seen in Europe before though worse, than they will give for a better Silk of ye same fashion worn ye former yeare.[38]

In fact, the EIC policy of catering to fashion opened a debate that addressed the foundations of mercantilism, industrial innovation, and social order. Those defending the Company were driven to justify consumption and fashion as ends in themselves, equally imperative for the individual, society, and economic prosperity. An anonymous East Indies tract of 1728 even maintained that the ascendancy of fashion resulting from the introduction of Asian manufactures must be regarded as a natural force, a response to vital human needs:

> Men's Tastes, like all other Parts of Nature, require Variety and Change; the very Air we live by would be fatal, without a fresh Succession, and a new Circulation. No part of the world can vie with the East-Indies, in the Variety and Goodness of its Product. . . . Fashion and Custom, and indeed the Nature of Things, having fix'd and set a Value on the East-India Goods, they are become necessary to all the nations of Europe; and Men can be no more restrain'd from them, than they can from their Food and Raiment.[39]

Such arguments struck Western manufacturers and craftsmen as merely hypothetical, if not transparently hypocritical. Faced with threats to their livelihood, they struck back. In 1680 London silk workers attacked the offices of the EIC in Leadenhall Street because of the influx of Indian cloth. Their alarm stemmed in part from the manifest superiority of the Indian cottons to their own product, for the former had water-resistant colors, stemming from use of mordants and resist-dyeing, whereas European colored fabrics faded easily when washed. France banned calico imports in 1686, and England followed suit in 1701; an English law of 1678 stipulated that corpses must be interred wrapped in woolen cloth. Legislation in 1720 even prohibited the use and wearing of calico, an interdiction that remained in place until 1774 and gave rise to sporadic episodes of "calico chasing." Mobs shouting "Callicoe, Callicoe, Weavers, Weavers!" harrassed women wearing the forbidden fabric.[40] James Gilray (1757–1815) expressed a like sense of indignation in his painted engraving, *The Advantages of Wearing Muslin Dresses!* (1802): a portly matron's Indian-print gown is accidentally set on fire, leading to panic among her companions that sends Chinese porcelain teacups crashing to the floor.

As soon as the first Chinese hand-painted wallpaper went on sale in London in the late seventeenth century, English "paperstainers" produced their own copies and pastiches of "paperhangings with fine India figures" (as a journal advertisement of 1693 boasted) and then immediately protested imports of the Chinese product.[41] Shipping lacquer furniture from East Asia provoked even more antagonism from European artisans. A natural polymer, lacquer is a sap obtained by tapping a tree *(Rhus verniciflua)* native to China and Japan. Both countries prized lacquer as a deep black, impermeable coating for materials such as silk, wood, and bamboo; works executed in lacquer often were incised, tinted, or inset with silver and mother-of-pearl.

Ricci may have been the first Westerner to suggest the advantages of obtaining lacquer technology: "The export of the product of this particular tree might well be the beginning of a profitable enterprise, but up to the present it seems that no one has given any thought to such a possibility."[42] Inasmuch as lacquer sap could not be transported to Europe and the lacquer tree itself could not be transplanted there, French and Italian Jesuits published accounts of Chinese methods of lacquer manufacture, just a few years before Dentrecolles departed for China. They recommended that European craftsmen employ shellac and varnish for veneer and lampblack for coloring.[43] More optimistic than the Dominican Naverette, the Jesuits believed that Westerners could imitate to perfection whatsoever they had seen brought out of China.

By the late seventeenth century, artisans in London and Paris were turning out facsimiles of lacquer furniture, as well as lobbying against competition from East Asian imports. In 1700 the Joyners' Company of London, whose members produced imitations of lacquer furniture, complained that "several merchants and others have procured to be made in London of late years and sent over to the East Indies patterns and models of all forms of cabinet goods and have yearly returned from thence . . . quantities of cabinet wares manufactured after the English fashion."[44] Soon after, the Joyners called for crippling duties on lacquerware since "the large quantities of Japan'd goods expected shortly to be brought from the Indies will not only tend to the Ruine of the Japan-trade here in England, but also obstruct the Transportation of our English lacquer to all Europe, which is a considerable Advancement to His Majesties Customes, whereas the Indian Lacquer being Exported from hence, draws back the Custom."[45]

The vessel *Amphitrite* carried 180 cases of porcelain to France in 1700, and it held so much lacquerware, mainly screens, boxes, and miniature cabinets, that Parisians took to calling the product "Amphitrite." French furniture makers persuaded government ministers to impose a ban on further imports of the commodity. The prohibition was not very successful, however, and smuggling was widespread. Still, some high-ranking individuals encouraged the French imitations. Louis-Henri de Bourbon (1692–1740), prince of Condè and cousin to Louis XV, financed a workshop for making japanned furniture. He also supported the porcelain manufactory of Cicaire Cirou (1700–1760), which experimented with applying lacquer decoration to pottery. Marie Antoinette (1755–93), the Austrian wife of Louis XVI (r. 1774–92), ordered japanned furniture and porcelains embellished with pseudo-lacquer. The royal Sèvres manufactory produced a pair of bravura vases for her with a black surface imitating lacquer and with Chinese-style dragons in gilt bronze rearing up on the handles.[46]

Daniel Defoe (1660–1731), the author of *Robinson Crusoe* (1719), was as well informed as any of his contemporaries about global trade. His description of a true-bred merchant as "a universal scholar" who "understands languages without books, geography without maps[,] . . . [and whose] journals and trading voyages delineate the world," anticipates Smith's high regard for "the great mercantile republik." But, although he believed Britain thrived on overseas commerce, he worried about its effects on domestic industry and denounced purchase of foreign luxury goods. In a 1713 tract for a popular journal, he describes London's streets as deformed by foreign exchange, swamped with "trade in baubles and trifles." The dismal transformation called to mind an apocalyptic passage from Ezra (3:12): "the ancient men who had seen the old temple wept when they saw the weak foundations of the new."[47] Long-established drapers and woolen wholesalers fled to back alleys, hounded out by peddlers of gilded lacquer boxes, teapots, tea-table utensils, and chinaware. In Defoe's view, the whirligig of fashion gyrated faster and faster every year, destroying solid industry, enthroning the ephemeral, and sapping the moral fiber of the nation.

Defoe stood out among his contemporaries in his denigration of China, even singling out its famed porcelain for derision. In *The Farther Adventures of Robinson Crusoe,* published a mere four months after the best-seller, the adventurer goes to China, where he belittles the natives as "a contemptible herd or crowd of ignorant, sordid slaves, subjected to a government qualified only to rule such a people." So degenerate a society cannot be expected to make better pottery than do Westerners, hence Crusoe's prejudices are confirmed when he is taken to see a palace reputedly made entirely of porcelain: "and when I came to it, it was nothing but . . . a timber house, or a house built, as we call it in England, with lath and plaster, but all the plastering was really China ware, that is to say, it was plastered with the earth that makes China ware." The legendary Chinese reputation for "performance in crockery-ware," he concludes, is no more than a fraud perpetrated on gullible foreigners.[48]

Defoe's vilification of China and porcelain testifies to the depth of his conviction that native manufactures should be preferred to imports for both moral and commercial reasons. The resourceful castaway of *Robinson Crusoe* was an exemplary figure inasmuch as he made all he required in his tropical island kingdom, including several porcelain-like pots "as hard burnt as could be desir'd; and one of them perfectly glaz'd with the running of the sand."[49] (As a sometime owner of a tile-making business, Defoe knew a lot about ceramics.) In a pamphlet on woolen production, Defoe maintained that honest English broadcloth is "better than even silk itself," so there is no need for his countrymen to "run to India and China for all the slightest and foolishist trash in the world."[50] His implausible boast represented a denunciation of what he saw every day on London streets. A visiting Frenchman observed in 1725 that the costumes of English gentlewomen are "made according to the season either of rich silk or of cotton from the Indies. Very few women wear woolen gowns."[51]

In terms of import substitution and product innovation, porcelain was unique. As Defoe realized, it indeed was necessary to run to India and China for cotton and silk since the plants needed for those textiles could not be grown in England—though in the late eighteenth century, it began importing unfinished cotton and weaving its own cloth. Lacquer and tea necessarily also were foreign commodities. England remained dependent on China for tea into the nineteenth century, when Assam plantations in northeastern India began producing it, and European artisans found only second-rate substitutes for lacquer. Porcelain, then, represented the only significant Chinese product that Westerners could aspire to contest on their own grounds, not only producing imitations of it, but even making a near-equivalent with native materials. That consideration helps explain both Defoe's exasperation with the high status of porcelain in Europe and persistent attempts to create a European version of the ceramic.

THE PORCELAIN KING: AUGUST II OF SAXONY

Paying silver for exotic textiles and porcelain attracted the most attention from critics of Asian manufactures because these goods were more extensively imported and publicly displayed than wallpaper and lacquerware. With millions of porcelains pouring into Europe, Ehrenfried Walther von Tschirnhaus (1651–1708), a German man of letters and scientist, declared that "the purchase abroad of so great a quantity of goods as the Chinese porcelain represented was a national loss to be averted."[52] He condemned the Chinese as "Saxony's porcelain bloodsuckers" *(porzellanene Schröpfköpfe)* because of the riches, mostly silver from Spanish America, that went off to China.[53] He was familiar with the problem, for he served August II ("the Strong") (1670–1733), Elector of Saxony and King of Poland, a ruler passionate about amassing porcelains and discovering how to manufacture the ceramic.

The Polish king was the most illustrious victim of what critics of Asian luxuries termed "porcelain disease" *(la maladie de porcelaine* or *Porzellankrankheit),* a feverish desire to possess the ware. His aristocratic agent in Holland purchased huge amounts of Chinese and Japanese porcelain at VOC auctions and sent it by cart and river barge to Dresden. August dreamed up projects well beyond the capacities of porcelain, such as a palace built entirely of the ceramic, including walls, bed, throne, and a chapel with a porcelain pulpit, organ, and altar; he also posed for sketches for a life-sized porcelain equestrian statue of himself. He commissioned life-sized statues of the Twelve Apostles, a task that required using extensive internal and external supports for the pieces in the kiln so they would not collapse.[54]

August's potters made four hundred life-sized porcelain figures of animals, based on those in the royal menagerie. In 1753, during the reign of August III (1733–63), who inherited his father's passion for porcelain, an English visitor to Dresden described the ceramic beasts:

> The next curiosity is the *Chinese* palace, so called from the taste of the building, and the intention of furnishing it with porcelain. . . . Here are a great number of porcelain figures of dogs, squirrels, monkeys, wolves, bears, leopards, etc. some of them as big as the life; also elephants and rhinoceros's of the size of a large dog; a prodigious variety of birds, as cocks, hens, turkeys, peacocks, pheasants, hawks, eagles, besides parrots and other foreign birds. . . . The long gallery in the second story had already two marble chimney-pieces, each adorned with near 40 very large pieces of porcelain, of birds, beasts, and vases, ranged to the height above 20 feet in a most superb taste, the figures being all made so natural, that I could conceive no idea superior in this kind.[55]

A handful of *Fantasievögel* sprang up amid the flock of birds, including one that combined the features of a golden pheasant and a Chinese phoenix. Reflecting on this zoomorphic wonderland, Postelthwayt noted in his *Universal Dictionary* that "His Majesty of Prussia [Frederick II, "the Great," r. 1740–86] sometimes calls his brother of Poland the Porcelain King."[56]

In one of the best-known episodes in the European career of Chinese porcelain, August II in 1715 traded 600 Saxon dragoons (or mounted infantry) to King Frederick William I (r. 1701–40) of Prussia for 151 blue-and-white vases of the Kangxi period. Lidded and almost a meter high, such pieces henceforth were known as "Dragoon vases." In the Prussian (and later German) army, the soldiers pawned off for chinaware formed the Saxon 94th Infantry Division, though it informally styled itself the "Porcelain Regiment." The unit saw its last significant action at the battle of Stalingrad in 1942–43, still bearing the emblem of crossed swords that denoted August II's rank as a marshal in the Holy Roman Empire and that was the hallmark inscribed on Meissen porcelain.[57] (See figure 5.)

By the time he died, August II possessed no less than 35,798 pieces of porcelain, roughly half from his Meissen manufactory, half from China and Japan (lumped together in a royal inventory as "East Indian Porcelain"). He housed the collection in Dresden in his Japanese Palace, a building with a porcelain bell in its belfrey. Originally known as the Dutch Palace (Holländisches Palais), it had been purchased from Count Jacob Heinrich von Flemming (1667–1728), a leading minister of Saxony. When von Flemming in 1726 offered to sell August the Orangerie of his Übigau Castle as well, the king replied with a well-nigh clinical description of psychological obsession: "Do you not know that it is the same with oranges as with porcelain? Namely, that those who have one sickness or the other never believe that they have enough but always feel that they need to have more?"[58] Happily for the Elector, he managed to indulge his two addictions simultaneously when he planted orange trees in Chinese porcelain tubs in the garden of his Dresden palace.

August II's passion for porcelain matched the scientific interests of Tschirnhaus. A man of formidable intelligence and broad interests, Tschirnhaus seems to have known all the intellectual lights of Europe. He began his advanced education by studying natural science and medicine at Leiden in the Dutch Republic. In the 1670s he corresponded with the Dutch philosopher (and lens grinder) Baruch Spinoza (1632–77), who lived in Amsterdam, and he told Leibniz about the radical religious views set forth in Spinoza's *Ethics*. Tschirnhaus conducted scientific experiments in London with members of the Royal Society, and in Paris soon after, finance minister Colbert hired the budding savant to tutor his son in mathematics. Colbert helped direct Tschirnhaus to the question of applying scientific learning to the development of new technology.

Christiaan Huygens (1619–95), the Dutch physicist who discovered Saturn's rings and pioneered the use of the pendulum in clocks, instructed Tschirnhaus in the grinding and polishing of glass lenses, one of the most exacting craft skills in science. He turned that knowledge to good account by heating potter's clay using large focused mirrors, an accomplishment that led to his enrollment as the first German member of the French Academy. In 1682, building on his growing reputation, he joined Leibniz in putting out the journal *Acta Eruditorum* (Learned Deeds). The question of the secret formula for porcelain, known as the *arcanum*, first gripped Tschirnhaus during his studies in Leiden, situated in the middle of the United Provinces, the Western country most inundated with chinaware. Working for August II from the early 1690s, he experimented with optic lenses to achieve the temperatures needed for creating porcelain. He then devoted years to investigating the nature of the ceramic itself, experimenting with various minerals and clays to come up with the right formula.[59]

In 1694 Leibniz's secretary visited the Dresden laboratory of Tschirnhaus and admired a waxy, porcelain-like material he had concocted. A few years later, the scientist spied out the techniques used in the Delft potteries of Holland and those in the Saint-Cloud pottery manufactory, located to the west of Paris on the road to Versailles. Like many other well-connected gentlemen, Tschirnhaus no doubt took the opportunity to tour the nearby country estate of Monsieur (the Grand Dauphin, duc de Bourgogne), Louis de France (1661–1711), the son of Louis XIV. The heir to the throne owned some porcelains that a celebrated legation from Siam (Thailand) had presented to the king in 1686. His collection included about four hundred pieces of blue-and-white and some porcelain figurines (known as "Siam Pagods"), as well as agate vessels, rock crystal, and Japanese lacquer furniture. Monsieur had the blue-and-white porcelains set out amid chairs and couches covered in fabric of blue and silver.[60]

On top of huge purchases of porcelain by August II, Tschirnhaus's travel and research naturally cost a fortune. The Elector's finance minister, distraught at his master's uninhibited spending, moaned that "China is the bleeding bowl of Saxony."[61] Significantly, alchemists hired by August to discover the Chinese secret referred to porcelain as the new "white gold" *(weissener Gold)*.[62] But despite the king being as extravagant and feckless as any monarch in Christendom, he nonetheless proved more forward-thinking than all his more enlightened contemporaries in financing systematic research on porcelain. A few years after Johann Friedrich Böttger (1682–1719), an alchemist and Tschirnhaus's associate, told August in March 1709 that he had succeeded in creating porcelain—he could claim sole credit for the achievement because Tschirnhaus had died five and a half months earlier—the king had himself portrayed on a vase viewing his Meissen manufactory in the guise of a Chinese emperor.[63]

August ordered that Asian porcelains be used to decorate the walls of his Japanese palace, set against "a background of wall coverings of Eastern splendor, great Indian figures painted onto silk or golden dragons embroidered on silk, or Indian gilded paper onto which various Chinese pagodas would be painted, clearly and neatly."[64] He also commissioned a ceiling painting for his gallery of porcelain beasts that, in effect, trumpeted the virtues of product innovation and import substitution by asserting the superiority of Meissen porcelain over that from Japan, a principal source of his early purchases. As he envisaged it, the first scene of the painting "will depict Saxony and Japan who quarrel over the perfection of their porcelain manufactories. . . . The goddess [Minerva] will graciously bestow the award of the struggle into Saxony's hands. Jealousy and dismay will prompt Japan to load their porcelain wares back onto the ships that once brought them here."[65] Only in the realm of fantasy could August the Strong let a single piece of porcelain escape his grasp. Still, he intended his painting to convey basically the same message that the Lisbon guild of potters had addressed to Philip III in 1619 when they boasted about their "pilgrim art": by fabricating a facsimile of porcelain, Westerners had created pottery that supplanted the finest wares hitherto imported at crushing expense from Asia.

Headstrong, compulsive, and fabulously rich, August II anticipated the modern recognition that science and industry necessarily are linked and, furthermore, that their coordination calls for strong state support. He followed the same course of action advocated later by Postelthwayt. The English writer lamented the hemorrhage of New World silver to Asia and extolled manufactures from there, notably Bengal linen, Japanese lacquer, and Chinese porcelain. He believed that the most effective means to improve Western commerce was "the patronage of princes," that is, cash rewards for new discoveries. His counsel to monarchs and entrepreneurs regarding pottery was based on his reading of du Halde's *Lettres édifiantes* and on his understanding of the excellent use already made in his own country of Dentrecolles's reports:

> Upon the whole, in whatever mechanical or manufactural arts other nations may excel Great-Britain, our artists should be upon the watch, not only to imitate, but surpass, if possible. Throughout this work, those particulars appear wherein other nations do excel us . . . and [which] may well deserve the attention of our artists. But those which are imported, and which they can see, handle, and minutely examine, they are the most likely to imitate or excel. As we have arrived at a great perfection in the China ware, why may we not in divers other eastern arts and manufactures?[66]

An anonymous pamphlet of 1697–98, *The Advantage of the East India Trade to England, reconsidered,* advanced the same general argument: "The *East India* Trade, by putting Persons upon Invention, may be the Cause of doing Things with less Labour. . . . [I]t is therefore very likely to be the Cause of the Invention of Arts and Engines, to save the Labour of Hands in other Manufactures; these are the Effects of Necessity and Emulation."[67]

Along with other Western rulers committed to economic self-sufficiency, Louis XIV believed that appropriating the manufacturing techniques of China would result in effective competition with its imports. He therefore heeded arguments by Bouvet and others that one of the advantages of sending Jesuits to China might be disclosure of the secrets of porcelain. In addition, the years around 1700 witnessed the high tide of mercantilist thinking in French royal circles, a forceful concern with the role of bullion, commodity production, and trade balances in the welfare of kingdom. With the creation of the Council of Commerce in 1700, the chief officials of the French state for the first time began directing economic policy in consultation with merchants and manufacturers.[68]

Beyond official strategy, however, ruling by example remained important in an age of personal monarchy. In the Nine Years' War (1688–97), in which France fought almost the whole of Europe to a standstill, Louis XIV tried to make up for financial shortfalls by having his silver table service and silver furniture melted down and coined at the mint, a patriotic gesture that failed to inspire the French elite to follow suit. The king also started using a porcelain-handled walking stick in place of one with silver, a relic "carefully preserved and shown to the curious" in the private royal apartments until the outbreak of the French Revolution in 1789.[69]

The last great war of the reign broke out in 1702 when the grandson of Louis became heir to the Spanish crown as Philip V (r. 1700–46). The Spanish succession raised the same sort of alarm throughout Europe that Philip II's union of the Spanish and Portuguese crowns had in 1580: it radically upset the balance of power, especially since, as a political observer remarked, control of American silver would provide the Bourbon dynasty "an inexhaustible resource for . . . the conquest of the world."[70] Yet rather than win Louis control of Spanish silver, the ruinous War of the Spanish Succession (1701–14) instead compelled him to order his nobles to turn over their silver plate to defray army costs, a measure that netted the treasury the huge sum of ten million *livres*. According to the duc de Saint-Simon (1675–1755), a high aristocrat and memoir writer, the courtiers of *le Roi-Soleil* rushed from the palace of Versailles to boutiques in Paris for faience imitations of their table settings. Indeed, Saint-Simon numbered among them, for he ordered a serving dish from Rouen decorated with his coat of arms.[71]

Coming about the same time that August II made his breakthrough in producing porcelain, the wholesale destruction of silver plate in France meant there was more urgency than ever in the French search for the secrets of porcelain. Not only did France want its own recipe for porcelain, but an unintended consequence of the melting down of silver plate was that more of the white metal than ever ultimately went off to China to pay for porcelain. Moreover, tin-glazed earthenware could not satisfy privileged circles accustomed to tableware of precious metal and newly acquainted with gleaming chinaware. A French comedy portrayed a woman smashing Dutch delftware (termed *ces porcelaines d'Hollande*) and proclaiming she henceforth would use only Chinese porcelain.[72] Seeking to staunch the torrent of silver to Asia, French ministers aspired to "make porcelain like the Chinese," as the royal letters patent for a pottery in Rouen declared around 1700, and thereby safeguard the economic health of the kingdom.[73]

"TO TELL OF THE PORCELAIN MADE THERE": THE SECRETS OF JINGDEZHEN

Dentrecolles wrote his letters from Jingdezhen with these ambitions in mind. One of the obstacles he faced, however, was that his mission to reveal the secrets of porcelain took place against a background of confusion and disagreement regarding the nature of the ceramic. Introducing Dentrecolles's first letter, du Halde remarks that most explanations of porcelain are "mere Inventions of such Writers, who have boldly palm'd their own silly Conjectures on the World."[74] No doubt he was thinking of authorities such as Duarte Barbosa (d. 1545), a Portuguese writer, who informed his readers that the ceramic was made from "fish ground fine, from eggshells and the whites of eggs and other materials."[75] Around the same time, one of Barbosa's compatriots told an Ottoman naval commander that porcelain actually was a stone "just like lapis lazuli," closely resembling rock crystal and passed down like an heirloom from father to son.[76] Basing themselves on Barbosa's account, Girolamo Cardano (1501–76), a mathematician and astrologer, and Justus Caesar Scaliger (1484–1558), a philologist and physician, debated the nature of porcelain in the mid-sixteenth century. Cardano argued that the ceramic was the same as an ancient, glasslike pottery, "for it is certain that porcelain is likewise made of a certain juice which coalesces underground." For his part, Scaliger maintained that "eggshells and the shells of umbilical shellfish (named porcelains, whence the name) are pounded into dust, which is then mingled with water and shaped into vases. These are then hidden underground. A hundred years later they are dug up, being considered finished, and are put up for sale. . . . The best are brought from the region of China."[77]

Gaspar da Cruz (d. 1570), a Dominican friar, scorned those who believed porcelain was made from oyster shells or from "dung rotten for a long time." In his *Tractado* (1569–70), the first European book devoted to China since *The Travels of Marco Polo,* he reported that "porcelain is a white and soft stone, and some is red which is not so fine; or in better speaking, it is a hard clay, the which after well beating and grinding it, and laying in cisterns of water . . . of the cream that remaineth on the top they make the very fine porcelain."[78] Although this was accurate enough, erudite speculation continued to link porcelain to agate, seashells, lobster shell, gypsum, mother-of-pearl, and precious minerals long into the seventeenth century. A 1617 publication retailed a cluster of popular notions:

> A large mass of material composed of plaster, egg- and oyster shell, of sea-locusts and similar creatures, is well mixed until it is of one consistency. It is then buried by the head of the family, who reveals the hiding place to one of his sons. It must remain in the ground for eighty years without seeing the light of day. After this time is elapsed, the heirs must take it from the ground, and use it to make the beautiful translucent vases of such perfect form and colour that no critic could find fault with them.[79]

Sir Francis Bacon (1561–1626) voiced a similar opinion in his *Novum Organum* (1620): discussing the transformation of natural substances by burying them in the earth, he cited the practice of the Chinese "with their china, masses of which, made for that purpose, are said to remain in the ground for forty or fifty years, and to be transmitted to their heirs as a sort of artificial mine." In his *Pseudodoxia Epidemica* (1646), a work debunking vulgar myths and folklore, Sir Thomas Browne (1605–82) devoted pages to the claims of scholars such as Bacon, Barbosa, and Scaliger, "which when the Author had seen he could hardly contain from laughter at the common opinion[s] . . . rejected by us."[80] More than sixty years later, Dentrecolles recorded that the silly conjecture still persisted in Europe that porcelains had to be buried underground for many years to attain perfection.

Like Browne, most informed investigators sensibly concluded by the late seventeenth century that porcelain was made from clay, though many continued to believe that it must be of a recondite sort, closer to such marvels as exotic gems, the chambered nautilus, rhinoceros horn, and ostrich eggs than to common earth. Before the letters of Dentrecolles, the most thorough account of porcelain appeared in the *Itinerario* (1595–96) of Jan Huyghen van Linschoten (1563–1611), a popular Dutch report on China:

> To tell of the porcelain made there, is not to be believed, and those that are exported yearly to India, Portugal and Nova Hispania and elsewhere! But the finest are not allowed outside the country on penalty of capital punishment, but serve solely for the Lords and governors of the country and are so exquisite that no crystalline glass is to be compared with them. These porcelains are made inland of a certain earth which is very hard, which is pounded to pieces or ground, and they leave it to soak in troughs cut out of stone, and when it is well soaked and frequently stirred as milk is churned to make butter, they make of that which floats on top the finest work, and after that somewhat lower the coarser, and so on, and they paint them and make on them those figures and likenesses they want, and then they are dried and baked in the kiln.[81]

Regrettably, Linschoten's prosaic description did not reveal the formula for porcelain. Desired almost as much as the philosopher's stone, the fabulous magical substance that transformed base metals into gold, the arcanum seemed every bit as elusive and enigmatic. Measured against the goal of plumbing the secrets of porcelain, European suppositions seemed feeble and faltering, scarcely more than confessions of ignorance. Alvarez Semedo, a Portuguese missionary monk who wrote a book on China, was typically unhelpful: "In this work there are not those mysteries that other people have written about, neither in the matter, the form, nor the manner of working. They are made absolutely of earth, but of neat and excellent quality. They are made in the same time, and in the same manner, as our earthen vessels; only they make them with more diligence and accurateness."[82] In the late 1690s a

French expert on glass and pottery could do no better than to suggest that porcelain was "a prepared clay that is glazed with a white glaze."[83]

In addition to lacking useful information from his contemporaries, Dentrecolles faced daunting obstacles in his espionage. Potters everywhere maintained intense vigilance regarding their techniques. Italian artisans were typical. In his *Three Books of the Potter's Art* (1557), the first major European work on ceramics, "wherein is treated not only of the practice thereof but in brief of all its secrets," Cipriano Piccolpasso observes that among his fellow craftsmen are "many who till the last day of their lives keep [a trade secret] hidden from their own sons and then, knowing themselves to be on the point of death, among the other goods they leave behind, summon their oldest and wisest son, and reveal to him this secret." Browne noted that Chinese potters are "so reserved concerning that Artifice [of porcelain], that 'tis only revealed from Father unto Son."[84] At times, of course, that was a risky practice: the patriarch of the Wei family in Jingdezhen died just as he was about to pass on to his son the secret of the molasses-like mortar that had made his lineage the principal kiln repairers in the city.[85]

Beyond the secrecy of potters, Dentrecolles had to deal with Chinese bureaucrats, who regarded foreigners with grave suspicion; hence their policy of keeping Western merchants penned in Canton and visitors to Jingdezhen sequestered on riverboats at night. Moreover, he had to learn potters' colloquial expressions for their techniques and materials, slang and abbreviations so baffling that "even the natives have to make inquiries before they get the meaning."[86] Not surprisingly, he often met with concealment, as when he asked about the proportion of ingredients in a certain red glaze but only could report that alum, unspecified oils, and "the urine of a young boy" made up the bulk of the formula.

Still, Dectrecolles learned a remarkable amount about porcelain production "by wandering through the workshops and instructing myself with my own eyes or by asking questions of the Christians who are involved in this work." He apparently won cooperation from some non-Christians as well, including shopkeepers, merchants, and even a couple of magistrates, most notably Tang Ying. He checked the responses of his informants against ancient chronicles and the official annals of Fuliang, the county magistrate's headquarters, five kilometers north of Jingdezhen. But he was disappointed in the latter since they said nothing about the origins and history of porcelain. Instead, they recounted incredible local sensations, such as the sow that supposedly gave birth to a tiny elephant with a perfectly formed trunk, though "there has never been an elephant in the country." (Right about the sow, he was wrong about elephants.) Furthermore, since one could purchase the honor of being cited in the annals, local luminaries filled the volumes with self-serving comments. Dentrecolles correctly concluded that the Chinese simply had no concern for the sort of information about porcelain desired by Europeans.

From the perspective of Dentrecolles, this problem was compounded by others.

He inferred that crucial information about porcelain had been lost inasmuch as China had experienced "all sorts of revolutions" in the past. Not only that, he believed that loss of knowledge went hand in hand with decline in manufacturing expertise. He regarded Jingdezhen as waning from its golden age, although the potters still could fashion marvelous things. He attributed the deterioration to the porcelain city's mushrooming production: as kilns and workers multiplied, standards plummeted and quantity substituted for quality. In the past, he argued, the potters "had less regard for the expense [of production] than for the perfection of the work." Since kilns used to be much smaller in Jingdezhen, more individual attention was given to select wares. He cited an old chronicle as claiming that sometimes an entire furnace would be dedicated to producing a single urn worth the extraordinary sum of fifty-nine taels. He asserted that broken saggars tied with reeds are used to save on materials, an economy never practiced in bygone times. In addition, wares used to be removed from the kiln only after it had slowly cooled so they would not crack with a sudden change of temperature; but in contemporary Jingdezhen, to save on wood for a subsequent firing, vessels are taken out while the furnace is still smoking. Nowadays, Dentrecolles concluded, kiln owners are much greedier for profits, workers are impoverished, the court requisitions burdensome quantities of wares, raw materials are more expensive, and potters lack the technical skills of their forebears. In short, Jingdezhen's very success had led to prolonged decline in the excellence of its products.

Since Jingdezhen had experienced rapid change and expansion in the previous two centuries, Dentrecolles's pessimistic analysis held some truth, especially regarding the increased cost of wood and clay. Yet he was too quick to assume that the porcelain city had gone substantially downhill and that inarticulate or evasive responses to his queries confirmed a decline in functional proficiency. As a Jesuit, he perhaps placed too high a premium on lucidity and coherent exposition. His gloomy perspective also smacks of hackneyed laments voiced by elderly potters and shopkeepers about the falling away of the younger generation from the exalted standards of the good old days. In addition, Dentrecolles identified the history of porcelain with the career of Jingdezhen itself, seemingly unaware that the ceramic had been produced elsewhere in China for many centuries before the city on the Chang became the dominant production center.

Dentrecolles applied the same blinkered perspective to the prosperity of the porcelain city: "According to the History of Jingdezhen [at Fuliang], the profits that were once made were greater than those produced now; but that is difficult to believe, since porcelains for Europe were not made at that time on so large a scale." Long before the arrival of Portuguese ships off the Chinese coast in the early sixteenth century, however, the profits of Jingdezhen and other pottery centers from commerce with the Asian region of the ecumene were at least as great as those made later from trading with Europe. As a Frenchman, Dentrecolles focused on Western commissions for chinaware, some of which he evidently facilitated, without recognizing that most of Jingdezhen's production went to areas other than his own homeland. This was true for Chinese commerce in general. Only 10 percent of customs duties collected in Guangdong ports in 1735 came from European ships, which were vastly outnumbered by Chinese vessels. Indeed, Western trade with China would not reach the level of Chinese trade with continental and maritime Southeast Asia (the Archipelago) until the end of the eighteenth century.[87]

"THE WHITE PORCELAIN SHELL": FROM MARCO POLO TO ROCOCO

Questions about the origins of porcelain baffled Dentrecolles. "One cannot say," he writes, "who the inventor was nor to what endeavor or accident one owes this invention." Since European artisans, alchemists, and scientists at the time were trying to devise a formula for porcelain, Dentrecolles naturally thought in terms of "invention" of the substance in China rather than of long-term, incremental discovery. This assumption combined with puzzlement regarding the origin of the term *porcelain.* He points out that its source is mysterious, "for the name of the material . . . is so little a Chinese word, that none of the syllables of which it is composed can even be pronounced or written in Chinese." The same thing troubled Ricci, who acknowledged, "It is not quite clear to me why it is called porcelain in the West."[88]

As Dentrecolles explains, the Chinese phrase for the ceramic is *tseki,* apparently his rendering of the characters for *ci* (high-fired pottery, i.e., stoneware and porcelain) and *shi* (stone). He speculates that *porcelain* may derive from the Portuguese language, though he confesses his suggestion is at odds with the contemporary Portuguese custom of using the term *porcellana* to signify a cup or bowl while referring to chinaware simply as *louça* (pottery). The latter usage actually originated in the early sixteenth century, for porcelain was known as *louça da India* (pottery from India) ever since Vasco da Gama brought chinaware to Lisbon on returning in 1499 from his great voyage around Africa to India.

After the Portuguese conquered the key port of Melaka in Southeast Asia in 1511, they commissioned Chinese traders there to purchase porcelains and other merchandise for them. Portuguese ships departing India for home generally carried as much as a third of their cargo in porcelain.[89] Within a short time, the Rua Nova dos Mercadores in Lisbon had seventeen shops selling *a louça da India.* An Italian in Portugal declared, with considerable hyperbole, that porcelain there was "esteemed more highly than gold and silver."[90] In 1563, during a dinner in the Vatican, a Portuguese cleric commended blue-and-white chinaware to Pope Pius IV (r. 1559–65):

> In Portugal we have a kind of tableware which, being made of clay, may be compared advantageously to silver both in its elegance and its cleanliness, and I would counsel all princes to use it in preference to any other service and to banish silver from their tables. In Portugal we call it porcelain. It comes from India and is made in China. The clay is so fine and transparent that the whites outshine crystal and alabaster, and the pieces which are decorated in blue dumbfound the eyes, seeming a combination of alabaster and sapphires.[91]

With comparable enthusiasm, the poet Paul Scarron (1610–60) lauded the principal attractions in the shops at Saint-Germain in Paris:

> Bring me to the Portuguese,
> There we shall see something new,
> The merchandise from China.
> There we shall see gray amber,
> Beautiful works of varnish,
> And of fine porcelain
> From this majestic country
> Or, rather, from this paradise.[92]

Although Dentrecolles could not have known it, he went to Jingdezhen seeking the secrets of a material whose name derived from Europe itself, not from China. The designation goes back two centuries before da Gama, for it first was applied to Chinese pottery by an even more famous European traveler to the East, Marco Polo.[93] *Porcelaine* is the rendering of *porcellane* in the Italian-French dialect of Languedoc, the tongue in which Rustichello of Pisa, Polo's scribe, recorded the Venetian's adventures in 1298 when they both languished as prisoners of war in Genoa. In *The Travels,* Polo refers to cowry shells *(Cypraea moneta)* as *porcellani,* and he employs the same word to identify white pottery he saw near the city of Quanzhou in Fujian province. Italians had long called cowries *porcellani,* supposedly because they perceived a resemblance between the humped back of the shell and a piglet *(porcellus,* diminutive of *porco).* Scaliger evidently had an intimation of this connection inasmuch as he claimed that shellfish, "named porcelain, whence the name," constitute a principal ingredient of the ceramic.

Harvested in the Maldive Islands, southwest of Sri Lanka (Ceylon) in the Indian Ocean, cowries were employed as currency in small-scale transactions throughout East Asia, India, and Africa. They served as the foremost means of payment in China in the Shang and early Zhou period, when people deposited large quantities of the shells in tombs. They remained in circulation long after the emperor Qin Shihuang (r. 221–210 B.C.E.), founder of the Qin, China's first empire, prohibited their use. People employed them far and wide by Polo's time, even in the mountain valleys of Tibet and the jungle hinterlands of Borneo. Inasmuch as no government would consider it financially worthwhile to mint a coin with so paltry a value, the shells served as a handy economic lubricant in everyday dealings, above all, in peasant communities.[94]

Large-scale merchants in Asia relied on coins and ingots of tin, lead, and silver, as well as on swatches of silk, for legal tender in international trade; common people used Chinese copper-cash or cowries, often supplemented by packets of rice. Ibn Battuta, who traveled widely and wrote an account of his adventures, noted that one gold dinar (the premier Arab coin, 4.25 grams in weight) could be exchanged for 400,000 cowries in the Maldives, the cost of a beautiful slave girl in Bengal. In his day, a small Bengalese silver coin purchased 10,520 of the shells. A century later, according to Tomé Pires (ca. 1468–1540?), the Portuguese author of the *Suma Oriental* (ca. 1513), the Burmese paid 500 cowries for a chicken while the Javanese used copper-cash and cowries for "small money" and nuggets of native gold for substantial purchases. In the late eighteenth century, the Chinese even made porcelain replicas of the shell to export to Southeast Asia, where they took their place alongside the sea-bred kind.[95]

Polo knew about cowries before his journey to China because of the Mediterranean trade carried on in his city. Venetian merchants acquired them in Cairo as ballast for their galleys and then sold them in Portugal and Morocco. According to João de Barros (ca. 1496–1570) in *Décadas da Ásia:*

> With these shells for ballast, many ships are laden [in Lisbon] for Bengal and Siam, where they are used for money, just as we use small copper coin for buying things of little value. And even to this Kingdom of Portugal, in some years as much as two or three thousand quintals [i.e., over 100 million cowries] are brought by way of ballast; they are then exported to Guinea, and the kingdoms of Benin and the Congo, where they are also used for money.[96]

Portuguese captains traded the cowry shells to coastal peoples in West Africa for ivory and slaves. When Ibn Battuta visited the Mali kingdom of Mansa Sulayman (r. 1341–60) in 1352, he noted that the "buying and selling of its inhabitants is done with cowry-shells."[97] In fact, it was the same throughout the continent, for the shells made an arduous journey inland from the west coast, eventually meeting up in central Africa with other Maldive cowries that had passed through the hands of traders on the Swahili coast. In China and maritime Southeast Asia, Polo often saw "the white porcelain shell" *(porcellane bianche)* being used for payment in markets.[98] He applied *porcellane* to Chinese pottery evidently because the latter had a milky quality, glossy sheen, and fracturing characteristic that reminded him of the shells.

Polo was not original in identifying seashells and pottery; many cultures also saw similarities between the two.[99] The flowing shapes, ridged patterns, and stippled colors of mollusks seemed to call out for replication in baked clay. Some Japanese ceramics from the late Jomon period (1500–1000 B.C.E.) are shaped like shells, and Minoan pottery of the Bronze Age (3000–1000 B.C.E.) is decorated with images of shells, along with those of seaweed, dolphins, and octopus tentacles. In ancient China and the Khmer region of continental Southeast Asia, potters simulated conch shells in clay and decorated pottery with depictions of them. Khmer potters even copied the interior chambers of the large shell that produced its booming tone. Indian craftsmen also replicated the conch in clay: they associated it with the lotus, the seat of deities, and they regarded the sound of the shell, blown like a trumpet, as the voice of Shiva, destroyer and renewer of worlds. In ceremonies of the Hopi of the Western Pueblo in the American Southwest, the Plumed Serpent (Paalölöqangw), a monstrous deity associated with the underworld and earthquakes, emerges from a ceramic vessel with the blowing of a conch to ravage cornfields, represented by corn sprouts set in clay containers.

Above all, there was a powerful association between the chambered nautilus *(Nautilus pompilius)* and porcelain, since the nacreous inner layer of the spiral shell of the cephalopod, from which comes mother-of-pearl, bore a striking likeness to the lustrous ceramic. Chinese aristocrats in the Tang period used vermilion nautilus shells, known as "Parrot Snails," as sumptuous wine goblets. A few nautiluses reached the West as early as the thirteenth century, where they fetched high prices and, like porcelains, usually were set in mounts of silver or gold. In the sixteenth century, the Portuguese imported large quantities of the shells to Europe. The Dutch also traded in them, especially after 1605 when the VOC seized control of the island of Amboina in the East Indies, whose warm, shallow waters abounded in the shells. In 1603 the governors of the Dutch Republic presented Henry IV of France (r. 1589–1610) with porcelain tableware and a valuable collection of shells, all looted from the *Santa Catarina,* a Portuguese carrack. In his *Still Life with Shells* (1659), Abraham Susenier (ca. 1620–ca. 1667) depicted a nautilus looming over a score of other exotic shells, and a Dutch compatriot of the painter owned a collection of shells valued at almost 100,000 guilders (about $128,000 in present-day terms).[100]

Amsterdam merchants treasured nautiluses and had themselves depicted holding them in paintings. They also commissioned still lifes featuring the shell alongside porcelain, as in *Flowers in a Wanli Vase with a Blue-Tit,* by Johannes Goednert (1617–68). Willem Kalf (1619–93) painted several such works, including *Still Life with a Porcelain Ewer, Dish, and Bowl* and *Wine Glass on a Gilded Silver Foot and a Bowl of Fruit.* An anonymous Dutch painting in the dazzling, jeweled technique of Kalf, *Still Life with Chinese Sugar Bowl, Nautilus Cup, Glasses, and Fruit,* presents a Wanli bowl, decorated with the Eight Immortals (Ba Xian) of Daoism, next to a nautilus supported by a gold statue of Neptune, with the mouth of the shell refashioned into the cavernous jaw of a gilded marine monster. In *A Richly Laid Table with Parrot and Macaw,* by Jan Davidsz de Heem (1606–83), the silver and amber tones of nautilus shells are in vivid contrast to the reds of a lobster and tropical birds. Christoffel van den Berge's *Bouquet of Flowers in Glass* (1617) includes seashells from the western Pacific, as well as porcelain wine cups. In *Still Life with Gold Goblets and Collectibles,* Clara Peeters (1589–ca. 1657) portrays a delicate Chinese bowl next to an assortment of shells from East Asia and the Caribbean, a juxtaposition that highlights the correspondence between crafted and natural objects.[101]

Bernard Palissy (ca. 1510–89), royal potter to Catherine de' Medici (1519–89), the Florentine mother of three French kings, was fascinated by the nautilus. Having won fame for works skillfully imitating natural things, especially for his creation of elaborate grottoes populated by clay replicas of frogs, lizards, and salamanders, he suggested that the nautilus shell poses a powerful, perhaps insuperable challenge to human creativity: "Why do [craftsmen] not observe from what the fish has formed such a beautiful house, and take similar materials to make a beautiful vase? The fish that made the said shell is not as glorious as man—it is an animal which has little form—and yet it is able to do what man is unable to achieve."[102] João de Mascarenhas (1632–81), the first marquis of Fronteira, decorated the walls of a grotto in the gardens of the Fronteira Palace near Lisbon with blue-and-white porcelain, shells, and colorful pebbles. When the future Pedro II (1648–1706), regent for his insane brother, Afonso VI (r. 1656–67), attended a banquet at the palace, the marquis ceremonially smashed the grotto's chinaware in tribute to the princely visit.

Many notable European purchasers of porcelain also collected exotic shells, and some regarded both commodities as possessing magical properties. Of course, others believed that the essential secret of porcelain consisted of some rarified blend of seashells and clay. A patron of alchemists, as well as a zealous collector of Chinese pottery, Francesco I de' Medici, Grand Duke of Tuscany (r. 1574–87), owned a magnificent double-nautilus cup, mounted in silver and engraved with Chinese motifs of plants and animals.[103] August II displayed his large assortment of Asian shells, hanging from imitation pine branches made from New World silver, alongside his porcelains, and his Meissen porcelain manufactory produced dinner services with elaborate shell molding. The potters who produced one of the most famous Meissen dinner ensembles, the Swan Service *(Schwanengeschirr),* a set of more than three thousand pieces commissioned by Count Heinrich von Brühl (1700–1763), a Saxon minister under August III, based their designs on shells from the royal collection. Shell-shaped tureens boasted mermaids for handles, and cruets took the form of cherubs astride swans. Potters also bedecked the service with motifs of snails, dolphins, the sea nymph Galatea, Venus rising from the waves, and Neptune flourishing his trident.

Pottery manufacturers at Capodimonte (near Naples) and Plymouth (in England) decorated dishes and vases with designs taken from shells. The vessels often had scallop-shaped edges or were painted with feathered rims suggesting shells. Louis XIII of France (r. 1610–43) added to the collection of shells begun under Henry IV, and Louis XV, the reigning sponsor of Sèvres porcelain, owned the largest assembly of shells in Europe, some of which he sent as models to his potters. Tsar Peter the Great (r. 1682–1725), who began the Russian royal practice of purchasing porcelains in western Europe, also bought shells in Holland, as did the monarchs of Denmark and Sweden. Like other amateur scientists and learned entrepreneurs, Wedgwood collected shells, and his Etruria manufactory produced shell-shaped desert services christened "Nautilus wares." (See figure 7.)

In the early eighteenth century, pottery ensembles like Wedgwood's catered to the vogue for rococo style (perhaps from the French *rocaille,* "shellwork"), which emphasized elegant and charming effects, such as elaborate ornament, coiling outlines, and asymmetrical curves, the serpentine lines characteristic of grottoes, plants, and shells. Meissen turned out a splendid porcelain clock, with a statuette of Venus sitting atop gilded shellwork and vignettes of gallant men and modish women in chinoiserie settings. Indeed, porcelain played a key role in stimulating enthusiasm for rococo fashion, especially in the decorative arts; chinoiserie designs on the wares emphasized the natural world as exemplified in meandering motifs. Westerners tacitly agreed with the Chinese that shells, whether the nautilus or the cowry, had a special affiliation with porcelain.[104] (See figure 8.)

To be sure, there is a world of difference between the regal nautilus and the plebeian cowry, the one traveling first class, the other in steerage. Still, Polo's use of *porcellane bianche* for the Chinese ceramic as well as the shell is an instance of the long-standing association between the two materials. Piglets aside, porcelain was well named.

NEPTUNE AND VULCAN: EARLY MODERN SCIENCE AND THE SECRETS OF PORCELAIN

According to Venetians, a Chinese bowl in the treasury of the Basilica of San Marco had been brought to the city by its most renowned citizen.[105] Although the traveler almost certainly had nothing to do with it, the "Marco Polo bowl" was a rarity in the fourteenth century, perhaps the earliest authenticated porcelain in the West. It is, however, a nondescript vessel, the sort of ware turned out by Fujian kilns for sale in the unsophisticated markets of Southeast Asia. Polo records that he traveled in Fujian province, but he apparently did not visit Jingdezhen in neighboring Jiangxi, a pottery center that became prominent in the late Song, a generation before he reached China. The city had begun producing porcelains seven centuries earlier. In the opening years of the Tang, local officials even sent gifts of pottery to the imperial court at Chang'an in northern China. At that time, the leading pottery centers were in the north, and Jingdezhen prospered only to the extent that it imitated them.

Dentrecolles remained ignorant of this background, and even if he had extended his investigation beyond Jingdezhen, he would have faced an impossible task collecting information from documentary sources. The history of porcelain attracted virtually no attention from the literati, and it would not be studied in systematic fashion in China until the twentieth century. Jingdezhen bears some responsibility for this inasmuch as its eventual, sweeping triumph led to the decline of older, dominant pottery centers whose very existence was lost to memory long before Dentrecolles arrived in China.

Dentrecolles surely learned a great deal about porcelain manufacture from Tang Ying, "the mandarin who . . . honored [him] with his friendship." Tang's interest in the details of pottery production was exceptional for a scholar-official. He served in the porcelain city for over twenty years in connection with the imperial kilns and became a master potter in his own right, even creating vessels known at the time as "Tang ware" (Tangyao). Remarkably, Tang carried out the first excavations at Jingdezhen in the 1740s, almost two hundred years before serious archaeological fieldwork of any sort began in China. The early Ming kilns of Jingdezhen were not discovered until the 1980s, when gas and water pipes were laid in the garden of the old imperial workshops.[106]

Having worked and lived with the potters, Tang erected stone tablets next to the imperial kilns with inscriptions of things he had learned. In 1743, at the command of the Qianlong emperor, he wrote his commentary on Jingdezhen's industry, *Description of the Twenty Illustrations of the Manufacture of Porcelain.* With pictures on silk in ink and color, it is an invaluable document in the history of porcelain, though it depicts a charmingly pastoral version of the grimy toil. A similar album sent to the Sèvres manufactory from Beijing by Joseph-Marie Amiot (1718–93), a Jesuit, may have been modeled on the illustrations commissioned by Tang.

Zheng Tinggui, the early-nineteenth-century editor of Lan Pu's *Potteries of Jingdezhen,* singled out Tang's writings on porcelain for praise; but he also complained that

> there has never been any book devoted exclusively to an account of pottery. What one finds in the books are either references to particular vessels introduced to give point to an incident or incidents introduced to give point to some vessel; or a name or two extracted from a verse of poetry. . . . In the present compilation, most of my knowledge of pottery affairs and local products has been obtained by making personal inquiries.[107]

A century earlier Dentrecolles likewise had no choice but to rely on interrogating the potters and shopkeepers of Jingdezhen. Although he did not discover the secrets of porcelain that would enable his countrymen to replicate it, that failure stemmed from no fault of his own: he and his superiors had no comprehension of how intractable a task they faced.

Among the crucial obstacles confronting Westerners was the conceptual and technical inability of early modern scientists to grapple effectively with the question of the nature of porcelain. Dentrecolles sent a sample of clay to Louis-François

Orry, who passed it along to René-Antoine de Ferchault Réaumur (1683–1757), an innovator in the study of heat and the nature of rock. Once renowned in intellectual circles, his name would be known to everyone today if his novel system for measuring temperature, which he developed from researching porcelain, had not lost out to the scales devised by Gabriel Fahrenheit (1686–1736) and Anders Celsius (1701–44), inventor of the centigrade thermometer. Réaumur also pioneered new ways to look at the history of the planet: his examination in southern France of sedimentary rock with substantial concentrations of seashells was an important step toward considering the Earth on a time scale dwarfing that prescribed by the Bible. Deeply involved in what his contemporaries called the "furor for porcelain-making" and inspired by his analysis of Dentrecolles's sample, Réaumur in 1715 helped direct a search for suitable clay materials in all the provinces of France. According to the savant, "Europe envied what China had too much not to attempt to create something similar."[108]

Réaumur successfully identified the principal components of the clay sent by Dentrecolles, and his experiments in sedimentary petrology (known at the time as mechanical mineralogy) suggested that transformation of the material in the kiln could be explained by the growth of crystals. But a correct analysis of porcelain eluded him because he shared the contemporary assumption that crystals could develop only from an aqueous solution. The Genesis story of Noah's Flood lay at the bedrock of that hypothesis, influencing Western notions of how the surface of the Earth took form. The biblical fable also served to account for seashells and fossilized fish being embedded in rock on the summit of mountains. That bewildering discovery had attracted the attention of ancient Greeks, most notably Aristotle (384–322 B.C.E.), and, bracketing Réaumur's life, Nicolaus Steno (1638–86) and James Hutton (1726–97) made it central to their geological speculation.[109]

A great admirer of Steno, Leibniz put forward the dominant educated view, known as Neptunism, when he asserted that floodwaters came from an abyss deep inside the Earth, with the result that "enormous bubbles . . . have burst here and there, so that certain portions [of rock layers] subsided to form the trough of the valleys, whereas others, more solid, have remained upright like columns and, for that reason, constituted mountains."[110] Since Réaumur also assumed that water was the most significant force shaping the planet as well as the basis of solid matter, he concluded that at the molecular level (as it is now called), some kind of "lapidifying juice" melted stony substance, thereby giving rise to crystalline extensions.[111]

Later in the century, such views became obsolete. Hutton and James Hall (1761–1815) experimented with volcanic emissions and determined that crystallization takes place in basalt, a fine-grained, black igneous rock. As geologists determined in the late twentieth century, the continents, largely made of granite, float on a sea of hot basalt five kilometers deep, and a similar volume of basalt also underlies the slim blanket of sediment that constitutes the ocean floor.[112] When the basalt undergoes certain chemical changes over millennia, liquid magma (molten lava) is created. The magma sometimes acts likes it is escaping from a pressure cooker, thrusting to the surface through fractures in the Earth's mantle, resulting in a spectacular pyroclastic event—that is, volcanic eruption. Since basalt is abundant on all continents, the conclusions about crystallization by Hutton and Hall placed volcanoes at the center of a momentous debate on the natural history of the planet.

Hutton's research was influenced by *Observations on Mount Vesuvius* (1772) by Sir William Hamilton (1730–1803), the British plenipotentiary at Naples for thirty-six years (1764–1800) and an ardent collector of so-called Etruscan classical pottery, many pieces of which he illicitly dug up in the foothills of the volcano.[113] (Although most experts regarded the vessels as Etruscan, that is, from central Italy, Hamilton himself correctly argued for their Greek origin.) The most famous eruption by Vesuvius took place in 79 C.E., wiping out Pompeii and Herculaneum. Except for a few outbursts, it remained silent until the late seventeenth century. It became especially active once again during the years Hamilton lived in Naples, giving rise to much speculative literature, most notably his *Observations.*

Hamilton's book inspired Hall to travel to Italy in 1785 to witness the eruption of Vesuvius and collect lava samples. Vulcan, the god of fire, who gave the Italian form of his name to volcanoes in the late seventeenth century, thus vanquished Neptune: the emerging science of geology established that the world's surface derives from the fiery depths of the Earth, not from giant crystals precipitated from a primordial Ocean or as a result of torrents of water from a subterranean chasm. One of Hamilton's correspondents speculated that recognition of volcanoes as a geological force would call for reevaluation of all earthly life since they are "part of the great system of dissolution & renovation which seems the Universal Law of Nature, by which a sparrow, a city, a region or a world . . . have their beginning & end."[114]

VOLCANOES AND VASES: CHARLES DARWIN AND JOSIAH WEDGWOOD

The most far-reaching reevaluation of earthly life began when Charles Darwin (1809–82) set off in 1831 on his five-year journey around the world in the HMS *Beagle.* After experiencing a volcanic eruption and an earthquake off the coast of Chile in 1835, Darwin pointed to the events as definitive, contemporary proof that a geology based on Scripture, as in the story of the Flood, must be abandoned. Discovering the Chilean shoreline had risen by several feet, he recalled the frontispiece of *The Principles of Geology* (1830) by Charles Lyell (1797–1875), which displays an engraving of the so-called Temple of Serapis at Pozzuoli on the Bay of Naples. Lyell's examination of black discoloration on the columns of the temple had persuaded him that they had been underwater for centuries and that eruption by Vesuvius, followed by earthquake, had raised them to the surface once again.

In *Voyage of the Beagle* (1839), Darwin declared that taking into account the role played by volcanoes in elevating the Andes Mountains, "we can scarcely avoid the conclusion, however fearful it may be, that a vast lake of melted matter, of an area nearly doubling in extent that of the Black Sea, is spread out beneath a mere crust of solid land."[115] Accordingly, Darwin recognized that massive segments of the planet had emerged over eons, that volcanoes represented part of a piecemeal process of destruction and creation happening in geological time—indeed, in time enough for evolution to take place in what he later called nature's "manufactory of species."[116] In *On the Origin of Species* (1859), he represents the natural world as a self-improving "workshop" characterized by pitiless competition, teeming productivity, and a "division of labor" founded on specialization.

Darwin's family heritage and the early stages of the industrial revolution formed the backdrop for the development and expression of these ideas. His voyage on the *Beagle* had been underwritten by the Wedgwood family fortune inasmuch as he was Josiah's grandson and his wife, Emma (1808–96), was a Wedgwood first cousin with a handsome trust fund. Darwin hung Josiah's portrait in his study, which was well stocked with books on the economy and industrial production, and he remained close to his Wedgwood cousins who managed the Etruria pottery. Wedgwood dedication to cutthroat competition in the commercial world, the astonishing productivity of industrial workshops, and the specialized division of labor in the Wedgwood manufactory helped direct Darwin toward his revolutionary theory of natural selection.[117]

English entrepreneurial culture in general, and the Wedgwood pottery in particular, provided Darwin with instinctive confirmation for his theory of evolution. Wedgwood's enterprise also played a clearly identifiable role in the development of earlier innovative ideas in chemistry and geology. For almost a century after Réaumur's analysis of the clay sent by Dentrecolles, progress in the new sciences was coupled to investigation into porcelain and furnace transformation.[118] Pottery kilns served as handy laboratories for the study of volcanic action, and Wedgwood's invention of a pyrometer for measuring temperatures above 1,000°C proved useful not only for baking clay but also for researching volcanic flare-ups. According to a contemporary observer, Lazaro Spallanzani (1729–99), a celebrated scientist who measured the heat radiating from Vesuvius in degrees Fahrenheit, "very happily applied [Wedgwood's pyrometer] to determine the degree of fire necessary to fuse the lavas of volcanos."[119]

Wedgwood discussed volcanoes and antique pottery with Hamilton, corresponded with Hutton about the latter's pioneering theory on the formation of the Earth's crust from lava flows, and gave advice about kilns and ceramics to Hall for his experiments in melting basalt. Wedgwood and Erasmus Darwin (1731–1802), the scientist's grandfather, naturally knew that the Peak District of Derbyshire, lying between the Trent and Mersey Rivers and immediately to the north of their homes in the West Midlands, held outcroppings of basalt. Both men enjoyed "fossilling" there for trilobites and bivalve shells, and Erasmus (an outspoken atheist) had the motto *E conchis omnia* (Everything from Shells) added to his family crest of three scallop shells.[120] In 1767 Wedgwood wrote to his partner, Thomas Bentley (1731–80), that digging the Trent and Mersey canal in the vicinity of his pottery had disclosed an extraordinary fossil: "Sometime last month was found under a bed of Clay, at a depth of five yards from the surface, a prodigious rib, with the vertebre of a monstrous sized Fish."[121] With such learned geologic preoccupations shared by many in an increasingly educated and secular public, Wedgwood and other Staffordshire potters catered to the interest by producing teapots decorated with patterns of fossils and shells.

In more ways than one, volcanoes and vases went together. As Wedgwood wrote, "I apprehend the Etruscan body owes its lightness to a mixture, either natural or artificial, of volcanic ashes."[122] When he opened his new manufactory in 1769—named Etruria after the classical wares made famous by Hamilton—he sat at a pottery wheel and ceremonially threw six stoneware vases made from an original ceramic formula he christened Black Basalt. He chose that name because of the connection he saw between the black rock forged by volcanoes and the new black ware fired by the Etruria kilns. In a fashion, as Wedgwood saw it, his manufactory replicated the actions of both Vesuvius and the potters of antiquity. He had the black-basalt vases inscribed with the legend *Artes Etruriae renascuntur* (The Arts of Etruria Are Reborn) and decorated with classical motifs from engravings in a book on Hamilton's pottery collection. When he later produced a ceramic portrait plaque of Hamilton, with the background painted to give it the look of an ancient Greek ("Etruscan") vase, he naturally had the figure done in black basalt.[123]

It seemed to Wedgwood that the world spun like a pot on a wheel while the bowels of the Earth poured forth new raw material for molding by natural forces. He received advance copies of Hutton's *Theory of the Earth* (1785), one of which he sent to James Watt (1736–1819), a fellow amateur scientist who experimented with liquefying igneous rocks and porcelain constituents.[124] Friends and business colleagues, Wedgwood and Watt both owned potteries, and their common interest in steam power, furnace transformation, mineral crystallization, and volcanoes made them receptive to the idea that the interior of the planet functioned like a titanic kiln, a heat-generating engine disgorging molten lava that threw up mountains and reshaped continents.

Yet it was not until a century later, with the invention of X-ray diffraction around 1900, that crystallography finally would reveal the internal structure of the clay minerals that made porcelain such a unique substance. And it was not until the mid-1960s that chemistry and physics would unite in the groundbreaking theory of plate tectonics, explaining the actual role of volcanoes in the inexorable deformation and reconstruction of the planet's surface. The same theory also would illuminate the substantive link between the lava expelled by volcanic eruption and the clays fired to make porcelain.

In effect, in his letters from Jingdezhen, Dentrecolles focused on details of porcelain production because questions regarding the fundamental nature of the ceramic remained well beyond the scientific understanding of his time. As it turned out, however, those details proved highly instructive. Fascinated by Dentrecolles's description of how one piece of pottery passed through the hands of more than seventy workers during production, Wedgwood copied the passage into his Commonplace Book in 1743, and a quarter of a century later he made it the basis for the organization of his Etruria pottery, the first industrial establishment founded entirely on the division of labor and the first directly to apply the steam engine to manufacturing.[125]

Since Etruria provided the model for the modern factory system, the triumphs of the industrial revolution thus owed something to the potters of Jingdezhen. In his mission to make the West competitive with the most renowned works of China, Dentrecolles ultimately proved more successful than he ever could have dreamed. Both his painstaking investigation and the accomplishments of the porcelain city won recognition from Wedgwood. Incorrigibly given to grandiloquent verse, Erasmus Darwin praised Wedgwood's manufactory—"A new Etruria decks Britannia's isle"—and declared that ancient Etruria rivaled China itself in the antiquity and excellence of its pottery.[126] Wedgwood, however, responded by suggesting that Darwin read Dentrecolles's letters in du Halde's *Description.* He would learn from the French Jesuit that "the Chinese employ their art in the formation of works of great magnificence such as the building, ornamenting & covering of entire pagodas, & other superb edifices." Justice should be done, the pottery baron told his old friend, to "my distant brethren in the plastic art."[127]

3

The Creation of Porcelain

China and Eurasia, 2000 B.C.E.–1000 C.E.

While a modern-day Dentrecolles certainly would find it substantially easier to learn about the history and nature of porcelain, problems of definition and interpretation still puzzle a newcomer to the subject. A central difficulty is that China and the West categorize porcelain differently in relation to earthenware and stoneware. Based on a Western taxonomy, the contemporary view regards pottery as encompassing the three types, ranked according to the ware's material composition and the temperature at which the clay is fired. The Chinese, however, have traditionally recognized only two groupings, *tao,* or earthenware, and *ci,* which includes both stoneware and porcelain. The Western classification of stoneware and porcelain makes sense in terms of chemical and physical analysis inasmuch as there are genuine differences in mineral ingredients and material formation between the two ceramics; but when the categories are employed in historical narrative, they lead to considerable confusion. Depending on which modern account one consults, a material comparable to porcelain emerged in any of the periods of Chinese history from the Zhou to the Yuan, a span of more than two thousand years.

EARTHENWARE, STONEWARE, AND PORCELAIN

Clay generally is associated with substances such as silt and sand, but it actually is decayed or "rotten" rock. A fine-grained, earthy material that is plastic when wet and hardens when heated, clay is formed from the breakdown of granite, a coarse-grained igneous rock found mainly in the continental crust. It is a relatively soft, sedimentary composition, a complex, varied substance, not an inert material. Many of its microscopic particles carry a negative electric charge and join in chemical reactions, a feature that makes it a likely source for the origins of life on Earth, a template or prototype for proteins and nucleic acids.[1]

Earthenware, made from many types of clay, forms at between 600°C and 1,000°C; if the temperature is higher than 1,000°C, the vessel will bloat and collapse in a liquid melt. The fired clay, which is red, brown, or buff, is soft and not wholly fused; because it is porous, it must be covered in a second firing with a glaze, essentially a thin layer of glass, to become relatively impermeable. Stoneware is produced at 1,100°C to 1,250°C, resulting in a product whose hardness lies between that of earthenware and porcelain; it is vitreous, almost nonporous, resonant when struck, and varies in color from light gray to black, depending on the amount of iron in the clay. Whiteness and translucency are the chief distinctions between porcelain and stoneware. The porcelain produced in northern China from the early seventh century was made from kaolin (china clay), whereas that made at Jingdezhen at the same time was made from china stone (or porcelain stone, *cishi*).[2]

The term *kaolin* derives from Kaoling (High Ridge), a modest hill northeast of Jingdezhen where miners worked china clay from ancient times. Kaolin not adulterated with impurities, such as that found near Jingdezhen, is scarce. Europeans employed it for pottery from the early eighteenth century, but before that time they used it chiefly for cosmetics and wig powder. *Petuntse* (or *baitunzi*, "little white bricks"), which stems from Chinese potters' slang, appears in most descriptions of porcelain as the commercial term for the processed china stone used in the manufacture of porcelain. According to Dentrecolles, potters always had to purify the china stone shipped to Jingdezhen because "from a people who roll little grains of paste in pepper powder to cover them and then sell them with real peppercorns, there is no safeguard from the sale of petunse without it being diluted with some dregs."

From the late thirteenth century, Jingdezhen potters mixed kaolin with china stone: fusing under heat into a kind of natural glass, china stone gives translucency and hardness to the porcelain paste but is difficult to work with by itself; kaolin softens the paste, providing plasticity, smoothness, and whiteness. As recognized by Réaumur in the early eighteenth century, the two components combine to form a single mass when heated to about 1,350°C, undergoing a process of vitrification that renders the finished product resonant, wholly impermeable, very white, and translucent when thin. China stone also provides the glaze of Jingdezhen porcelain; mixed with a bit of lime and wood ash to aid in fusing, it makes for a glassy, rock-solid bond with the clay body. Song Yingxing describes the liquid glaze as having "the appearance of a clear rice broth."[3]

In the early twentieth century, X-ray crystallography revealed that both kaolin and china stone contain alumina, a mineral oxide that produces microscopically fine crystals of mullite at high temperature. Referring to "little silvery particles that are sprinkled in the kaolin," Dentrecolles unknowingly identified grains of mica that pointed to the presence of the mineral oxide. Chinese potters similarly described the clays as "sprinkled with spots like stars of silver."[4] After William Cookworthy (1705–80), an apothecary and potter, read Dentrecolles's reports in du Halde's *Lettres édifiantes* in the 1740s, he noted specks glinting in the granite walls of the church of St. Columb in Cornwall, a clue that contributed to his discovery of deposits of china clay and china stone in that southwestern English county. To the present day Cornish clay mines are bordered by small white mounds made up of the residue left behind when the two materials were processed.[5]

In kaolin and china stone, crystals of mullite are piled up in stacks set at right angles to other stacks. When heated in a kiln, the crystals glide over one another, extending like shafts into the matrix of the clay, forming a glassy, latticelike structure that fuses the substance. In terms of composition and temperature, kaolin and china stone fired below 1,000°C make earthenware; fired above 1,000°C but below 1,300°C, they produce stoneware. Because porcelain is impermeable, glazing merely improves its appearance or protects pigments added between initial firing and final return to the kiln.

The terms *proto-porcelain, quasi-porcelain,* and *porcellaneous stoneware* are used in many modern accounts to identify pottery made with some of the ingredients of porcelain and sharing some of its physical characteristics centuries before the emergence of northern porcelain in the Tang dynasty or of Jingdezhen porcelain in the Song period. The long-established Chinese category of *ci* for high-fired pottery (stoneware and porcelain) has the virtue of avoiding such perplexing terminology, albeit at the price of ignoring real distinctions between stoneware and porcelain.

The two classifications of porcelain point toward significant differences between the ceramic experiences of China and the West. Stoneware defined the ceramic history of China for centuries before Tang or Jingdezhen porcelain appeared. Whether in the early seventh or late thirteenth century, the Chinese grouped porcelain with stoneware inasmuch as the new material resulted from incremental change within an established tradition, mainly a matter of adjusting the proportion of known ingredients. And since porcelain vessels, in their solidity, resonance, and impermeability, seemed virtually indistinguishable from stonewares, potters and connoisseurs had good reason to regard the new product as just another, slightly improved variety of *ci.*

Earthenware defined the ceramic history of the West until the arrival of Chinese porcelain in the sixteenth century. Northern German stoneware developed late in the history of European pottery, though it was highly valued, especially in England and Holland. At least ten million pieces of the Rhenish pottery entered the port of London between 1600 and 1640, just before the deluge of Chinese porcelain began. The stonewares gained popularity as drinking vessels, pharmacy jars *(albarelli),* and chamber pots and as nearly impermeable containers for transporting beer, wine, mineral water, mercury, and Dutch gin.

Attractive and pleasingly decorated, Rhenish stoneware figured in altarpieces by

Hans Memling (ca. 1433–94) and in paintings of peasant festivities by Pieter Bruegel the Elder (1529–69) and David Teniers the Younger (1610–90). Artists of the Netherlands often featured stoneware jugs in Marian devotional paintings, showing them filled with flowers or embellished with the sacred trigram of the Holy Name of Jesus *(De nomine Jhsesu)*. Dutch taverns served beer in the jugs, and the vessels also found their way into wealthy Dutch households. Floris van Schooten (1590–1655) painted *A Kitchen Scene* (ca. 1625), which depicts shelves of pewter dishes, blue-and-white porcelain bowls, and potbellied stoneware jugs topped with pewter lids. In *Young Woman at Her Toilet* (1667), by Gerrit Dou (1613–75), an ermine-draped lady is surrounded by expensive furnishings, including a Persian carpet, upholstered chairs, a silver-gilded metal pitcher, and a stoneware wine jug resting in a massive cooler of black marble. Still, the spread of stoneware in northern Europe from the fourteenth century did not inspire attempts to replicate or analyze the product because good, well-glazed earthenware was close to the Rhenish commodity in price and distinction. Chinese porcelain was a different matter altogether, for Europeans regarded it as superior to their ubiquitous earthenware in all respects.[6]

Eager to compete with China, Westerners from the late seventeenth century—soon after porcelains began to arrive in massive quantities—investigated the composition of the ceramic and experimented with formulas that approximated it. Dentrecolles shared the scientific mentality on which this enterprise was founded. Observing that the manufacturers of Jingdezhen were ignorant of chemical analysis, such as the use of nitric and hydrochloric acids to dissolve solid materials, he declares that "their inventions [are] all . . . of extreme simplicity." He seems to share Leibniz's notion that the Chinese were oblivious to the mental abstraction and critical method that are the essence of Western experimental science. He suggests that "by taking the necessary risks and expense of various tests, it might be possible to discover the art of making with certainty what had once been the consequence of chance." This presages a major thrust of early-twentieth-century research in traditional industries such as cotton mills, breweries, tanneries, and potteries: that is, investigating the scientific grounds for practices until then inarticulately embedded in traditional handicraft procedures.[7] Dentrecolles recommends that just as Europeans might employ experimental techniques to recover the lost secret of making stained glass, so too "perhaps someone in Europe can invent what the Chinese now do not know" about porcelain.

In the service of August II, who financed what is arguably the first research-and-development project in history, that is precisely what Tschirnhaus and his associates did in the very years Dentrecolles carried out his investigation in the workshops of Jingdezhen. Early-nineteenth-century European analysis of the mineral and chemical composition of ceramics resulted in the contemporary classification *earthenware, stoneware,* and *porcelain*. The category of stoneware emerged by a kind of back-formation: having identified the physical characteristics of porcelain, Western scientists inescapably recognized those of stoneware as well.

This taxonomic achievement stemmed from European admiration and craving for chinaware, though lack of that achievement in China reveals little about mental abstraction and critical method there. Dentrecolles believed that the potters of Jingdezhen used only rudimentary techniques, yet he actually witnessed them employing some that have a distinctly experimental air, such as producing a component of copper-oxide glaze by sprinkling water on molten copper and then removing the oxide scale for examination.[8] Dentrecolles describes this procedure in his letter of 1722, just before he left Jingdezhen for his new post in Beijing. As it happens, pottery experts in the two cities embarked on experimental and systematic studies of ceramics in the 1720s, leading to innovative techniques that transformed porcelain aesthetics by the introduction of brilliant enamel colors and unusual effects. The Chinese, however, did not focus on the composition of porcelain itself simply because they regarded the material as a natural extension of their ancient craft of high-fired pottery. For their part, Westerners investigated the nature of the ceramic because of the novelty, excellence, and expense of the exotic commodity.

TECTONIC PLATES, VOLCANOES, AND YELLOW EARTH

The most striking contrast between China and the West in the history of ceramics is that the former was more than three thousand years ahead of the latter in the manufacture of high-fired wares. The foundation for this precocious achievement lies in the unfathomable past, as far back as the geologic formation of what is now called China.[9] The supercontinent of Pangaea, "All Earth," began forming 600 million years ago during the Precambrian, assembled from landmasses riding on tectonic plates averaging one hundred kilometers in thickness. When plates meet at subduction zones, one is pushed beneath the other, with the heat of the collision melting the rock to magma. Made relatively fluid as a result of being mixed with gases and water, the magma rises from the base of the upper mantle, or lithosphere, of the Earth, where the temperature is 1,400°C, about the firing range of porcelain—a conformity that would not have surprised Hutton or Wedgwood.

When magma surfaces explosively, volcanoes are born. More commonly and less dramatically, as the magma cools and crystallizes, it forms granite that thrusts through cracks in the mantle, thereby propelling tectonic plates at an average speed of two centimeters a year, roughly the rate at which fingernails grow. North China (technically, the Sino-Korean craton) and South China (the Yangzi craton) collided with Siberia 250 million years ago, completing the consolidation of the continents and heaving up the Altai and Tian Shan mountain ranges of Central Asia.[10]

As a consequence of their different tectonic heritages, the northern and southern parts of China have contrasting topographies and natural resources, a fault line also accentuated by climatic divergence. The region of northern China—that is, the territory along the Yellow River and north of the Qinling mountains (or the provinces of Shaanxi, Henan, Shanxi, and Hebei)—with an average winter temperature below freezing, became a land of millet, wheat, and sorghum, a central plain of relative physical simplicity. Southern China—encompassing the east-central provinces near the Yangzi River (Hunan, Hubei, Jiangxi, Anhui, Jiangsu, and Zhejiang)—evolved as a jumbled, subtropical region of countless lakes, river-dissected hills, and highly indented coastline; having an average winter temperature above freezing, it became a land of rice, sugar, oranges, and cotton. Boundless fields and carts in the north, forested slopes and boats in the south. Much of the history of China may be regarded as a successful human endeavor to extend what tectonic collision set in motion—to wield together the disparate areas by deploying a universal written script, a common textual canon, an overarching civil service, and a man-made waterway (the Grand Canal) linking the two great tributaries that flow from the Himalaya mountains to the Pacific Ocean.

About 140 million years ago, during the dinosaur-dominated periods of the middle Mesozoic, huge volcanic eruptions took place in southern China, perhaps associated with westward movement of the Philippine tectonic plate as it slid over the Eurasian plate. For millions of years, massive clouds of volcanic dust and cinder specks periodically shrouded the region, settling out eventually as a sedimentary deposit identical in chemical composition to igneous rock. Compacted over eons, the pale-colored volcanic debris turned into the china stone that became the basis for the earliest porcelain of southern China. Pulverized by trip-hammers and leached through tanks of water, the stone experiences a simulated weathering that telescopes epochs into hours. Fired in a kiln, it fleetingly reaches toward its primordial origins as molten lava. A phenomenon of furnace transformation, the white porcelain of Jingdezhen is essentially igneous material that has been melted once again and then solidified into artificial rock, a substance re-created in human fire and shaped to human purposes.[11]

Long after the volcanoes of southern China stopped belching volcanic ash, northern China received its own accretion of new soil. Some 45 million years ago, 20 million years after extinction of the dinosaurs by an asteroid, the last great convulsion of the Earth's crust began. The bulk of what is now the Indian subcontinent split away from Antarctica and headed north, propelled by movement of the Indian-Australian tectonic plate. Thirty-seven million years later, a slow-motion smashup took place as India plowed into southern Eurasia, raising the Tibetan Plateau and Himalaya mountains, a range of peaks more than three thousand kilometers long. These constitute the youngest and largest uplands on the planet, called "the roof of the world" since Marco Polo made them known to Europeans in his travel narrative. India still drives north today, raising the mountains an average of ten centimeters every year, making the region the most dangerous earthquake zone in the world. Erosion whittles down the uplift just as inexorably, thereby maintaining the peaks at a general elevation of five kilometers above sea level, apparently the maximum height that mountains can reach on the planet.

The Indo-Eurasian cataclysm had manifold consequences, even proving decisive for the evolution of human life. The tectonic collision displaced southern China eastward, farther into the Pacific, and it thrust out what became the great peninsula of Southeast Asia. The Tibetan Plateau cut off the flow of rain from the south, the influx of which had nurtured Central Asian grasslands and replenished the Tethys Sea, a body of water extending from Europe to East Asia. As the inland sea dried up, vast deserts expanded in the rain shadow of the mountains; the region lost its mild climate and fertile soil, turning into the desolate wasteland that later afflicted travelers on the Silk Road. The Tibetan upheaval led to creation of the monsoon systems of South and East Asia, the colossal natural engines that were unfailingly to speed the passage of ships between India and China. The drainage pattern of East Asia's great rivers also emerged, with the Yellow, Yangzi, Hong, Mekong, and Salween all flowing from the ice and snows of the gigantic uplands, separated in their headwaters by fewer than seventy kilometers and in their ultimate destinations in the engulfing sea by thousands.

Formation of the Himalaya mountains and the Tibetan Plateau resulted in cooling of the Earth, since exposure of such a huge expanse of bare igneous rock—the size of Texas, California, Montana, and New Mexico combined (or 1,817,400 square kilometers)—led to removal of carbon dioxide, a greenhouse gas that traps solar heat, from the atmosphere. Monsoon rains that drench the rock particles produced by tectonic collision and erosion also flush out more carbon dioxide from the atmosphere, thereby accelerating global cooling. In central Africa, lower planetary temperatures led to shrinking of the rain forest and expansion of savannah grasslands. Faced with a new and more difficult environment, tree-dwelling primates explored the advantages of walking upright on the plains in search of food, launching the evolution that would culminate millions of years later in their most successful descendant, modern *Homo sapiens.*

Cooling of the Earth spawned glaciers up to 3.2 kilometers thick that ground up the continental bedrock in a succession of ice ages. Glacial pulverizing and both mechanical and chemical weathering of the granite peaks of Central Asia produced immeasurable amounts of loess (from German *löss,* "loose") a yellow rock dust as fine as flour, composed mainly of quartz, a crystalline mix of silicon and oxygen. Some 2.4 million years ago, loess began blowing into China from the Central Asian desert belt, including the Gobi, blanketing the highlands of the north during long periods of desiccation, finally burying the luxuriant forest of the broad continental plain and its clay-rich surface. Carried southeast on winter winds from Mongolia, the quartz powder forms a layer as much as three hundred meters deep in places, the thickest topsoil on the planet.

Loess, known in China as *huangtu* (yellow earth), is the sediment that gives the Yellow River (Huang He) its name and turns it into the muddiest waterway in the world: at times as much as a third of its mass is silt, and a tributary once was recorded as reaching almost twice that.[12] This is reflected in the Chinese adage that "the waters of the Yellow River become clear only once in a thousand years." The river, Ricci remarks, has "no respect for Chinese law and order," frequently ravaging whole districts when it breaks its loess banks, seemingly changing its course at will.[13] Unlike the Yangzi, the Yellow River is navigable for only a short distance from its mouth, making it a dismal channel for trade and communication. Pouring out over a billion tons of silt every year, the river saturates the Yellow Sea with its ocher deposit.

An exceptionally fertile soil by virtue of its capacity to fix atmospheric nitrogen and bring up mineral nutrients by capillary action, loess provided the agricultural basis for the emergence of civilization in northern China, despite the sparse vegetation of the central plain. Millet agriculture supported a population of four hundred per square kilometer, and the high productivity of cereal farming gave rise to a host of independent, small-scale cultivators, the tax base of all future centralizing bureaucracies. To some extent, China owed its lack of feudal magnates and its perennial imperial integration to the rich topsoil swept in from Central Asia. Fittingly, as early as the Zhou period, the Middle Kingdom itself came to be symbolized by a yellow dragon, and writers described the Yellow River as flowing with the supple motion of a mighty dragon.

Potters in ancient China placed their workshops in loess deposits, and tens of millions of Chinese to the present day, notwithstanding calamitous earthquakes, live in roomy caves carved out of cliffs of yellow earth. Loess comprises the bulk of the raw material used in making the seven thousand life-sized, unglazed terra-cotta (earthenware) statues of warriors and horses buried in the famed mausoleum of Qin Shihuang, the first emperor of China. Around the same time, laborers constructed the first sections of what came to be known as the Great Wall from a rammed-loess foundation faced with bricks of fired loess. Centuries later, in the reign of the Ming, loess gave its tint to the yellow tiles on the roofs of the imperial Forbidden City in Beijing, and officials used yellow pottery for rituals at the Altar of the Earth.

By itself, because it is difficult to shape, loess has limited use as a pottery material; but after being washed by water over millennia, it becomes more malleable as a result of increments of clay. Kaolin, sedimentary clay washed out from granite rock and often found in association with loess, proved most useful for making pottery. As early as the Neolithic, artisans mined suitable clays, establishing their kilns where the loess layer was thinnest so they could gain ready access to the underlying kaolin. That clay became the material basis for the white vessels of the Shang, and for more than a thousand years, northern potters shaped it into high-fired wares.[14]

POTTERY AND METALLURGY IN CHINA

Employment of loess determined the development of Chinese pottery in the Bronze Age (1700–500 B.C.E.).[15] Since it consists mainly of quartz, loess has a high melting point, making it excellent material for constructing high-temperature kilns. In addition, the low clay content of loess means that it does not shrink during drying and firing, thereby proving ideal for making ceramic piece molds for casting bronze ritual vessels, the artistic masterworks of the Shang era. The shapes of the vessels usually came from pottery, though the designs they bore—animal heads, imaginary beasts, a pattern of whorls—generally appeared in low relief or incised rather than as painted decoration. A common motif was the *taotie,* a mystifying, masklike face of a creature with horns, bulging eyes, and gnarly mouth.[16] Potters carved the ornament on clay molds, sometimes on as many as eleven separate pieces for cauldrons *(ding),* large round vessels standing on three legs. Artisans assembled the pieces around a clay core and poured in molten copper and tin to form the metal vessel; after they pried off the clay sections, they polished the bronze. The excellent consistency and minimal shrinkage of the loess molds ensured that the final product preserved its meticulous embellishment.

Retaining their shape under intense heat, the molds made possible the creation of bronze pieces of superb quality, characterized by detailed ornament, dense patterns, and complex interlocking forms. Used as ceremonial, feasting and burial utensils, the bronze containers were objects of desire and distinction. As items central to status assertion as well as ritual order, royal courts controlled their manufacture, and their possession signified high rank and political authority. The Shang tomb of Fu Hao, a consort of King Wu Ding (ca. 1200 B.C.E.), held a treasure of more than 400 bronze objects, including 53 wine goblets, along with pottery, carved ivory, jade jewelry, a lacquered coffin, and 7,000 cowry shells.[17] Improvements in smelting in the Spring and Autumn period of the Eastern Zhou led to large-scale production of bronzes; archaeological excavation of a single foundry has unearthed tens of thousands of molds.

The ritual vessels had a substantial influence on the history of Chinese art and culture, a heritage of the Bronze Age that remained vigorous until the fall of the Qing dynasty in the early twentieth century. In 113 B.C.E. workers digging a pit discovered a bronze cauldron, reportedly enveloped in a supernatural, shimmering mist as it came to light. When scholar-officials presented the vessel to the emperor, they explained that in antiquity there existed only "one divine *Ding*-cauldron," which became the model for the actual bronzes employed in imperial ceremonies honoring Heaven, Earth, and sacred spirits.[18] In some measure, then, all ancient *ding* shared in the auspicious nature of this elemental cauldron, endowing them with awesome prestige. From the Han through the Qing, emperors collected *ding* and other ancient bronzes since they regarded the vessels as embodying the legendary past, spiritual power, and political legitimacy. In turn, potters traded on the reputation of the bronzes, for as early as the Shang they copied the metal pieces in their own medium.

As a consequence of innovations in metallurgy, ceramics, which had enjoyed the highest status for ritual and funerary purposes during the Neolithic, lost first place to bronze objects during the Shang.[19] And when wares in gold and silver came into wider use in the Han, pottery took second place to them as well. Still, its capacity for skeuomorphic reproduction rescued pottery from being relegated to the category of mere utilitarian crockery. By simulating costly metal wares in clay, potters brought preferred shapes, hues, and ornament within reach of prosperous social groups outside the royal courts. During the Shang, artisans replicated bronze containers in earthenware and stoneware, with glazes that mimicked the patina on bronzes and with spiral patterns that roughly imitated the dragon designs on them. In the Warring States period, ceramic surrogates of ritual bronzes gained widespread acceptance among the upper stratum as *mingqi* (radiant artifacts), tomb furnishings for the spirits of the dead. Inasmuch as a bronze vessel cost the equivalent of a worker's income for two years, a clay substitute had obvious attractions, in particular when the economy went downhill. That was the case in the period of disunity following the fall of the Han empire in 220 C.E.: bronze production plummeted, and glazed vessels took on a broader role in privileged households.[20]

Employment of loess for ceramic molds meant that pottery and metallurgy developed together in China. The same individuals certainly labored as both bronze casters and potters since each craft depended on the other. Until the molten copper and tin actually flowed into a mold, potters controlled almost all stages of production of the bronzes. As a result, effective kilns, clay molds, and ceramic artistry stimulated high-temperature smelting, bronze casting, and imitation of pottery forms in bronze. In western Eurasia and elsewhere, workers first made metal objects with a hammer and anvil as part of a smithy tradition. Hammering metal, however, was not a traditional Chinese technique. Rather, China stood out among cultures in that its earliest metal artifacts emerged from a ceramic context, with bronze vessels modeled on pottery and with ceramic molds determining the shape and embellishment of bronze implements.

This established a significant technological pattern: the reciprocal relationship of ceramics and metallurgy continued long past the Bronze Age. Increasingly powerful smelting ovens encouraged the development of kilns for making high-fired pottery in northern China by the late Shang. By 500 B.C.E. greatly improved furnaces led to the use of iron, which melts at 1,535°C, higher than copper (1,083°C) or tin (231.89°C). Most significantly, Chinese craftsmen relied on their long experience with mold technology to produce iron: they made cast iron—heating and pouring the metal into molds, as with bronze—rather than wrought iron, as in the smithy tradition everywhere else. While casting techniques are more wasteful of the ore than hammering, China possessed plentiful iron. It also had reserves of coal to fuel furnaces when northern China became largely deforested during the Tang.

Cast-iron manufacture had enormous rippling effects in many parts of the Chinese economy, especially in forest clearance, farming, engineering, mining, and salt production. Using ceramic molds, ironmasters forged thin pans (for evaporating brine), axes, plowshares, chains, hoes, saws, pots, and armaments generations before the beginning of the common era. In the early Tang, the notorious sovereign-empress Wu (r. 691–705), an enthusiastic sponsor of Buddhism, ordered construction of a three-story pagoda of cast iron, topped by a towering cast-iron phoenix (her symbol) sheathed in gold plate. During the Northern Song (960–1126), builders erected an iron pagoda in Kaifeng, thirteen stories (fifty-four meters) high and clad in glazed tiles. In the eleventh century, coke, a solid residue obtained from coal, fueled blast furnaces in northern China, raising output of cast iron to over 113,000 metric tons a year. The state armaments industry of the Song equipped an army of one million men with iron weapons, including spears, arrowheads, swords, and body armor. Europe did not use such large amounts of cast iron until five centuries later.[21]

A mutually reinforcing relationship between potters and metal-smiths, coupled with natural resources of first-rate ceramic materials, coal, and iron ore, gave China a key advantage over other regions of the ecumene. Civilization—as defined by the appearance of cultivated plants, domestic livestock, village settlements, formal religious cults, written script, and territorial polities—first came into existence around 4000 B.C.E. in Sumerian temple communities on the lower Tigris and Euphrates Rivers in Mesopotamia (present-day Iraq). Within a few centuries, Sumerian-Akkadian villagers had 350 names for ceramic vessels, and guilds of potters used mass-production techniques.[22] In contrast, China came relatively late to civilization, for the same set of defining breakthroughs did not appear on the Yellow River until more than two thousand years after the Mesopotamian achievements.

Since some communication took place across the breadth of Central Asia long before the advent of recorded history, it is unclear which of those breakthroughs to civilization in China resulted from Mesopotamian influence and which were indigenous developments. In any event, China swiftly surged past the accomplishments of Southwest Asia in ceramics and metallurgy as a consequence of the remarkable fashion in which the two technologies reinforced and stimulated each other. Although several considerations, including relative geographical isolation and early imperial unification, set China on a unique trajectory among Eurasian civilizations, the millennia-long symbiosis of clay and metal contributed substantially to the triumphs of China so much admired and coveted by Leibniz and his contemporaries.

THE CREATION OF TANG PORCELAIN: POTTERY FROM THE BRONZE AGE TO THE TANG

Over the course of centuries, a proficient kiln technology emerged from the combination of loess soil, the early predominance of ceramics in metallurgy, and considerable experience developing furnaces for making pottery, bronze, and iron.[23] No matter what sort of clay goes into making a vessel, if the kiln does not reach a sufficiently high temperature it merely will turn out earthenware. According to Piccolpasso, Italian potters "look on the manner of making the kiln as an important secret and say that in this consists the whole art" of creating superior wares.[24] Of course, Chinese potters shared the same perspective. They had the advantage, however, of more efficient and durable kilns than those of Southwest Asia. The clays of that region have poor refractory properties, firing adequately in the 1,000° to 1,100°C range but melting above 1,170°C. This limitation contributed to early Mesopotamian and Egyptian potters modeling their kilns on smelting furnaces, in which the smithy strives to keep metal and fuel in close contact with each other. They built furnaces of brick from the ground up, squat bottlelike structures with fireboxes directly beneath the pots, a technology that produced a uniform but modest temperature.

Potters achieve the best results, however, by separating vessels from the source of heat in the kiln. In China, they had long experience with refractory clays in building furnaces, and their downdraft kiln served as the model for the reverberatory furnace used for making cast iron, in which flames are deflected downward from the roof on the metal. Artisans employed loess soil to build such furnaces: they simply excavated a chamber in rising ground, tamped the walls, and dug a vent to the surface; an effective chimney gave a good draft and strong flame. Chinese potters had the additional benefit that their heat-resistant kilns could handle at least sixty firings before needing repairs, whereas those of Southwest Asia could not withstand much more than a dozen because of their construction from clays readily damaged by high temperature.

From early times, Chinese kiln builders put their structures partly underground, using the quartz-rich soil to provide efficient insulation. The small, boxy kilns of the Neolithic period reached 800°C, about the temperature obtained by putting raw wares in a bonfire, the earliest means of baking clay; at that heat only porous earthenware can be produced.[25] As early as the Shang, however, kilns achieved temperatures of 1,200°C, well within the range at which stoneware forms. In the Warring States period, northern kilns partially built into hillsides of loess soil took on a horseshoe shape, with two chimneys side by side, and fired as high as 1,350°C, the temperature at which porcelain ingredients fuse. In the lower Yangzi region during the Zhou era, "dragon kilns" *(longyao)* used the slope of hills to produce a natural draft that raised the firing temperature higher than European furnaces could obtain before the nineteenth century. Resembling a radically inclined chimney, the multi-chamber tube relied on the use of downdraft circulation, with fire at the mouth of the kiln preheating the air rising in the structure. Zhou potters fired stoneware at temperatures of 1,200° to 1,300°C and exported them as luxury items from the Yangzi valley to the region of the Yellow River. By the Song period, the largest dragon kilns of southern China stretched up slopes as much as 140 meters, at a 20-degree angle, and could fire tens of thousands of porcelains simultaneously.

At Jingdezhen toward the end of the Ming dynasty, the egg-shaped kiln *(zhenyao)* came into extensive use. Potters called it that because they thought it resembled half an egg lying on its side, though Tang Ying saw the oven as evocative of a tipped-over water jar. The switch from the dragon kiln stemmed from thinning timber supplies in the vicinity: since the egg-shaped furnace heats up rapidly and is exceptionally well insulated, it requires less fuel than the dragon kiln, making it cheaper to use. Described by Dentrecolles as typically thirty-six meters high and seventy-three meters long, the egg-shaped kiln fired various types of wares and glazes at different temperatures in a single operation. Perhaps bringing to mind his own experience, Dentrecolles tells his readers that the arch of the kiln is "thick enough that one can walk on it without being troubled by the fire inside."

Many pages of Dentecolles's letters concentrate on glazes, an aspect of pottery technology that depended on employment of sophisticated kilns. His report of 1722 is devoted almost entirely to recipes for glazes, surely because he grasped that French potters could more easily replicate the surface coatings of porcelain than they could its clay fabric. High-fired glazes first appeared in China in the Yangzi basin during the Bronze Age, fully three thousand years before Europeans started applying them.[26] Wood-fueled kilns develop a coating of ash glaze when fired to temperatures needed to make stoneware; some of it falls on the vessels being fired, producing a blotchy, uneven veneer. Glaze crystals also form on the wares themselves, an effect augmented as kilns grow larger and cooling takes longer. In all likelihood, potters as early as the Shang noticed these phenomena and thereafter deliberately sought to replicate and manipulate glazes during kiln operations.

In the Han period (206 B.C.E.–220 C.E.), potters added lead to glazes as a flux, that is, to lower the temperature at which the glaze is fired. This was a turning point in technology, for when iron-oxide pigments were covered by the lead glaze, colors appeared bright and clear on the smooth, reflective surface of the fired vessel. Lead glazes thus encouraged increased recourse to painted decoration, a hallmark of Tang ceramics. Excellent glazes on Chinese porcelain help account for the extraordinary success and prestige of the commodity; they enabled the potter skillfully to imitate metalwork and gemstone surfaces, to create a variety of shades reminiscent of other substances, such as jade, bronze, and tinted lacquer, and to paint vessels with precisely delineated, polychrome adornment.

Such artistry depended on comprehensive knowledge and control of firing techniques, especially regulating the amount of oxygen in the kiln. In reducing conditions (i.e., with restriction of oxygen during firing), grayish bodies and cool (or bluish green) tones dominate in the fired clay; in oxidizing conditions (with ample oxygen), brownish bodies and warm (or yellowish) shades dominate.[27] Tang potters accidentally discovered the effect of various kiln atmospheres, and Song artisans later developed techniques to control the process more surely. It all called for remarkable ingenuity and proficiency, however. Without precise measuring tools, potters had to learn by trial and error how to gauge oxygen levels in the kiln to produce desired hues on the fired vessel while simultaneously taking account of how the temperature affects glaze viscosities, assorted pigments, and the clay itself. As Dentrecolles emphasized, making porcelain was an exceptionally volatile, costly business as substantial wastage invariably occurred when potters aimed for subtle and complex effects employing large kilns and high temperatures.

Given the natural resources Chinese potters had available, from loess soil to kaolin clay to china stone, creation of porcelain by the Tang period would have been impossible without prolonged experimenting with raw materials, improving furnace design, developing glazing techniques, and learning how to manage the kiln environment. The achievement also owes much to the existence of the large number of pottery centers that sprang up as China slowly recovered from the fall of the Han empire in 220 C.E. From then to the rise of the Sui dynasty in 581, imperial China fractured into several petty kingdoms, the north fell into the hands of nomadic invaders, and multiple disasters—drought, floods, and epidemics—resulted in population decline. Scores of refugees fled the north, including potters, some of whom rebuilt their workshops just south of the Yangzi River. With recovery of the economy and centralized political control under the Sui and Tang dynasties, pottery manufacturers flourished as never before, contending for business within a growing domestic market, producing wares for new export clients, and struggling to win the all-important patronage of the imperial court.

In the Tang, kiln complexes cropped up in at least fourteen provinces (and fifty counties), five times the number of centers that existed in the Sui period, with most manufacturers producing an assortment of wares.[28] Intense competition placed a premium on the purest clay, the silkiest glaze, the stylish shape, the gleaming color, the innovative embellishment. As was always the case, potters quickly adopted—that is, spied out and copied—the best practices of their rivals, thereby impelling overall ceramic standards to higher levels. It is not surprising that porcelain, prepared for by two thousand years of craftsmanship, first appeared during the reign of the Tang.

A transformation in China's relationship to Eurasian culture also spurred the ultimate transition to porcelain in the early seventh century. During the Sui period, merchants began importing Southwest Asian commodities to China in unprecedented amounts. Silver vessels from the eastern Mediterranean and Persia ranked among the most prized merchandise, eagerly embraced at the imperial court and reproduced by Chinese artisans in bronze and lacquerware. Novel and appealing, the silver objects and their Chinese counterparts became models for the ceramic industry within a generation. Northern Tang potters copied the pale shades, slender bodies, and lobed shapes of silver ewers, bowls, and dishes.[29]

Of course, they employed kaolin as their primary clay, a mineral substance that satisfied all the requirements for such replication since it fired a striking white, shaped to fine tolerances, and withstood the hottest portion of the kiln without melting. Tang potters thus created the first true porcelain, a new material that nevertheless emerged naturally, all but invisibly, from the long evolution of high-fired pottery, inspired at the last moment by new cultural forces linking China to the rest of the ecumene.

YELLOW CHINA AND BLUE CHINA: LAND AND SEA IN EAST AND WEST

Extensive cultural contact with western Eurasia represented only one aspect of widespread change in early Tang China (618–755). Based on Sui precedents, the Tang modestly extended the system of staffing government posts by examinations based mainly on Confucian texts, one of the Chinese practices most admired by European philosophes a thousand years later. Another Sui achievement, building the Grand Canal to connect the basins of the Yellow and Yangzi Rivers, became equally important in time. At its greatest extent in the fourteenth century, the Grand Canal attained a length of 1,794 kilometers (or roughly the driving distance from New York City to Memphis, Tennessee). In contrast, the largest such undertaking in Europe was the Canal de Languedoc, a waterway of 240 kilometers, completed in 1681 to connect the Mediterranean to the Atlantic by slicing through southwestern France. In England, the canal linking the Trent and Mersey Rivers, built in the decade after 1766—and running past the front gates of Etruria Hall, Wedgwood's Georgian-style country house—was acclaimed by patriots as the "Eighth Wonder of the World."[30] Making possible a barge journey from Hull on the North Sea to Liverpool on the Irish Sea, with a canal connection running down the Severn River to the Bristol Channel, it stretched for 150 kilometers, only 8 percent the length of the Grand Canal.

China's new man-made river functioned as the vital artery by which the surplus rice of the south reached the center of power in Chang'an (Xi'an), the capital of the Tang dynasty in the northwest. The grain fed Chang'an's population of one million, as well as armies guarding the frontier against nomadic attack. By the eighth century, the lower Yangzi region had emerged as the indispensable granary of the empire: the riches and resourcefulness of the south secured the bureaucrats and bulwarks of the north. Ricci records that "the southern provinces provide the King with everything needed or wanted to live well in the unfertile province of Pekin [Beijing]: fruit, fish, rice, silk cloth for garments, and six hundred other things, all of which must arrive on a fixed day."[31] Using a simile already commonplace for centuries, a Ming official declared that the Grand Canal "may be likened to a man's throat. If food cannot be swallowed for a single day, death ensues at once."[32] At the same time, the waterway also benefited southern China: it lowered transport costs by 80 percent for agricultural and manufactured products. Porcelain from Jiangxi featured among the earliest merchandise carried north on barges on the Grand Canal.

In general, the Middle Kingdom experienced chronic tension between the forces of what has been called "yellow China" and "blue China."[33] In simplified terms, the former stood for the Yellow River, the Great Wall, agrarian predominance, continental priorities, a command economy, Confucian bureaucracy, and disregard for the maritime world; the latter signified the lower Yangzi, a market economy, local self-sufficiency, cultural interaction, long-distance trade, and orientation to the sea. Fear of marauding pastoral nomads in the north made for a static frontier, swept by loess winds and crisscrossed by camel caravans. The south contended with a fluid frontier, with Tang administrators and merchants moving into the hot lands of Guangdong, Guangxi, and northern Vietnam. As a northern writer observed after the fall of the Han dynasty, the south is "a distant place with different customs, and the climate is not the same."[34] In the late thirteenth century, Marco Polo looked on northern and southern China as virtually separate kingdoms, the latter called Manzi and the former Cathay.

If construction of the Grand Canal had never taken place, an enduring blue China might have emerged in the south, an independent state that could have reached out to encompass coastal Vietnam and parts of insular Southeast Asia, areas bound together by a common maritime orientation and commitment to long-distance exchange. The Chinese elite, however, remained focused on command from the north and on a continental point of view.[35] For them, the sea signified the merchant's realm, dominated by men given over to profit rather than to principle or position; it stood for the alien and the ungovernable, viewed with apprehension and to be shunned whenever possible. The VOC and the EIC had limited cultural influence in Asia before the late eighteenth century because of unbreakable ties to their metropolitan countries and the unrelenting focus of the merchants on realizing earnings for their companies and shareholders. Seagoing Chinese merchants had limited cultural influence abroad because Chinese officialdom regarded them as loosening ties to their metropolitan country in the service of an unseemly focus on garnering profit. As a consequence, the "Asian Mediterranean," encompassing the expanse from the South China Sea to the Gulf of Siam, played a limited role in the Chinese worldview.[36]

In contrast, western Asia always had a pronounced maritime outlook. For

Mesopotamian empires, control of the Persian Gulf and access to the sea lanes of the Indian Ocean were crucial for commercial prosperity and state security. For ancient Egypt, Phoenicia, and Greece, the eastern Mediterranean represented a highway of trade and a cockpit of war, considerations that impelled kingdoms and coastal city-states to maintain harbors and finance fleets. For Rome, the whole of the Mediterranean bound the empire together: it provided for the movement of troops, as well as for indispensable, never-ending relays of grain-filled galleys from the Nile to Italy. More than anything else, control of the Mediterranean accounts for imperial Rome's exceptional stability and resilience, making it the greatest of all Iron Age empires of the West. The countries of northwestern Europe—Portugal, Spain, France, Holland, and England—succeeded to this maritime legacy inasmuch as they looked out on the Atlantic and traded by water with the Mediterranean.

In the West, then, commerce, war, and the sea were entwined from birth. Columbus brought the traditions of the Mediterranean and the Atlantic together inasmuch as he was a native of the port of Genoa, spent many years in Portugal, appealed for financial backing from European monarchs, and made his great discovery by voyaging across the Ocean Sea. As warrior-merchants serving trading empires, the men of the VOC and EIC hearkened back to institutions and perspectives forged in Genoa and Venice, most conspicuously a lethal joining of warfare to trade, power to profit.[37] The VOC followed a policy of *coophandel met force* (trade by force of arms).[38] In 1718 an English governor of Bombay counseled the directors of the EIC of the need for sea power to prevent Indian monarchs from controlling commerce: "if no naval force, no trade, if no fear, no friendship."[39] According to Ricci, when Portuguese ships reached the ports of southern China in the early sixteenth century, Chinese customs officials expressed alarm at the cannon they carried and at the swift recourse to violence of Portuguese commanders determined to win access to trade.[40] The officials at first opposed allowing the foreigners entrance to the kingdom, though they soon gave way before the lure of illicit profit. In this encounter, Confucian bureaucrats and Western mariners represented their respective traditions, the former oriented toward continental, land-based centers of power, the latter toward maritime enterprise backed by armed force.

Ricci believed that a difference in character distinguished the two great regions of China: "The northern Chinese are more warlike and courageous than those from the south but less alert mentally." The claim made its way into Montesquieu's *Spirit of Laws,* which drew heavily on du Halde's compilation of Jesuit reports: "The peoples of northern China are more courageous than those of the south."[41] Indeed, even the Chinese recognized an age-old contrast between northern martial valor and southern commercial acumen (or avarice). Still, tensions between yellow and blue China relaxed in the early phase of the Tang dynasty, with administrators in the north and merchants in the south finding common ground. For the first and last time, blue China came to clear dominance during the Southern Song period (1127–1279), when nomadic conquest of the north resulted in the capital and court being relocated to the city of Hangzhou, just south of the Yangzi.

Building the Grand Canal extended both imperial control and Chinese culture to the south, a vast region that began to be incorporated into what is now called China only from the time of the Han. From the Tang period to the eleventh century, the population soared, increasing by over 500 percent.[42] Migration from the north during the post-Han troubles, coupled with expansion of crop cultivation thereafter, especially of more productive strains of rice, explains the demographic explosion, while the latter sheds light on the increasing importance of maritime dealings in the southern provinces from the Tang through the Song. The hilly coastal provinces of Fujian and Guangdong are hemmed in by mountains, resulting in a congested environment that effectively oriented the population to the sea. Xie Li, a Quanzhou official and poet of the thirteenth century, described the problems and opportunities of the area:

> The Quanzhou population is dense.
> The mountains and valleys are barren.
> The people want to till, but there is no land to open up.
> To the south lies the ocean, vast and limitless.
> Every year they build ships and go to foreign lands.[43]

Inhabitants of the coast turned to the sea in one capacity or another: they cultivated foodstuffs for export (sugar, wine, salt, wheat flour), made export merchandise (pottery, paper, silk, iron wares), imported foreign goods (pearls, sapanwood, sulfur, coral, kingfisher feathers, Indian cotton cloth), and even migrated to Chinese settlements in Taiwan, Korea, Vietnam, and maritime Southeast Asia. Merchants in China shipped pickled plums and lychee nuts to maritime Southeast Asia, metal utensils and sides of cured pork to India, silk fabric and porcelains to Japan and Persia. A Song scholar observed that "the entire province of Fujian makes its living by the practice of seaborne commerce."[44]

Merchant shipping and overseas trade increased considerably during the early Tang; Canton and Quanzhou emerged as important ports for the first time. A visitor to Canton recorded seeing "unknown numbers of Indian, Persian, South Sea and other boats, laden with incense, drugs, and precious things piled up like mountains."[45] The rise of Buddhism in China from the first centuries C.E. fostered trade in aromatic woods and other forest products of Southeast Asia needed for religious ceremonies.[46] During the Tang, however, trade shifted to spices from the Moluccas and pepper from India. As luxury products, the Buddhist commodities made for handsome proceeds when transported on ships of slight tonnage; but pepper had a low value-to-volume ratio, which meant carrying it in bulk on larger ships resulted in higher profits. As a consequence of the advantages obtained from economy of scale, sizable Chinese junks came to dominate trade to India after the ninth century, displacing smaller vessels from the Indian Ocean. This would be especially lucrative for the pottery industry.

Chinese seaborne enterprise during the Tang met with a similar impulse from Southwest Asia.[47] The rulers of Sassanian Persia (226–651) promoted trade in the Indian Ocean and farther east, as far as Sri Lanka for the collection of silk, spices, and sandalwood. The Arab ruling class that conquered and replaced the Sassanians in the seventh century pursued a similar commercial policy. Proclaimed in 618, the outward-looking Tang dynasty coincided with the foundation of Muslim power in Arabia in the time of the Prophet Muhammad (ca. 570–632). Arab-Muslim conquest of Iraq (637), the Levant (640), Mesopotamia (641), Egypt (642), and Persia (651) reconfigured the Southwest Asian trading zone, uniting a region previously divided by language, religion, and warring empires. Ships from Southwest Asia began arriving in Canton in the eighth century, and large numbers of Arabs and Persians settled in the city. During the early period of the Abbasid dynasty (750–1258), Persian vessels made high profits taking goods on the 16,000-kilometer round-trip to China. A diaspora of Muslim merchants from Southwest Asia in the entrepôts of the Indian Ocean, Southeast Asia, and China later would prove crucial for the development of Jingdezhen blue-and-white porcelain.

The Tang dynasty, however, traced its descent from Turkic-Mongol peoples of Central Asia, so the imperial court in Chang'an showed more interest in northern caravans than in southern argosies. Indeed, the early Tang period witnessed the most extensive cultural and political contacts between Southwest Asia and China before establishment of the Mongol empire some five hundred years later. The Eastern Market of Chang'an displayed merchandise from all provinces of the Middle Kingdom; the Western Market featured goods from Central Asia and Southwest Asia, as well as six Zoroastrian and Manichaean temples.[48]

Central Asian artisans fashioned vessels in silver and carved figurines from jade. Southwest Asian entrepreneurs imported purple-dyed woolen cloth from Syria, deep-pile carpets from Persia, rock crystal from Samarqand, and lapis lazuli from the mountains of northern Afghanistan. A thousand precious items buried near Chang'an in 732 suggest the scale and variety of the caravan trade: the hoard included coral, gold dust, fabrics, Sassanian coins, silverwork, Baltic amber, and blocks of jade.[49] The goods may have been carried to Chang'an by Sogdians, perhaps the most numerous Southwest Asian residents of the city. Many came from Samarqand, a city in the region the Romans called Transoxiana (now mainly Uzbekistan and Tajikistan). A Tang chronicler noted that Sogdians "excel at commerce and love profit [and] . . . they go wherever there is profit."[50] Sogdian merchants calculated prices in silver ingots and copper-cash, quantities of which they carried to Transoxiana. Significantly, *sabao,* the name of the Tang office appointed to supervise foreigners, derives, by way of the Sogdian language, from a Sanskrit term meaning "caravan leader."

THE SILK ROAD: CHINA AND SOUTHWEST ASIA IN THE TANG

Pottery statues of two-humped Bactrian camels figure among the paraphernalia deposited in Tang tombs, along with ceramic models of dogs, horses, houses, and granaries. The pieces commonly are painted yellow, green, and brown. Depicted with lively realism and dramatic movement, the camels bear sacks of merchandise and cooking equipment. Pilgrim flasks and haunches of meat dangle from the saddles, provisions for the dead on their ultimate journey. Some of the camels improbably carry eight-man orchestras playing stringed instruments and woodwinds; dwarfs and jugglers cavort on others. Familiar animals in northern China for hundreds of years, camels are represented in mortuary sculptures as far back as the fourth century B.C.E. In Central Asia, a winged camel symbolized good fortune, and by the time of the Han, the Chinese regarded the beasts in the same way because of the rich cargoes they brought from distant lands.[51] But whatever their original spiritual function, the Tang pottery camels inescapably evoke the caravan trade of the Silk Road.

Camels travel no faster than four kilometers an hour, so journeying on the Silk Road meant a long, grueling trek through the desiccated heart of Asia.[52] Caravans of camels and carts departed from Chang'an, passed through the long corridor of Gansu province, and trudged west to skirt the southern edge of the Gobi Desert, where a line of modest Chinese towns serviced travelers. At Dunhuang, the route split into northern and southern passages around the Taklimakan desert of the Tarim basin, a stony badlands between the Tian Shan to the north and the Kunlan range of the Tibetan Plateau to the south. Bleak and cruel, the terrain took many lives; travelers stacked up the bleached bones of animals as landmarks. Most caravans took the northern route, which passed by fewer sand dunes and salt crusts, relics of the primeval inland sea. On the southern route, steep mountain trails led by way of Tibet and Nepal into the Ganges valley of northeastern India.

Here and there along the Silk Road, the greenery of crops and pastureland, the legacy of glacial runoff and loess soil, relieved the desolation of rock and sand. A series of polyglot oasis communities—including Turfan, Kucha, Aksu, Khotan, and Cherchen—lived off the passing caravans. They were not merely neutral waystations for travelers, however, for their residents acted as cultural brokers, adapting material objects and religious ideas, as well as ornamental designs and symbols, which filtered in from the great centers of agrarian civilization. On his sixteen-year, roundtrip journey to India, Xuanzang (ca. 596–664), a famous and influential Buddhist monk, received hospitality in Silk Road settlements from Uighur (Turkic) khans and warlords. Of course, his fellow monks welcomed him as well. In the Kucha area alone, some one hundred Buddhist monasteries with five thousand monks offered lodging to travelers while also serving them as banks, storehouses, and caravanserais.[53]

The two tracks of the Silk Road joined again at Kashgar, the point at which pilgrims such as Xuanzang headed south through the towering peaks of the Pamirs and the Hindu Kush to the Buddhist shrines and reliquaries (stupas) of India. The main route, however, continued west through Transoxiana to the prosperous metropolis of Samarqand, from which paths ran southwest to Persia and west to Mesopotamia. After crossing the Tigris and Euphrates Rivers, a relatively short journey brought the camel-weary traveler to the Levant or Egypt. The trek from westernmost China to ports on the eastern Mediterranean traversed some 7,000 kilometers, a distance covered at no more than 32 kilometers a day. Polo may have counted himself fortunate to have journeyed to China in only one year since his father and uncle previously had spent three years on their return because of political turbulence along the route.

Although regular travel on the Silk Road went back centuries, probably as far as the early Bronze Age, the Chinese still perceived western Asia before the Tang as a treasure-house of legends and fantasies, a land of monstrous and marvelous animals, the realm of the Daoist Queen Mother of the West (Xiwangmu), the setting of a fabulous mountain of jade, the most magical and sacred of substances. The Tibetan Plateau seemed to be the edge of the world: Daoists believed the paradise of the Immortals lay beyond the wasteland of deserts and mountains; Chinese folktales imagined the distant west as a region of quicksand, killing frosts, and everlasting night, plagued by elephant-sized red ants and wasps big as bottle gourds.[54]

This cloud of unknowing began to lift in the early Tang as China for the first time since the Han period looked far beyond its familiar northern periphery. In the mid-seventh century, Tang military forces even seized control of the oasis communities around the Tarim basin, extending Chinese sway almost to Persia. To maintain its armies and alliances, the Tang regime sent a million bolts of silk every year to the Sogdians of Transoxiana in exchange for 100,000 warhorses, imports needed to keep the army's stock at the required number. After 180,000 horses reportedly perished from disease, a Tang chronicler stated, "Horses are the military preparedness of the state; if Heaven takes this preparedness away, the state will totter to a fall."[55] Purchasing Transoxianian horses represented one of the heaviest drains on the treasury inasmuch as a million bolts of silk amounted to 84 million square meters of cloth. Export of such fabulous quantities resulted in pieces of silk being used as common currency throughout Central Asia. Traders in Persia and Mesopotamia paid twice the price the fabric cost in China and carried it on to the Levant to sell at high profit. According to reports by travelers, Chinese silk overflowed the market stalls of Southwest Asia.

Under the early Tang regime, Buddhist pilgrims, Syrian merchants, Persian priests (Manichaeans, Zoroastrians, and Nestorian Christians), Sogdian craftsmen, Jewish physicians, Arabian gem merchants, Tibetan mercenaries, and Uighur horse dealers traveled the Silk Road. The flourishing Tang economy, especially the teeming metropolis of Chang'an, acted like a magnet, drawing acrobats, painters, dancers, magicians, and musicians along the camel trails. Fleeing from Muslim conquerors, some of the elite of Sassanian Persia, including the imperial family, sought refuge in China. Within a generation, Persian traders recently converted to Islam crowded the streets of Chang'an. Diplomats, missionaries, and merchants moved along the corridors of Central Asia, their collective activity drawing distant regions closer together than ever before. The Tang court sent four embassies to India in the mid-seventh century, and northern Indian states dispatched envoys to Chang'an to plead for military help against Tibetan warlords.

The western focus of the early Tang meant that the seventh century was a great age for the expansion of Buddhism in East Asia. Chinese merchants imported Indian commodities needed for Buddhist ceremonies, including coral, pearls, incense, lapis lazuli, cotton cloth, and glass vessels. Chinese monks, who sometimes acted as diplomatic emissaries, exchanged large amounts of silk fabric for precious Indian relics, such as the supposed parietal bone of the Buddha (Siddartha Gautama, d. ca. 483 B.C.E.). With a train of twenty horses, Xuanzang brought back 650 manuscripts of Buddhist lore, statues of the Buddha, and enshrined bits and pieces of the Buddha's flesh. Regrettably, his only elephant plunged into a river gorge in the Pamirs when bandits attacked his caravan.[56]

At the request of the emperor Taizong (r. 626–49), Xuanzang wrote an account of his travels, and in subsequent generations his exploits became incorporated into Chinese popular culture, including oral tradition, folktales, poems, and short stories. Most notably, Wu Cheng'en (ca. 1500–1582) made them the basis for his influential comic fantasy, *The Journey to the West,* in which the trickster character Monkey takes center stage rather than the pious pilgrim. As a consequence of the novel's popularity, the supernatural adventures of Monkey and Xuanzang (called by the Buddhist name Tripitaka) became common themes for porcelain decoration in the seventeenth century.

Xuanzang's pilgrimage took him to numerous shrines of Buddhist bodhisattvas, enlightened beings who forgo nirvana to help others. Near the Ganges in northern India, he prayed before a sandalwood statue of the bodhisattva Avalokitesvara (Perceiver of the World's Sounds). With the Buddha himself having attained nirvana and thus having transcended human affairs, believers regarded bodhisattvas (or Buddhas-in-the-making), as accessible and altruistic beings who aided mortals in daily troubles and in reaching Buddhahood. By the early centuries of the common era, Avalokitesvara had emerged as the most significant personification of compassion in Buddhism, a model of perfection in a hallowed pantheon. In due course, Buddhist communities throughout Asia paid worship to him. He became a revered figure to wayfarers on the Silk Road because of his response to cries of pain and suffering. According to the Lotus Sutra, the most popular Buddhist scripture in the Tang and Song periods, "This bodhisattva can grant fearlessness to living beings.

If you call upon his name, you will be delivered from these evil-hearted bandits [who rob traveling merchants]."[57]

Xuanzang promoted the holy man's cult when he returned home, a cause later taken up by the empress Wu. The career of Avalokitesvara subsequently took a curious turn, however: by the tenth century, under the impact of Chinese folk religion, he became transformed into Guanyin, the spirit of compassion and fecundity, who herself later would take on the guise of the mother of the Christian god. At all stages of the bodhisattva's cultural pilgrimage, porcelain figurines represented the principal means by which the sacred personage became tangible and proximate to most believers, whether as Avalokitesvara reclining on a lotus blossom, Guanyin cuddling a baby boy, or the Virgin Mary fingering a rosary. (See figure 13.)

EURASIAN CULTURAL EXCHANGE IN THE TANG

Porcelain and other Chinese arts responded to the Tang opening to the west.[58] Oasis communities on the Silk Road played the role of middlemen, conveying their versions of Indian and Persian pictorial methods to China, such as rhythmic patterns, rotating arabesques, stylized flowers, geometric shapes, molded relief, interlaced designs, and exuberant colors. A number of plant patterns, including the acanthus, palmette, and peony-like blossoms, entered the mainstream of Chinese art after monks and artisans copied them from hundreds of Buddhist cave temples and monumental tombs at Dunhuang, near the eastern terminus of the Silk Road. The quintessential Buddhist motif of the lotus, journeying from South Asia to the Middle Kingdom by way of Persia and the caravan track, embarked on its triumphal progress in Chinese art and architecture. Foreign influence led to one of the most enduring Chinese decorative devices on porcelain plates and bowls, a swirling band of flowers surrounding a central medallion, such as a sketch of a carp, duck, or blossom—a format whose remote descendant is the conventional pattern still frequently painted on modern dinner plates.

Auspicious Buddhist objects, such as the bodhi tree, parasol, wheel, and shell, became fixtures in Chinese ornament. Figures of peacocks, stags, wild sheep, and eagles entered the Chinese ornamental repertoire by way of decoration on Southwest Asian silverwares. Transmitted by designs on imported textiles, so too did winged horses, griffins, and chimera, fanciful counterparts of the actual novelties—peacocks, cheetahs, date palm, fig tree, jasmine flower, almond tree, lettuce, polo, and chess—that arrived in China from Southwest Asia around the same time. In return for these bequests, China sent the apricot tree, the rhubarb plant, and silkworm cocoons in the opposite direction.

Southwest Asian glass and metal vessels also reached China, where in due course potters replicated them. Silver bird-headed ewers came from Persia (or possibly the eastern Mediterranean) to China, where elite consumers called them *huping,* "barbarian bottles," and placed them in their tombs. Potters copied the ewer, helping to make it a fashionable item. In a centuries-long roundabout typical of ceramics, merchants exported Tang bird-headed porcelain ewers to Southwest Asia, where both earthenware potters and metalworkers duplicated them in the eleventh century. Traded across Central Asia, the Southwest Asian pilgrim flask became conflated with the leather canteens of pastoral nomads, and Chinese potters then translated the flask, known as a *bianhu* (flattened vase) in China, into porcelain. Often depicted on the funerary camels of the Tang, the pilgrim flask, sometimes with simulated leather stitching on its sides, carried decorative motifs deriving from the Hellenistic kingdoms (323–30 B.C.E.) of the eastern Mediterranean, including reliefs of dancing girls, winged lions, and mounted archers. Porcelain pilgrim flasks became high-status relics in Tang tombs, and merchants exported the flasks to Southwest Asia along with other vessels.[59] (See figure 13.)

Archaeologists have excavated shards of Tang ceramics from many locations in Southwest Asia, although some black-glazed, fifth-century pottery fragments discovered beneath the floor of a mosque at Siraf, on the northern coast of the Persian Gulf, represent the earliest dated evidence for Chinese artifacts in the region. The pottery almost certainly came to Persia in ships, always the preferred mode of transport for that merchandise. The first archaeological confirmation of direct maritime trade between China and Southwest Asia came as recently as 1989 from investigation of an early-ninth-century Arabian or Indian shipwreck off the island of Belitung, between Sumatra and Borneo. Chinese ceramics made up all but a small fraction of the cargo and included fine northern whitewares, polychrome bowls, and large storage jars made in Guangdong province. The whitewares are the first porcelains known to have been exported from China.[60]

Just as commerce on the Silk Road depended on knowledge of oases and mountain passes between Dunhuang and Samarqand, maritime trade between Canton and the ports of the Persian Gulf depended on understanding the monsoon winds of the southern seas. Known to the Chinese as "ocean ship-driving winds" *(bozhao feng),* they constituted the climatic regimen that determined when ships could set sail.[61] Since at least the first century C.E., Roman, Southwest Asian, and Hindu merchants had a practical grasp of the schedule of monsoon winds in the Indian Ocean. Around that time, an anonymous writer, perhaps a Greek businessman in Egypt, produced a small handbook on the subject, the *Periplus of the Erythrean Sea,* which spelled out when vessels could safely voyage between Arabia and the western coast of India.[62] During the Tang and Song, merchants from various countries and trading groups also gathered information on the operation of East Asian winds. Seagoing traders in the Tang period based their enterprise on this accumulation of knowledge. Relying on steady, predictable ship-driving winds from China to the Indian Ocean, they established the first substantial and enduring maritime connection between the Middle Kingdom and Southwest Asia. Chinese merchants in particular benefited from that achievement when seaborne commerce expanded during the Song.

Still, the Silk Road, rather than the maritime porcelain route, remained most important in the early Tang, the leading indicator of the dynamism and affluence of the period. Around the mid-eighth century, though, the dynasty ran into momentous problems, disrupting its homeland as well as interactions with the rest of Asia. The An Lushan rebellion (755–63), led by a Tang general of Turko-Sogdian descent, plunged the empire into disorder, stymied the flourishing economy, and provoked violent xenophobia. With tens of thousands of persons killed, parts of Chang'an became a wasteland. All but a handful of the city's numerous Buddhist monasteries and convents disappeared; so too did foreign dress and entertainments. Buddhism lost its cultural cachet among the aristocracy and gentry; vandals destroyed Buddhist wall paintings and statues throughout the empire in the 840s. The regime turned against alien religions, even proscribing them in 845, a legacy the Jesuits still struggled with centuries later. Fighting, rebellion, and disease devastated the richest provinces, including the lower Yangzi region. Mobs in Canton massacred thousands of Muslim merchants in 755, and when trade revived decades later, subsequent pillages of the port by pirates and rebels dealt additional blows to overseas connections. Many seaborne merchants, both Chinese and Southwest Asian, took refuge in northern Vietnam and carried on their business from there. Losing control of Korea and the frontier region between Sichuan and Burma, the Tang regime found itself besieged on all fronts.[63]

An event far from China foreshadowed the disasters that would strike the empire with the An Lushan rebellion. At the battle of Talas in 751, somewhere to the east of Samarqand and south of Lake Balkhash, an Abbasid army defeated Tang forces. The victors took several thousand Chinese men into captivity in Baghdad, the Abbasid capital on the Tigris-Euphrates, where they may have passed on knowledge of paper making, silk weaving, and pottery glazing. An obscure clash at the outermost limit of two great empires, Talas nonetheless represented a turning point in the fortunes of Eurasia. As a consequence of the An Lushan rebellion a few years later, the Tang treasury fell bankrupt just when the cost of warhorses from Transoxiana soared to the prohibitive price of forty bolts of silk each. Facing overwhelming force as well as financial crisis, the Chinese military pulled back from their strongholds on the Silk Road, leaving Islam to expand and Buddhism to face a long decline.[64] The ceramic horses modeled by Tang potters, lean racehorses and muscled warhorses, ceased to be produced, replaced in the Song by smaller, less magnificent animals characterized with round noses and stubby legs.

The battle of Talas marked the end of an era of Chinese dominance in Central Asia; the empire would not reassert its power there until the nineteenth century. The Tang, however, sought good relations with the Abbasids, welcoming several of their embassies to Chang'an and gaining Arab military help in fighting the An

Lushan rebels. Although diminished, trade continued on the Silk Road, and Abbasid envoys to China returned home with gifts of porcelain. A Persian governor presented Caliph Harun al-Rashid (r. 786–809) with twenty pieces of imperial porcelain, "the like of which had never been seen before," along with two thousand more commonplace vessels.[65] When Abbasid caliphs constructed immense palaces on the Tigris-Euphrates in the ninth century, they also stimulated demand for porcelain and other Chinese products.

No matter how unstable and dangerous conditions became on the Silk Road, some merchants always found it profitable to make the trek. But the battle of Talas, the An Lushan rebellion, Tibetan seizure of the Gansu corridor, and the rise of powerful nomadic confederations to the north of China transformed political relations between the Middle Kingdom and Central Asia during the Song dynasty. As a consequence, China looked toward the Silk Road solely with apprehension, no longer beguiled by the myths and fantasies of earlier days. Beyond the tracks of camel caravans lay the steppe lands of tented nomads, the horse warriors who would redraw the map of imperial China. For its own security, the Northern Song necessarily maintained diplomatic contact with powers in Central Asia until the Liao state of the Qidan (907–1125) conquered northern China, bringing the first phase of the dynasty to an end. Subsequently, the Jin state of the Jurchen (1115–1234) defeated the Liao, cutting off imperial China from almost all contact with Central Asia.[66] The Jurchen even extracted an annual tribute from the Southern Song of 250,000 taels of silver and as many bolts of silk. From 1126 to 1279 (and conquest by the Mongols) the monarchs of the Middle Kingdom governed their rump empire, encompassing two-thirds of the Chinese lands, from a provisional capital at Hangzhou in Zhejiang province, south of the Yangzi. Blocked from the Silk Road by powers to the north, the Southern Song turned decisively toward the sea.

4

The Culture of Porcelain in China

Commerce, Confucians, and Connoisseurs, 1000–1400

During the Song period, a Chinese writer exulted, "The ships which sail the Southern Sea and south of it are like houses. When their sails are spread, they are great clouds in the sky."[1] Government officials and private entrepreneurs had reason to look upon the huge junks with satisfaction, for their voyages contributed substantially to a flourishing economy. Despite costly, relentless threats from nomadic confederations, Song China experienced growth and riches surpassing that of the early Tang.

COMMERCE

Most regions of China, down to the village level, were drawn into a commercialized, monetary network.[2] Silver ingots in circulation rose from less than one million around the year 1000 to ten million by 1300. Copper coins issued by the government—200 million strings (or 200 billion coins) under the Northern Song—streamed out of China to become the international hard currency of East Asia. A government magistrate complained in 1074 that junks ventured abroad loaded with nothing but copper-cash, as if mariners were pouring all the currency of the Middle Kingdom down a hole in the South China Sea. The next year, another official wrote, "There are none among the ship captains who do not take cash away with them. Cash was once a treasured commodity in China, but today it is commonly used among the barbarians."[3] A customs official complained that to trade for pepper with Java, merchants "are in the habit of smuggling copper-cash [out of China] for bartering purposes."[4]

To increase the amount of cash in circulation, the Gaozong emperor (r. 1127–

62) of the Southern Song prohibited making domestic utensils from the metal, and trying to set an example for privileged ranks, he took the dramatic step of melting down fifteen hundred bronze vessels in the imperial collection. Government declarations of the early thirteenth century specifically called for export of ceramics to halt the flow overseas of cash and costly metals. As a government document of 1219 stated, "It is lamentable that officials often use gold and silver for trading abroad, thus losing valuable hard currency. Trading with silk textiles, porcelain, lacquer and other commodities is urged as an alternative."[5] A few years later, an official noted that pottery went to a quarter of all the foreign places with which China had contact.

Trying to staunch the hemorrhage of precious metal, the Southern Song put out the equivalent of 400 million strings of copper-cash in the form of paper money, an expedient that contributed to galloping inflation that persisted ino the early Ming period. Accompanying this, an economic revolution in credit mechanisms, agriculture, water transport, and metallurgy made China the richest, most economically dynamic region on Earth. With a population of perhaps 140 million in 1100, China had 6 million people living in cities, more urban residents than the rest of the world put together. Seaborne trade became a mainstay of the government, accounting for 20 percent of its revenue. The court rewarded port officials who collected substantial taxes on imports, and with such incentives, custom duties, which the government set at 10 percent, soared from 500,000 strings of cash in the late tenth century to 63 million a century later.[6]

For the first time, China began treating the sea as a commercial zone, and the term *haishang* (maritime merchant) was coined to identify those who made a living from seaborne exchange.[7] China emerged as a sea power, with a navy for coastal defense and a transportation fleet for carrying grain and tribute from south to north. Seagoing craft entering the Yangzi River numbered in the tens of thousands, and nine major shipyards built over six hundred vessels a year. Timber for ship construction called for fifty thousand trees a year, a harvest that contributed to deforestation along the coast. Duanhong, a Buddhist monk, bemoaned the destruction in 1282: grief spread through the mountains, he wrote, when "tens of thousand trees were chopped down."[8]

Advances in maritime technology, such as the magnetic compass, large lighthouses, and sailing charts, reduced the perils of the ocean; pigeons carried messages between ship and shore about prices and purchases. The merchant fleet grew in numbers and capacity; some junks accommodated a thousand passengers and boasted six masts, four decks, and watertight compartments. The largest vessels the world had ever seen, they linked the flourishing economy of the Grand Canal and lower Yangzi with the markets of Southwest Asia. Muslim traders of the Song period described the Chinese as "the only people in the world with two eyes"; Arabs supposedly had only one, Europeans none.[9] A Chinese poet described a coxswain singing as he steered his ship out to sea: "I have limitless power in my hands, [so] who is afraid of the towering waves!"[10]

Having lost all revenue from the northern provinces of his realm, the Gaozong emperor encouraged overseas trade to reduce the tax burden of the population in the south. He declared that "the profit of maritime trade contributes much to the national income. Therefore, pursuing the former custom, the people of far-away countries should be encouraged to come to sell an abundant supply of foreign goods."[11] From government shipyards in Fujian and Guangdong, officials provided vessels to private entrepreneurs, with 70 percent of trading profits going to the court and the rest to merchants. Members of the imperial clan took refuge in Quanzhou, where they became involved in shipping as investors and traders, despite laws excluding them from such activities. Incorporated into the Agency for the Southern Branch of the Imperial Clan, affiliates of the clan rapidly increased from 300 to 2,300 individuals, making them a formidable trading presence in a city with a population of 200,000.[12] Zheng Xie (1044–1119), a Song official, described Quanzhou: "Maritime merchants crowd the place. Mixing together are Chinese and foreigners. Many find rich and powerful neighbors."[13]

Above all, population growth and foreign trade stimulated the ceramics industry. The Song period accounts for 75 percent of all pottery kilns ever established in China's history; porcelain was produced in 130 counties in nineteen provinces.[14] In Hebei province, a major producer under the Northern Song, one excavated site revealed pottery debris fifteen meters deep over an area twice as wide. In Longquan county in Zhejiang province, whose potteries dominated production under the Southern Song, more than five hundred furnace complexes, many using the great dragon kilns, have been identified by archaeologists. In Guangdong province, where output mainly served overseas markets, 100 kilns turned out wares. In southern Fujian, 10 percent of a population of slightly over three million owed their livelihood to the manufacture and export of ceramics. With some 150 kilns in the region, the pottery industry generated one million strings of cash annually. Quanzhou displaced Canton as China's busiest port, in part as a result of the enormous growth of the ceramics export industry.[15] Coastal deforestation probably owed more to consumption of wood by porcelain kilns than to construction of ships.

Expansion of seaborne commerce during the Southern Song especially benefited the pottery industry. Commodities high in value and low in bulk (such as jade, emeralds, amber, and jasmine oil), as well as those high in both value and bulk yet relatively straightforward to transport (silk, cotton fabric, and carpets), were always the principal merchandise on the Silk Road. In exceptional cases, taking just a single piece of pottery from China could bring a fortune: during the Tang, an Arab merchant carried from China a black-glazed porcelain with a gold lid encasing "a golden fish with ruby eyes, garnished with musk of the finest quality."[16] Of course, most chinaware was considerably more mundane. It traveled in caravans as large as eight hundred camels and carts, such as the one seen in Samarqand by Ruy Gonzalez de Clavijo in 1405. The Castilian observed that goods from China comprised "the richest and most precious of all those brought thither from foreign parts, for the craftsmen of Cathay are reputed to be the most skillful by far beyond those of any other nation."[17] Silk made up the largest and most valuable part of the caravan's cargo, which also included sacks of diamonds, pearls, spice, and dried rhubarb. The caravan carried pottery, but a report by a functionary of the Ming dynasty points to the difficulty of taking that commodity on the Silk Road:

> In the capital of the Ming Empire, Peking, carts from Mongolia, Manchuria, Persia and Arabic countries were fully packed with Chinese goods to carry back to the countries from which the tributaries had come. Goods were stacked as high as thirty feet, including great quantities of porcelain loaded on dozens of carts. Each piece of porcelain for the long journey was filled with soil and beans. Then the pieces of porcelain were tied together with string. They were carefully placed in damp conditions and sprayed with water, until the beans grew and their roots spread around the porcelain to maintain the tight packing. Porcelain sent in this way to inner Asia was sold at ten times its original price.[18]

Although the high profits realized by selling chinaware in Southwest Asia plainly justified such meticulous, labor-intensive handling, the quantities transported necessarily had to be limited. Decorative influences from India and Persia moved more easily across the Silk Road to China than did porcelains in the opposite direction. But with inauguration of a maritime porcelain route from China, a seafaring equivalent of the ancient caravan trail, chinaware became a preferred, convenient bulk cargo of high value. Writing about ships leaving Canton, a Song merchant reported, "Among the goods exported, porcelain remained the main item. The big porcelain pieces enveloped the small ones, and there was hardly any space left between them."[19] Even undersized pieces contained profitable items, such as iron needles, carved rhinoceros horn, and tortoiseshell combs. As the Spanish in the Philippines discovered when they had their Chinese stevedores stock the Manila galleon for its annual Pacific voyage to Acapulco, the Chinese had a deserved reputation for exceedingly skillful packing. A sea chest packed in Manila typically held double the amount of merchandise as a similar one readied in Seville.

Maritime traders had a particular liking for porcelain as a cargo: since the ceramic is both heavy and impermeable to fluids, it proved practical as ballast, weighty material packed deep in the ship's sodden hold to enhance stability in rough seas. Ballast cargoes otherwise were low-profit commodities: cowries, rock salt, lead and tin ingots, black pepper, stone, sand, timber. Porcelain, though, provided a hefty ballast of high value, thereby increasing shippers' proceeds.[20] Transporting porcelain as ballast was so lucrative that Western merchants speedily adopted the practice when they entered Asian trade networks. Agents of the EIC in Vietnam reported to London in 1672 that "coarse porselon here for ballast [is] very reasonable," a convenient freight to ship to the Philippines and Thailand.[21]

VOC carracks sailing home from Batavia in Java sometimes carried large Chinese jars filled with ginger preserved in sugar syrup, five hundred of which nicely trimmed a ship. Spanish galleons going from Seville to the New World habitually used tiles and olive jars *(tinajas)* for ballast, with the jars being sold and recycled as roof vaulting for churches on arrival. Captains of the VOC packed 160 chests of porcelain, each eighteen meters square, into a moderate-sized frigate. In 1702 a director of the EIC instructed a supercargo:

> It must be your first care to provide Kintlage commoditye [i.e., ballast as heavy as pig iron] to stiffen your ship and make her sail worthy, and which will be first wanting to lade on board her . . . [above all] china ware of the usefull sorts, especially plates and dishes which stow close. Buy us some large, some middling, some smaller punch bowls, some great china flower potts, fitt to sett orange Trees in and other lesser dimensions fitt for smaller Trees and flowers. . . . What hollow chinaware you buy fill up with sago [palm starch] or other more profitable commodity.[22]

Both the VOC and the EIC packed tea in lead-lined chests to keep it fresh and placed the chests on top of crates of porcelain: the pottery kept the tea dry; the latter provided a shock-absorbing layer that curtailed damage to the chinaware. In VOC ships, an average of only 5 percent of the porcelain was broken on its journey to Amsterdam, though even the shards could be sold as masonry material. An additional virtue of the ceramic was that it lacked odor and therefore would not affect the flavor of the tea.

Transporting porcelain by the maritime route, rather than by the Silk Road, meant that greater quantities of the ceramic reached Southwest Asia than ever before, resulting in lowered costs to consumers there. In China, increase in pottery production had a similar effect on prices. Potters responded to government promotion and domestic demand with technological innovations that swelled output to unprecedented levels. They began to use carved molds in great quantities: by pressing a mold engraved with ornament over clay, they substituted mechanical techniques for laborious decoration by hand. In addition, by employing stacked saggars in the kiln, potters could fire four times as many wares simultaneously than with the old method of using a single saggar for every item on the furnace floor.

Along with reduced prices, however, there was a certain decline in the quality of wares. Enormous numbers of pots, plates, and jars were turned out, much of it coarse and run of the mill, for unrefined foreign and domestic customers. The use of carved molds meant that wares often were dully uniform, and, for technical reasons, the use of stacked saggars meant that the rims of vessels went unglazed. Nonetheless, superlative porcelains also were produced in quantity and attracted the attention of the imperial court. The latter was especially important, for official patronage proved crucial for conferring enhanced cultural stature on the ceramic. Imperial catalogues and workshops set standards for the elite—for sophisticated study by antiquarians, for appreciation by connoisseurs, and for acquisitions by the trendy wealthy.

Ever since the early common era, kiln owners had sent modest shipments of pottery to the court, but the quantities escalated so much after 1000, amounting to fifty thousand pieces at a time, that the government established a ceramic storehouse in Kaifeng, the Northern Song capital in Henan province. An early-eleventh-century emperor so valued his porcelains that he ordered their unglazed rims to be adorned with ten thousand ounces of gold and silver.[23] Given imperial patronage and widespread use, pottery for the first time became a focus of connoisseurship. Since Song porcelains were at the root of this erudition, they remained highly prized by collectors thereafter, ranking just below venerable works in bronze and jade. Later generations adhered to the perceptions of the Song elite so comprehensively that well into the twentieth century Song antiquarian notions determined how archaeologists in China classified and interpreted the ceramic objects they unearthed.[24]

CONFUCIANS

While the pottery statues of Bactrian camels represent the ecumenical perspective and imperial reach of the early Tang regime, the vases of the Song exemplify the spirit esteemed and promoted by that dynasty—elegant, subtle, classical, and self-contained. The porcelains still convey those qualities to modern sensibilities, as shown by T. S. Eliot's *Four Quartets* (1944), which epitomizes artistic transcendence as a Chinese jar that "still moves perpetually in its stillness," a triumph of pure form over inexorable, annihilating time.[25] Indeed, elite styles changed dramatically in all respects from the Tang to the Song. In particular, the spread of Buddhism in the Tang had made the exotic bodhisattva a kind of fashion model for privileged ranks, sparking a rage for elaborate hair ornaments, silver jewelry decorated with Buddhist lotuses, and belts set with jade and lapis lazuli.

In the Song, however, restrained styles asserted themselves, with an emphasis on modest dress and unassuming motifs of vines and leaves on jewelry and silks. So too with pottery. Whereas the hallmarks of Tang ceramics are exotic motifs, vigorous naturalism, robust profile, and painted adornment, Song porcelains are characterized by reserved decoration, balanced proportions, fluid shapes, and cool-toned, monochromatic shades. Embodying consummate integration of contour, glaze, and ornament, they rank among the finest achievements in the history of ceramics.

During the Song, porcelain came to be a fixture in dining, domestic decoration, and the scholar's studio. Contemplated, commissioned, and collected by connoisseurs, it became incorporated into the social rituals, cultural outlook, and self-perception of the upper stratum. In the eyes of professional scholars and educated gentry, it came to embody urbanity and refinement, epitomizing the abstemious life of reclusive Daoists as well as the unpretentious reserve of Confucian gentlemen. Song Yingxing wrote in the late Ming that porcelain vessels "sparkle in quiet retreats or at festive boards, a concrete sign of civilized life."[26] For those of middling social status, who lacked the wherewithal to acquire expensive replicas of ancient bronze vessels, owning porcelains functioned as an affirmation of social respectability and upward mobility, not unlike the English blue-and-white bone china that still graces the armoires of many middle-class American homes. For common villagers and city dwellers, porcelain also very slowly became part of daily life, as is clear from an anonymous poem of the Song period:

> By the side of the road, three or four families run a rural inn.
> It is clean and bright, but there is no broth let alone tea.
> These are laborers and farmers, no connoisseurs.
> Yet there is a porcelain vase with a sprig of purple bindweed.[27]

Porcelain entered the life of the privileged in significant ways just as the class of Confucian scholar-officials emerged as the dominant status group of government. The Song regime greatly extended Tang precedents by formally establishing that all administrative offices must be staffed through examinations based on Confucian texts. This system, which formed the basis of the civil bureaucracy until its abolition in 1904, ensured that varieties of Song moral philosophy would prevail in intellectual life until the end of imperial rule.[28] Even more important, the system led to creation of a clerical-governmental class with a corporate identity that in time became independent of the turnover of dynasties and court patronage. Yuan, Ming, and Qing—all those who ruled the empire after the Song found themselves compelled to rely on a self-perpetuating Confucian bureaucracy, the elite men who represented sanctioned values, revered tradition, and managerial know-how. There was a pragmatic dimension to this, captured in the Chinese stock scene in which the alien conqueror acknowledges that though he has seized the empire by the sword, he necessarily must rule it by the pen. The ideological dimension was equally significant, however, for the Confucian literati came to see themselves as representing doctrinal unity over time, even exercising moral authority that was grounded in the cosmic order itself and therefore, ideally, independent of transient political authority.

By promoting the examination system as a tool for bureaucratic recruitment, the early Song emperors successfully undercut the power of the landed military aristocracy that had dominated politics and government in the Tang period. Under the new dispensation, men of the book and culture took precedence over men of the sword and lofty pedigree. This attracted talented individuals from a broad social base as, in principle and significantly in practice, the examination system was based on merit. Competition was fierce. In 1090, at the initial, prefectural stage of examination, 4,000 candidates in Fuzhou prefecture in Fujian province competed for just forty positions as "selected men" *(juren),* those moving forward to the provincial level of tests. More than 14,000 men took the final exam in 1002 at the capital level in Kaifeng, each candidate hoping to win one of the fifteen hundred advanced scholar *(jinshi)* degrees on offer, possession of which guaranteed an official post, usually in county administration. By the early twelfth century, 400,000 men sat for examination at the prefectural level, though there were no more than 25,000 officeholders in the empire—a modest number for governing 140 million people spread over some five million square kilometers. Feeling in need of supernatural help, candidates appealed to Wenchang and Kuixing, Daoist deities regarded as instrumental in winning a degree. Not surprisingly, the idols became popular motifs on blue-and-white porcelains in the Ming period.

Producing more educated men than the government bureaucracy could absorb, the examination system effectively directed failed place-seekers into alternative careers in scholarship, private academies, tutoring, lineage organizations, local tax collection, charitable institutions, manufacturing, literature, and the arts. By the late Song, those who possessed status as scholar-officials were complemented by an extensive penumbra of educated and semieducated men who had aspired to join the Confucian administrative elite. For scholar-officials, a sense of corporate identity arose from shared values and a sinecure in the bureaucracy. For those who failed to win entry to office, self-esteem and corporate identity could be achieved by participation in the urbane, learned activities ideally characteristic of officeholders. Like Victorian gentlewomen learning to play the piano as a voucher of middle-class refinement, members of the Chinese intelligentsia excluded from bureaucratic circles arrogated to themselves the right to be part of the cultural realm of the specialist and professional.

Shaped by a humanistic Confucian education, scholar-officials and educated gentry developed a cult of self-expression that manifested in a variety of cultural pursuits, most notably the "Three Perfections": calligraphy, poetry, and painting. The three went together inasmuch as poetry took form in written words, and painting employed calligraphic techniques.[29] When porcelain bowls and silk scrolls embodied all Three Perfections, they held special appeal for the educated. The Three Perfections went hand in hand with the Four Gentleman's Attainments: expertise on the lute *(qin)* and in chess, painting, and calligraphy. Subtlety in these pursuits corresponded with the somber ink painting that came to dominate Song landscape art and with the muted tones of Song porcelains that displaced polychrome embellishment. To be accomplished in the Perfections and Attainments identified one as a refined individual and (at least for men) provided cultural credentials for advancement within the Confucian establishment. For those unsuccessful in administrative careers, the expressive pastimes gave solace in retirement retreats, certifying that failure in the examination round did not rule out partaking of the finer things in life.

"COPYING THE ANCIENT"

In their contrast with earlier ceramics, Song porcelains suggest the reorientation of elite perspectives in the period. A sweeping change of style took place with the advent of the new dynasty, a thoroughgoing abandonment of Tang approaches to shape, ornament, and foreign subject matter.[30] The rejection stemmed from the xenophobia and turmoil that shook China from the mid-eighth century. Threatened by nomadic powers and anxious to bolster imperial legitimacy, the Song dynasty lacked the poise and self-sufficiency that admirers perceive in the renowned porcelains of the era. After building the most powerful military force in history, the Song failed to hold back their enemies to the north, and paying them huge sums to purchase peace contributed to imperial bankruptcy and ultimate disaster. The most splendid achievements of the period served to mask harsh realities; a deeply conservative culture arose from lack of confidence, the result of failure to resist alien incursions and maintain imperial unity.[31]

Emperors and literati sought revitalization and legitimacy in ancient tradition. Renewed interest in Confucian philosophy, leading to the first complete printed edition of the Confucian classics in the mid-tenth century, helped direct attention to the material legacy of the past. In 1143 the Song revived the Han custom of building the Jiaotan, a temple on a raised-earth platform to hold the imperial ceremony of worshiping Heaven. In another conservative reflex, the Song increasingly focused on indigenous Buddhist traditions. Since arrival of the new faith centuries earlier, Chinese Buddhism had defined itself and become intellectually self-sufficient; hence, though Chinese monks still made pilgrimages to India and brought back valuable items, they no longer looked to the subcontinent as an obligatory (and superior) source for texts, rituals, and relics.

Emphasizing the Confucian classics and immemorial tradition, Song monarchs asserted equivalence between their own governance and that of fabled rulers of past times. The Lizong emperor (r. 1225–64) had himself portrayed in paintings as a legendary cultural hero, and he identified himself as "the ancestor to the Hundred Kings" of mythic ages.[32] In like fashion, other emperors and their courtiers promoted a coherent style in the arts that had its roots in antiquity. Song artists produced silk-scroll paintings of the *Eighteen Scholars of the Tang* to show that the dynasty revered ancient values, sponsored elite culture, and provided exemplary leadership.[33] The Huizong emperor (r. 1101–25), an accomplished painter and authority on antique objects, ordered that "ancient vessels should be sought for throughout the empire," since their use in ritual sacrifice manifested and certified the virtue of his rule.[34] He brought together a collection of 1,257 paintings by old masters from the Han to the end of the Tang, works regarded as conveying "the spirit of lofty antiquity."[35] He also launched a series of major cataloguing projects that used the new technology of woodblock printing to present visual proof that his dynasty incarnated the values of the honored past.

Drawing on the vast imperial art collection, scholars produced *Illustrations of Antiquities of the Xuanhe Era* (i.e., the reign of Huizong), annotations to images of eight hundred objects in twenty categories, with Shang ritual bronze vessels given pride of place. After the collapse of the Northern Song (and consequent loss of part of the imperial collection), the Gaozong emperor even more forcefully accentuated his regime's fidelity to antiquity. The son and brother, respectively, of the last two Northern Song emperors, he had fled south to the Yangzi to declare himself ruler of what is now identified as the Southern Song. As such, he bore heavy responsibility for abandoning not only the heartland of the empire but also the sacred tombs of his imperial predecessors. He bolstered his claim to legitimacy by presenting himself as an unrivaled champion of Confucian scholarship. After personally copying out classic Confucian texts used for the civil examinations, he had his transcriptions engraved on stone tablets and rubbings made of them sent to government schools and learned academies throughout the realm. He ordered stone tablets engraved with portraits of Confucius and his supposed seventy-two disciples, and he commissioned the court workshop to produce silk scrolls illustrating the 305 verses of *The Book of Odes* (ca. 1000–600 B.C.E.), a volume believed to have been compiled by Confucius. Gaozong paid a posthumous price for his fulsome public reverence for Confucius, however: commentators in the Ming period criticized the emperor for manipulating the image of the sage for his own political ends.[36]

Imperial commissions such as those ordered by Huizong and Gaozong had an immense formative influence on the course of Chinese culture, sustaining a continuity of inspiration, pious and programmatic, unique in world history. Put in a Western context, it was as if kings of the Capetian dynasty (987–1328) of France supplied published models for successive cultural renaissances down to the French Revolution (1789–99) based on objects created by the megalith builders of Brittany, part of a Neolithic culture whose most famous relic is Stonehenge, a monument whose construction began around 2600 B.C.E.

Even this analogy fails to capture the extraordinary cultural continuity of China, much of which depended on the language employed for literary purposes by the Confucian elite of the empire. The prehistoric people who built the stone circles flourished some four millennia before the language of Montesquieu and Voltaire existed in any form, and the French tongue of the Capetian period—when the literary medium for those of high status was Latin—was quite different from that spoken in the Enlightenment. The elite of Qing China, however, expressed themselves in a language that effectively became ossified in the last century before the common era.

Classical Chinese took shape between 500 and 100 B.C.E., and what is called literary Chinese, which took the former as its model, became stabilized in later generations, especially during the Tang and Song, as the Confucian canon emerged as the basis of learned discourse, authorized ideology, and the civil examinations.[37]

Ricci observed that "from time immemorial, [the Chinese] have devoted most of their attention to the development of the written language and did not concern themselves overmuch with the spoken tongue"; hence, like the ancient Greeks, all their eloquence is found in literature and not in public speaking.[38] Increasingly isolated from the linguistic forms assumed by colloquial Chinese, both verbal and written, literary Chinese remained the language of the educated and propertied, giving them direct access to the writings of the remote past. In contrast, by the eighteenth century Western vernacular tongues largely ousted Latin as the educated lingua franca, except for the writing of scientific papers—a convention that led Linnaeus to employ it for the specialized function of botanic nomenclature.

Confucian officials and intelligentsia of the Qing employed a stylized, almost autonomous language that had been cut off from common, everyday life for well over a millennium. It was as if Montesquieu's *Spirit of the Laws* comprised annotations in classical Latin on concepts of government found in Gallo-Roman manuscripts written around the time Julius Caesar (100–44 B.C.E.) campaigned in Gaul—and with those manuscripts regarded as having a linguistic and cultural lineage stretching back more than a thousand years into the Celtic mists. Yet it was a Chinese reality that ideographic inscriptions on Zhou bronzes about dynastic honor and governance shaped later Chinese concepts of ancestor veneration and imperial legitimacy as a consequence of the texts being first transcribed in the Han era. At that time, emperors effectively created the first classical renaissance by casting themselves in the role of ancient sage-kings, and writers grounded political authority in an imagined antiquity, thereby periodically reconsidering the past in the service of the ruling state.[39] Written in classical Chinese, the *Annals* of the Zhou and the *Book of Changes* could be read easily by Confucius, and his *Analects* set forth venerable notions of ritual and ethical conduct that generations of scholars discussed and debated in literary Chinese until the end of imperial China in the early twentieth century.

Despite later Western perceptions of an "unchanging China," however, substantial continuity of tradition did not amount to millennia of monolithic inertia. In reality, Chinese intellectuals recast their culture by turning to perceived high points in their past at least as often as did the Western elite. Although educated ranks committed to cultural restoration generally claim they are revitalizing an authentic past, they in fact always are involved in an act of creative imagination. Whether Confucian literati in twelfth-century Suzhou or Renaissance humanists in fifteenth-century Florence, they select from a largely imaginary, constructed tradition that seems most useful or most appealing in the present. More often than not, when they claim to be transmitting mores and values from antiquity, they are reinterpreting their meanings, doing something novel while claiming to be entirely conservative.

Chinese tradition persisted as a dynamic, variable force, renewed and reinvented across eras. The great historian Sima Qian (ca. 145–ca. 87 B.C.E.) said that noble men who had failed to accomplish what they wished "wrote of past affairs in order to pass on their thoughts to future generations."[40] The past to be revived varied according to contemporary preoccupations and political contingencies; nomadic conquerors and alienated literati looked to different traditions to validate, respectively, their mandate to rule and their obligation of passive resistance. Writers on art and literature perceived tradition in terms of interaction and mutation, marked at periodic intervals by "copying the ancient" *(fang gu),* a return to exemplars.[41] It became commonplace to say that just as a tree or stream had its starting point in the soil or on mountain peaks, so too tradition remained vital by sympathetic, creative interpretation of ancient models by later generations.

Zhao Mengfu (1254–1322), a Yuan official and an artist famous for his depiction of horses, looked to Tang and Northern Song painters for inspiration. He declared that "the lack of the spirit of antiquity will create so many faults that the result will not be worth looking at."[42] But since all things in the world were seen to exist within the dynamics of a specific configuration *(shi),* the educated ranks assumed that the very disposition to revive antiquity must be modified in response to shifting circumstances. In *Literary Minds and the Carving of Dragons* (Wenxin diaolong), Liu Xie (ca. 465–ca. 522), a founder of Chinese literary criticism, stressed the value of employing canonical traditions in the service of originality: "When a writer casts and moulds his works after the pattern of the Classics . . . he will be able to conceive new ideas and carve original expressions."[43]

In practical matters, this flexible perspective on tradition facilitated cultural change. During the Song, the focus in the arts was on antiquarian reconstruction by the imperial court for the purpose of renewing ancient values. While this emphasis was never wholly abandoned, the growth of a market economy by the Ming period resulted in artists and craftsmen, in response to an elite consuming public, using archaic forms in diverse media for expressive and decorative purposes. Propelled in the first instance by imperial and political dictates, the focus on archaic tradition took on a vitality of its own, with objects in metal and clay valued as much for their aesthetic qualities as for what they were seen to represent of the remote Chinese past.

Confucianism displayed the same sort of adaptability, thereby helping to sanction changes in the arts. During the Song period, Confucius came to be viewed as a model of creativity for his supposed editing and compilation of earlier texts, such as *The Book of Odes:* he was seen as actively refashioning past thought as he transmitted it to the future. Zhu Xi (1130–1200) was the most influential exponent of what became known as Neo-Confucianism, for his commentaries and those of his epigones on the canon formed the substance of the examination system. (Like *Confucius* and *Confucian, Neo-Confucianism* is of Jesuit coinage; in the Song, the thought of Kongzi, or "Master Kong," was known simply as "Learning of the Way," *Daoxue,* or as "Learning of Principle," *Lixue.*) Zhu himself believed that a rupture with tradition had taken place, making it impossible for antiquity to serve as an ideal model for the present. Rejecting inflexible adherence to tradition, he committed himself to creative revision of archaic texts, believing that "great doubts lead to great progress."[44] He reformulated the relationship between revered predecessors and contemporaries when he declared that "at bottom, descendants are the force of ancestors." This shifted emphasis from conciliation of awesome forebears through ritual and sacrifice to ancestors being grateful to the current generation for their enduring spiritual authority. As in the change from antiquarian reconstruction to aesthetic appreciation, the effect was to sustain links with ancient traditions while at the same time crucially renewing them.

The mental reflex of justifying innovation by looking to the past proved invaluable, for it meant that alien influences could be absorbed and accommodated without a sense of cultural shock. Thus Chinese pottery painters in the Yuan and Ming periods readily adapted to the very different Islamic design tradition of Southwest Asia by circumscribing it within their own long-established modes of decoration. So too Chinese artists responded to the pictorial realism of European engravings, introduced by Jesuits in the late sixteenth century, by copying the ancient, in this case, by reviving Song painting techniques regarded as similar to the Western ones.[45] Far from a monolithic burden, the past was a living presence for artists, antiquarians, and connoisseurs, a source of strength and inspiration, perpetually reworked and reevaluated.

Tradition also could be exploited by the cognoscenti for contemporary political ends. Tao Qian (d. 420) retired from office in the period of the Six Dynasties (265–589) to write poetry and preserve his integrity. Landscape scenes from his "Peach Blossom Spring," a poem about a utopian land free from dynastic upheaval and poverty, were commissioned by literati for Yuan and early Qing pottery because they carried a subtle message about the need to preserve one's virtue under alien dynasts. Literati of the mid-seventeenth century charged potters to paint vessels with motifs taken from the *Water Margin,* a novel by Shi Nai'an (ca. 1290–ca. 1365) and Luo Guanzhong (ca. 1330–ca. 1400) about Song heroes battling illicit authority. In the light of late Ming political corruption, the old swashbuckling tale had a contemporary relevance for estranged Confucians. Recognizing the censure, the Chongzhen emperor (r. 1628–45), the last Ming ruler, banned printing of the novel. Similarly, soon after the Qing came to power, estranged members of the literati class commissioned potters to decorate wares with motifs showing Shang officials choosing death over serving the triumphant Zhou dynasty—obvious stand-ins for the despised Manchu rulers.[46]

The past, then, represented a treasure-house of praiseworthy actions and resources. Song potters produced tripod incense burners modeled on archaic bronze food containers and decorated with symbols from the *Book of Changes.* The same antiquarian impulse became more intense and widespread in the Yuan period after the Northern Song imperial collection of antiquities was dispersed among literati in southern China; it seemed more important than ever to maintain cultural traditions, especially since many Confucians refused to take office under the Mongol dynasty. In the early Ming, Song archaic-style porcelains served as models for numerous bronzes cast to furnish the palace of the Xuande emperor. In *Essential Criteria of Antiquities,* Cao Zhao (fl. 1388) declared that "a wide knowledge of antiquities is the first requisite of a gentleman."[47] Potters of the Xuande period put reign marks on the bottom of porcelains in imitation of signatures on ancient bronzes. Copying illustrations from the Huizong catalogues, Ming artisans turned out Song-style porcelains, as well as their own variations on Song ceramic imitations of Shang and Zhou bronzes. They produced porcelains with panels depicting archaic bronze vessels adorned with flower sprays, as well as porcelain incense burners modeled after archaic bronzes and decorated with the *taotie,* the monster-face motif of the Shang era. Ming dynasts adopted porcelains for ceremonies at the Altar of Heaven in Beijing, with the pieces modeled on Zhou bronze cooking utensils for grain and meat as "in the Worship of Heaven and Earth, simplicity is valued beyond everything, and pottery is used in imitation of ancient methods of manufacture."[48]

Qing potters created vessels with Bronze Age ornament, Tang shapes, and Song-type glazes; gilt paint and turquoise glaze replicated the appearance of ancient bronze. In Jingdezhen, imitating works by Song artisans, Tang Ying fashioned green-glazed porcelains embellished with dragons and script in archaic styles. Potters in the Qing imperial workshops in Beijing modeled vases on ancient bronze vessels, adding a touch of modernity by decorating them with bright enamel colors (rose, peach, and lemon yellow) introduced to China by Jesuit missionary-craftsmen. Some Qing connoisseurs claimed that as a result of magical transmutation in the kiln, Song ceramics exhaled a "strange fragrance after long years," even changing their scent with the turnover of the imperial ruling house.[49]

As scions of the conquering Manchu from north of the Chinese frontier, the Qing emperor Yongzheng (r. 1723–35) and his successor, Qianlong, established their legitimacy on lines laid down by the Song rulers. Sponsoring the practice of copying the ancient, as had Huizong and Gaozong, they published illustrated catalogues of antiquities as guides to contemporary forms and images in porcelain and other arts. The most famous tome comprised a record from the mid-eighteenth century of 1,400 imperial possessions, including many Shang bronzes featured in catalogues published six hundred years earlier. Other compilations included representations of 6,000 paintings, calligraphic art, inkstones, jades, and coins. One scholar produced the *Illus-*

trated Catalogue of Bronzes in the Qing Imperial Collection, a study of 1,500 objects.[50] Published to memorialize the past and ensure cultural continuity, the catalogues sometimes instead resulted in novelty. Engravers of the woodblock illustrations mistakenly depicted some bronze altarpieces with an elongated shape foreign to the originals; hence when Qing metalworkers and potters used the illustrations as models, a new form, supposedly sanctioned by antiquity, entered the Chinese repertoire. (See figures 14 and 15.)

Qianlong studied the edition of *The Book of Odes* published during the reign of Gaozong, and he composed poems to accompany illustrations he specially ordered for the text. He commissioned white porcelains in the form of a bronze tripod cup known as a *jue,* examples of which had been found at sites of the Shang period. He also paid reverence to famed paintings by Guanxiu (832–912) of the Buddha's sixteen *luohan* (venerable disciples) by commissioning jade carvings of the figures.[51] The emperor treasured Song ceramics, and his collection included hundreds of green-glazed porcelains made for his imperial predecessors, as well as many contemporary copies of them. He had his poems praising the vessels carved on the underside of some. One such poem, composed in 1786, read:

> On its bare [unglazed rims] alone, Ding ware is not liked.
> But the secret ware of the [Southern Song] official kilns is highly treasured.
> It stands alone in the world, rarely seen, and although blemished,
> The connoisseur's desire is still very keen.[52]

Anxious to show they warranted allegiance as a "glorious dynasty transmitting antiquity," Qing emperors ordered porcelains produced for the court to be decorated with calligraphy and landscape scenes taken from art of the Han and Tang reigns.[53] Porcelain brush pots for the scholar's studio were inscribed with texts from the Han, such as "How the Supreme Ruler Begets Virtuous Officials." This had the shrewd effect of focusing attention on past glories while also elevating Qing achievements. In 1742 Qianlong commanded Tang Ying to make a type of hanging vase suitable for a sedan chair, with the emperor's poem inscribed on it: "Guan wares and those of Ruzhou are famous classes [of ceramics], / Yet the shapes of the new vases are even more admirable."[54]

Court-promoted archaism from the Song to the Qing spread veneration for the remotest traditions among the educated and affluent of society. Ricci noted that high esteem is "always accorded to those who display any knowledge of antiquity."[55] The painter Qui Ying (ca. 1502–51) captured the sophisticated preoccupations of connoisseurs in his *Ranking Ancient Works in a Bamboo Court,* in which three men of letters examine bronzes, fan paintings, and porcelains on an elegant terrace framed by decorated lacquer screens.[56] Garden scenes of literati engaged in examining antiquities also featured as a common motif on pottery in the early Qing period, around the time Dentrecolles arrived in China. He observes that though modern porcelains attract many admirers, "there is no doubt that China has its antiquarians, those who revere old things."

Dentrecolles's judgment is nicely illustrated by *The Story of the Stone* (better known as *The Dream of the Red Chamber*), the great Qing novel by Cao Xueqin (ca. 1715–63) about an illustrious family in Nanjing. They lived in a mansion whose reception hall boasted an imperial inscription in "archaic seal-script" on the wall, beneath which stood

> a long, high table of carved red sandalwood, ornamented with dragons. . . . In the centre of this was a huge antique bronze *ding* [cauldron], fully a yard high, covered with a green patina. On the wall above the *ding* hung a long vertical scroll with an ink painting of a dragon emerging from clouds and waves, of the kind often presented to high court officials in token of their office. The *ding* was flanked on one side by a smaller antique bronze vessel with a pattern of gold inlay and on the other by a crystal bowl.[57]

An inner apartment presented a less showy display, a small, four-legged *ding* and a rare Song ceramic, "a narrow-waisted Ru-ware imitation *gu* [antique bronze goblet] with a spray of freshly cut flowers in it."

Zhu Yan remarks that "a genuine old piece [of porcelain] is beyond price."[58] During the Song period and after, individuals who could not afford to purchase treasured objets d'art commissioned bronze casters and potters to replicate them, a practice that resulted in a thriving market for counterfeiters. Indeed, bogus antiquities represented the tribute fraud paid to time-honored cultural values. Complaining of the difficulty of evaluating Song porcelains, a Ming connoisseur said that "the genuine and the false remain intermingled"; a Qing expert cautioned buyers that "crowds of fakes jostle the genuine."[59] And Ricci wrote, "Counterfeiters of antiques are numerous, and they are clever at cheating the unwary by selling them utterly worthless things, taking advantage of the fact that they are poorly posted in value." Sometimes it sufficed if an object simply looked old, as Ricci discovered when a statue of the Virgin Mary fell to the floor and smashed: "That would have ruined its value in Europe, but served only to increase it in China. When the pieces were put together again, the statue took on an appearance of antiquity, which made it more valuable here than when it was whole."[60]

Zhou Danquan, a well-known Jingdezhen potter of the late sixteenth century, specialized in reproducing historic pieces and selling them to collectors. In *Potteries of Jingdezhen,* Lan Pu observes, "In making vessels, [Zhou] was a famous hand of the times. . . . Every time a famous vessel-type was produced, everyone wrangled to buy them."[61] Zhou's creations were regarded as far from worthless: he secretly replicated a Song ceramic censer owned by a connoisseur, and even after revealing it was a copy, he received forty ounces of gold for it because the man longed to match it with his authentic piece. A generation later, a top government official, presumably unaware of the provenance of the reproduction, reportedly persuaded a descendant of the connoisseur to sell it for one thousand ounces of gold.

CONNOISSEURS AND THE CULTURE OF TEA

With great wealth in their hands, the burgeoning upper classes of the Song era stimulated demand for luxury ceramics, both antique and contemporary. Production of silver vessels could not keep pace with demand, thereby widening the market for sales of fine replicas in clay.[62] Although silver and lacquer utensils dominated the tables of the privileged early in the Song, porcelain came into increasing use, with wealthy consumers motivated in part by the novel notion that eating from gold and silver plate harmed one's health. New standards of decorum also came into play. "Set a high value on porcelain and a low one on gold and silver," one writer advocated. "Esteem elegance!" As Ricci observed, porcelain vessels became "highly prized by those who appreciate elegance at their banquets rather than pompous display."[63] Not everyone shared this view, however, for some members of the upper class remained faithful to their vessels of precious metal and regarded endorsement of pottery as a display of smug puritanism. "A gentleman of true taste, talents, and sentiment," a Ming critic sniffed, "thus takes vows not to mention 'elegance.'"[64]

Porcelain gained wider acceptance during the Southern Song, just as Chinese cuisine took on its distinctive characteristics. Rice became the chief grain—conquerors controlled the millet of the north—and soy sauce, bean curd, steamed buns, sesame oil, vinegar, and stir-fried vegetables became standard fare. Taverns, tea shops, and stew kitchens sprang up in the empire's twelve hundred walled cities, providing a more relaxed and diverse environment than existed in the Tang period. In the latter, cities were predominantly bureaucratic outposts and gathering spots for the landed aristocracy; in the Song, they became centers of leisure pursuits and social interaction for the scholar-gentry class and well-to-do entrepreneurs.[65]

In the early twelfth century, Zhang Zeduan painted a scroll five meters long, *Peace Reigns over the River*, an immensely detailed panorama of what is probably Kaifeng, including a merchant's shop advertising porcelain wares alongside noodle stalls and cafés. Fancy restaurants emerged, with porcelain and lacquer vessels the customary tableware. A journal writer in Kaifeng in the Song period retailed information about food and dining, emphasizing the many regional cuisines available in over seventy named restaurants. He explained that eateries provide "first-class ceramic bowls with shallow rims. . . . In the old days we used only spoons, but nowadays they use chopsticks."[66] One multistory restaurant, built with a cast-iron frame, served a thousand persons at a time and provided entertainment in the form of skits, bawdy songs, and dramatic verse. Conferring prestige on the new establishments, the Huizong emperor sometimes ordered take-out from his favorite eatery.

Rather than sit on mats or cushions, diners used chairs, a form of furniture introduced from India to Chinese Buddhist monasteries by the fifth century.[67] The Buddha and associated bodhisattvas often are depicted seated on lotuses or on lotus-shaped stools and couches. The court eventually took to raised seating, and in the early twelfth century emperors ordered chairs and tables of imported sandalwood for their palaces. Chairs replaced mats in well-to-do households, impelling a move from large eating utensils to smaller cups, bowls, and plates. In turn, small wares set on a table surrounded by chairs prompted a decline in communal consumption, in which diners shared large bowls and goblets, and a move toward place settings for each person. The same evolution toward individualized dining would not take place in Europe until the late seventeenth century and then largely due to the introduction of Chinese table services. Europeans had used chairs for centuries, and Ricci in fact thought it remarkable that in certain everyday customs the Chinese "alone of all nations outside of Europe are in accord with the West. Their use of tables, chairs, and beds is wholly unknown to any of the peoples of the states that border on China, all of whom place straw mats on the ground or floor and use them in place of chair, bed, or table."[68]

In Song China, caterers served the rich in their mansions, setting up banquets complete with porcelain tableware, napkins, drapery, and furniture. Most important for the enhanced reputation of porcelain and potters, the ceramic also graced the imperial table. The Huizong emperor painted *Literary Gathering,* a hanging silk scroll portraying eight gentlemen (perhaps including the artist himself) seated on chairs around a black-lacquered table set with an array of floral bouquets. Each diner has a place setting consisting of a porcelain cup, a stand, a dish, and chopsticks. Attendants stand nearby with wine carafes, and a portable cupboard holds additional tableware. In the foreground, servants prepare tea.[69]

By the reign of the Song, polished soirées such as that depicted by Huizong customarily featured tea *(cha)* served in porcelain cups or bowls.[70] An enthusiast for tea, as well as porcelain, the emperor wrote a treatise praising the virtues of the drink. Tea is derived from *Camellia sinensis,* an evergreen shrub of southwestern China, named by Linnaeus in honor of Georg Joseph Kamel, the Jesuit who sent dried specimens of the plant from the Philippines to Europe. Tea was used first for making a medicinal drink in the early Han period, often boiled with ginger, orange peel, or peppermint. By the early common era, drinkers abandoned those ingredients, and tea started to be regarded as more than a health aid or flavoring agent.

Around the same time, drinking tea played a role in the replacement of the traditional "ear-cup"—an oval vessel with two winglike projections, generally used for sipping wine—with a round, relatively high-sided cup, a shape much better suited to hold hot liquid safely.[71] Of course, porcelain proved especially practical for drinking tea as it has minimal thermal conductivity, unlike vessels made from metal and glass. Dentrecolles made that point in discussing the advantages of Jingdezhen's pottery: "They find their porcelain is more useful [than glass]: it bears hot liquids, and one can hold a cup of boiling tea without being burned . . . which one cannot do with a cup of silver of the same thickness and shape."

Wandering adepts of Chan Buddhism (later known as Zen in Japan), whose only permitted sustenance during their long, sleepless nights of meditation was tea, spread its use in the Tang period. According to legend, Bodhidharma, an Indian monk who came to China in the early sixth century, stayed awake until his eyelids fell onto the ground and sprouted into the first tea plants in the Middle Kingdom. The Buddhist poet Gao Shi (d. 765) wrote that "reading books cannot compare to reading scripture, and drinking wine is no match for drinking tea."[72] A popular subject in Tang paintings was that of monks studying religious texts while sipping tea from white porcelain bowls, white being considered auspicious by Buddhists. The theme entered into Chinese high culture, as shown in Qui Ying's hand-scroll *Zhao Mengfu Writing the Heart Sutra in Exchange for Tea,* a mid-Ming work that reflects the lasting association between Buddhist monks and the culture of tea.[73]

Tea won over the imperial court in the Tang period. Feng Yan's ninth-century *Record of Things Heard and Seen* states that "the way of tea is in great fashion, and among the imperial court and nobles, there is not one who does not drink it."[74] As early as 821, a court official claimed, "Tea is no different from rice and salt. People far and near alike depend on it."[75] The state established a monopoly on tea in 793, by which time taxes on the tea trade amounted to 400,000 strings of cash (40 million coins). The drink became common refreshment throughout China only in the Song, however. Preparation advanced from grinding and then boiling slices of prepared tea bricks in the late Han to the practice of pouring boiling water over dried tea leaves in the Southern Song (and still today). This entailed a change of implements. In the earlier method, powdered tea simply was tossed into a heated cauldron of water and the liquid scooped out and poured into a bowl; use of dried leaves demanded creation of a pot for stewing the tea.

Taste also evolved: consumers favored red tea in the Tang and white tea in the Northern Song; green tea, sometimes scented with floral essences and musk, became most popular in the Southern Song and remained so thereafter. During the Song, tea and sociability forged an enduring association, eclipsing the drink's origins in monkish meditation. As Gaspar da Cruz later described it, "Whatsoever person or persons come to any man of quality's house, it is customary to offer him on a fair tray in a porcelain cup . . . a kind of warm water which they call *cha* . . . made from a concoction of somewhat bitter herbs."[76] The Jesuit Rodrigues reports that some well-heeled tea votaries in the late Ming took the edge off the bitterness by dropping morsels of apricot and almond into tea; when the drink was consumed, the fruit was eaten with silver or bronze spoons. According to another Jesuit, sometimes the Chinese prepared a hasty meal by beating two egg yolks with heaps of powdered sugar, then pouring the concoction into infused tea. Many tea drinkers, however, frowned on making the beverage into a kind of sweet or spicy soup. A

Ming writer pronounced that "adding fruit, seeds, salt, pepper, ginger, bitter orange, and so on to tea is a calamity."[77]

Whatever the preparation, color, or tidbit, everyone agreed that tea must be accompanied by porcelain. In fact, it is no coincidence that the camellia leaf entered the diet of the moneyed classes on a wide scale at the same time porcelain came into wide use. Zhu Yan remarks that during the Tang, various kinds of cups were used for drinking liquids, especially wine—including "ruddy gold, white jade, engraved silver, rock-crystal, and glass, beautifully carved and designed"—but in the Song, cups typically were made of porcelain, "the material most esteemed at the tea-tasting parties of the period."[78] A comparable transition took place in eighteenth-century Britain, when tea challenged ale, beer, and gin in popularity, with pottery teacups and teapots jostling aside mugs, tankards, and beakers on the table.

In becoming a defining aspect of Chinese culture, tea gathered a cluster of associations in which porcelain played a central role, especially among the elite. The Tang poet Lu Yu (ca. 730–804) wrote *The Classic of Tea,* a work so influential that it led to his memorialization as the god of tea. He taught generations of tea drinkers to cultivate an urbane sensitivity to the natural world. He informed his readers that when boiling water is poured into a pot, the frothy patches that bubble up from the tea leaves should bring to mind "eddying pools, twisting islets or floating duckweed at the time of the world's creation. They should be like scudding clouds in a clear blue sky and should occasionally overlap like scales on fish. They should be like copper *cash,* green with age, churned by the rapids of a river, or dispose themselves as chrysanthemum petals would, promiscuously cast on a goblet's stand."[79]

Lu obviously would have thought that dumping egg yolk and powdered sugar into his cherished beverage spelled calamity. He laid down the dictum that "moderation is the very essence of tea," and he specified what he called "the Twenty-Four Implements," from water filters to bamboo pincers to porcelain bowls, all of which he considered indispensable for preparing a superb drink.[80] He also was the first to express a true interest in ceramics, classifying and ranking tea bowls by regions, just as he did for certain teas. By the Song era, according to Zhu, "detailed rules [for tea drinking] were laid down for every step, from the gathering, the preparation, and the storing of the crop up to the boiling of the water for the infusion. . . . The utensils employed were also carefully selected, and only adopted after repeated trials, nothing being taken at random."[81] As a consequence of Lu's influence, the culture of tea also complemented upper-class devotion to antiquity as potters modeled the biggest pieces of tea equipment, such as water jars and cauldrons, on bronze vessels from the Shang and Zhou periods.

Competitive tea-tasting parties placed special demands on the proper utensil. As Zhu explains, "The rule was that he whose tea trace disappeared first lost, and that the one whose trace lasted longest won." Inasmuch as detecting the trace was the point of the competition, it was essential that the tea bowl have an impermeable surface to prevent staining by tea tannin and to ensure the clarity of the liquid. Many experts followed Lu in regarding green-glazed porcelains as superior for enhancing the color of tea, bringing to light a green akin to "the bright tint of distant well-wooded mountains."[82] Others preferred thick, black bowls with curious effects on the glaze, such as "partridge feathers" (white pearly spots) and "hare's fur" (orange-brown streaks), decoration characteristic of Zhejiang and Fujian kilns. In his treatise on tea, Huizong argued that green tea looked best set off against dark porcelain: "The black-hued tea bowls are to be preferred. Those with distinctive hare's fur glaze are the best." In the early Ming, many drinkers preferred teacups "pure white in colour, like jade, so that one can observe the colour of the tea."[83]

Teapots became indispensable accessories when consumers turned to brewing tea leaves during the Southern Song. Zhu states that the cognoscenti considered porcelain pots obligatory "so that the fragrance may not escape and the aroma not be lost."[84] By the Ming period, those most admired by connoisseurs came from Yixing, a town on the shores of Lake Tai in Jiangsu province, immediately to the east of Anhui.[85] In the twelfth century, Yixing craftsmen, perhaps under Buddhist influence, adapted the shape of wine ewers to create a pot for infusing tea. Employing a stoneware clay that fired a soft rose-brown, the potters molded elegant vessels with a pear-shaped form, curved spout, ring-shaped handle, burnished (and unglazed) surface, spare decoration, and simple, clean lines—in other words, what eventually came to be regarded everywhere as the classic teapot. Singling out another leading feature, the Portuguese introduced the ware to Europe in the sixteenth century under the name *boccarro* (big mouth). The Japanese, who had little use for a teapot since they did not stew tea, nevertheless admired the Yixing product, which they called "vermilion ware."

Buddhists may have been responsible for development of the Yixing vessels, but, like the camellia leaf centuries earlier, the teapots soon transcended a monastic or religious context. Tea enthusiasts avidly sought them out, paying as much as six taels of silver for one and commissioning customized pieces with calligraphic inscriptions. Some owners even ordered their teapots placed in their tombs. Su Dongpo (1037–1101), a magistrate, poet, and connoisseur, designed one of the most enduring pot shapes in the Yixing repertory, and an eminent Yixing potter received the dubious honor of having his signed teapots counterfeited in his lifetime.

An enthusiast for Yixing ware claimed that furnace transformation magically affected his pot so that when hot water spilled into it, "the stored tea turns to rosy clouds like shimmering silk."[86] Porcelain and tea made for a potent brew. Reading omens in tea leaves (tasseomancy) goes back a long way, perhaps as far the Tang dynasty, and the practice was transmitted to (or independently conceived in) Europe, as illustrated by the English poet Charles Churchill (1731–64), who describes "matrons, who toss the cup, and see / The grounds of fate in the *grounds* of tea."[87] When a prominent Song general gave a teacup to a poor man, his charity received supernatural acknowledgment, for when tea splashed into the vessel "there immediately appeared a pair of cranes, which flew out of the cup and circled round it, and only disappeared when the tea was drunk." When Su Dongpo ceremonially offered tea to images of the Buddha's venerable disciples, the liquid became "transformed into white milk, or solidified into flowering trees, among which the peach, plum, and peony were in turn clearly identified by him."[88]

PORCELAIN AND DOCTOR GOOD TEA

Jingdezhen was closely connected to Qimen in Anhui province, a major tea-growing area, so the porcelain city naturally turned to manufacturing tea vessels from early in the Tang. It also created imitations of the Yixing ware in blue-and-white in the Ming, but they could not compete with the real thing. When the Yixing teapots reached Europe in the mid-seventeenth century, at the same time the VOC began importing tea, they proved as popular as in China.[89] Rising to the challenge, Western silversmiths soon copied the novelty and produced variations on it: the well-known Queen Anne silver teapot derives from a pear-shaped Yixing vessel. One Dutch potter who made facsimiles of the Yixing pots boasted in a 1678 newspaper advertisement that "he has achieved such perfection in the manufacturing of red teapots that they are in no way inferior in colour, purity, and durability to the Indian teapots."[90] Yixing pots were celebrated in China for their naturalism, with pieces often shaped like lotuses, melons, pomegranates, gourds, and bamboo. Western potters copied such exotic items because the organic forms charmed their customers, though of course they remained unaware that in China the plants conveyed symbolic significance rooted in Buddhism and Daoism. (See figure 11.)

August II purchased many Yixing pots at VOC auctions in Amsterdam, and Tschirnhaus believed the red stoneware held the key to a formula for porcelain. He investigated the work of Ary de Milde (d. 1708), a potter in Delft, a town forty-eight kilometers southwest of Amsterdam, who styled himself "Mr. Teapot Maker" because of his outstanding imitations of Yixing pots. His workshop, the Crowned Teapot, obtained a fifteen-year patent for his version of the ware in 1680. Dutch painters also memorialized the Chinese teapots.[91] Pieter Gerritz van Roestraeten (ca. 1630–1700), a Haarlem painter, produced *Still Life with Tea Things*, which depicts a black-lacquered table on which stands a Yixing teapot, an imposing tea caddy, blue-and-white porcelain cups, and a chunk of rock crystal. His *Still Life with Silver and Ebony Casket* features a silver-mounted nautilus shell next to a Yixing teapot with a gold cupid fastened to the lid, evidently a reference to the supposed aphrodisiac effect of the popular Chinese drink.[92]

While not everyone agreed on the efficacy of tea as a sexual stimulant, there was a consensus wherever the drink became widely consumed that it improved one's health. Lu Yu wrote that "its liquor is like the sweetest dew of Heaven," and if taken five times a day, it alleviated arthritis, eye strain, constipation, and melancholia.[93] In an essay on herbs in 1578, Li Shizhen claimed that drinking tea would "promote digestion, dissolve fats, neutralize poisons in the digestive system, cure dysentery, fight lung disease, lower fevers, and treat epilepsy."[94] Perhaps inspired by payments from the VOC, Dr. Cornelius Decker (1648–86) of Holland, the so-called Doctor Bontekoe (Doctor Good Tea) who wrote the *Treatise of the Excellent Herb Tea* (1679), assured his readers that the drink cured an astonishing range of ailments, including scurvy, sore throat, colic, gout, bad breath, and inflamed eyes. He recommended that "all men and women drink it every day and at all hours, beginning with ten cups a day and increasing the dose as much as the stomach can stand and the kidneys can expel."[95] Mockers, however, claimed that "his joints rattled like castanets" from the two hundred cups he sometimes drained in a single day. Some tea drinkers nearly fulfilled Decker's prescription: an acquaintance told Montesquieu that housewives in Holland drank some thirty cups every morning.[96] A popular verse of the 1670s also endorsed the drink:

> Tea that helps our head and heart
> Tea medicates most every part
> Tea rejuvenates the very old
> Tea warms the piss of those who're cold.[97]

According to Postelthwayt, tea strengthens the brain and stomach, promotes digestion and perspiration, and provides a "remedy against chronic [ills], from its admirable quality of sweetening and diluting the blood."[98] The Japanese shared European and Chinese estimations of tea as a panacea. Yosai (1141–1215, also known as Eisai), a Buddhist monk and pilgrim to China, brought tea seedlings back home to Kyushu, the southernmost island of Japan. In 1211 he wrote *Record of Curing Disease with Tea* (Kissa Yojoki), in which he asserts that tea wonderfully fortifies the body against illness:

> Whereas the Chinese drink tea to provide the heart with the taste it likes, in Japan we do not drink tea. Therefore, the Chinese have healthy hearts and live long. . . . The heart is the chief of the Five Organs and the [bitter] taste of tea is the most important. In other words, bitter is the chief among the [Five] Tastes. This is why the heart likes bitter tastes. It is healthy, thanks to the bitter foods it partakes, and secures the health of the other subordinate organs.[99]

Ricci and Rodrigues, however, came closest to proposing the most significant health benefit from drinking tea. In China, as Ricci points out,

> their drinks, which may be wine or water or the drink called Cia, are always served warm, and this is so even in the hot summer. The idea behind this custom seems to be that it is more beneficial for the stomach and, generally speaking, the Chinese are longer-lived than Europeans and preserve their physical powers up to seventy or even

> eighty years of age. The custom might also account for the fact that they never suffer from gallstones, so common among the people of the West who are fond of cold drinks.[100]

After declaring that tea relieves hangovers, reduces fever, and safeguards chastity (by purging the kidneys of hot blood), Rodrigues observes:

> It has various other advantageous properties. Thus both China and Japan are densely populated and the people, especially in China, are greatly crowded together. Yet there is usually no plague in these two kingdoms as in Europe and other places, and pestilence is very rare. Many people maintain that this results from *cha,* which evacuates all superfluous matter that causes evil humours. They drink *cha* continuously both day and night, and *never touch cold water* for hot *cha* is their ordinary drink summer and winter, and normally they always drink it at the end of a meal.[101]

In fact, the principal health benefit conferred by tea was that it required consumers to boil their water, thereby reducing the incidence of diseases arising from contamination of wells and streams—a cause of some three million deaths annually even today.[102] Though tea was not a cure-all, then, its preparation certainly acted as a prophylactic. At least in Song China, the popularity of tea also may have promoted broader concern for untainted water. Zhuang Chuo, an early-twelfth-century physician, commented that "when the common people are traveling they take care to drink only boiled water."[103] Because brewed tea cost very little, it came within reach of the lower classes, thereby ensuring that much of the urban population had greater protection from lethal infection in congested, disease-rich environments.

Beyond its association with tea, porcelain itself also contributed directly to reducing disease: its impermeable surface led to a decrease in bacterial infections stemming from particles of decayed food being retained in the pores and scratches on plates and bowls made of wood, earthenware, pewter, and precious metal. Traveling in Italy in 1581, Michel de Montaigne (1533–92) admired faience dishes, "which are so white and neat that they seem like porcelain," but he recoiled from the grimy, scratched pewter vessels placed before him in hostelries.[104] A mid-seventeenth-century Italian manual on household management pointed out that the rich considered high-quality pottery "safer than tin [i.e., pewter], not picking up bad odors, and cleaner, just as one sees princes use crystal for drinking although they have cups, glasses and other dishes in gilded silver."[105]

Moreover, after the Tang period Chinese glazes contained no lead, whereas low-fired, lead-glazed pottery used everywhere else released minute amounts of the metal into cooked and stored food, exposing consumers to significant health risk, even death. In particular, acetic acid in wine leaches the mineral from lead-glazed ceramics. In 1760 the Society for the Encouragement of Arts, Manufactures and Commerce in Birmingham heard a report warning against use of common pottery:

> For the Delfware is clumsy and rotten, and its Glaze dangerous to Health, on account of the Quantity of calcin'd Lead and Tin employed in it. Matters that, even after Vitrification, submit to the milder Acids. The same bad Consequences attend the other Common Potteries that are glaz'd with Lead.[106]

An anonymous engraving of 1799 shows skeletal Death seizing an Englishman who is spooning soup from an earthenware bowl, and in *The Lady's Assistant for Regulating and Supplying the Table* (1773), Charlotte Mason counseled that only Chinese porcelain should be used to hold acidic foods.[107]

Song China, early modern Japan, and eighteenth-century Europe all experienced striking population growth. Though obviously impossible to measure or prove, it is likely that widespread use of tea in the three regions and increasing use of porcelain vessels in food preparation, consumption, and storage contributed to improving the health of the general population.

CONNOISSEURS AND THE CULTURE OF PORCELAIN

Along with imperial patronage, restaurant dining, and private banquets, the culture of tea guaranteed that porcelain would come to occupy a prominent niche in the mental universe of the Confucian intelligentsia. The ceramic offered agreeable occasions for demonstrating that the taste of a refined gentleman matched the delicacy of his finest vessels. Aficionados claimed that the simplicity of porcelain elevated it above vulgar gold, the substance of choice of parvenu merchants and military commanders. Often articulated with intimidating hauteur, restraint and modesty were keynotes of Song connoisseurs. Still, as one writer made clear, the supposedly straightforward virtues did not rule out exquisitely nuanced feelings when a treasured vessel came to grief:

> When a good piece of porcelain is broken, you know there is no hope of repairing it. I then hand it to the cook, asking him to use it as any old vessel, and give orders that he shall never let that broken porcelain bowl come within my sight again. Ah, is this not happiness?[108]

A Song connoisseur maintained that pottery should be preferred to opulent containers and that "pouring from silver vessels into cups of jade only dazzles the eyes." Verses by Su Dongpo indicate that porcelain cups held a place of honor on the table equal to that of vessels of precious metal: "From the silver vase flows out in oily stream the ant-wing wine: / On the brown bowls float up the grains of coiling-dragon tea."[109] Scholars and gentry regarded porcelain boxes as de rigueur for preserving the odor of incense and for keeping their vermilion seal colors bright, far better than traditional jade receptacles. They reclined their heads on porcelain pillows because those are "most efficacious in keeping the eyes clear and preserving the sight, so that even in old age fine writing can be read."[110] The Song art of flower arranging, later elaborated by Zan Qiande (1577–1643) in the *Treatise on Vase Flowers* (Pinghua Pu), called for an assortment of porcelains and other accessories, such as vases to display each flower type and furniture to match the vessels. Experts agreed that porcelain vases superbly exhibited the hearty blossoms of summer and autumn, while those of winter and spring flourished best in bronze containers.[111] (See figure 12.)

Calligraphers preferred porcelain inkstones, and they judged that their porcelain brush holders, unlike those of copper, did not make brushes brittle by poisoning the water. Porcelain incense burners, modeled after ancient bronze vessels, only needed to be "submitted to thorough friction to remove the new gloss of the fire from their surface," though pieces with "tortoise-shell markings scattered all over the surface . . . are too elaborately ornamented to be fit for a simple scholar." Like-minded gentlemen believed that vases for the library or study should be modest and restrained, as well as shaped into natural forms, fashioned "like aubergine fruit, like flower-jars, or flower-bags, or in the form of divining-stalks, or bulrush heads."[112] According to one authority, porcelains lacking elegance and sophistication are "only fit for use on the ladies' dressing table. They are not the refined garniture of the scholar's study."[113] From the Song to the end of the Qing, decoration of vases made for the scholar's study commonly included representations of accessories, such as flowerpots, stands, hanging scrolls, brush holders, and bronze antiquities, that customarily adorned the study of the scholar. Natural forms in porcelain dominated paraphernalia on the scholar's desk: an inkstone in the appearance of a peach, a pomegranate-shaped water dropper, a peach-shaped censer, brush holders of bamboo design, a brush rest fashioned as a pea pod, a seal-paste box in the guise of a scroll, and a scroll weight wrought as a dragon cresting waves.

During the Song, connoisseurs favored porcelains from certain kilns. They identified "the Five Great Wares" of Ru, Ding, Jun, Guan, and Ge, labels derived from the county locations of kilns patronized by the court. Despite its limited utility, this taxonomy still dominates discussion of Song ceramics. In China, however, five is a number with which to conjure. Confucian and Daoist scholars of the early common era elaborated a system of mutual correspondences to classify and understand reality. They originally identified five elements (earth, fire, water, wood, and metal) and later extended the symbolically weighty number to identify five of everything significant in nature and society, including, inter alia, organs, tastes, planets, spices, colors, classical texts, virtues, and human aspirations.

In short, rather than provide a literal inventory, allusion to five of anything represents rhetorical shorthand for a complete amount.[114] The Yuan court mandated that the imperial dragon depicted on porcelains and other objects must have five claws, and porcelain brush rests for scholars were molded to stand for the five Sacred Peaks, epitomized as the spines of five subterranean dragons. The Qianlong emperor relished "the exquisite state of repose produced by tea, that precious drink, which drives away the five causes of sorrow," those aspects of life inimical to "fivefold happiness."[115] The five-petaled flowering plum *(mei hua),* a symbol of good luck, was chosen in 1930 as the official emblem of China, whose national flag displays five stars. The notion of "the Five Great Wares," then, says less about the principal kilns of the period, many of which still await archaeological investigation, than it indicates that by the Song porcelain had achieved a stature sufficiently elevated in the eyes of the literati to assign it a canonical fivefold listing.

Besides enjoying imperial patronage, the Five Great Wares displayed luminous glazes and a range of colors, including blue-green, dark green, sea green, olive green, amber, duck-egg blue, blue-white, ivory, and pale silver. In the late Tang, potters summed up the most important distinction in ceramics with the phrase "green in the south, white in the north."[116] In the Song, the most prestigious kilns producing whitewares were located in Hebei province in northern China, and those manufacturing the best-known greenwares *(qingci)* were at Longquan in Zhejiang province. In modern nomenclature, greenware usually is called *celadon,* a term derived from the name of a character that an anonymous writer attached to a shepherd in *L'Astrée,* a French pastoral drama based on a novel by Honoré d'Urfé (1568–1625). The character Celadon always appeared clad or decked out in ribbons of grayish green, a shade that became identified with shepherds in masques at the court of Louis XIV. The shepherd and his bucolic comrades often appeared on eighteenth-century French pottery, and in the nineteenth century French collectors applied *celadon* to the dominant hues of monochromatic Song porcelains. Of course, no one in Song China employed the term; "the color of natural things" sufficed to describe porcelains whose tints ran the spectrum from blue-green to amber.[117] (See figure 10.)

CONNOISSEURS AND THE CULTURE OF JADE

Jade (*yu,* or nephrite) came in the same range of colors as the Five Great Wares, the shades varying according to the iron content of the semiprecious mineral.[118] Potters pleased their patrons by consciously playing on the likeness of porcelains to jade. Indeed, a driving force behind ceramic development in the Song was the ambition to replicate the feel, look, and timbre of jade, a substance of great ceremonial, symbolic, and artistic significance. Whereas gold was believed to derive from the power of the sun, jade was regarded as "the shining essence of moonlight," a supernatural treasure that formed and revealed itself in the nighttime glow. Scarce in China, jade was imported from Central Asia, especially from mines and riverbeds at Khotan on the Silk Road. Merchants termed good-sized pieces without pits or fractures "the jade worth the value of cities."[119]

Costly even in its crude state, jade became costlier when carved by an artisan: because it is harder than any metal, shaping and engraving the mineral with abrasives and cords is a labor-intensive, time-consuming task. In 1735, soon after Asian jade arrived in Europe, John Barrow, in *Dictionarium Polygraphicum,* a lexicon of the arts, wrote that jade is "much esteemed for its hardness . . . and is only to be cut with the powder of Diamond."[120] This makes a specific achievement of Qianlong's craftsmen more extraordinary. In 1780 they carved a jade boulder 200 centimeters in height and weighing more than 5,000 kilograms: depicting scores of laborers and engineering tasks in minute detail, *Da Yu Controlling the Floods* was inscribed with a poem by the emperor suggesting that jade artworks have value equal to painting since they last virtually for eternity.[121] In the realm of the decorative arts, the achievement of the palace craftsmen was as monumental as the hydraulic projects memorialized by their jade masterpiece.

As a consequence of its seemingly imperishable nature, jade became identified with the quest for immortality as early as the Neolithic, an era when perforated jade axes were buried with the honored dead, whose bodily orifices were sealed with jade plugs. Jade seemed endowed with Heaven's creative and preservative powers, making it the most appropriate material for cult objects in rites for the deceased. By the Shang period, specialists in sculpting and polishing jade had emerged in the lower Yangzi region.[122] The tomb of Fu Hao of the Shang held seven hundred pieces of jade, including some Neolithic ones that probably already were regarded as precious heirlooms. Kings of the Bronze Age ritually presented ring-shaped jade objects at altars to pay homage to Heaven and Earth. Incised animal masks of jade covered the faces of royalty in tombs. Jade scepters, with shapes based on bronze weapons, signified hallowed authority, and jade ornaments, such as pendants, buckles, and jewelry, denoted high rank.

By the Warring States period, the authority of Confucius buttressed the spiritual prestige of the gemstone, for one of the texts attributed to him states that "the noble man will never appear without jade ornament, and his virtue is like his jade ornament."[123] The sage's judgment became the warrant for elaboration later in the period:

> Anciently superior men found the likeness of all excellent qualities in jade. Soft, smooth and glassy, it appeared to them like benevolence; fine, compact and strong, like intelligence[,] . . . its flaw not concealing its beauty, nor its beauty its flaw, like loyalty; with internal radiance issuing from every side, like good faith.[124]

The association of Confucius and jade appeared in the earliest Western portrait of him, included in the first Latin translation of his supposed writings: in the *Confucius Sinarum Philosophus,* published in Paris in 1687 with a subvention from Louis XIV, Confucius prayerfully holds a jade tablet *(gui),* a symbol of political investiture.[125]

Princes of the Han dynasty were buried in full suits of jade plaques stitched with wire because it was believed that the soul could return to the body only so long as it remained preserved. Jade cicadas, symbols of reincarnation, were placed in the mouth of the deceased, jade pillows cushioned their heads, and jade pigs, symbols of wealth, were clutched in their hands. Jade discs placed in tombs were decorated with dense, interlaced motifs of zoomorphic masks that resemble the *taotie,* the ogre face engraved on bronzes in the Shang and Zhou periods. The meaning and function of the discs remain uncertain, however.

Chinese medical manuals recommended swallowing powdered white jade for internal disorders and using jade needles (also replicated in porcelain) in acupuncture. In Daoism, the Jade Emperor, the link between the heavenly realm and the gods of the empire, is the counterpart of the worldly ruler, and the Jade Gate (Yumen) marked the Chinese entrance to the Silk Road. The Tang emperor Xuanzong (r. 712–56) so desperately wanted some renowned "five-colored" jade from Khotan that he supposedly dispatched an army of forty thousand men to seize it. The most momentous political ritual of the Tang dynasty involved the Son of Heaven climbing Mount Tai in Shandong province to sacrifice to Heaven and Earth, a ceremony affirming his mandate to rule and climaxing with him burying a stone coffer containing jade tablets wrapped in gold cord.

Jade thus represented perpetuity, nobility, supremacy, and the sacred. In addition, as a consequence of their exceptional molecular density, jade and high-fired pottery are resonant when struck. From ancient times, the chiming of the materials conveyed spiritual significance. Pottery bells were used for ritual occasions in the early third millennium B.C.E., to be replaced by bronze bells several centuries later. Musicians in the Shang cajoled ancestral spirits to communal feasts by playing melodies on bronze and pottery bells. Music also mollified the spirits of the privileged dead: in the fifth century B.C.E., an aristocrat from Hubei province went to his tomb with a magnificent array of sixty bronze bells.[126] Musicians playing on pottery and bronze bells sometimes were accompanied by others performing on jade chime stones. Some wealthy Tang monasteries possessed jade bells, as indicated by one poet's tribute to a temple in Zhejiang: "The mountain sings—with bells of Khotan."[127] Zhao Xigu (fl. 1220), a Song connoisseur, wrote that the tinkling of jade pendants is akin to being transported to "the Bright Garden's Jade Pool" in the Land of the Immortals.[128]

Lute players attributed the exquisite "jadelike" sound of their instruments to the powder of the stone that was used to mend cracks in them. The art of playing on celadon cups from Longquan, "the tones of which surpassed those of the hanging musical stones of jade," became a stylish diversion among the upper stratum in the Tang and Song.[129] Literati of the Ming delighted in the melodious chime given forth when one porcelain chess piece struck another, an effect otherwise possible only with exorbitantly costly jade. Dentrecolles describes an instrument called a *yunlo:* a set of nine small, slightly concave porcelain plates suspended in a frame on which one plays with mallets like a dulcimer. Soon after porcelains started to reach Europe in large numbers, Westerners discovered a like amusement. John Evelyn (1620–

1706), a member of the Royal Society and a connoisseur of the arts, reports in his diary that in an Amsterdam tavern in 1641 he saw displayed as a rarity "a chime of Purselan dishes, which fitted to clock-worke, sung many changes and tunes without breaking."[130]

Porcelain was the only material that could replicate the sound, shades, sheen, and unctuous appearance of jade. Pouring, dipping, and brushing, Song potters applied numerous glaze layers to their wares, creating a shell as much as ten times thicker than was customary in the Tang. Remarkably, sometimes they made the glaze thicker than the body of the porcelain vessel itself. Milled coarsely, the burnt-lime glaze imparted a translucent, jadelike texture to the piece; milled finely, it created a more transparent effect, deepening the color of the fired clay. In either case, the glaze layers, which enfolded countless quartz particles and tiny gas bubbles, gave a luminous glow to the porcelain as a consequence of the scattering and bending of light.[131]

Song connoisseurs likened such effects to that created by shimmering ice, snow, or silver; they extolled the vessels for having "deep pure colour like beautiful jade."[132] They called jade of the finest quality "mutton fat," so they naturally described porcelain glazes as having the glossiness of pig or chicken fat, "like lard dissolving but not flowing."[133] In his poem *The Secret-Colored Bowl Remaining from the Gifts to the Emperor,* Xu Yin (fl. 890) struck a loftier tone in paying tribute to an outstanding celadon vessel:

> Newly glazed in auspicious jade-like colors,
> The finished bowl was first offered to my lord.
> Skillfully molded like a full moon dyed with spring water,
> Deftly turned like a swirl of thin ice holding green clouds,
> Like a moss-covered ancient bronze mirror present at this occasion,
> A tender, dew-soaked lotus leaf parted from the river's edge.[134]

THE CREATION OF JINGDEZHEN PORCELAIN

As early as establishment of the Tang dynasty, kiln owners in Jiangxi sent gifts of what they called "imitation jade" *(jiayu)* to the imperial court.[135] According to a Song writer, artisans there excelled at making porcelains with shades that could "be likened to those of icy jade."[136] The ceramics also were called "jade from Rao" because wares shipped down the Chang River reached Lake Boyang at that town. During the Tang, Jingdezhen was known as Xinping, and it thrived as a market center rather than as a maker of pottery. It received its permanent designation under the Song emperor Zhenzong (r. 998–1022), who renamed it Jingde (his reign title for several years) and officially declared it a trading town *(zhen).* The river port acted as a staging post for pottery produced in more than a dozen outlying villages, some as far away as sixty kilometers. During the Song, a visitor to the district recorded seeing "kilns connecting one village to another, kilns firing everywhere."[137] Production at the time involved seasonal labor, with men farming half the year, tilling land belonging to the kiln owners, and devoting themselves to potting and firing from April to November.

Porcelain from northeastern Jiangxi attracted the attention of the imperial court during the Northern Song, leading to establishment of an official ceramic repository in Jingdezhen.[138] The area specialized in white ceramics that were glassy and sugary in texture, jadelike wares so fashionable that competing kiln centers on the southern coast copied them. The commodity received more substantial backing from the court during the Southern Song, when potters also began to focus on overseas markets. With the court calling for an increase in exports, artisans began to make some wares exclusively for foreign customers. Small jars in the form of a star-shaped fruit *(balimbing)* were made for the Philippines, and *kendi,* bulbous jars originally based on an Indian vessel, became popular items throughout maritime Southeast Asia, where shamans incorporated them into divination rituals.[139]

Imperial patronage and maritime trade boosted Jingdezhen's importance as a marketplace and carrier at the same time the city started to lay the foundations for its transformation into an industrial center. In the Southern Song, brokers set up shop there, taking advantage of an improved canal network to supply distant clients. In the Yuan and Ming periods, kiln owners began to shut down in outlying villages and relocate to Jingdezhen. The increased volume of output made river carriage essential, while hundreds of years of mining and forest removal made it difficult for local resources to sustain high production levels. Clay and timber increasingly had to be floated downriver from Anhui province to the porcelain city. Demands for large amounts of pottery by early Ming emperors led to establishment of an imperial depot in Jingdezhen for supervising the manufacture and shipping of wares to the court. With the rise of the city as a porcelain producer from the Southern Song to the Ming, villagers migrated there looking for employment, potters devoted themselves to their craft on a full-time basis, government supervisors organized employees into guilds, kilns began to specialize in certain vessels, and workshops adopted mass-production techniques across the board.

In the course of this long industrial transformation, the basic ware associated with Jingdezhen also changed. From the late tenth century, the main line of production in Jiangxi was known as *qingbai* (blue-white). According to a Tang scholar, *qingbai* originally referred to "the colour of white jades with delicate undercurrents of pale blue." In the Song period, the term exclusively identified Jingdezhen ceramics with a white body and blue-tinted glaze. A Qing connoisseur praised antique *qingbai* for being "blue as the sky, bright as a mirror, thin as paper, and tinkling like a chime."[140] By the last years of the Song dynasty, however, the pottery was losing those distinctive traits. Miners had reached deep deposits of china stone that lacked alumina, the mineral oxide that permitted the ceramic to withstand high firing. To maintain the quality of their product, potters began adding kaolin (or china clay) to the china stone, thereby restoring the percentage of the mineral required.[141] (See figure 12.)

It was a turning point in the history of pottery, for the compound recipe proved vastly superior to anything before. The fusion of china stone and kaolin generally is still regarded as the defining characteristic of "true" porcelain, compared to which other formulas are considered mere imitations. By the end of the Southern Song, this also became the Chinese perspective, especially as other pottery centers, such as the Guangdong and Longquan kilns, took second place to Jingdezhen or closed altogether. A seventeenth-century Chinese treatise declared that porcelain is made from both "glutinous rice" (china stone) and "non-glutinous rice" (china clay), the former bestowing a fair complexion and the latter delicate bones.[142] Dentrecolles relates that a Chinese porcelain vendor told him that a Dutch trader once obtained a sample of china stone to send to Amsterdam for analysis, but since he did not include china clay as well, his countrymen naturally failed to make porcelain: "The Chinese merchant said to me with a laugh, 'They wanted to have a body without bones to support its flesh!' "

Adding kaolin to china stone permitted potters to raise the kiln temperature to above 1,300°C. This led to fusion of the china stone and kaolin, greatly augmenting the translucency and solidity of the ware. Around the same time, potters enhanced the whiteness of the final product by creating a new, colorless glaze that eliminated the bluish tinge typical of *qingbai*. The brilliant white surface of the porcelain thus presented new potential for decoration, and the superior strength of the clay body meant that larger, more elaborate vessels now could be produced. The workers of Jingdezhen had created a material that would transform ceramic traditions around the world. Those changes began, however, only after they had adopted a significant innovation from the potters of Southwest Asia.

5

The Creation of Blue-and-White Porcelain

Muslims, Mongols, and Eurasian Cultural Exchange, 1000–1400

From the perspective of Dentrecolles, the most celebrated wares of Jingdezhen presented something of a mystery. Learning from local annals that "people here in times past made only white porcelain," he wondered how it came about that in his day "one hardly sees any in Europe except those which have a vivid blue on a white background." When he questioned his parishioners about the origins of the coloring, they related a tale about a porcelain merchant who was shipwrecked on a remote island, where he discovered that "rock fit for making the most beautiful blue was quite common; he carried away with him a big load, and they say that Jingdezhen never had seen so lovely a blue. He later searched in vain for the coast where luck once had sent him."

Dentrecolles did not subscribe to this fable, and the source of Jingdezhen's "beautiful and celestial blue" (as local annals described it) remained as obscure to him as other aspects of the history of porcelain. The answer to this mystery is suggested by Xiang Yuanpian, a calligrapher and artist of the late Ming, who described a fifteenth-century porcelain cup modeled on a jade vessel of the Han dynasty: "The glaze is of a uniform translucent white, like mutton-fat or fine jade, rising in minute millet-like tubercles, and the blue so pure and brilliant as to dazzle the eyes, being painted with Muslim blue."[1] The pigment, then, had associations with the world of Islam. The Chinese term for the religion is Huijiao (the Hui sect) and Muslims are known as Huihui or Huizu. The words came into Chinese usage in the Song period, perhaps deriving from the Chinese name for the Uighurs of Central Asia, a people who converted to Islam in the years after the Tang lost control of the Silk Road. "Muslim blue" is *huihui qing,* sometimes called *sumali* or *suponi qing,* variations on the Chinese transliteration of the Arabic *samawi,* "sky-colored" or "azure."[2]

"Muslim blue" points to the cultural and economic interdependence of the ecumene from the time of the Song. Blue-and-white porcelain developed over time as a consequence of the impact of Song ceramics on Southwest Asia, and Muslim merchants from that Islamic heartland shipped cobalt pigment to Jingdezhen, where, during the last generation of the Yuan dynasty, potters began using it to decorate wares for domestic and international markets. This stimulated a far-reaching development in art: during the late Yuan and early Ming, the design traditions of China and Southwest Asia came into greater contact than ever before, with the free-flowing spatial organization of the Chinese tradition and the geometric precepts of the Southwest Asian tradition meeting and learning from each other. Blue-and-white porcelain achieved its full artistic development in the Ming period, during which it comprised the bulk of the export trade in ceramics. It went on to triumph far and wide, reshaping (and sometimes destroying) pottery traditions in virtually every society it touched, from the Philippines to Portugal.

The creation and success of the new porcelain style demonstrate that long-distance exchange around the turn of the first millennium integrated the ecumene into a world system, a network of overlapping economies. The most significant part of that system was China, while Islam provided the cultural foundation for the principal trading diaspora that circulated the products of the system. After making his extraordinary journey from Morocco to the Middle Kingdom, moving chiefly from one Muslim community or kingdom to another, Ibn Battuta concluded that "there is no people in the world wealthier than the Chinese."[3] Given its population of 140 million in the Song, as well as the attractions of its manufactured products, China exerted enormous leverage when it turned to the wider world, having an effect on markets as far away as the commercial fairs of northern France and Burgundy.

CHINESE TRADE AND SOUTHWEST ASIA

During the Song, Chinese overseas trade and Islamic communities in Chinese ports recovered from the sharp decline triggered by the An Lushan rebellion in 755. Song emperors encouraged overseas exchange, and their counterparts on the Tigris and Euphrates Rivers, the Abbasid caliphs, launched gigantic construction projects that spurred the economy and attracted long-distance commerce. Al-Mutasim (r. 833–42) moved the court to Samarra, 130 kilometers upstream from Baghdad, where he built a palace complex many times larger than Louis XIV's Versailles. Within a few decades, the new capital sprawled along the Tigris River for thirty-five kilometers.[4] International maritime traffic revived, conducted mainly by Chinese merchants, who supplanted Arabs as the dominant businessmen trading between the Middle Kingdom and the Indian Ocean. Yet the future of the reconstituted cosmopolitan order depended principally on Muslim merchants from diverse ethnic groups—Egyptian, Arabian, Persian, East African, Indian, and Southeast Asian. Muslim and Chinese merchants did not constitute entirely separate categories, however, for Chinese converts to Islam and Muslim families resident in China for generations also took an active role in seaborne trade.

Although Muslim communities in the ports of China organized under their own religious law, in their overseas exchange they fell under imperial regulations enforced by maritime trade commissioners *(tiju shipo),* an office created under the Tang. The commissioners collected customs duties, directed tribute collection, managed state monopolies (on imports such as coral and ivory), and superintended merchant vessels. In Canton, Muslims built a grand mosque with a minaret from which flags and fire beacons guided the arrival and departure of oceangoing junks. In Quanzhou, they had six substantial mosques, including one financed by a merchant from Siraf. A shrine known as "the Holy Tomb" supposedly contained several of Muhammad's disciples, believed to have arrived in the city a few years after the death of the Prophet.[5]

According to a Quanzhou magistrate of the eleventh century, "this prefecture has developed overland communication routes leading to every corner of the empire. Maritime merchants crowd the place. Mixing together are Chinese and foreigners."[6] Muslims made up most of the latter, and the size of their community in the Song era has been disclosed by excavations of a large cemetery outside the city, where Arabic phrases in Kufic script are inscribed on tombstones. Muslim merchants, most notably the Pu family, controlled fleets of ships and dominated the office of maritime trade commissioner. Chinese copper-cash found by archaeologists in the ports of the Persian Gulf probably often reached there in vessels owned by Muslims living in Chinese coastal cities.[7]

A Song geographer described the Gulf as "the Sea of China," testimony to the enterprise of Chinese and Muslim merchants who brought commodities to Siraf and nearby harbors, transfer points for goods taken inland through the Zagros Mountains to towns in the Persian provinces of Fars and Kirman. Ships also went 355 kilometers farther north to Basra, located on the delta of the Tigris-Euphrates, the route to the great cities of the caliphate. The historian Al-Tabari quotes a caliph boasting, "This is the Tigris; there is no obstacle between us and China; everything on the sea comes to us on it."[8] Shards of Song porcelain, mainly from tableware and perfume bottles, have been excavated in the ruins of palaces and harems at Samarra, and earthenware fragments of local wares reveal that Mesopotamian potters copied the shapes and glaze tints of the Chinese imports.[9]

Great fortunes could be earned shipping Chinese products to Southwest Asia. According to a contemporary Islamic account, Ramisht of Siraf (d. 1140), a wealthy merchant based in Aden (in southwestern Arabia), "removed the silver water-spout of the Ka'aba and replaced it with a golden one, and also covered the Ka'aba with Chinese [silk] cloth, the value of which cannot be estimated."[10] Entrepreneurs such as Ramisht typically had several ships at sea, and their profits depended on painstaking attention to the fortunes of dozens of ports and polities from China to the Persian Gulf. But few merchants actually traveled the entire distance, a voyage of some one hundred days with favorable winds. Rather, short-range passages and local markets were woven into a trading network connecting the littorals of maritime Asia.

In the decades around 1000, long-distance merchants shifted their resources from the Persian Gulf to the Red Sea. Destroyed by an earthquake in 977, Siraf never regained its former eminence because flagging commercial revenue made reconstruction impossible. Economic crisis and bankruptcy plagued Basra, Baghdad, and Samarra; caliphal weakness and political turmoil made the passage of caravans uncertain and hazardous. Muslim Turkish warriors from Central Asia invaded Mesopotamia and established their own regimes, the most outstanding of which was that of the Seljuqs (1038–1194). Ports on the Persian Gulf fell victim to the inland disorder, losing their key role of importing goods destined for the Mediterranean, the Tigris-Euphrates, and Transoxiana. Government revenue from import fees declined, and no regime in the Gulf area minted silver coins for two centuries after 1060. A crushing blow came from invading Mongols, who finally toppled the Abbasid Caliphate in 1258.[11] The dynasty of the Mongol Ilkhanids of Persia (1258–1353) eventually restored stability and a measure of prosperity, attracting seaborne trade back to the region; but an immediate result of Mongol devastation was to reinforce the advantages of the Red Sea over the Persian Gulf. With the sacking of Baghdad, some merchants and artisans migrated to Cairo.

Establishment of the Fatimids (969–1171) and Ayyubids (1171–1250) in Egypt further impelled a shift in trade from the Gulf to the Red Sea. As much as Egypt benefited economically from instability in Iraq and Persia, however, its own swelling population and its government's need for revenue would have attracted the commerce of the Indian Ocean in any event. Far more than the Abbasid caliphs, the Egyptian sultans relied on income from customs duties; hence they promoted the expansion of commerce, a policy the successor regime of the Mamluks (1250–1517) pursued even more energetically. Egypt also profited from an upsurge of activity in the merchant republics of Italy, such as Venice, Genoa, and Pisa. Rising to prosperity and power on trade in Asian spices purchased in Egypt, they were pacesetters in the commercial revolution of the European Middle Ages.[12] They led the way in the revival of sophisticated urban life and long-distance exchange, restoring the Mediterranean to its ancient role as a major commercial highway, a position it had lost with collapse of the western Roman Empire in the fifth century. By the eleventh century, then, as a consequence of economic activity booming at both ends of Eurasia, Egypt became the great intermediary for exchange in the ecumene, the pivot of trade between the Mediterranean and East Asia.

THE EXPANSION OF ISLAM IN MARITIME ASIA

Creation of a world system after 1000 was a consequence of Chinese commercial expansion to the Indian Ocean and the spread of Islam into India and maritime Asia. In terms of world history, the latter development was more important and enduring. Chinese junks did not sail beyond Melaka after the early fifteenth century because that port, whose emergence at the time owed everything to Chinese patronage and protection, came to function as a site where traders from the Indian Ocean and the Middle Kingdom exchanged merchandise. *Melaka* derives from the Arabic word for "rendezvous" or "meeting place," terms that fittingly apply to the gathering of either merchants or monsoons. Located in the straits formed by the Malay Peninsula and the long, narrow island of Sumatra, the port is protected from the brunt of storms from the southwest and northwest; it also enjoys a lengthy period of equatorial calm between monsoon seasons. Ships from the Indian Ocean arrived there in April, around the time the Chinese departed; when the latter returned six months later, the Indians and Arabs set off for home. For Chinese merchants, Melaka served as a convenient emporium, sparing them the need to make the long, costly voyage to India.

In contrast to the retreat of Chinese vessels from the Indian Ocean after 1433, adherents of Islam pushed on to the east and south from the homeland of their religion, making the five centuries after 1000 the greatest period in the geographic extension of the faith. The Ghaznavids (1001–1186), Muslim Turks from Afghanistan, conquered northern India, and their successors established the Delhi Sultanate (1206–1526), which ruled the Indo-Gangetic plain by the early thirteenth century.[13] Islamic expansion elsewhere proceeded with considerably less belligerence. From around 1000, communities of Muslim merchants established themselves in ports along the Swahili coast of East Africa and on both the western (Malabar) and eastern (Coromandel) coasts of India, creating a diaspora of trading communities. Their expansion represented a search for more secure sources of profit inasmuch as the Abbasid caliphate and northern India were experiencing severe political upheavals that disrupted long-distance commerce.

By the fourteenth century, Muslims dominated overseas trade in Indian harbors, especially as prohibitions against sea travel, with its attendant caste pollution, gained greater force among Hindu merchants involved in maritime affairs.[14] In Calicut on the coast of Malabar, the principal center of trade in black pepper, the Hindu ruler styled himself "Lord of the Sea" (Zamorin) and worshiped Kapalotta, the goddess of navigation, but he merely monitored and taxed the Muslim merchants who actually controlled the day-to-day business of shipping. Ludovico di Varthema (ca. 1465–1517), an Italian merchant, observed that "the pagans [i.e., Hindus] do not navigate much, but it is the Moors who carry the merchandise; for in Calicut there are at least fifteen thousand Moors, who are for the greater part natives of the country."[15] Tomé Pires likewise recorded that "all the merchants of Malabar who trade on the sea are Moors, and they have the whole of the trade."[16]

Muslims from Gujarat (in northwestern India), Malabar, Coromandel, and Bengal ventured into maritime Southeast Asia in the thirteenth century, bringing their trading contacts and religious devotion. Within a hundred years, a substantial Muslim population in the Archipelago made it profitable enough to import tombstones—a funerary memorial not part of indigenous traditions—made of fine cream marble and decorated with panels of Kufic calligraphy, from as far away as Gujarat. The oldest Islamic tombstone in northern Sumatra dates from 1320, the oldest in Java from 1376. Muslim traders from Calicut introduced the black pepper plant *(Piper nigrum)* to Sumatra and Java, and foreign Muslims held office as harbormasters *(shabandars)* in Southeast Asian ports.[17] Ibn Battuta noted that the piety of Muslims in Sumatran courts compared favorably with that of fellow believers in the long-established Islamic communities of Southwest Asia. Influenced by both Sumatran precedents and Muslims from China, the Hindu-Buddhist ruler of Melaka converted to Islam in 1414. Expansion of the religion thereafter went along with extension of Melaka's trading connections in the Archipelago. According to Pires, one early-sixteenth-century Melakan ruler even proposed that his city "be made into Mecca, and that he would not hold the opinion of his ancestors about going [on pilgrimage] to Mecca."

Chiefs of the northern ports of Java converted to Islam in the early fifteenth century. Pires notes that Muslims from Arabia, Gujarat, Persia, and Bengal

> began to trade in the country and to grow rich. They succeeded in way of making mosques, and mollahs came from outside, so that they came in such growing numbers that the sons of these said Moors were already Javanese and rich, for they had been in these parts for about seventy years. In some places the heathen Javanese lords themselves turned Mohammedan, and these mollahs and the merchant Moors took possession of these places.[18]

Tradition maintains that Malik Ibrahim (d. 1419), the first Muslim apostle of Java, earned a living as a merchant and served as harbormaster on the northern coast. A century later, a coalition of Muslim coastal polities launched an attack on Majapahit (ca. 1290–1528), the Hindu-Buddhist kingdom in the interior of Java, eventually forcing the royal family to flee to the island of Bali, immediately to the east of Java. Unlike the disunited and vulnerable kingdom of Majapahit, Bali in the early sixteenth century experienced political consolidation and religious reform under powerful Hindu-Buddhist rulers; they used their connection with Javanese royalty to shape an ideology justifying strong monarchy.[19] As a consequence, the wave of Islam swept past Bali, leaving it the only sizable outpost of the old faith in Indonesia.

Around the same time Islam began making advances into the heartland of Java, Muslim holy men and merchants from the sultanate of Brunei on Borneo brought their faith to Mindanao in the Philippines, the southernmost point of that insular chain. The Spanish arrived in the Philippines in 1565, however, and they plundered Muslim settlements on the central island of Luzon soon after.[20] The profusion of porcelain discovered by Spanish soldiers in the home of a Muslim chieftain astounded them and seemed to herald future riches. Legazpi reported that on Chinese junks captured near the island of Mindoro "the decks of both vessels were full of earthen jars and crockery; large porcelain vases, plates, and bowls; and some fine porcelain jars, which they call *sinoratas*."[21] With a secure base of operations on the American mainland and subsidized by American silver, Spanish forces headquartered in Manila subjugated the Philippines and Christian friars began to convert the native peoples. The Spanish governor warned the Brunei sultan to stop propagating Islam in Luzon and the southern Philippines.

Thus by the narrowest of margins, in an unwitting race with a hostile faith across half the globe, Islam failed to win over all of maritime Southeast Asia. As Juan González de Mendoza (ca. 1540–1617) wrote in his *History of the Great Kingdom of China* (1585), "Our Lord has had great mercy, sending them the remedy for their souls in so good season; for, had the Spaniards delayed a few years more, all the natives would now be Moors."[22] The Philippine Islands came onto the global stage as the remotest satellite of Christendom, a bastion of Western power and commerce. The better part of a century after Columbus set sail from the Gulf of Cádiz for the fabled land of Cathay, inspired by *The Travels of Marco Polo* and intent on outflanking the Muslim powers of Southwest Asia, the Spanish at last had drawn within reach of what he described as the riches and "manifold marvels" of China.[23]

Despite the Spanish advance to the northern margin of the Archipelago, however, Islam eventually became the faith of the majority in maritime Southeast Asia. Muslims adapted as readily to the harbor principalities of Sumatra, Java, and Borneo as they did to the inland dominion of Majapahit and the bureaucracy of the Middle Kingdom. Their religion made for a practical fusion of cultural flexibility and confessional solidarity, a combination that enhanced their commercial enterprise in unfamiliar settings. All Muslims regarded themselves as belonging to the *umma*, the universal community of the faithful. Everywhere believers gathered, they shared social norms and practices, such as alms-giving, dietary restrictions, and public prayer; they took pride in having a lettered high culture, a common canon of learning, and a collective background that transcended parochial loyalties.[24]

Pilgrimage to Mecca, or hajj, one of the obligations of the faith, joined believers from diverse cultures in a stirring collective ritual. As witnessed by Ibn Battuta, a host of pilgrims came to worship in the holy city, "so many that the earth surged with them like the sea and their march resembled the movement of a high-piled cloud."[25] They carried banners embroidered with sacred texts in Arabic calligraphy, the all-embracing script of the faith. Pilgrims used the hajj as an occasion to take part in an immense, short-term market fair at Jedda, the port not far from

Mecca on the coast of the Red Sea. Muhammad had been a trader before taking the mantle of a prophet, and believers quoted him as saying that "the merchant enjoys the felicity both of this world and the next."[26] The high esteem accorded traders in Islam and a body of commercial law shaped by the Qur'an bestowed coherence and legitimacy on a widespread maritime culture.

In going from his hometown of Tangier to Beijing, Ibn Battuta traveled almost entirely within what his fellow believers called the "Abode of Islam" (Dar al-Islam). The intercontinental reach of the Islamic *umma* made for a remarkable flow of communication, commodities, and pilgrims. Ibn Battuta met the same religious scholar from Ceuta, a town close to Tangier, in both northern India and coastal China; he later encountered the man's brother near the Niger River in sub-Saharan Africa. He considered it providential that an expensive robe of goat hair that he owned in Bengal ended up in the hands of a Muslim holy man in Beijing. The pilgrim from Morocco served as a judge in Islamic courts in the Maldive Islands, relied on gifts and loans from Muslim merchants in Malabar ports, received hospitality from Muslim notables on the Swahili coast, and recited the Qur'an with worshipers everywhere he went. Significantly, only in China, which had promised to be the climax of his journey, did his passion for novelty finally peter out:

> The land of China, in spite of all that is agreeable in it, did not attract me. On the contrary I was sorely grieved that heathendom had so strong a hold over it. Whenever I went out of my house I used to see any number of revolting things, and that distressed me so much that I used to keep indoors and go out only in case of necessity. When I met Muslims in China I always felt just as though I were meeting my own faith and kin.[27]

Although Ibn Battuta condemned China as a land of infidels, at least its southern ports made him feel at home. He regarded Quanzhou as a cosmopolitan city when he stayed there for a time in 1342. Its residents used Persian as a lingua franca, and Chinese administrators identified the substantial population of foreigners, especially Arabs, Persians, and Indians, as "people with colored eyes" *(semuren)*.[28] Muslims controlled their own quarter of the city, and Ibn Battuta remarks that calls to prayer from the minarets of mosques sounded through the streets. According to the traveler, Muslims of Quanzhou became so elated when a fellow believer arrived from abroad—"They say 'He has come from the Abode of Islam' "—that "they make *him* [rather than orphans or the poor] the recipients of the tithes on their properties, so that he becomes as rich as themselves."[29]

In Canton, also a harbor of diverse cultures, Ibn Battuta strolled through the markets, the largest of which was the porcelain bazaar. He expressed astonishment that high-quality vessels could be purchased for less than the cost of lackluster earthenware in Tangier. Naturally, he knew from personal experience that merchants of his faith circulated porcelain throughout the ecumene. It is exported, he wrote, "to

India and other countries, even reaching as far as our own lands in the West, and it is the finest of all makes of pottery."[30]

SONG PORCELAIN AND SOUTHWEST ASIAN POTTERY

Chinese whitewares and celadons came as a revelation to Southwest Asia, especially as the region previously knew only utilitarian, generally unglazed pottery. The foreign vessels seemed to have a jewel-like, magical quality. The earliest extant Southwest Asian description of porcelain accentuates its unique character: in *Accounts of India and China* (Akhbar al-Hind wa'l Sin [851]), an Arab merchant named Sulayman, who had traveled to the Middle Kingdom, wrote that "there is in China a very fine clay from which are made vases having the transparency of glass bottles; water in these vases is visible through them, and yet they are made of clay."[31] Chinese pottery retained its reputation for the marvelous down the centuries. In Samarqand, a generation after Ibn Battuta's visit, a Chinese envoy to the court of Shah Rukh (r. 1405–47) of the Timurid dynasty (1378–1506) of Persia, observed of Persian imitations of porcelain that "they are very beautiful but they do not match the light, blue, clear and sparkling ones of China. If such a vessel is hit, it makes no sound. The nature of clay is like that."[32]

Of course, it is more accurate to say that the nature of Southwest Asian clay is like that. In striving to imitate porcelain, potters of the region fell back on an ancient tradition of compensating for inferior resources with artistry and ingenious substitution. Lacking good stone for monumental architecture, builders instead employed mud-bricks and carved, gilded, and painted stucco.[33] Lacking abundant timber for firing kilns, potters turned to dried grass, weeds, straw, and animal dung. Lacking kaolin and china-stone for pottery, craftsmen in ninth-century Basra developed a pottery body made of crushed quartz, white clay, and ground glass (frit). The new ceramic recipe, now known as fritware, turned out an unusually hard earthenware product that did not require firing at a very high temperature. Brittle, stiff, and difficult to work with, the material made possible very thin vessels and, like porcelain, could be molded into fantastic shapes.[34] It became a standard material for fine pottery in many parts of the Islamic world.

Another innovation proved among the most significant in the history of pottery. Since the low-firing earthenware of Southwest Asian potters could not duplicate the lustrous white surfaces of porcelain, artisans also developed a novel glaze technology around the same time fritware emerged. They simulated the effect of Chinese whitewares by adding tin oxide as an opacifier to a clear glaze: a fine cloud of tin-oxide particles diffused through the lead-glaze coating, covering up the brownish tones of the earthenware with a soft, matte white.[35] Practical and profitable, the technology of tin glazing shaped the development of pottery in the ecumene for centuries. It became the standard technique in Southwest Asia for hundreds of years. From the thirteenth century, it swept through Europe, generating the emergence of maiolica, faience, and delftware; it reigned supreme there until the advent of Chinese porcelain and its Western counterparts in the period between Vasco da Gama and August the Strong.

With the decline of trade in the Persian Gulf in the tenth century, Fustat (Old Cairo) in Fatimid Egypt emerged as one of the largest cities in the Islamic world and a center of trade and industry. The geographer Al-Muqaddasi (d. 1000) declared that "the City of Peace [Baghdad] cannot compare to it in greatness. It is the treasure-house of the West and the emporium of the East."[36] Fustat paid high prices for porcelains as they had to be carried overland 1,100 kilometers from Red Sea ports, a trek that inescapably raised costs. Egyptians referred to the imports as *hindi* (India-ware), a usage similar to *a louça da India,* the phrase used by the Portuguese centuries later. Twenty percent of 700,000 pottery shards excavated at Fustat are Chinese in origin, with Longquan celadons dominating through the thirteenth century, after which Jingdezhen blue-and-white becomes most common. Craftsmen repaired tens of thousands of Song whitewares with copper wire and iron clamps, an indication of how much owners valued the pottery. Virtual mountains of earthenware fragments also reveal the extent to which Egyptian potters ingeniously tried to imitate the Chinese white vessels.[37]

Because tin-glazed pottery lacked the gleaming appearance of white porcelain, Southwest Asian potters enhanced it with painted decoration. Indeed, monochromatic Song ceramics ran counter to a venerable regional tradition of color and decoration on glass and pottery. Around 4500 B.C.E., Egyptians imitated the prized mineral lapis lazuli by using cobalt oxide and copper to dye soapstone blue, a color deemed magical, representative of life and resurrection. Centuries later, they also replicated the bright red of carnelians and the rich blue-green of turquoise stone. When Egyptians discovered how to make glass around 2000 B.C.E., they naturally turned to cobalt as a coloring agent. In the Eighteenth Dynasty (ca. 1570–1293 B.C.E.) of the New Kingdom (ca. 1570–1070 B.C.E.), artisans used cobalt oxide to decorate pottery as well as tiles. Nebuchadrezzar II (r. 604–562 B.C.E.) of the Neo-Babylonian Empire (625–539 B.C.E.) of Mesopotamia built the turquoise-colored Gate of Ishtar in Babylon, the grandest ancient achievement in the art of glazed brickwork. The succeeding empire of the Achaeminids (559–330 B.C.E.) continued the tradition of using colored brickwork for palaces and monuments, though the regime's conquest by Alexander the Great (336–323 B.C.E.) of Macedonia and Greece ended the practice until its revival in the Abbasid period. In the fifth century C.E., Egyptian and Mesopotamian glassmakers applied colors to their wares derived from copper and silver, but potters seldom glazed works other than those used for ceremonial purposes.[38]

Fustat potters in the tenth century adapted the ancient glass-tinting methods to their novel tin-glazed vessels. Known as luster-glazing, the technique involved painting decorations on the surface of a glazed vessel with a compound of silver and copper; after a brief, third firing in a smoky kiln, the metals diffused in an extremely thin film on top of the glaze. Subsequent polishing of the vessel produced an iridescent sheen of hues ranging from a coppery-red bronze to a pale lemon yellow.[39] In a treatise written in 1301, Abu'l Qasim al-Qashani, scion of a prominent tile-making family of Kashan (Persia), declared that the glaze of lusterware "reflects like red gold and shines like the light of the sun."[40] In the Seljuq period, when precious metal became scarce due to a worsening economy, lusterware appealed to moneyed customers who desired table settings with familiar, glittering sun colors. Moreover, Qur'anic tradition condemned those who ate and drank from silver and gold, so lusterware functioned for some Muslims as a stylish substitute for plate of precious metal. Scholars quoted Muhammad as saying, "He who drinks from a silver vessel will have hell-fire gurgling in his belly."[41] They also counseled abstaining from costly metal to thwart lower-class indignation, for when the poor witness the opulence of the rich, "they get desperate at the thought that these have gold and silver plate when they cannot even get earthenware."[42]

At some Muslim banquets, Ibn Battuta noted, servants brought delicacies such as pomegranates "in vessels of gold and silver with golden spoons, and others in vessels of glass with wooden spoons."[43] He recorded that Muhammed bin Tughlaq (r. 1325–54) of the Delhi Sultanate used porcelains at dinners and receptions. Many scrupulous Muslims steered clear of precious metal utensils. Although the throne room of Süleyman I ("the Magnificent") (r. 1520–66) of the Ottoman Empire (1288–1918) boasted a stunning display of inlaid gold and jewels, the sultan and his viziers dined on blue-and-white porcelains and used wooden spoons. A century later, a French visitor reported that in the holy month of Ramadan, when believers had to fast during the day, high Ottoman officials used yellow porcelain bowls for their food when they feasted after sunset.

Still, most wealthy Muslims scorned severity at the table as Qur'anic proscriptions failed to eradicate long-standing associations of gold and silver with nobility, luxury, festivity, and wine drinking. Nor did privileged ranks fret that their conspicuous consumption might inflame the pottery-deprived masses. Despite his piety, Ibn Battuta clearly regarded it as a mark of his own consequence that powerful men of the Islamic world entertained him at sumptuous feasts served on gold and silver platters. When he rebuked a local potentate in Persia by pointing to his gold goblet—"there is nothing to be brought against you as a ruler but this!"—it was not because the man drank wine from a forbidden vessel but because he had become shamefully drunk.[44]

In addition to elite attachment to tableware of precious metal, practical considerations limited the scope of lusterware. Luster-glazing required costly metal compounds, but the technique proved difficult to control; hence wastage ran high, making the final product commensurately expensive. More important, the ware also proved virtually impossible to clean without damaging the very surface that made it attractive to purchase in the first place.[45] As a consequence, luster-glazed vessels never entered into everyday use but remained restricted to celebratory and ceremonial occasions, such as regal feasts and religious observances. In such settings, the shimmering wares, often decorated with images of the sun and stars, evoked identification with divinity and brought to mind the much-cited "light verse" of the Qur'an (24:35–36): "God is the light of the heavens and the earth. His light may be compared to a niche that enshrines a lamp, the lamp within a crystal of star-like brilliance."

THE MONGOLS AND EURASIAN UNIFICATION

As it turned out, the color of the sky rather than that of the sun and stars held the greatest promise for the future of ceramics. At the same time Egyptian potters developed luster-glazing, Persian craftsmen experimented with painting designs in cobalt blue on the white face of their tin-glazed products. They concentrated on blue as a coloring agent because cobalt oxide was readily obtainable in central Persia, where it appeared near the surface of the ground in flowery masses of pink or metallic-black crystals. Elsewhere, cobalt could be obtained only by mining, followed by hazardous methods of processing. *Cobalt* derives from the German *Kobald,* "goblin," because smelting of the ore produces toxic fumes of arsenic, regarded by silver miners in August II's Saxony as an emanation of malevolent, subterranean spirits. Persian artisans called cobalt *lajvard* in reference to the radiant blue characteristic of lapis lazuli.[46]

Persian potters had difficulty achieving good effects with cobalt, however, since it tended to run in the tin glaze and blur the painted designs. This technical problem became the stimulus for bringing the two great ceramic traditions of the ecumene into closer contact than ever before. In fact, significant change had been taking place in both traditions during roughly the same time: in the last decades of the Southern Song, craftsmen at Jingdezhen experimented with a new formula for their *qingbai* wares, while in Southwest Asia, under the stimulus of Song imports, potters created an innovative glaze and explored several original techniques of decoration. One of the most extraordinary events in world history provided the context in which these developments came together to create blue-and-white porcelain and foster the encounter of Chinese and Islamic design traditions.

In the early thirteenth century, the Mongol leader Genghis Khan (ca. 1167–1227) united the mounted archers of the Central Asian steppes into a powerful confederacy and launched their force against the centers of urban civilization. The Jin state in northern China fell to the Mongols in 1234, though the regime of the Southern Song held out for another generation. Campaigns in the west between 1218 and 1241 led to establishment of Mongol khanates in Central Asia and Russia, territories loosely integrated into what the conquerors termed the "Great Mongol Empire" (Yeke Mongol Ulus). In the early 1250s Hulegu (d. 1265), third grandson of Genghis, led a Mongol army of 200,000 men into Persia and Iraq, a force spearheaded by Mongols and Turks but also including Russian, Armenian, Georgian, and Chinese troops, as well as some European experts in siege-craft. Hulegu sacked Baghdad in 1258, killed the last Abbasid caliph, and founded the dynasty of the Ilkhanids, "the vassal khans."[47]

Mongol invasion meant pillage and destruction on a frightful scale, including extermination and enslavement of some urban populations; Samarqand lost three-fourths of its population of 100,000 households. When Marco Polo and Ibn Battuta passed through northern Persia many years after Hulegu's devastation, they found parts of the region still desolate and abandoned. Reflecting the political realities of the new Eurasian empire, Persia became oriented more toward China than toward Southwest Asia, while Iraq survived as a weak, vulnerable territory wedged between Egypt and the Ilkhanids. Except for Mamluk Egypt, which fought off Hulegu's troops in 1260, the people and culture of Southwest Asia suffered terribly. Of course, potters and ceramic traditions were no exception: old workshops disappeared, those that survived declined into provincialism, and sophisticated pottery such as lusterware died away.

Kublai (1215–94), the fifth Great Khan of the Mongols and the second grandson of Genghis, proclaimed the Yuan dynasty of China in 1272. After the Southern Song finally fell in 1279, the descendants of Genghis Khan ruled a loose-jointed dominion that stretched from Korea and Vietnam to Hungary and Russia, the greatest land empire in history, extending across an area of between 26 million and 31 million square kilometers, roughly the size of the African continent. They created a postal system across Eurasia that sped communications as never before, with 10,000 staging posts and 200,000 horses for couriers. Mongol troops guarded the network, guaranteeing secure passage for traders on the Silk Road, such as the Polo family. Francesco Pegolotti, a well-traveled Florentine who wrote *La practica della mercatura* (The Practice of Trading) in the late 1330s, around the same time Ibn Battuta journeyed to China by sea, declared that "according to reports by merchants who have used it, the route from Tana [at the head of the Sea of Azov] to Cathay is absolutely safe by day as well as night."[48]

Unprecedented cross-cultural exchange flowed from Eurasian unification under the Mongols, dwarfing that of the early Tang period. This was not a predictable result of the so-called Pax Mongolica, for the nomadic warriors had trouble establishing long-term peace, especially among themselves. Rather, cultural interaction stemmed from Mongol strategy inasmuch as the conquerors used the skills, technologies, and traditions of their subject peoples to advance their own command and riches. A Chinese writer in the fourteenth century boasted that "loyal, virtuous, brave and talented men from a multitude of places and myriad countries all willingly enter the emperor's service."[49] Perhaps that was true; but given the small Mongol population at the time of Genghis—a total of 700,000, including 100,000 horse warriors—in relation to the overwhelming numbers they conquered, the ruling elite of the Eurasian empire actually had no alternative but to depend on the self-interested collaboration of their subjects.

Ogodei (r. 1229–41), the second Great Khan, drew on the resources of Eurasia to construct and provision Karakorum, his capital city in central Mongolia, built from scratch on the Orkhon River, a good distance from trade routes. He commanded that five hundred wagons loaded with food and drink arrive in the city every day from China, a task that required sixty thousand teamsters working full-time. In his *History of the World Conqueror* (1260), Ata Malik al-Juvayni (1226–83), an Ilkhanid minister, declared that Ogodei's extravagance created a boom for traders from far and wide, "and whatever goods they had brought, whether good or bad, he would command them to be bought at the full price."[50] Not only that, the khan ordered that the merchants receive a 10 percent bonus on top of their handsome charges. Naturally, porcelain manufacturers counted among those who profited from this bonanza.

Mongols transplanted 100,000 artisans, including 20,000 captured at the sack of Samarqand, from Transoxiana to Karakorum and China. Chinese millet farmers moved to Azerbaijian, and groups of Central Asian Turkic speakers settled in both Persia and China. An account of the journey of Zhang Chun (1148–1227), a Daoist monk, from China to Southwest Asia noted that "Chinese craftsmen are found everywhere" in Samarqand and that the vessels of the inhabitants are "usually of brass or copper; sometimes of porcelain."[51] German and French workers seized in Hungary smelted metal and manufactured weapons in Karakorum. Persians served as garrison troops in Chinese harbors. Chinese cooks, engineers, physicians, and administrators traversed the Silk Road to Southwest Asia; Persian scribes, translators, architects, and carpet weavers went in the opposite direction. Chinese experts brought printing and gunpowder to the west; Southwest Asian craftsmen introduced distillation and sugar refining to China. Rashid al-Din (ca. 1247–1318), a minister of the Ilkhanids and a historian of the Mongols, asserted that as a consequence of God bestowing empire on the dynasty of Genghis, "philosophers, astronomers, scholars and historians from North and South China, India, Kashmir, Tibet, [the lands] of the Uighurs, other Turkic tribes, the Arabs and Franks, [all] belonging to [different] religions and sects, are united in large numbers in the service of majestic heaven."[52]

Distancing themselves and their people from the Confucian ideology of the Chinese elite, Kublai and his successors patronized Tibetan Tantric Buddhism, whose magical and supernatural aspects they found compatible with Mongolian folk religion. Determined to assert control over their imperial bureaucracy, Yuan emperors suspended civil service examinations and barred Chinese from office for decades. The Persian and Mongolian languages joined Chinese as official tongues of government. Muslims from Southwest Asia dominated financial and commercial policy, and they monopolized the office of maritime trade commissioner. Most strikingly, Sayyid Ajall Shams al-Din (ca. 1210–79), a Muslim from Bukhara in Transoxiana, won appointment from Kublai to govern the newly conquered region of Yunnan in southwestern China, a frontier province to which countless Muslims migrated. Sayyid displayed a religious eclecticism and tolerance typical of Mongol rule, founding mosques, subsidizing Buddhist monasteries, and supporting Confucian education. His son Nasir al-Din, who commanded the first Mongol invasion of Burma in 1278 (and thereby won mention in Polo's *Travels*) succeeded Sayyid as governor of Yunnan.[53]

After establishment of Mongol power across much of the ecumene, sovereigns began looking upon it as an arena for diplomatic stratagems, especially as Latin Christendom for the first time established direct contact with Central Asia. In the mid-thirteenth century Franciscan envoys of Roman popes journeyed to Karakorum, where they argued for an alliance of Mongols and Christians against the Mamluk regime in Egypt and Syria. In 1260 the father and uncle of Marco Polo traveled to China, where they joined a thriving community of Italian merchants. Kublai asked the Polos to have the pope send a hundred learned Christians to his capital—the kind of imperial suggestion that later would lead to Dentrecolles being sent to China.[54]

Kublai also sponsored a pilgrimage from China to the Holy Land by Rabban Sauma (ca. 1225–94), a Nestorian Christian monk; the Ilkhanid ruler Arghum (r. 1284–91) dispatched him farther west on a diplomatic mission to the kings of France and England. In 1287 Rabban Sauma buoyed the hopes of the College of Cardinals in Rome by telling them that "today there are many Mongol Christians," including offspring of the Great Khan and their wives.[55] Ghazan (r. 1295–1304), Arghum's successor and the first of his dynasty to embrace Islam, dreamed of an alliance with Christian monarchs against the Mamluks. In the half century before 1312, the Ilkhanids sent fifteen diplomatic missions to Europe seeking an accord, a strategy that led to the Mongols being transformed in Christian eyes from the apocalyptic Gog and Magog to prospective redeemers of sacred Jerusalem.

EURASIAN CULTURAL EXCHANGE IN THE AGE OF THE MONGOLS

Eurasian cultural exchange in the Mongol period extended to the realms of design and art.[56] *Sini*, as China was known in Southwest Asia, came to represent the standard of excellence in painting and other media. By the end of the thirteenth century, Chinese textile workers transplanted to Transoxiana contributed to the diffusion of East Asian motifs and forms to western Asia. Chinese and Central Asian cultural fashions and materials influenced a school of Persian artists established in Tabriz (Persia) by Rashid al-Din. The painting traditions of the regions facilitated artistic interaction inasmuch as they all employed flat colors, without shading or chiaroscuro. Persian painters of manuscript miniatures depicted foliage, mountains, and horizons in a style derived from China, though without the sense of flowing motion and spatial depth characteristic of Chinese design. They also adopted Chinese standards of beauty for the human countenance, such as a round face, arched eyebrows, rosebud lips, and dark, almond eyes.

As represented on the brocades of Central Asian weavers, the Chinese dragon took on a violent demeanor it never had in its birthplace, transformed from an awe-inspiring imperial symbol into a fire-snorting monster.[57] In like fashion, the phoenix, which in China symbolized the empress, turned belligerent on Southwest Asian illuminations and bookbindings, shown locked in mortal combat with dragons and griffins. In an early-seventeenth-century Mughal painting, the bird tackles the Gaja-Simha, a monster with a lion's body and elephant's head. The *qilin,* a Chinese mythical beast with the body of a deer, hooves of a horse, and tail of an ox, became transmuted in Southwest Asia into a sort of winged unicorn with a flowing mane. Persian artists also modified aspects of the *qilin* to portray the Simurgh, a creature from ancient Southwest Asian mythology with a stag's body, goose's wings, and cock's head.

Ornate cloud patterns from China, which stemmed from decoration on bronze vessels of the Shang and Zhou periods, appeared on Persian pottery and stone reliefs, transmitted by designs on Central Asian and Chinese fabrics. Persian metalworkers supplemented their customary griffins and sphinxes with dragons and phoenixes copied from Chinese embroidered silks and Mongol saddle decoration. In the fourteenth century, *The Seven Beauties,* the great Persian poem by Nezami de Gandjeh (1141–1209), received its first illustrations by painters in Baghdad and Tabriz: Bahram-Gour, the hero of the epic, is depicted slaying a Chinese-style dragon, a motif that subsequently became a staple of artworks in Mughal India, Ottoman Turkey, Safavid Persia (1501–1722), and Latin Christendom. An ancient Chinese adage held that the dragon must be honored "because it cannot be taken alive."[58] In western Asia, however, the only good dragon was a dead one.

Chinese vegetal patterns, the lineal descendants of ornamental motifs introduced to the Middle Kingdom from Persia in the Tang period, migrated back across the Silk Road and took root in Ilkhanid territory, refashioned over the centuries and enhanced by respect for things Chinese among the Islamic upper stratum. The motif of the lotus flower blossomed everywhere, ceaselessly reproduced on carpets, metalwork, stucco, bookbindings, tiles, and pottery. Indeed, Persia under Mongol rule witnessed the first appearance of chinoiserie, an artistic style reflecting Chinese influence as manifest in fanciful representations of Chinese culture.[59] Centuries later, around the time of Louis XIV, chinoiserie, especially as it appeared on pottery, architecture, and interior design, had an even more brilliant efflorescence in the West. Both Persians and Europeans could admire China so unreservedly because it was far enough away not to pose a political threat, yet in close enough communication to inspire imitation. In the age of the Mongols, however, Central Asian and Chinese themes had less direct impact on European art than on Southwest Asian, chiefly because porcelain reached the West in such trifling quantities.

Genoese and Florentine merchants bought large amounts of Chinese silk on the shores of the Sea of Azov and then sold it in Europe for three times its purchase price in China. Christian weavers and stonemasons copied the Chinese dragons and phoenixes embroidered on the fabric for decoration on textiles and cathedral facades. A painting of a dragon looked down from the wall of the choir in the cathedral of Notre-Dame in Paris, while Giotto (ca. 1266–1337) employed Mongol *Phags-pa* script as a decorative motif in the famed frescoes of the Arena (or Scrovegni) chapel in Padua. In masterpieces of the International Gothic style, the Limbourg brothers (fl. 1410), in the service of the dukes of Burgundy, portrayed figures in Central Asian and Turkish costumes, and in the early fifteenth century, the Bedford Master produced *Festivities at the Court of the Grand Khan,* showing a scene bustling with courtiers in Asian dress.[60]

Italian painters, such as Ambrogio Lorenzetti (d. 1384) and Antonio Pisanello (1395–1455), incorporated Central Asian costumes and figures in their works. In Giovanni del Biondo's *Martydom of Saint Sebastian* (ca. 1370) in the Duomo of Florence, Mongol archers take aim at the holy victim. Simone Martini (ca. 1284–1334), an influential artist of Siena, painted an altarpiece, *Saint Louis of Toulouse,* with the central figure wearing Persian cloth decorated with rings of gold and seated on a chair draped in Central Asian silk adorned with tiny representations of animals. In Martini's *Annunciation* (ca. 1333), the angel Gabriel wears a Mongol white-and-gold robe that is strikingly similar to vestments held in the treasury of Pope Boniface VIII (r. 1294–1303).[61] Mongolian gold-threaded silk fabrics *(panni tartarici)* had a significant impact on the Sienese textile industry, and Martini evidently developed new painting techniques in an attempt to reproduce the glittering patterns and textures of the exotic cloth. When Marco Polo died in 1324, an inventory of his possessions included *panni tartarici* as well as Chinese silk cloth embroidered with "strange animals."[62]

Paolo Uccello (1397–1475) painted *Saint George Fighting the Dragon* (ca. 1455–60), showing the monster covered with gaudy rosettes, a motif inspired by Asian carpets or brocades. In the sixth century the holy warrior from Cappodocia typically appeared in Byzantine art as torturing and beheading tyrannical persecutors of Christians, such as the Roman emperor Diocletian (r. 284–305). The dragon entered the legend only in the twelfth century: recently arrived in Southwest Asia from China, the winged monster came to Europe with soldiers and monks returning from papal crusades to Syria and Egypt. In *The Inferno,* Dante Alighieri (1265–1321) describes the dragonlike Geryon, with his pointed tail and horrid stench:

> He had two paws, both hairy to the armpits;
> His back and breast, and both sides down to the shank
> Were painted with designs of knots and circlets.
> No Tartar or Turk has ever woven a cloth
> More colored in field and figure.[63]

Cangrande della Scala (r. 1311–29), the lord of Verona and Dante's patron, had himself buried in robes of Chinese silk, a fitting interment for a princeling who appropriated the title of the awesome Great Khan.

Cangrande obviously was captivated by China, almost surely from reading about it in Polo's *Travels,* a highly popular work that sometimes was reproduced with illustrations of Chinese dragons in European guise. It is likely, however, that the Veronese despot never laid eyes on a piece of porcelain, for chinaware remained exceedingly scarce in the West because of the limited quantities carried on the Silk Road. With few exceptions, the porcelains that came west by caravan remained in the Islamic world, the premier export market for Chinese commodities. Significantly, Pegolotti's comprehensive book on trade never mentions porcelain coming into the hands of his fellow Italians in their depots on the Sea of Azov and in the Levant.[64] As always, a great deal of porcelain came by ship to the Persian Gulf and Red Sea, but Southwest Asian markets handily absorbed it all. If Europeans wanted porcelain, they would have to journey to its source.

In 1291, a year before Polo returned from China and the same year the Mamluks captured the city of Acre, the last Christian stronghold in the Holy Land, the Vivaldi brothers departed Genoa in two trading galleys in an attempt to reach the Indian Ocean by circumnavigating Africa. After sailing through the Strait of Gibraltar and heading south down the Moroccan coast, they vanished forever.[65] If they had succeeded in their audacious venture, a maritime route linking Europe to Asia, with all its ramifications for the history of the world, might have been established two centuries before da Gama's voyage. Inasmuch as that did not take place, Europe had to wait until after 1500 to have the same access to porcelain (and many other commodities) that the rest of the ecumene already took for granted. The scarcity of porcelain in the West in the days of Polo and Dante thereby testifies to the peripheral position of that region in a world system dominated by Muslim commerce, Mongol power, and Chinese economic might. Even taking into account their sporadic diplomatic forays, the various rulers of the Great Mongol Empire regarded Latin Christendom as marginal territory, a land of paltry kingdoms, the remotest and most impecunious market for Asian merchandise.

What is now known as the "Gaignières-Fonthill vase" indicates the extraordinary value placed on the few porcelains to reach the West during the Mongol period. Made in Jingdezhen, the white porcelain bottle with a bluish-tinged glaze came to Europe by the Silk Road in the early fourteenth century, perhaps carried there by Nestorian Christians of China on their way to meet with the Roman pope in Avignon, France. Mounted in silver gilt and inscribed with gothic gold letters, the *qingbai* vase became a prized possession of the Anjou kings of Hungary and Naples in the fourteenth and fifteenth centuries. After further peregrinations and some heraldic embellishments in Burgundy and Germany, the vessel eventually ended up at Saint-Cloud near Paris, into the hands of the most august collector of porcelain in Europe, Louis de France, Grand Dauphin.[66] The illustrious credentials of the piece attracted him, as did its enduring rarity: Europeans in the reign of the Sun King, as Dentrecolles points out, rarely saw anything but blue-and-white porcelain.

In fact, the Gaignières-Fonthill vase is symbolic of the end of an era inasmuch as blue-and-white porcelain came to dominate Jingdezhen's export trade after the early decades of the fourteenth century, around the same time the *qingbai* bottle departed for the West. Although the early history of cobalt-decorated porcelain is obscure, potters evidently began experimenting with the new painting technique in the opening years of the century, at a time when Jingdezhen still concentrated on manufacturing *qingbai* and other whitewares.[67] Entrepreneurs sold the small, clumsily painted blue-and-white vessels produced at first to undemanding customers in the Archipelago. The porcelain city had no inhibitions about finding a market for its shoddiest wares.

A Chinese shipwreck indicates that Jingdezhen did not produce blue-and-white in quantity until after the first two decades of the fourteenth century. On a voyage from China to Japan in 1323, a trading junk sank off southwestern Korea. Underwater excavation in 1976 revealed that it held twenty-eight tons of copper-cash (equivalent to one year's supply of coinage in Japan) and a single icon, a green-glazed porcelain figurine of Guanyin, probably specially ordered for a family temple or shrine. The main cargo comprised eighteen thousand porcelains, including celadons from Longquan, black-glazed teabowls from Zhejiang, and *qingbai* from Jingdezhen.[68] Significantly, it carried no cobalt-decorated wares.

That state of affairs soon changed, however. Though the Longquan kilns continued to claim a portion of the export trade until they finally ceased production in the 1600s, they suffered a long, irreversible decline as Jingdezhen improved its output of blue-and-white after the first quarter of the fourteenth century. As it happened, Ibn Battuta arrived in China in the early 1340s, soon after the new azure-painted ware began to appear in quality and quantity. The pilgrim certainly witnessed evidence of a historic turnabout when he wandered through the porcelain bazaar of Canton—piles of blue-and-white plates and bowls being readied for shipment to the Abode of Islam.

THE ORIGINS OF BLUE-AND-WHITE PORCELAIN

Beginning in the final decades of the Yuan dynasty, Muslim merchants in Quanzhou and kiln owners in Jingdezhen engaged in a commercial venture unprecedented in world history: cobalt ore shipped eight thousand kilometers from Persia to China; pottery customized in bulk for Islamic customers and then transported to Southwest Asian markets. Quanzhou merchants acted as midwives for the birth of blue-and-white porcelain, and they continued to play a crucial role in the development of the ceramic into the era of the Ming. On the one hand, they knew about the market for high-quality pottery in Southwest Asia, as well as about the frustrating attempts by Persian potters to paint in cobalt on tin-glazed vessels. On the other hand, they recognized that the potters of Jingdezhen had developed an improved ceramic body that offered an ideal white surface for painted decoration and sufficient strength to replicate in pottery the massive metal basins and platters that Southwest Asians used as utensils.

Moreover, the Quanzhou merchants filled their homes with Southwest Asian metalwork that could serve Jingdezhen artisans as models, along with carpets, textiles, and leather goods as sources of Islamic design. To experiment with cobalt oxide for decoration, the merchants even could provide samples of "Muslim blue," which they sold as medicine in their apothecary shops.[69] For their part, Jingdezhen potters discovered that the viscosity of their colorless glaze prevented the *huihui qing* from diffusing during firing, thereby allowing intricate designs to be executed on their white vessels. They also gained a decisive economic advantage over the craftsmen of Longquan since blue-and-white proved cheaper to produce than celadons. Manufacturing the latter was a chancy proposition; it demanded precise control of the amount of oxygen in the kiln, failure of which resulted in many ruined pieces. Potters fired blue-and-white with greater precision, producing substantially less wastage in firing; hence profits were higher in Jingdezhen than among its Longquan rivals.[70]

Muslim officials and the court of Kublai supported the Quanzhou merchants in their enterprise. In 1278, a year before conquest of the Southern Song capital, the new emperor established the Fuliang Porcelain Bureau to oversee production at Jingdezhen, the first time a monarch of China appointed magistrates to such a task. The Porcelain Bureau, which increased fourfold in personnel by 1295, came under the direction of the Bureau of Imperial Manufactures, the agency charged with supplying luxury goods to the court. Foreigners, especially Mongols and Persians, dominated Imperial Manufactures. They sent motifs taken from tapestries, war banners, and imperial robes to Jingdezhen for decoration on ceramics, and they provided cobalt oxide to the imperial workshops.[71] The early development of blue-and-white porcelain owed a great deal to the patronage of the Yuan court.

Imperial interest in Jingdezhen was primarily financial and commercial rather than aesthetic, however. The Yuan organized the tax system to extract maximum profits from the output and sale of salt, tea, metals, and pottery. They levied heavy payments on Jingdezhen, assessing rates based on the size of kilns and the number of workers employed; shop owners and river porters had to pay taxes for the first time.[72] Along with generating returns from taxation, the court aimed to boost production. Even more than the rulers of the Southern Song, the Yuan dynasts promoted foreign trade as a means of increasing state revenue. They therefore encouraged export of Longquan celadons and Jingdezhen *qingbai* as well as production of the promising blue-and-white ware. According to government records, "upon his ascent to the throne, Kublai Khan sent envoys with ten edicts to countries in Southeast Asia to invite merchants to come to China to trade."[73] The Mongols looked favorably on merchants, an attitude shaped by the chronic desire of nomads to attract sellers of desirable commodities to the barren steppes. Basing themselves on Song precedents, the Yuan created a scheme, termed "Government-Invested Ships," that combined official vessels with private merchant participation in maritime trade. In 1285 the court devoted nineteen tons of silver to building ships for the venture. These measures bore fruit, for Chinese merchants and those of other nationalities made profitable use of the spacious domain opened up by the steppe conquerors. According to a writer of the Yuan period, "China traders who go forth among the different courts and various territories travel as if between the prefectures of the east and the west."[74]

In contrast to the Mongol upper stratum, Confucians traditionally viewed merchants with disdain. From the time of the Han dynasty, they ranked at the bottom of the social scale in Confucian estimation, below gentry, artisans, and peasants. The official ideology stigmatized merchants as a disruptive social element, perpetually in need of supervision and regulation, necessary parasites whose materialism ran counter to Confucian ethics and harmonious social order.[75] Too much can be made of such condemnations, however, for they were more routine flourishes of conventional opinion than faithful descriptions of reality. From the Han through the Song, ideological scorn for merchants never significantly hampered commercial activity, just as ecclesiastical prohibitions against usury never stopped interest payments and loan-sharking in Latin Christendom. As regimes have discovered throughout history, it is virtually impossible to stop people eager to make money. When the Ming tried to enforce draconian laws prohibiting overseas trade, a sixteenth-century Chinese writer pointed out that it was a fruitless enterprise: "China and the barbarian countries have their respective unique products; thus trade between them is difficult to terminate. Where there is profit, people will certainly go for it."[76]

In any case, the Yuan did away with the Confucian social assessment by endorsing merchants and introducing many new social categories. Mongol princes entered into partnership with merchants, and Mongol rulers placed foreign traders (such as Marco Polo) in administrative positions.[77] When Quanzhou Muslims considered how to employ the potters of Jingdezhen to serve the markets of Southwest Asia, they therefore could count on backing from the Yuan court for a venture that benefited both merchants and the imperial treasury. In addition, trade in blue-and-white porcelain attracted the court because it established a mutually profitable link with Ilkhanid Persia, Yuan China's most loyal ally.

From early in the fourteenth century, Quanzhou merchants imported cobalt oxide from Persia to China, sometimes in processed form, smalt (powdered glass, potash, and cobalt), more often as raw material. Chinese potters identified the best-quality *huihui qing* as "Buddha's head blue."[78] After it was roasted in an oven for twenty-four hours, special workers, recruited from among the aged and lame, ground the cobalt ore, Dentrecolles records, "in an unglazed porcelain mortar using an unglazed porcelain pestle" for several weeks until it was reduced to a dust-like powder. According to Zhu Yan, Jingdezhen manufacturers considered the ore worth "twice the value of gold": every sixteen ounces of the unrefined oxide yielded just six-tenths of an ounce of pure pigment. A little bit of the cobalt went a long way, however, as it possesses extraordinary coloring power: just one part in 500,000 produces a perceptible tint, and one part in 5,000 creates a brilliant blue.[79]

Although by the mid-sixteenth century artisans exploited native Chinese sources of cobalt oxide and combined it with the foreign product, the Persian material remained so expensive that Dentrecolles recommended to his superiors that if the same ore were discovered in Europe, it could be shipped to China to exchange for "the most beautiful porcelains."[80] He noted that workers reused even tiny dried drops of the paint after scraping them up from paper placed under the vessel being decorated. Given the value of the substance, craftsmen in the imperial workshops routinely stole it to sell to private kilns. In the early Ming, the Bureau of Imperial Manufactures tried to stop the larceny by doling out exact portions to the workshops. A Fuliang official in the mid-Ming stopped the pilfering for a time by checking the weight of the cobalt before and after use. Some supervisors even compelled porcelain decorators to paint with their arms through wooden screens to restrict their opportunity for filching the precious pigment.[81]

By the middle of the fourteenth century, Jingdezhen potters were making large vessels in "vivid blue on a white background," as Dentrecolles described it, embellished with elaborate designs. An illustrious pair of porcelains, inscribed with the date 1351 and made for a Daoist temple in Jiangxi province, provide a benchmark. Known as "the David vases," they stand slightly over sixty-three centimeters tall and are decorated in eight horizontal bands with images of lotuses, plantain leafs, chrysanthemums, cloud scrolls, phoenixes, and dragons. In scale, layout, and quality of workmanship, the vases are striking evidence of the level of sophistication Jingdezhen had reached by the mid-fourteenth century in producing the novel ware.[82]

Just when blue-and-white porcelain had attained this level of maturity, however, the decline and fall of the Yuan regime, impelled by Mongol infighting, hard economic times, and epidemic disease, halted further development of the commodity for two decades. Peasant forces rebelling against the Mongols occupied Jingdezhen in 1352, forcing potters to abandon their kilns. In the same year, the Persian garrison in Quanzhou, led by the Pu trading family, rose up against the Yuan. The rebellion plunged Fujian into tumult for a generation, in the course of which Chinese soldiers and mobs massacred Quanzhou Muslims and destroyed their mosques and shrines.[83]

From midcentury until the flight of the last Yuan emperor to the steppes in 1368, China was wracked by rebellion against the Mongol rulers and by conflict among competing warlords. Fighting convulsed the lower Yangzi region, the power base of Zhu Yuanzhang (1328–98), the future Hongwu emperor, the "Grand Progenitor" (Taizu) of the Ming dynasty. His last major battle against a rival warlord, a monthlong naval clash, took place on Lake Boyang, not far from Jingdezhen. In January 1368, when the Hongwu emperor made the obligatory sacrifices to Heaven and Earth at his enthronement ceremonies in Nanjing, the entire Middle Kingdom came under native Chinese rule for the first time in four hundred years.

THE TRIUMPH OF BLUE-AND-WHITE PORCELAIN IN CHINA

Following the custom of all new Chinese sovereigns, the Hongwu emperor devoted attention to providing his court and palaces with suitable trappings. According to a contemporary account, in his first year on the throne, he "decreed that vessels for the Imperial Ancestral Temple should be changed and made of gold. . . . In the second year it was decided that all sacrificial vessels be made of porcelain."[84] He commanded that ritual wares be glazed monochromes of auspicious color, shades that mirrored the cosmological ordering of the imperial realm: white for official rituals at the Altar of the Moon, blue for use at the Altar of Heaven, red for that of the Sun, and yellow for that of the Earth. He appointed magistrates to supervise the reconstruction of the Jingdezhen kilns, and he ordered his palace in Nanjing, the capital of the early Ming, roofed with white porcelain and embellished with red dragons and phoenixes on the end tiles of the eaves. Building the palace and renovating the city required tiles produced by seventy kilns.[85]

In an act weighty with symbolism, the Hongwu emperor designated red-glazed vessels as official ware for his palace. The reign name *Ming* (Bright) connoted red or fire, the element that represented the south of China, the region in the vanguard of the rebellion against the Yuan, whose centers of power naturally lay in the north. *Ming* also could be taken to stand for the Red Turbans, partisans of a Manichaean cult that had rallied to the future emperor because they regarded him as the millenarian Prince of Radiance *(ming wang)* destined to defeat the dark force of the Mongols. And, fittingly enough, the ideographic character for red is *zhu,* the same as that for the surname of the Grand Progenitor.[86] Acutely sensitive to political gestures that legitimized his rule, Hongwu played up such evocative correspondences when he selected his palace tableware.

The Ming founder's second son, who came to power as the Yongle emperor (r. 1403–25), overthrew Hongwu's successor and grandson, the Jianwen emperor (r. 1399–1402). The usurper declared, "It is deeply satisfying to one's spirits that the court should utilise familiar, bright, clean and unadorned Chinese porcelain vessels."[87] He favored white porcelain, perhaps because he and his wife, Empress Xu (d. 1407), paid special devotion to Guanyin, a figure almost always depicted in that ideally pure, achromatic tone. The empress even wrote a Buddhist sutra as a result of a dream vision in which she saw Guanyin standing on a thousand-petaled lotus and holding a bejeweled rosary. Most of the shards excavated at Jingdezhen from the Yongle reign come from vessels of "sugar-white," a term applied to the wares from the sixteenth century, when refined sugar became a significant commodity. As adherents of Tantric Buddhism, the Yuan emperors also had preferred white porcelain, so even though Hongwu and Yongle despised the Mongol conquerors and explicitly based their governance on precedents from the Han and Tang dynasties, their selection of the same ceramic for important occasions suggests continuity with their immediate predecessors at odds with their official proclamations. In fact, although the early Ming emperors upheld Confucian precepts, in their reliance on military force and their hostility to the Confucian establishment they resembled the alien rulers they had supplanted. In their attitudes to foreign trade, however, Hongwu and Yongle diverged sharply from the Yuan. As a result, Chinese overseas merchants faced significant obstacles in the early fifteenth century, and exports of porcelain and other commodities declined.

White also represented mourning and filial piety; hence when the Yongle emperor in 1412 ordered construction of the nine-story Buddhist Baoen Temple in Nanjing in memory of his parents, he ordered its masonry frame sheathed with L-shaped tiles of white porcelain made at Jingdezhen. His splendid tribute conveyed the political message that despite his violent seizure of the throne from his nephew, he nonetheless remained obedient to the will of his father. Octagonal in shape and eighty meters high, the Baoen temple had 100 bells suspended from its eaves and 140 lamps that glowed in its windows at night.[88] Dentrecolles extolled it as the highest and most beautiful of all towers in China, and kilns in Jingdezhen turned out porcelain models of it as tall as a man, a few copies of which reached Europe and America in the eighteenth century.

The building became renowned in the West after Johann Nieuhoff (1618–72) published *An Embassy from the East India Company of the United Provinces to the Emperor of China* (1665), a best-selling account of a Dutch legation to Beijing in 1656. It provided the first information to Europe on Chinese architecture, a topic prominently featured in the volume's one hundred engravings. In his narrative,

Nieuhoff highlighted a description and illustration of the Baoen Temple, christened "the Purceline Pagoda" in English translation. French and English potters used the illustration of the pagoda as a decorative motif on their wares, and artisans in Japan painted it on teapots commissioned by the Dutch for their home market. Based on Nieuhoff's description, Westerners acclaimed the pagoda as the eighth wonder of the world in the belief that it was built entirely of porcelain. When Wedgwood told Erasmus Darwin that Chinese potters, by ornamenting entire pagodas, "employ their art in the formation of works of great magnificence," he evidently was thinking of the Porcelain Pagoda.

Inspired by Nieuhoff's report, Louis XIV's most extravagant gesture of esteem for China was the Trianon de Porcelaine in the park of Versailles, designed by Louis Le Vau in 1670 and built for Madame de Montespan (1641–1707), the king's mistress. The first of numerous European buildings in chinoiserie style, the one-story Trianon, with a Doric facade and mansard roof, actually bore no resemblance to the Nanjing tower. Instead, the association with China appeared in numerous decorative elements: blue-and-white faience urns bordering the roofline, metal flower tubs painted blue and white, wooden window casements adorned with motifs taken from blue-and-white porcelain, and rooms outfitted in Chinese embroidered silks, Chinese lacquer screens, and blue-and-white earthenware tiles. Poorly clad in blue-and-white Dutch tiles on the exterior walls, the Trianon de Porcelaine fared so dismally in the clammy ambience of Versailles that Louis XIV ordered it demolished in 1687. Thirty-two years later, when Defoe, no lover of the French, in *The Farther Adventures of Robinson Crusoe,* mocked a supposed Chinese porcelain palace for being just timber "plastered with the earth that makes China ware," he most likely was making a sly allusion to the fabled Trianon.

Although Jingdezhen produced modest amounts of blue-and-white porcelain for Hongwu and Yongle, neither emperor cared for the ware, presumably because it had too close an association with the despised Mongols. In addition, Hongwu came to prominence as an ill-educated warlord, and Yongle, though not uncultivated, devoted much of his energy to leading armies into the steppes against the Mongols. The two monarchs showed negligible interest in art—yet altogether too much for the court painter put to death by Hongwu for portraying a jellyfish mauling a five-clawed dragon, the imperial symbol. He also ordered the execution of Zhao Yuan in 1375 when his unconventional paintings of ancient worthies, such as *Lu Yu Brewing Tea,* offended the emperor's conservative taste.[89] The new dynasty did not sponsor blue-and-white until the rule of the Xuande emperor (1426–35), Hongwu's great-grandson, a century after the new style first emerged and fifty years after the end of Mongol domination.

After the tempestuous reigns of Hongwu and Yongle, the Ming dynasty settled down into long-established modes, eager to buttress its political legitimacy through emulation of venerated predecessors. Modeling himself on the Huizong emperor of the Song, Xuande likewise became an accomplished painter and art patron. He ordered the casting of bronze vessels for the Ancestral Temple, including hundreds of antique-style pieces based on illustrations in Song catalogues. Metalsmiths modeled some of the bronzes on Song porcelains that were themselves approximate copies of ancient bronzes in the same catalogues. A zealous promoter of porcelain, the emperor had his reign mark *(nianhao)* written on imperial wares, a practice in imitation of inscriptions on old bronzes. He sent paintings to the imperial kilns as models for designs, and in some years he ordered over 400,000 pieces of pottery for his palace. Fond of cricket fighting, Xuande commanded a court official to provide a thousand of the insects every year, and he commissioned porcelain cages to keep them cool in summer.[90] (See figure 13.)

Although some well-off families placed blue-and-white porcelain in their tombs in the late Yuan and early Ming, indicating that it had a certain appeal, the elite generally looked down on the ware until it won Xuande's patronage. After centuries of connoisseurship devoted to stylish, understated whitewares and celadons, the cobalt-decorated vessels struck many as vulgar and ostentatious. In the eyes of their enthusiasts, Song porcelains epitomized excellence by virtue of their sculptural quality and the patina on their flowing bodies, a range of hues that made them seem in harmony with both the natural world of flowers and the polished world of jade. Painting pictures and designs on celadons and *qingbai* was out of the question. Potters employed only subtle, restrained decoration on monochrome vessels, slightly incising the glaze or ceramic body, a technique known as "secret ornament" *(anhua);* at their most delicate, the designs could be seen only by holding the vessel to the light or when liquid filled it.

Another decorative technique arose from the actual firing of the pottery: sometimes as the porcelain cooled in the kiln, the glaze contracted at a rate greater than that of the body of the vessel, causing crackling, or crazing, of the glaze shell. The effect attracted admirers, so potters by the late Song deliberately created the fine network of cracks as embellishment, a procedure that dictated exact control of kiln temperature. In praise of a Song vase, the Qianlong emperor wrote, "Despite the pattern of hundreds of intermingling crackle lines, its texture is fine and smooth to the touch. . . . One discovers that the value of these undecorated wares is the same as that of unpolished gems."[91] Song connoisseurs and literati regarded the crackling as in accord with their aesthetic principles inasmuch as the effect seemed natural, arising spontaneously from the materials and techniques being employed. In contrast, porcelains ornamented with designs in blue paint, the end product of a common artisan's routine toil, appeared contrary to the simplicity and moderation acclaimed so fervently by scholars and gentry.

With imperial support in the Xuande reign, however, blue-and-white porcelain made its crucial breakthrough among the elite. Potters abandoned their fastidious subtleties and adopted new painterly techniques; the death knell sounded for celadons, and the Longquan kilns lost their extensive market. Following the lead of the court, connoisseurs speedily revised their perceptions and presumptions. They learned to take pleasure in ceramics with vibrant blue designs, fresh painting styles, fanciful portraits of animals, illustrations of scenes from classical drama, motifs conveying auspicious messages, and landscape compositions on a three-dimensional surface. Imperial porcelains in the new style were so admired that they set a formidable standard for subsequent generations. Blue-and-white vessels of the Xuande period became so prized that potters later in the century illicitly inscribed the emperor's reign mark on their wares to boost their value or to pay homage to an early zenith of the art.[92] In the late Ming, the best Xuande porcelains reportedly fetched higher prices than fine jade figurines. (See figure 14.)

But the early Ming renovation in taste amounted to a segregation of old aesthetic principles rather than their wholesale rejection. Song monochromes took on elevated status as antiques, emblems of the revered past, treasures to be commended and collected. Commitment of the scholar-gentry class to archaism thus consigned Song porcelains to a privileged niche, along with ancient jade carvings and bronze vessels, even while new standards of appreciation turned against the principles informing them.

In any case, by the end of the first decades of the fifteenth century, elite taste came to regard painted decoration on ceramics as more significant and gratifying than body shape and glaze tone. This had significant ramifications for the relationship of pottery to other media. Emphasizing form and sheen, Song ceramics had taken slight notice of other decorative arts, with the outstanding exception of pottery replicating ancient bronzes in response to court-promoted archaism and upper-stratum demand. The new blue-and-white style changed all that; ceramics now entered into a close relationship with painting, printing, calligraphy, weaving, and carved woodwork. Potters increasingly adopted motifs and designs from silk scrolls, published texts, embroidered fabric, and lacquer panels.[93]

Porcelain thus joined in a long-term convergence of the arts in which ceramics borrowed patterns from other media while the latter looked to designs on ceramics for inspiration. The administrative procedures of the Ming and Qing encouraged this development since their centralized procurement system, first established by the Yuan, meant that the same designs could be dispatched from the Bureau of Imperial Manufactures to potters, print shops, weavers, painting studios, and furniture workshops serving the court. The assumption that aesthetic forms are basically interchangeable is seen in the Qianlong emperor's fondness for treasure boxes *(duobaoge),* artfully designed storage chests for small objects—carved jade, ivories, lacquer boxes, calligraphic texts, pottery—from different periods and contexts.

The triumph of blue-and-white porcelain in the early Ming represented as striking a break with conventional standards and perspectives as had ever taken place in Chinese art. It signified a greater rupture with aesthetic tradition than that which occurred in the transition from the Tang to the Song, when potters turned away from painted adornment and exotic motifs in favor of elegantly proportioned, jadelike monochromes. Whereas the transformation in the Song registered recoil from the outside world, an expression of xenophobia and cultural seclusion, the move toward blue-and-white in the Yuan and Ming arose from an embrace of foreign styles and techniques. In effect, by applying the standards of expressive painting to pottery and rejecting those of sculpture, the Chinese, under Muslim and Mongol influence, accommodated themselves to the traditional aesthetic values of Southwest Asia.

PORCELAIN ART AND CROSS-CULTURAL EXCHANGE

Though the Chinese had been converted to the Southwest Asian cause of painted decoration on pottery, they generally relied on their own tradition for compositions. They had a kaleidoscopic repertoire from which to draw: plant forms (white lily, magnolia, morning glory, camellia, peach, crab apple), animal figures (peacock, lion, hen, heron, *qilin,* cicada, crane, quail), symbols of longevity, endurance, and wealth (evergreen, bamboo, flowering plum, gourd, tortoise, deer, goldfish), representations of the seasons (summer–lotus, autumn–chrysanthemum, winter–plum tree, spring–peony), and scores of auspicious emblems from Buddhism and Daoism (the shell, lotus petal, wheel, scepter, furled umbrella, vase, flaming pearl, sacred fungus, sun disk, calabash, endless knot).[94] (See figure 15.)

When combined in ornamentation of a plate or scroll, these elements usually conveyed symbolic meaning by way of encoded designs based on rebuses, that is, representations of words in the form of pictures or pictorial puns, often presented as a puzzle. A tonal language, Chinese is exceptionally rich in homophones inasmuch as it does not make morphological distinctions between grammatical categories; no linguistic structures, such as inflections, derivations, and compound forms, make it possible to distinguish between verbs, adverbs, complements, subjects, adjectives, and attributes. In the absence of these syntactical markers, the sound and meaning attributed to an ideographic character are powerfully determined by context. As Ricci explained, "Many of the symbols have the same sound in pronunciation, though they may differ much in written form and also in their signification. Hence it results that the Chinese is probably the most equivocal of all languages."[95] Precisely that aural ambiguity encouraged the literati to indulge in complex puns and layered meanings in both literature and ornamental design.

Except for tonal pitch, the word for goldfish, *yu,* sounds the same as that for "abundance," while *tang* can mean either a pond or a hall: thus a portrait of a goldfish in a pond is a rebus denoting a wealthy household.[96] The word for a carp, *li,* sounds like that for "profit," so representation of the fish swimming upstream is taken to symbolize a scholar's effort to surmount obstacles in the grueling Confucian examination system. The same academic struggle is personified by depiction of an egret (*lu,* which also can mean "road") in a lotus pond (*lian,* which also indicates "success"), while the painting of a lotus with two crabs and a reed stands for congratulations on passing the examination with high marks. The image of a rooster crowing near a peony tree brings together a cluster of homophones that punningly translates as "an official heaped with riches and honor." The word for "vase," *ping,* has the same sound as "peace"; hence a vase on a table (*an,* which sounds like "tranquility") accompanied by the Buddhist symbol of a scepter (*ju,* which also stands for "wish") conveys the sentiment, "May you achieve your wish for peace and tranquility."

A quail is regarded as an auspicious bird because it too is pronounced *an,* while two quail, a butterfly, a peony tree, and a chrysantheum coalesce to form a rebus for "rich living and redoubled tranquility." The Yongzheng emperor commissioned a porcelain painted with nine quail and a long-tailed pheasant, an avian menagerie that cued the expression, "May the country be in peace and order forever." A homonym for "good harvest" *(fengdeng)* is provided by an ornament of a wasp *(feng)* flitting round a lantern *(deng).* The bat *(fu)* represents happiness or good fortune since the same sound also denotes those concepts; "red" *(hong)* sounds like the word for "vast"; hence the image of a red bat signals an expectation for "vast good fortune." Extending the conceit, five red bats correspond to the notion of "fivefold happiness" (virtue, progeny, riches, health, and longevity), and if a butterfly *(die),* the sound of which also evokes the word for "duplicate," flutters around the quintuple red bats, then the message is that the hope for those manifold blessings is doubled.

Comprehending the symbols and concealed meanings of such ornamental designs called for considerable verbal wit and erudition, an art of the cognoscenti. The same sophisticated circles indulged in it that delighted in calligraphic finesse, tea-tasting contests, flower arranging, and ranking ancient works of bronze and ceramics. Rebuses that spoke of success in taking examinations and in scaling the imperial administrative ladder naturally appealed to scholar-officials. The welfare and future of their families depended so critically on the student or bureaucrat that relatives saw his accomplishments as attended by prophecies, omens, spirits, and revelatory dreams.[97]

Still, the allure of decorated porcelains cannot be explained simply in terms of camouflaged exhortations to realize ambitions or advance a career. For the intelligentsia, a cardinal attraction of the ornamental repertoire resided in its very multivalence: a design could be appreciated in turn for its message regarding professional accomplishment, its appeal to religious sensibility, its allusion to classical literature, its witty adaptation of a conventional motif, and its engaging depiction of a scene from nature. Just as a high-ranking Chinese gentleman might assume roles as a Confucian civil servant, Daoist devotee, authority on antique objects, connoisseur of contemporary art, and admirer of natural beauty, so too he relished the sensation that the painted scene on a vase or platter could be contemplated on several levels of implication.

Of course, the repertoire's symbolic associations spoke to a relatively small social group in China. The same was true in Korea, Japan, and Vietnam, where refined court circles comprehended the Chinese floral and animal symbolism inasmuch as their culture was strongly influenced by that of the Middle Kingdom. But when the decorated pottery traveled far abroad, its designs necessarily lost their multifaceted allusions. In the Archipelago and Southwest Asia (as well as later in Europe), plant forms and other imagery, removed from their original settings, became mere ornamentation or took on new meanings germane to the indigenous culture.

Members of the Chinese elite termed the pine, bamboo, and flowering plum collectively the "Three Friends of Winter" *(suihan sanyou);* they regarded the plants, which endure through harsh weather, as symbols of resolve, candor, and purity. As decoration on an early Qing vase, they expressed covert loyalty to the ousted Ming dynasty, the refusal of "leftover people" *(ymin)* to take office under the steppe conquerors.[98] Yet the plant motifs had no political resonance beyond that context, and in distant lands they became mere botanical adornment. A porcelain vase shaped like a double gourd symbolized Daoist notions of the relationship between Heaven and Earth, and its decoration of red bats stood for abundant good fortune from the gods; outside of a Chinese context, neither the shape nor the motif conveyed significance, though their idiosyncrasy evidently possessed a certain allure. A porcelain parrot with a pearl in its beak represented a companion of Guanyin in China but only a vivid tropical bird elsewhere. Cranes were regarded as celestial beings and as symbols of longevity in China; depicted on porcelains with the Eight Trigrams—eight combinations of three horizontal lines taken from the *Book of Changes*—they communicated a complex message about Daoist cosmic order. When Southwest Asian artists copied the motif, however, their customers saw it merely as an eccentric geometrical design surrounding a familiar animal. Distant potters appropriated styles and motifs heedlessly, with no comprehension of the original significance of the exotic decoration. In Delhi, Damascus, and Dresden, an egret was just another bird.

Misreadings by cultures distant from China were inevitable, although the scale on which they happened always varied. Sometimes alien artisans took over shapes and motifs with no change; sometimes they creatively combined the original source with a native one. Faced with bewildering Buddhist and Daoist symbols on Chinese porcelains, Persian potters sometimes mixed them helter-skelter on their earthenware. They fused Chinese images of a flowering crab apple and a stalk of peony into a single hybrid species, portraying the imagined plant alongside a cypress tree, a Southwest Asian evergreen associated with the Qur'anic garden of Paradise. Not recognizing the Chinese lotus on porcelain as a water plant, Persian and Ottoman potters sometimes rendered it with a bulky trunk and minuscule leaves. On some Ottoman earthenware, dragons stalk through jungles of towering lotuses, Chinese cloud patterns metamorphose into garlands on a lion's head, and Chinese peonies are translated into a mixture of forms, sometimes emerging from the other side of the cultural divide as tulips and carnations.

An image of a naked boy on a Yixing teapot, symbol of a newborn soul in Buddhist iconography, insinuated something else altogether when laid out on an Egyptian platter. In Persian art, the holy figure of Guanyin is transformed into a pretty maid buying fish, though she unaccountably carries a rosary as well as a shopping basket. In China, decoration of a peach tree on a porcelain plate is a representation of spring, but on a Persian pot, it turns into a flowering rose shrub, a Sufi reference to mystic union with the godhead. A painting on a Chinese vase of a hoary Daoist sage with a gourd of magic, life-prolonging elixir becomes a portrait of a grizzled wine guzzler on a Southwest Asian bottle. A porcelain representation of Chinese scholars in a wintry landscape, a representation of integrity and passive resistance to Mongol or Manchu conquerors, is transmuted on Persian pottery into warrior-aristocrats hobnobbing at a princely court.[99] Cast adrift from their native moorings, Chinese motifs and symbols lost their original meanings and took on new significance. Not surprisingly, material objects crossed frontiers with far less difficulty than cultural suppositions.

THE ENCOUNTER OF SOUTHWEST ASIAN AND CHINESE DESIGN

In their production of blue-and-white porcelain for Islamic markets, Jingdezhen potters turned out many utensils foreign to Chinese taste, such as pot-bellied jugs, ewers with handles and curved spouts, tankards, fish baskets, washbowl stands, gourd-shaped bottles, rose-water sprinklers, flasks with loop handles, massive rectangular vases, and large, deep dishes. They modeled the pieces on Egyptian, Syrian, and Persian metalwork, some sent to China by Muslim merchants for copying, others supplied by Muslim households in Quanzhou and Canton. In a few cases, the model from which the potter worked can be identified. Brushwork on a porcelain stand of the early fifteenth century indicates that the painter had a fourteenth-century Mamluk brass stand in front of him, and a porcelain candlestick of the Xuande period clearly derives from a thirteenth-century Persian brass counterpart.[100] (See figure 16.)

Of course, the potters of Jingdezhen had to accommodate themselves to novel vessel shapes because the dining styles of Southwest Asia were so different from their own. Eating meals in communal fashion called for wide platters, deep-welled basins, and big pouring vessels. Moreover, unlike the Chinese, Southwest Asians dined while seated on the floor or carpet-covered ground; hence they required platter stands to raise their utensils to a convenient height. In the early stages of the development of Chinese blue-and-white, pottery painters decorated the large vessels with busy, cramped designs, paying little attention to overall layout or elegant composition. But they soon learned how to embellish the pieces with their own designs in a coherent, well-ordered fashion, separating different categories of ornament (as on the David vases) into horizontal bands surrounded by ample white surface.

Potters decorated many items with motifs from Islamic culture, such as renditions of Arabic calligraphy and arabesques, that is, ornate designs of intertwined floral and geometric elements. They also used naturalistic images of a general nature, especially abstract floral patterns that were known in both China and Southwest Asia through reciprocal influence across the centuries. In the century of production of blue-and-white before the imperial court adopted the new fashion, however, the export market most influenced ornamentation. Pottery painters working on official wares proved "unwilling to take risks" (as a Qing connoisseur wrote), with the result that designs on their vessels sometimes appeared formulaic and uninspired, even when the pieces otherwise were painstakingly executed and of high quality. In contrast, export wares included imperfectly produced items, with slightly warped bodies and glaze impurities, yet with lively decoration and more varied, inventive designs.[101] After the Xuande court adopted blue-and-white, official wares benefited from an injection of the vigor and imagination that derived from potters having directed their talents toward export markets.

Creation of blue-and-white porcelains for Islamic customers, combined with Southwest Asian replication of the wares, entailed a remarkable encounter and accommodation between the two great design traditions of the ecumene. Since those traditions had developed largely in isolation from each other, they naturally embodied different aesthetic values and perceptions, which themselves arose from contrasting views of reality.[102]

The Southwest Asian approach to design emphasized symmetry, mathematically structured space, rectilinear patterns, and meticulous enrichment of surface. From the seventh century, that style became synonymous with the Islamic tradition, which incorporated design patterns used by the Achaeminids and Sassanians. The focus was on repeated figures and elastic effects, an indefinite extension of two-dimensional ornament. Geometric pattern appeared on all kinds of surfaces—stucco, brickwork, floor mosaics, metalwork, manuscripts, saddles, costumes, coins, bookbindings, and pottery. Arabic calligraphy was similarly ubiquitous, gliding ineluctably into geometrics as letters were intricately embellished with plaits, stars, and stylized flowers. Inspired by Sufi mysticism, Persian potters and metalworkers concealed the fundamental letters of God's name, *lam-alif* (L-A, *Allah*), amid a lattice of cursive script, rosettes, and flower buds. Perhaps with some irony, they camouflaged maxims urging restraint in speech within calligraphic convolutions, even leaving out the diacritical marks on Arabic characters that would make them decipherable.[103] In a predominantly illiterate society, a pleasing aesthetic pattern trumped textual substance.

Ornament typically consisted of squares and lozenges, zigzag bands, overlapping circles, radiating polygons, dovetailed hexagons, sunburst disks, floral-filled quatrefoils, and stellar grids.[104] The principles of Southwest Asian design are most famously displayed on Persian carpets, works that exemplify the *horror vacui* (abhorrence of empty spaces) supposedly characteristic of Islamic art. Images of porcelains appear on sixteenth-century Persian "vase carpets" as central medallions around which are spread out mirror images on a longitudinal axis, sometimes with as many as six layers of detailed pattern superimposed on one another. Carpet weavers also turned to the formal framework of actual gardens to create designs. Viewers saw their works as representations of Paradise, with a vase, fountain, or flowering tree depicted in the center of the carpet standing for the axis of the four cardinal directions.[105]

Artists favored designs taken from vegetation since Paradise, lush and well watered, appears prominently in the Qur'an, a consideration that gave prominence to gardens, fountains, and floral decoration in Islamic cultures. A popular motif was the paradisiacal Tree of Life, a symbol in Southwest Asian religions since ancient Babylon. Islamic believers located it in the highest reaches of Heaven, its trunk so colossal that it would take a horseman a century to canter around it. The motif appeared on window grilles, prayer rugs, palace facades, city gateways, ceremonial robes, and earthenware platters.[106] On the tiles of mosques, passages from the Qur'an threaded through stems and branches of the Tree of Life, the integration of the texts into the labyrinthine maze expressing the notion that the transcendent truths of Scripture are implanted in the created world. Elaborating on Qur'anic passages, Sufi mystics regarded all of Creation as united in praising God, including the scent of flowers, the humming of bees, and the color of sunsets.

Southwest Asian potters painted vessels with images of Chinese peacocks and ducks, yet they so skillfully formed the silhouettes of the birds from vegetal patterns, such as petaled blossoms and undulating stems, that the figures all but vanish into the device. This represented a standard mode of adapting Chinese figural designs: the Islamic artist flattened them out, rendering them as abstract patterns, arrayed in endless succession or caught in a web of geometric embellishment. Chinese craftsmen portrayed the dragon as a dynamic creature, an emblem of primal energy, whirling amid clouds or chasing a flaming pearl (a Buddhist symbol of perfection) across the sky; but Mamluk potters and tile makers, oblivious to the Chinese tradition, employed the motif of the dragon as a recurring decorative feature, statically flanking a series of indistinguishable phoenixes.

Islamic tile makers produced individual pieces as components of a larger configuration, decorated so they could be joined edge to edge in a notionally infinite pattern, as when blue-and-gold tiles adorned with eight-pointed star medallions swathed the domes of mosques in Samarqand, Isfahan, and Istanbul. Indeed, decorative design in Islam seems modeled on the employment of textiles, as if representations of motifs, graphic symbols, and geometric elements were a brocadelike adornment to be wrapped around any object, large or small, fixed or mobile.[107] More significantly, however, the ubiquity of all-enveloping ornament and calligraphic inscription reflects the Islamic conception that transcendent creation informs the mundane world, made manifest in all aspects of daily life.

In contrast to the Southwest Asian tradition, Chinese aesthetic conventions stressed asymmetry, a flowing sense of space, naturalism, curvilinear patterns, and rotating directional motifs, such as cloud, wave, and peony scrolls.[108] Dense, unrelieved ornament on Shang bronzes conveys a sense of *horror vacui;* but that design technique gave way after the Han era to open spaces and atmospheric effects, techniques of landscape painting and ink drawing that capture the evanescent qualities of trailing clouds, rushing water, and cliff-clinging mist. A classic example of Chinese modes of representation is the influential Southern Song hand-scroll by an unknown artist, *Dream Journey over Xiao Xiang* (ca. 1170): executed in a style known as "apparition painting" *(wangliang hua),* the work shows man-made structures—boat, bridge, pavilion, and path—set amid a fog-shrouded landscape dominated by crags receding in the distance.[109]

Purely geometric designs were rare in Chinese art; repetition and alternation were unobtrusive, masked by rhythmic movement and plentiful room around ornamental figures. Significantly, tiles decorated with comprehensive, interconnecting designs played no role in Chinese architecture, though sometimes large ceramic statues of gods and sacred animals topped the eaves of palaces.[110] Usually white or yellow, tiles generally served as an unadorned ground against which other architectural elements, as well as the landscape setting, appeared to best advantage.

In fact, use of plain tiles suggests the bias in Chinese aesthetics against vivid color in favor of muted tones, a reflection of the dominance of monochrome ink in painting after the late Tang period. Retreat of bright, alluring color and concentration on monochromes brought painting into closer association with calligraphy. In his *Reports on Famous Painters for All Dynasties* (ca. 847), Zhang Yanyuan, a calligrapher and commentator on art, contends that "mountains are green without needing malachite, and the phoenix is iridescent without the aid of the five colors. For this reason one may be said to have fulfilled one's aim when the five colors are all present in the management of ink [alone]."[111] This point of view confounded Ricci, who wondered why landscapes by well-known artists were in great demand "despite the fact that Chinese pictures are only outlines, done in black rather than in varied colors."[112]

When Chinese potters and connoisseurs shifted their allegiance from monochromatic to blue-and-white porcelains, they of course embraced a vivid color; but they were aided in doing so by the new style, like calligraphic art, making use of a

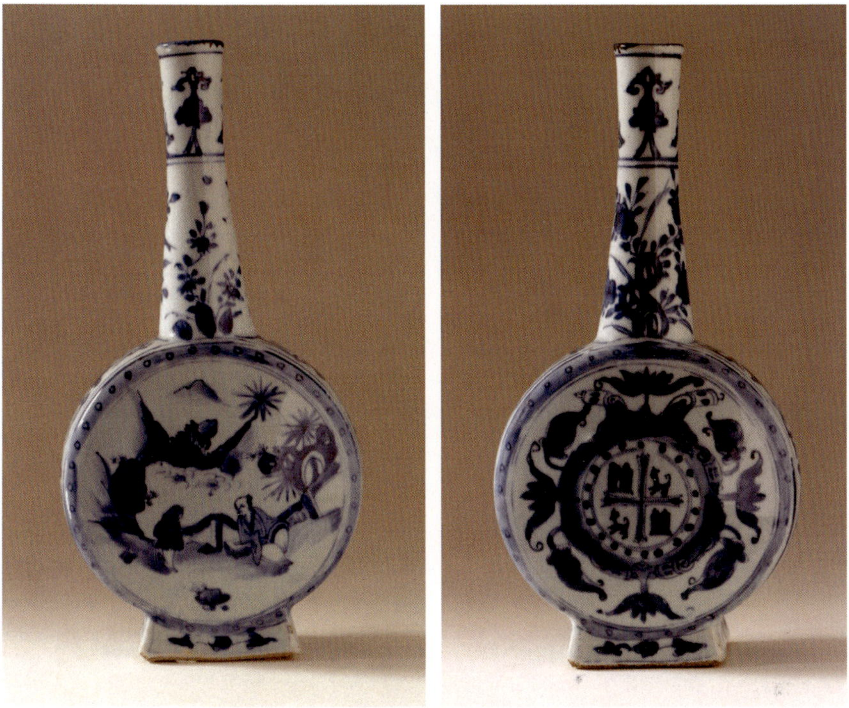

1. Chinese blue-and-white pilgrim flask. Ming dynasty (1368–1644). Jingdezhen ware, porcelain painted in underglaze cobalt blue. Peabody Essex Museum.

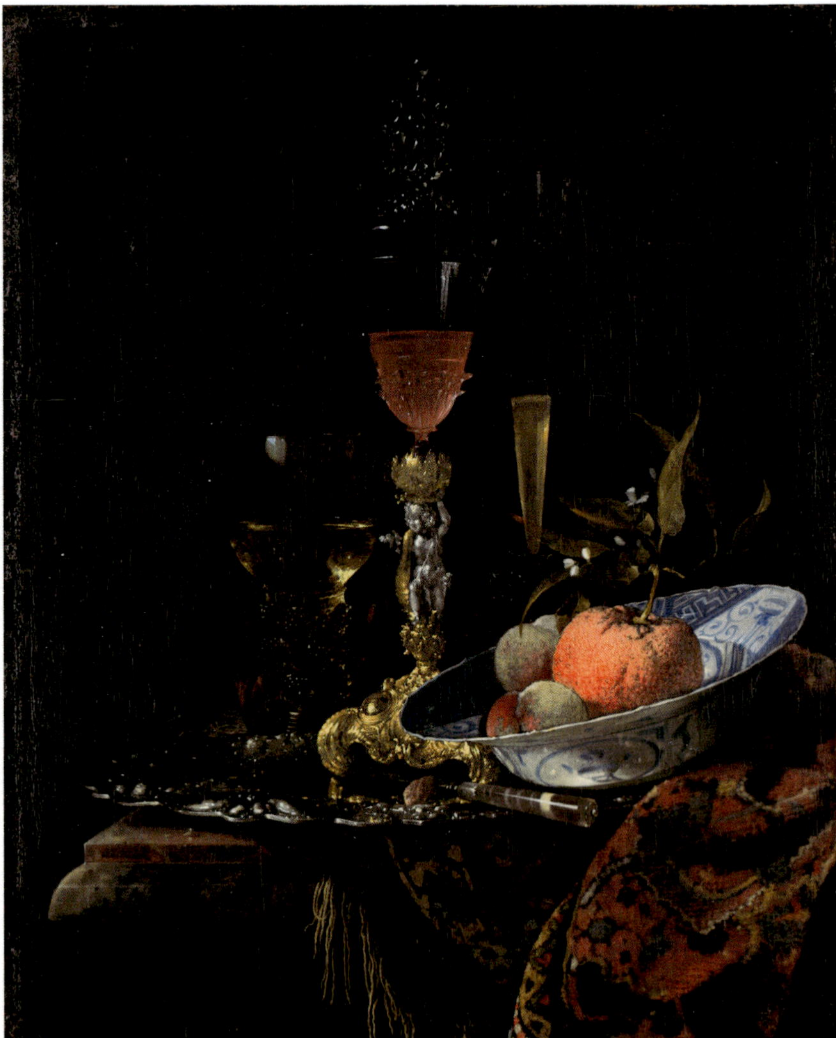

2. Detail of painting by Willem Kalf (1619–93), *Wine Glass on a Gilded Silver Foot and a Bowl of Fruit.* Oil on canvas. 23.7 × 27.7 in. (60.3 × 50.2 cm). © The Cleveland Museum of Art, Leonard C. Hanna, Jr., Fund, 1962.292

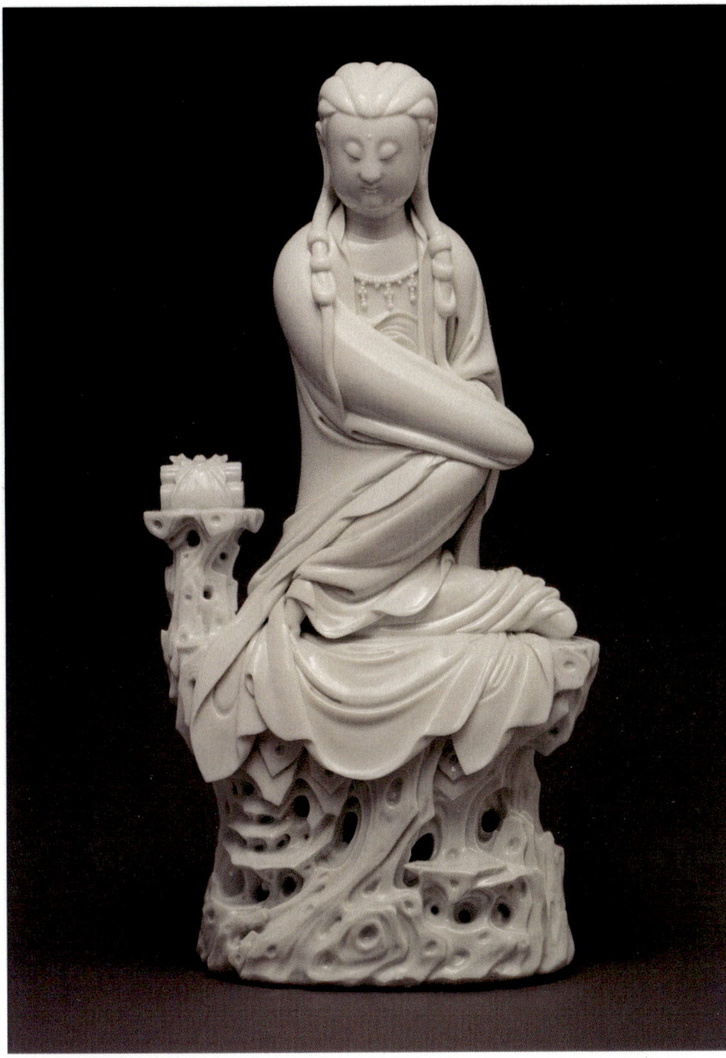

3. Chinese figure of Guanyin. Qing Dynasty, 18th–19th century. Dehua porcelain. H: 19 in. (48.3 cm). Eugene Fuller Memorial Collection, Seattle Art Museum. Photo: Paul Macapia. 43.42.

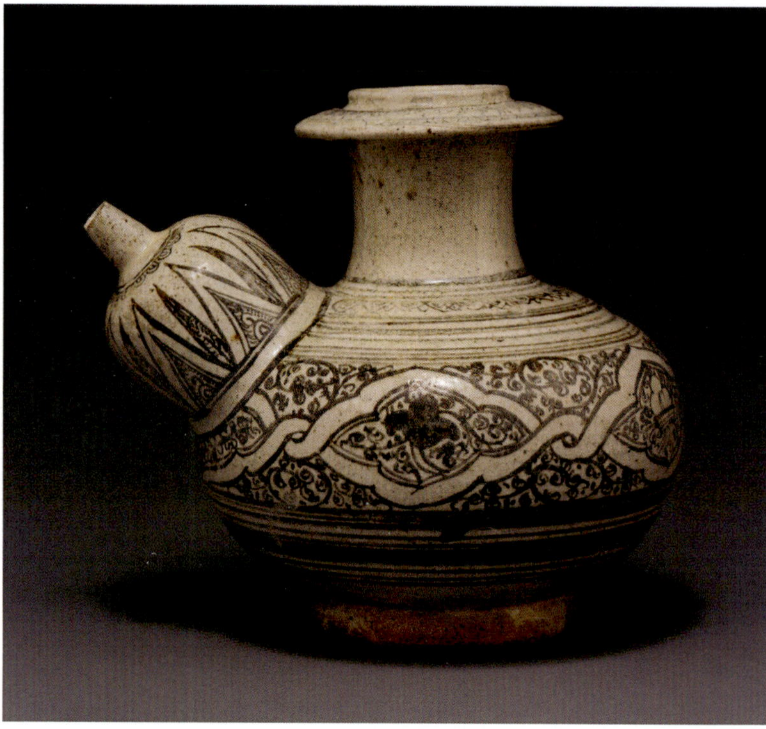

4. Thai *kendi,* 15th–16th century. Sistatchanalai ware, stoneware with underglaze black decoration. H: 8¾ in. (22 cm). Gift of Gloria Gunn Prince, Seattle Art Museum. Photo: Paul Macapia. 92.82.82.

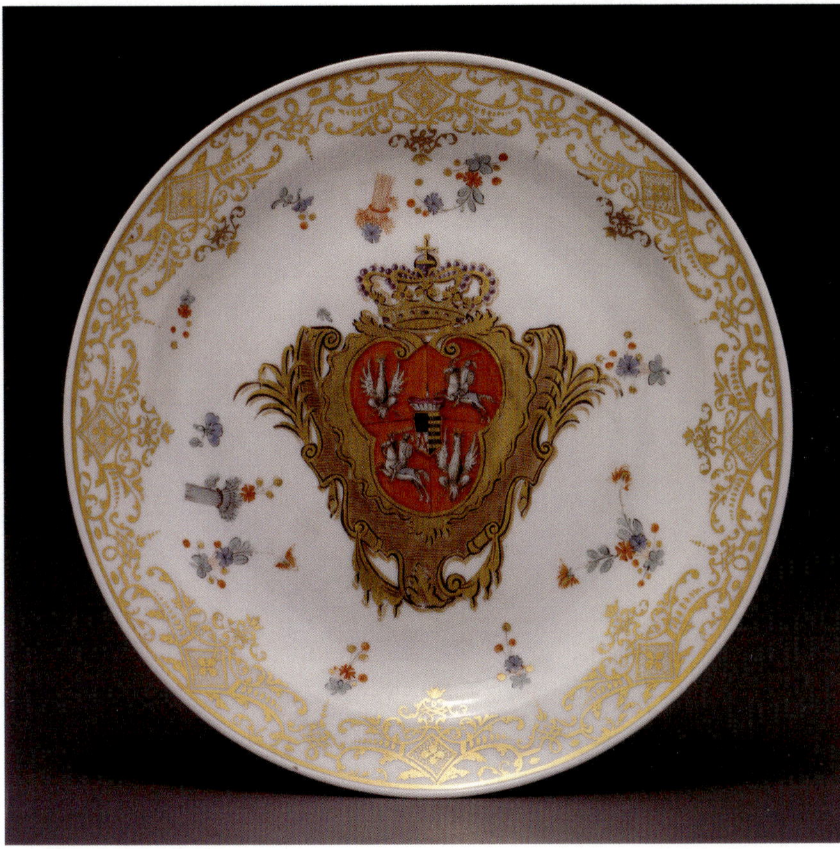

5. German plate. Meissen manufactory, ca. 1730–34. Porcelain with enamel colors and gilding. DIAM: 8⅞ in. (22.5 cm). Gift of Martha and Henry C. Isaacson, Seattle Art Museum. The crossed-sword emblem of the Elector of Saxony appears in the lower left corner of the coat of arms of the King of Poland and Elector of Saxony. The crossed-sword device was designated as the mark of the Meissen manufactory in 1723. Photo: Paul Macapia. 69.201.

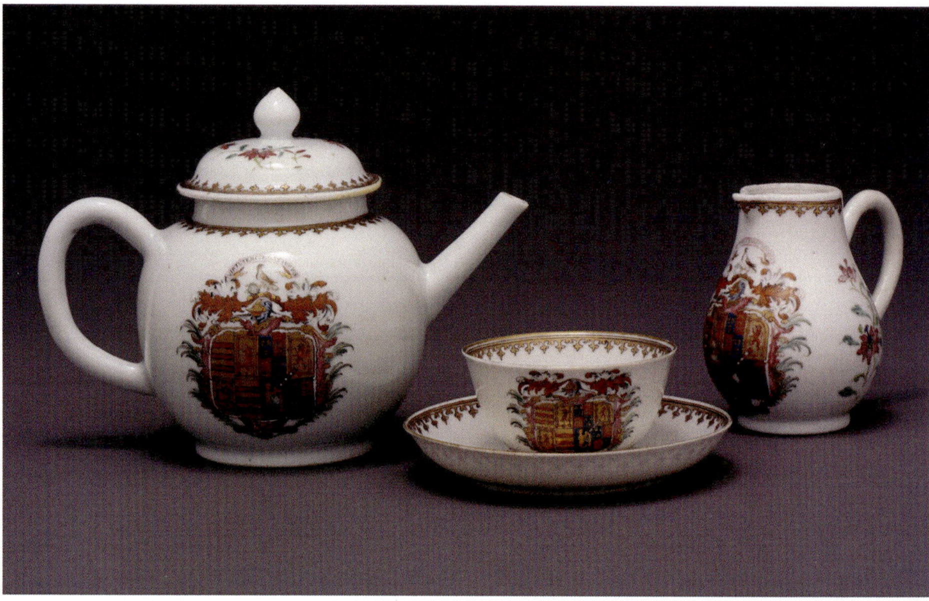

6. Chinese armorial teapot, teabowl, saucer, and milk jug. Qing dynasty, ca. 1744. Porcelain with overglaze enamels and gold. *From left:* Teapot: H: including lid 5½ in. (14 cm); teabowl: DIAM: 1½ in. (3.8 cm); saucer: DIAM: 4¾ in. (cm 12.1); milk jug: H: 3½ in. (8.9 cm). Gift of Martha and Henry C. Isaacson, Seattle Art Museum. Photo: Paul Macapia. 76.115.1.

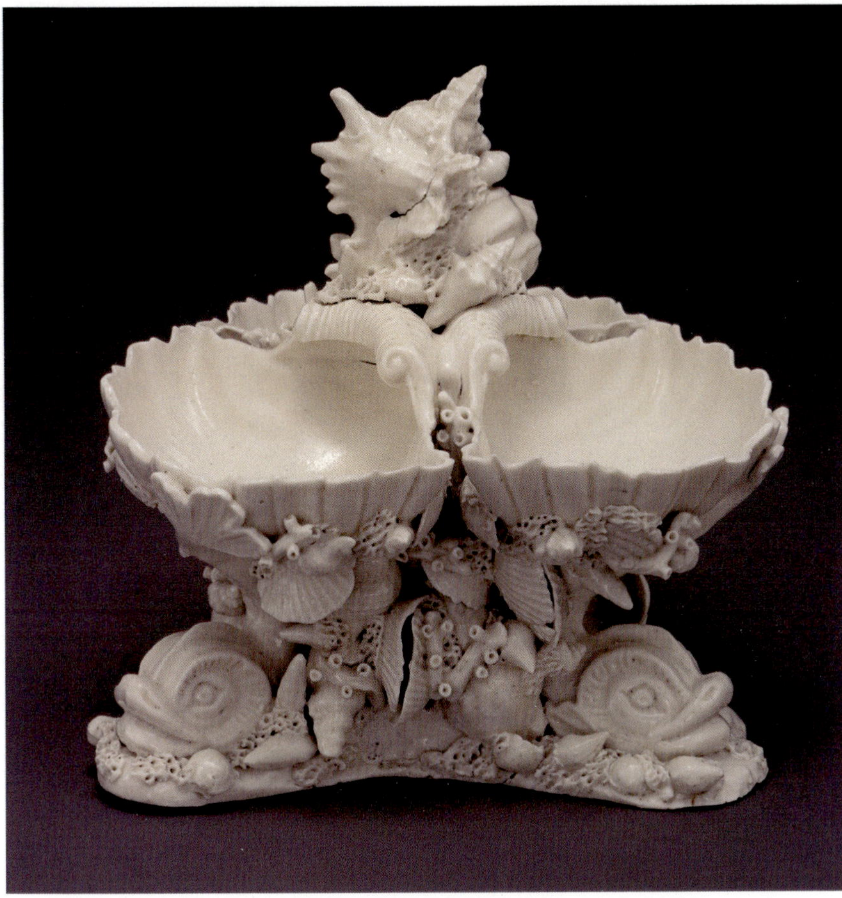

7. English triple shell dish. Plymouth-Bristol manufactory, ca. 1770. Porcelain. H: to top of bowl 6¾ in. (16.2 cm). Blanche M. Harnan Ceramic Collection, Seattle Art Museum. Photo: Paul Macapia. 57.85.

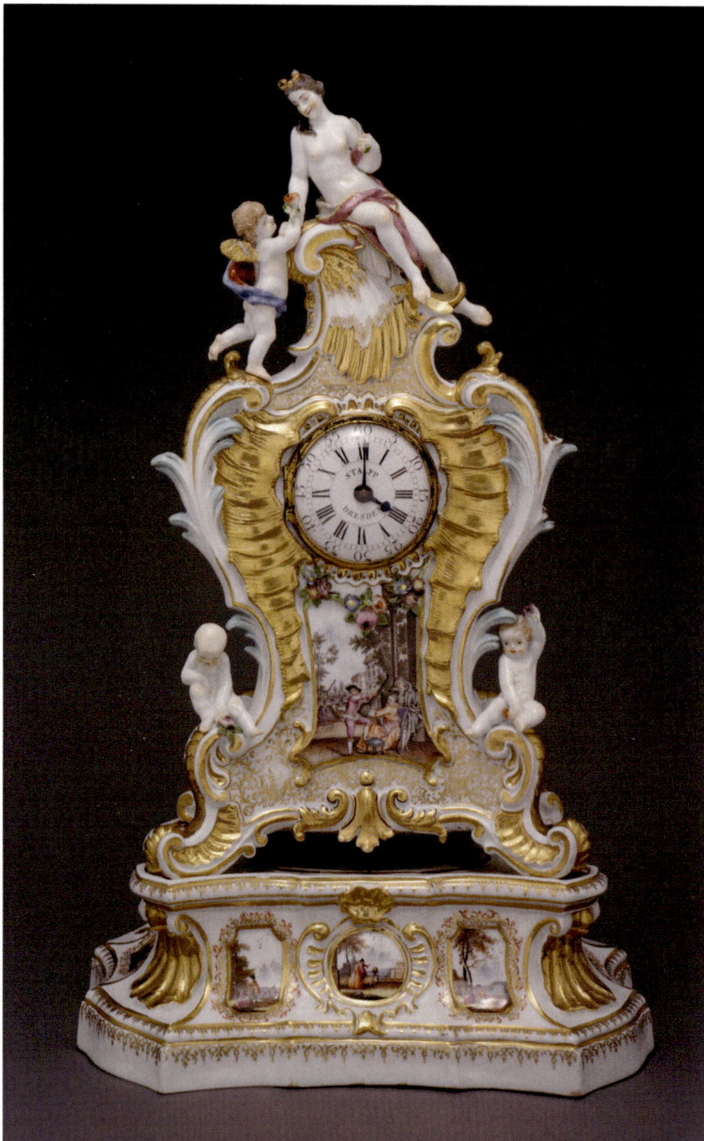

8. German clock. Meissen manufactory, ca. 1748. Porcelain with enamel decoration and gilding; metal clock face with white and black enamel. H: 15½ in. (39.4 cm). Gift of Martha and Henry C. Isaacson, Seattle Art Museum. Photo: Paul Macapia. 78.13.

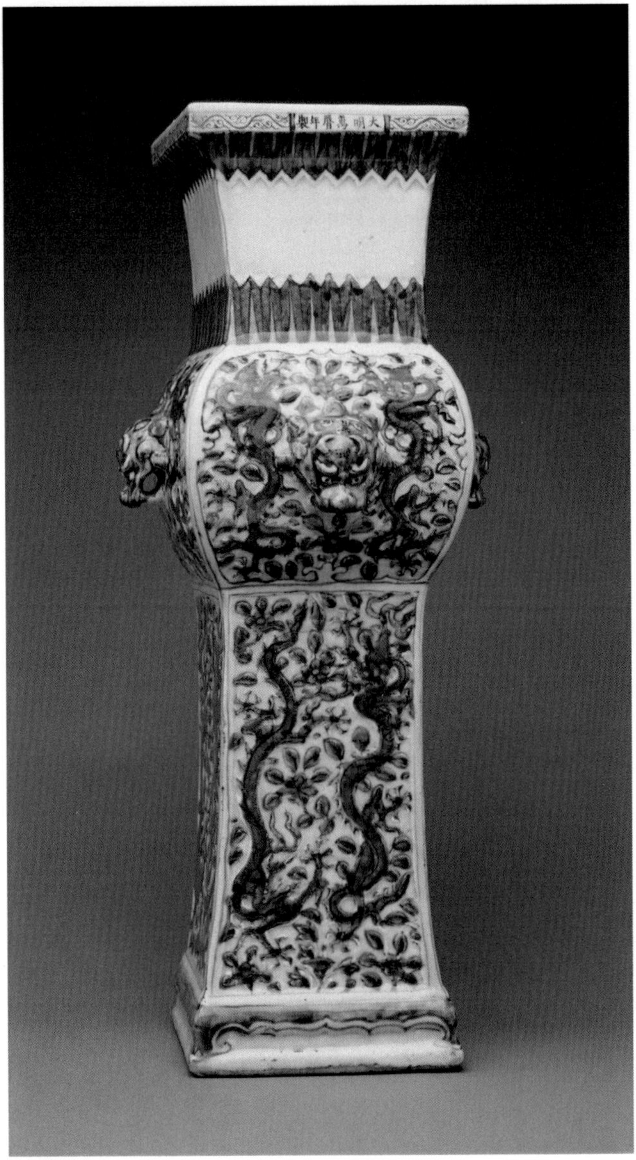

9. Chinese blue-and-white vase. Ming dynasty, Wanli period (1573–1619). Jingdezhen ware, porcelain painted in underglaze cobalt blue. H: 22½ in. (57.2 cm). Eugene Fuller Memorial Collection, Seattle Art Museum. The vase is based on an ancient wine vessel *(zun)* of the 12th–10th century B.C.E., with the lower half elongated, unlike the bronze prototype. The design includes five-clawed dragons, the emblem of the emperor. Photo: Paul Macapia. 54.120.

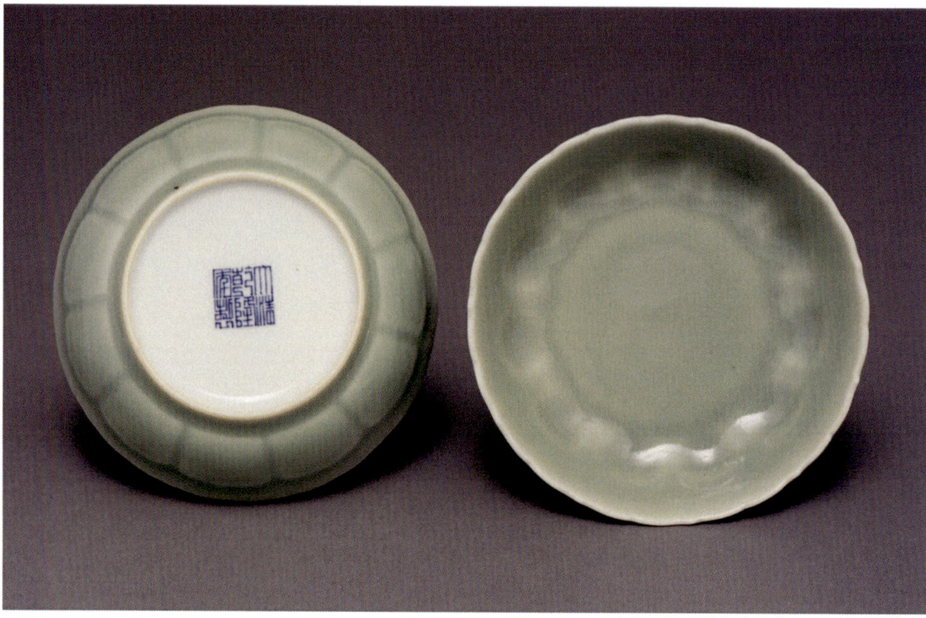

10. Pair of Chinese celadon dishes. Qing dynasty, Qianlong reign mark and period (1736–95). Jingdezhen ware, porcelain with celadon green glaze. DIAM: 4⅝ in. (11.8 cm). Eugene Fuller Memorial Collection, Seattle Art Museum. The celadon dishes are decorated with molded and incised petal designs. Photo: Paul Macapia. 44.118.1.

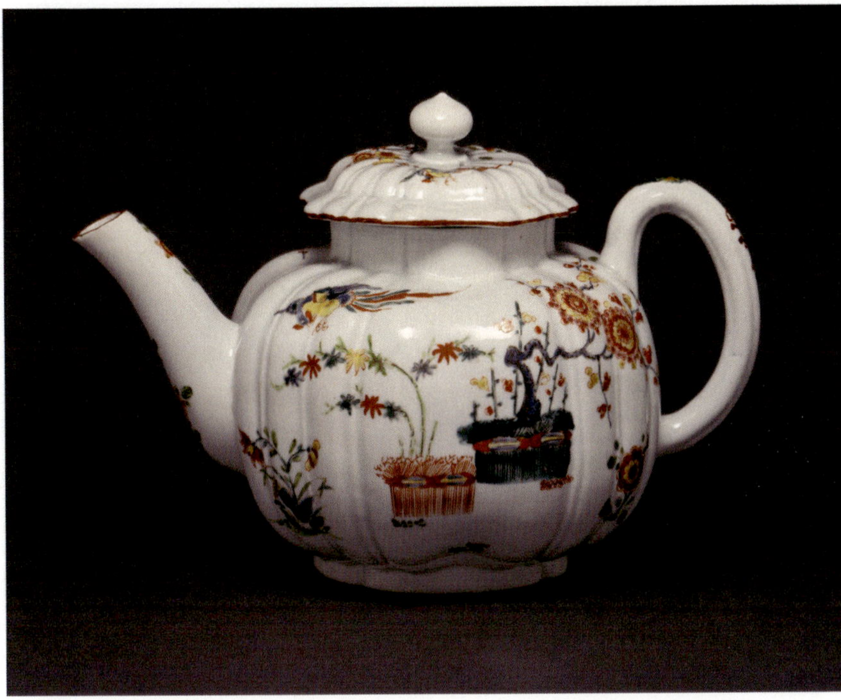

11. English teapot. Worcester manufactory, ca. 1753–54. Porcelain with enamel colors. H: 5½ in. (14 cm). Kenneth and Priscilla Klepser Porcelain Collection, Seattle Art Museum. Based on the red stoneware teapots of Yixing, the teapot has a lobed melon shape and is embellished with Japanese-influenced motifs. Photo: Paul Macapia. 94.103.5.

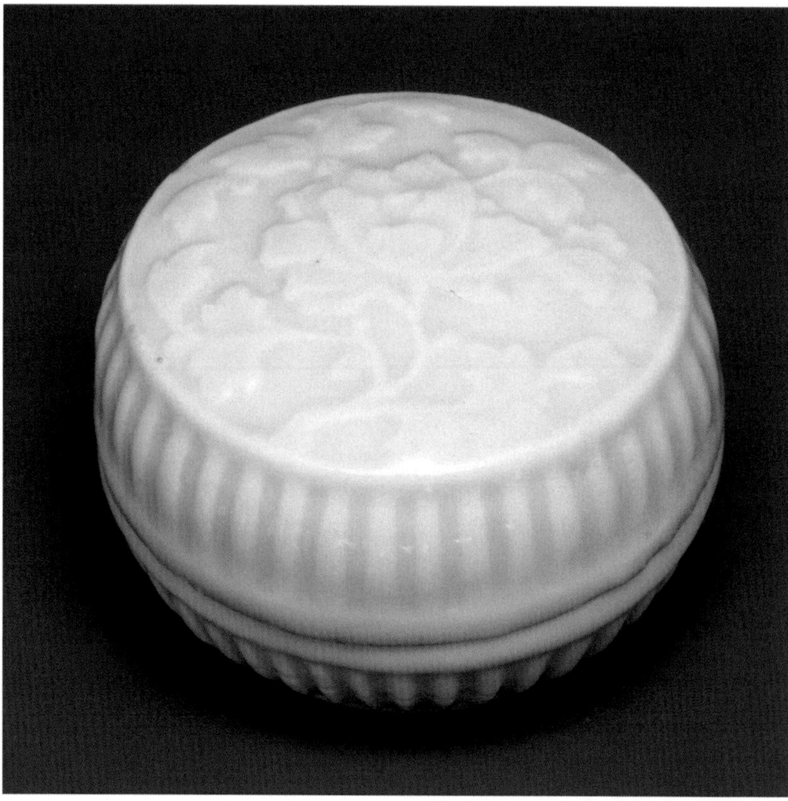

12. Chinese lidded box. Southern Song dynasty (1127–1279). Jingdezhen *qingbai* ware, porcelain with bluish-toned glaze and molded decoration. DIAM: 2½ in. (6.4 cm). Thomas D. Stimson Memorial Collection, Seattle Art Museum. Photo: Paul Macapia. 45.78.

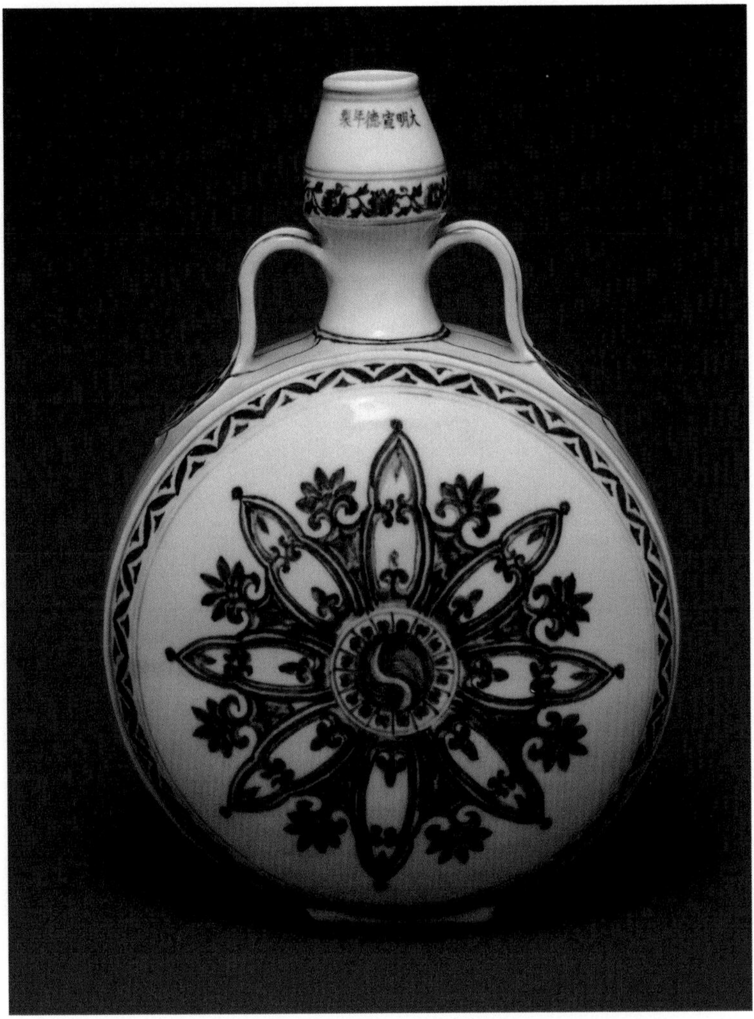

13. Chinese blue-and-white flask. Ming dynasty, Xuande reign mark and period (1426–35). Jingdezhen ware, porcelain painted in underglaze cobalt blue. H: 11½ in. (29.2 cm). Eugene Fuller Memorial Collection, Seattle Art Museum. The interlaced floral medallion on this flask is Islamic in origin. In shape, the vessel is a close relative of the pilgrim flask. The Xuande emperor's reign mark is just below the lip of the vessel. Photo: Paul Macapia. 48.167.

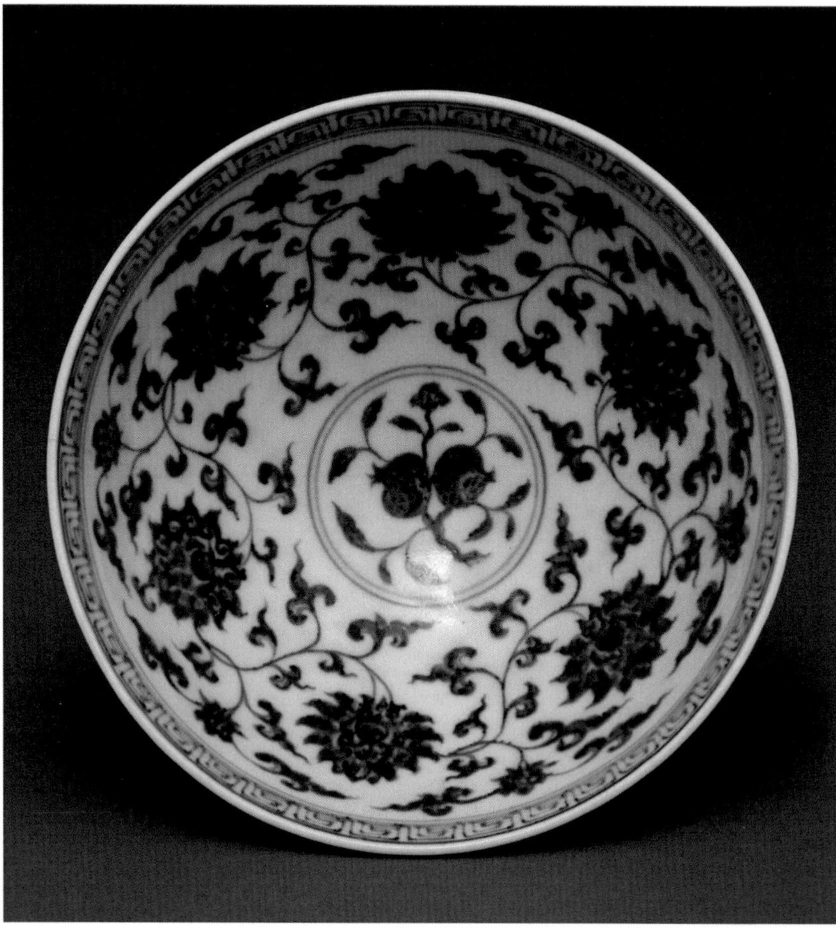

14. Chinese blue-and-white bowl. Ming dynasty, Xuande period (1426–35). Jingdezhen ware, porcelain painted in underglaze cobalt blue. DIAM: 8½ in. (20.6 cm). Eugene Fuller Memorial Collection, Seattle Art Museum. The bowl is decorated with lotus scrolls on the interior and double rows of lotus petals on the exterior, where the Xuande emperor's reign mark also appears. Photo: Paul Macapia. 49.154.

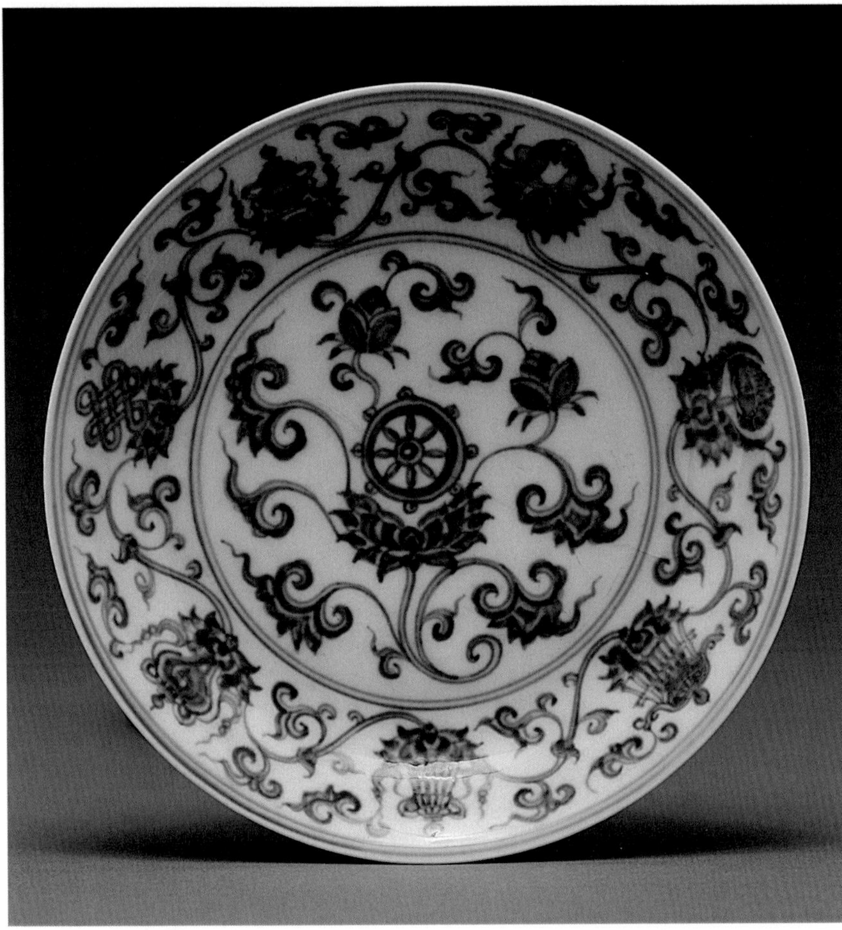

15. Chinese blue-and-white dish. Ming dynasty, Chenghua period (1481–87). Jingdezhen ware, porcelain painted in underglaze cobalt blue. DIAM: 7½ in. (19 cm). Eugene Fuller Memorial Collection, Seattle Art Museum. The dish is decorated with the eight auspicious emblems derived from Tibetan Tantric Buddhism. In the well of the dish, highly stylized versions of lotus buds encircle the Wheel of Law; the remaining seven emblems are painted on the rim of the vessel. Photo: Paul Macapia. 51.85.

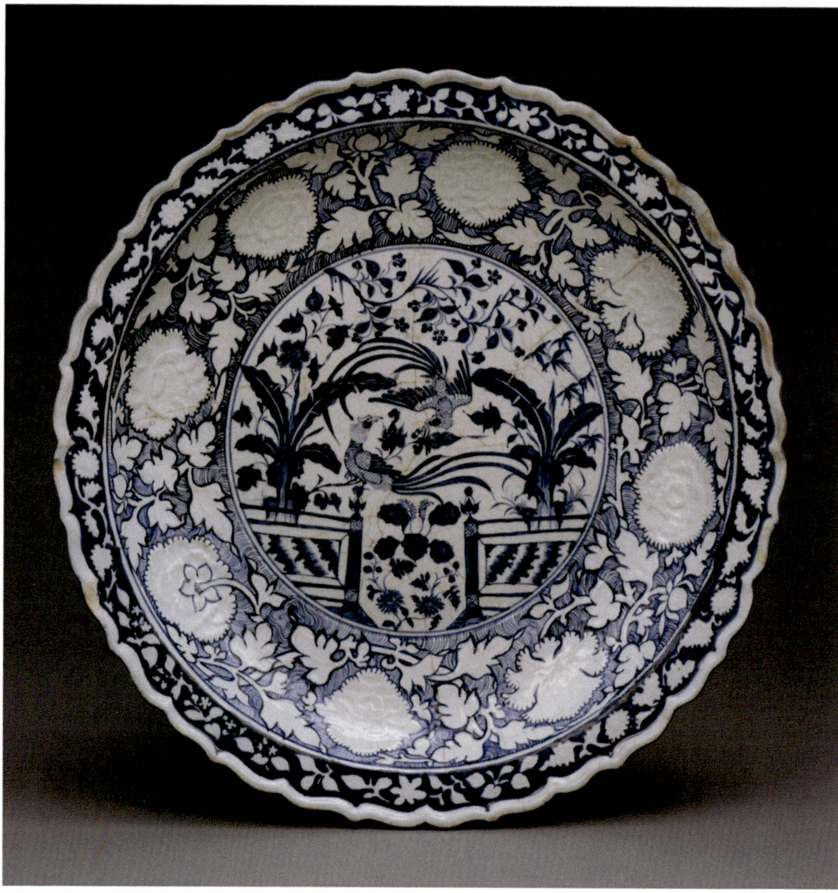

16. Chinese blue-and-white dish. Yuan dynasty, 14th century. Jingdezhen ware, porcelain painted in underglaze cobalt blue. DIAM: 18½ in. (47 cm). Purchased in memory of Elizabeth M. Fuller with funds from the Elizabeth M. Fuller Memorial Fund and the Edwin and Catherine M. Davis Foundation, Seattle Art Museum. Dishes such as this were made at Jingdezhen for Southwest Asia, where communal dining made large vessels necessary. Clay molds were used to produce such pieces. Photo: Paul Macapia. 76.7.

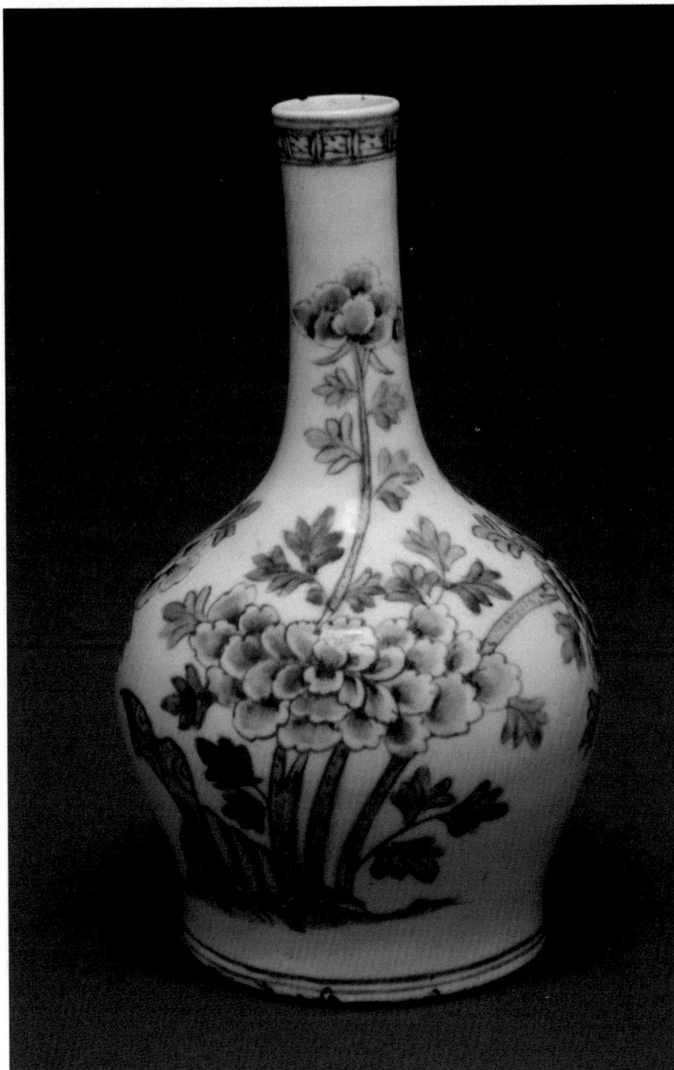

17. Korean blue-and-white bottle. Choson dynasty, ca. 1800–1850. Porcelain with underglaze blue decoration. H: 8 in. (20.3 cm). Gift of Frank S. Bayley III in Memory of Dorothy Stimson Bullitt, Seattle Art Museum. Photo: Paul Macapia. 82.127.

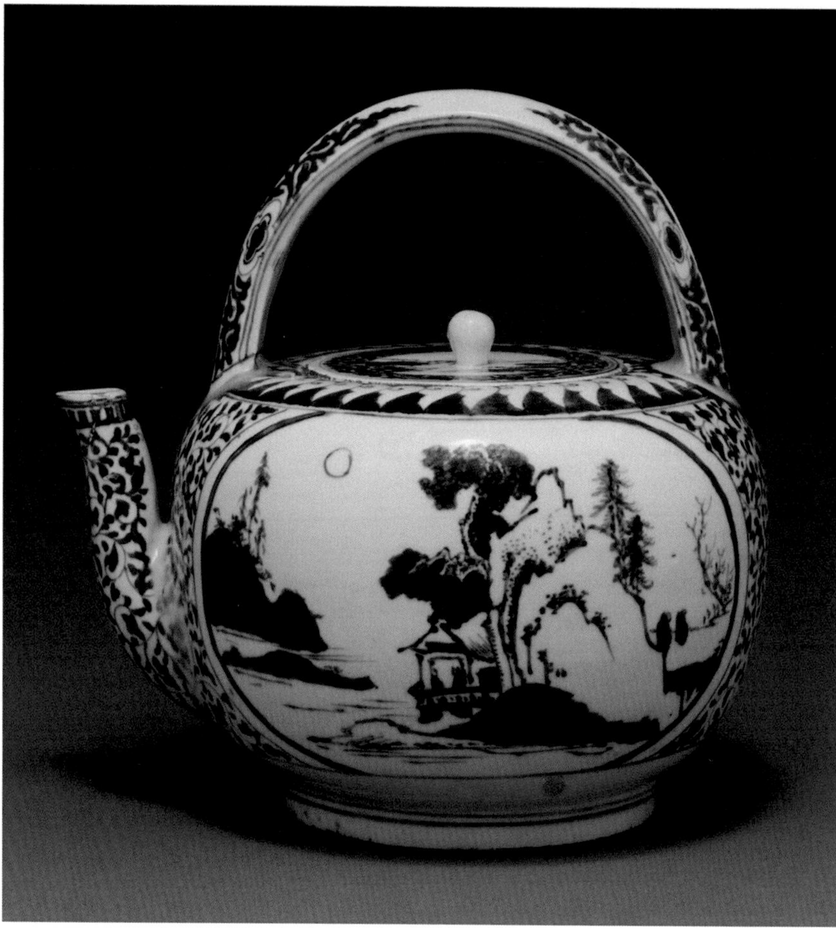

18. Japanese blue-and-white ewer. Edo period, ca. 1640–50. Imari ware, porcelain with underglaze blue decoration. H: 8⅞ in. (22.5 cm). Eugene Fuller Memorial Collection, Seattle Art Museum. Photo: Paul Macapia. 70.11.

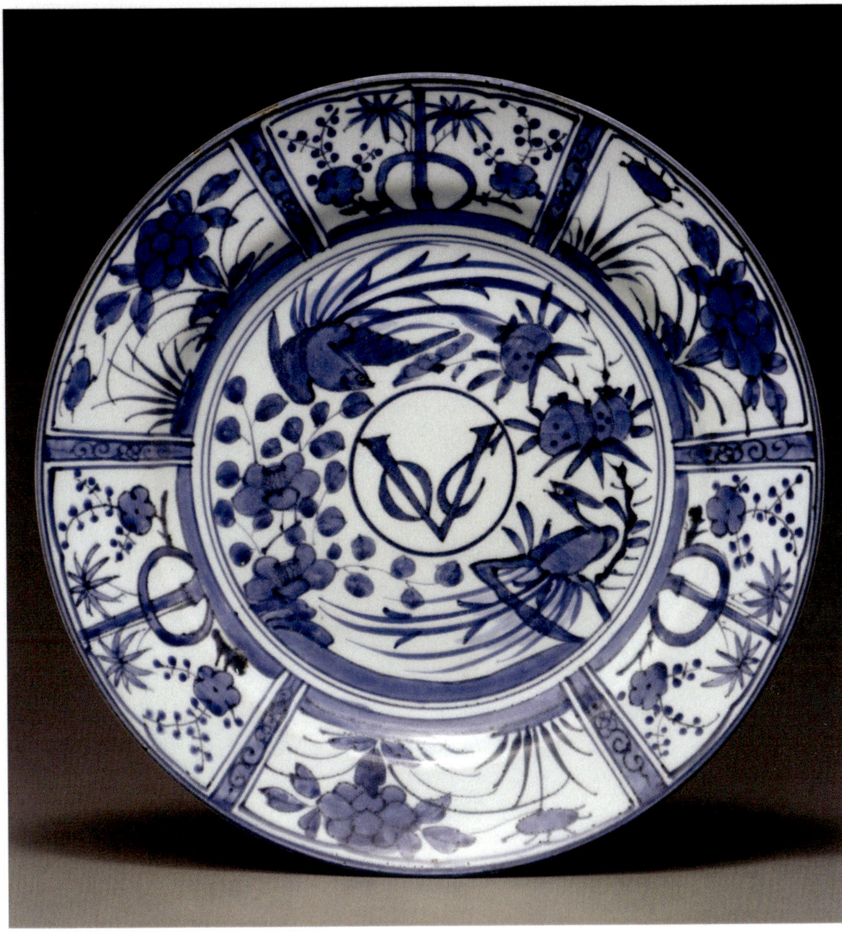

19. Japanese blue-and-white plate for the VOC, ca. 1660–80. Porcelain, Imari ware. DIAM: 15⅜ in. (38.9 cm) Floyd A. Naramore Memorial Purchase Fund, Seattle Art Museum. The monogram for the Vereenigte Oost-Indische Compagnie is in the center of this plate, encircled by two phoenixes, a flowering camellia, and pomegranates. The panels, in which decoration of bamboo plants alternate with peonies, are modeled on the style developed in Jingdezhen for Dutch *Kraakporselein.* Photo: Paul Macapia. 75.78.

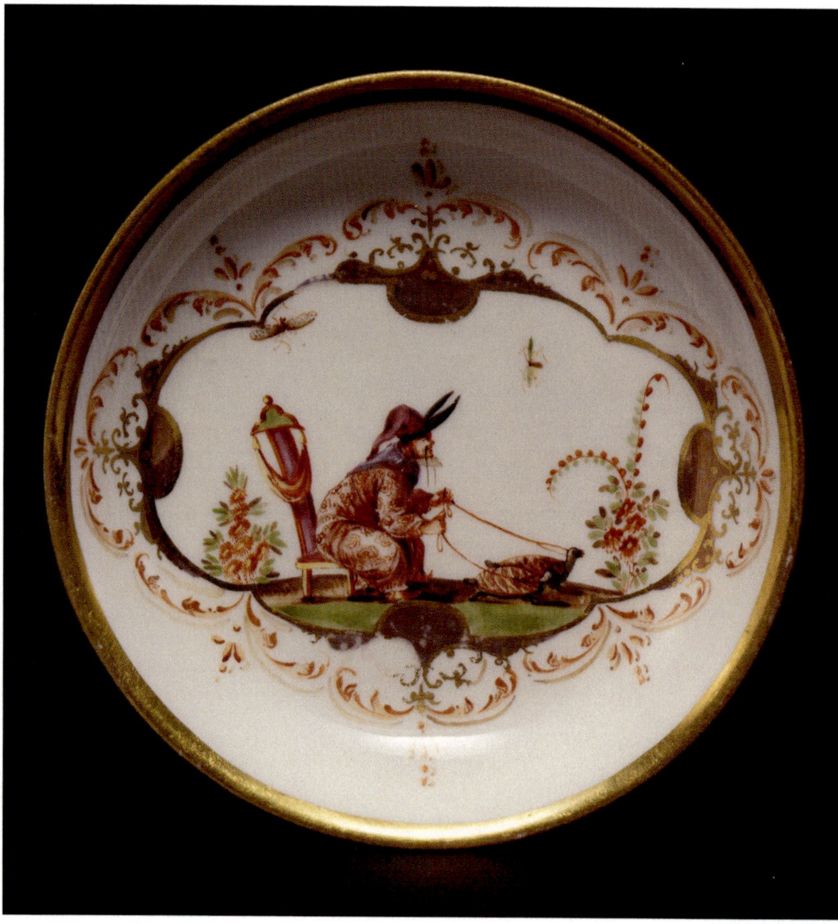

20. German saucer. Meissen manufactory, ca. 1725. Porcelain with enamel colors and gilding. DIAM: 5 in. (12.7 cm). Dorothy Condon Falknor Collection of European Ceramics, Seattle Art Museum. Photo: Paul Macapia. 87.142.101.

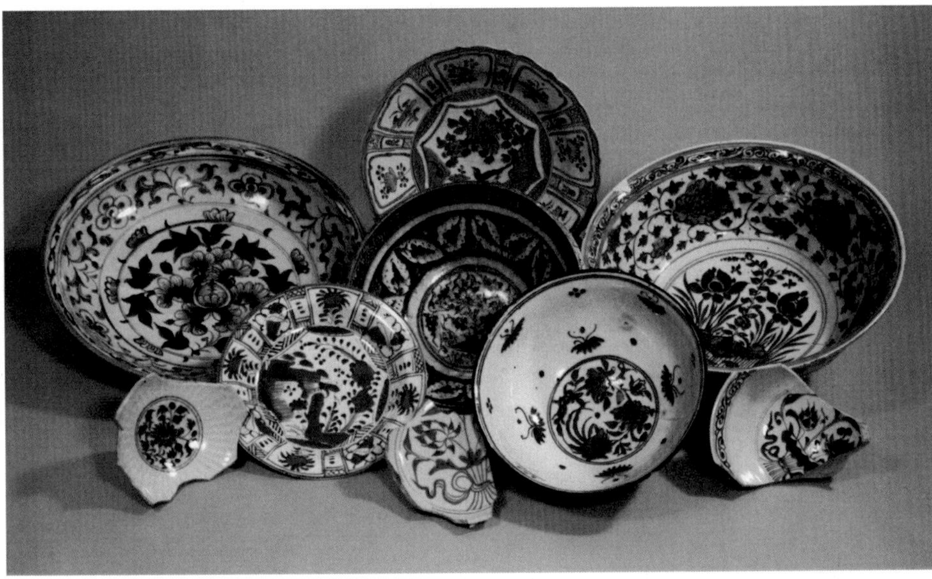

21. A collection of blue-and-white pottery. Large vessel on left: Vietnamese pottery, 15th–16th century. Large vessel on right: Chinese porcelain, 15th–16th century. Vessel at top center: Chinese porcelain, 17th century. Vessel in center, between two large vessels: Syrian pottery, 15th century. Vessel to left of Syrian piece: Japanese porcelain, 17th century. Vessel to right of Syrian piece: Persian pottery, 17th century. Three shards, Chinese porcelain, 14th century.

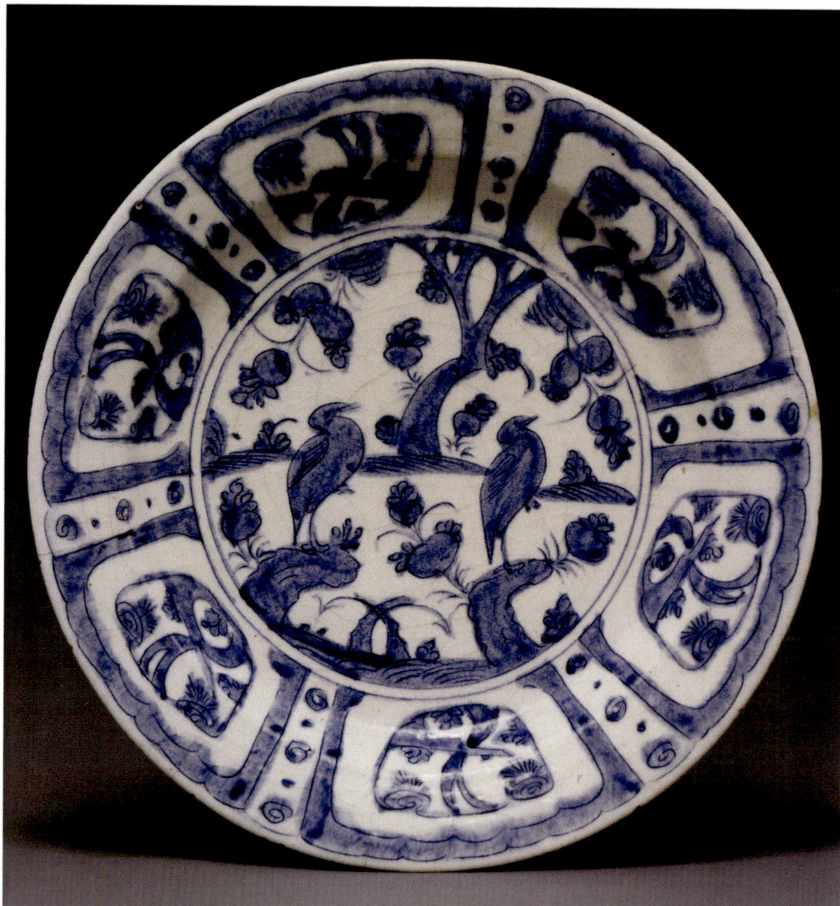

22. Persian plate. Safavid period, 17th century. Pottery with underglaze blue decoration. DIAM: 13⅞ in. (35.2 cm). Eugene Fuller Memorial Collection, Seattle Art Museum. Photo: Paul Macapia. 48.146.

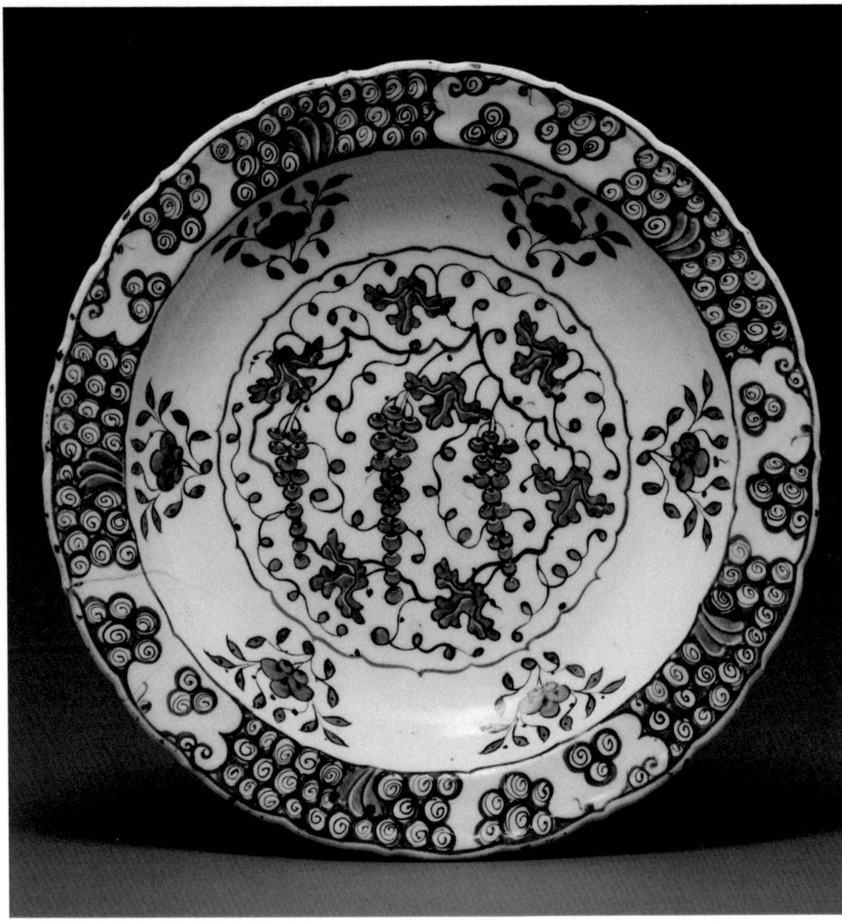

23. Ottoman plate. Late 16th century. Pottery with underglaze blue, black, and green decoration. DIAM: 12 in. (30.5 cm). Eugene Fuller Memorial Collection, Seattle Art Museum. The ornament mixes motifs used on vessels produced at the Iznik pottery and Jingdezhen. Photo: Paul Macapia. 57.17.

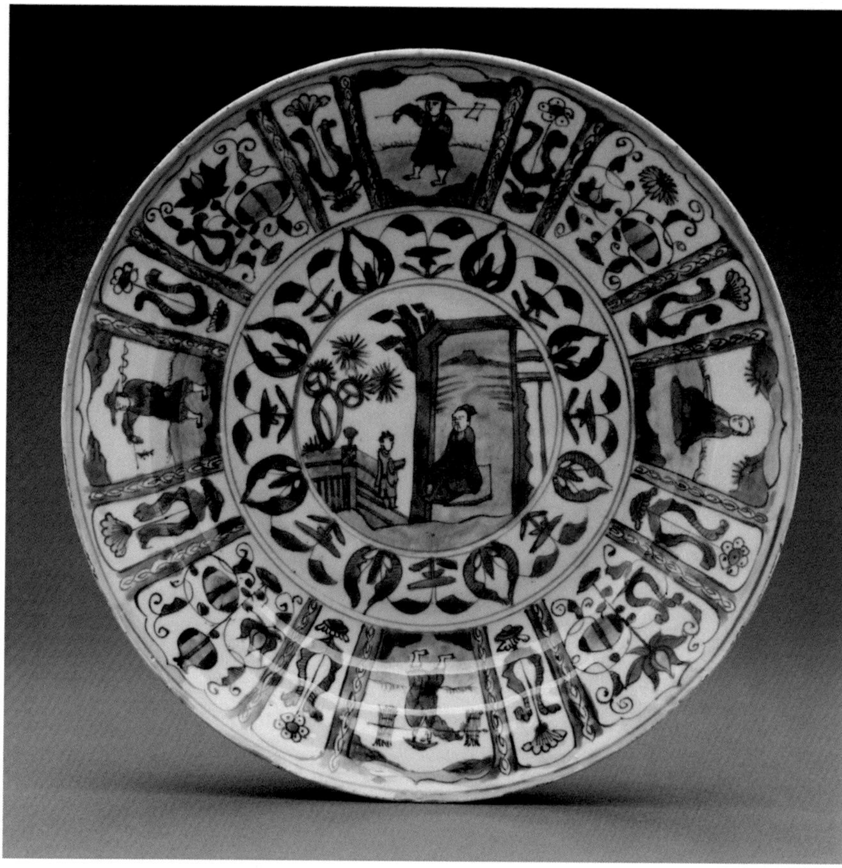

24. Chinese blue-and-white plate. Ming dynasty, ca. 1625–50. Jingdezhen ware, porcelain with underglaze blue decoration. DIAM: 14¼ in. (36.2 cm). Gift of Mr. and Mrs. Ford Q. Elvidge, Seattle Art Museum. The plate is *Kraakporselein,* probably modeled on a wide-rimmed Dutch wooden platter, with decoration of tulip buds and radiating panels with stylized flowers; the four panels represent down-to-earth rural labor and scholarly meditation, activities that stand for ideal pursuits among Chinese literati. Photo: Paul Macapia. 75.51.

single hue to delineate contour and volume. Moreover, spontaneity, which the literati enormously valued in calligraphy, came into play even in the routine labor of decorating blue-and-white. As with painting on silk, porcelain offered an unforgiving surface, permitting no second thoughts for those wielding brushes; it was essential that "their minds are not distracted," as Tang Ying puts it. And even more than the calligrapher and silk artist, the painter of porcelain had to master the tricky potentialities of the solution being applied: while the cobalt pigment looks murky black when applied to clay, after firing it is varied shades of blue, much as (Dentrecolles says) "the natural warmth of the sun makes the most beautiful butterflies, with all their tints, come out of their eggs."

Focusing on contour and volume, Chinese artists evoked the sense that the indeterminate field occupied by the design unfolds in three dimensions; they valued the impulsive, impressionistic, and undomesticated. Whereas Southwest Asian space functioned as an infinite, homogeneous expanse to be packed with fixed elements, Chinese space constituted an opening into limitless distance, a milieu through which things progress. Islamic painters rendered a pine tree as an elegant geometric pattern, its branches sinuously and meticulously entwining, all elements presented in a flat, symmetric outline; it is seen head-on, a still point of uniform order. Chinese painters depicted a pine tree as an embodiment of dynamic tension and natural movement, with a gnarled, twisting trunk, shaggy leaf clusters, and zigzag branches drooping to the ground and thrusting to the sky; it is seen from all sides, a manifestation of inexhaustible vitality. In *Chunjuan's Compilation on Landscape,* Han Zhuo (d. 1125), a functionary at the court of the Huizong emperor, explained that some pine trees possess

> the force of angry dragons or frightened young dragons, and some have forms like ascending dragons or crouching tigers; and some seem proud and haughty, or humble and modest; some spread out and lean over a bank as if to drink from the middle of a stream; some lean over from precipices of lofty mountain ranges with their trunks bending upward. These are the attitudes of pines, whose aspects are manifold and whose transformations are inestimable.[113]

According to Han, the painter strives to capture a fleeting moment in the life of a great pine tree, "the venerated ancient" of the world of flora, "a noble gentleman among lesser mortals."[114] Wu Zhen (1280–1324), a prominent Yuan artist, added an inscription to his *Crooked Pine,* a painted silk scroll: "You may hang [my pine tree] on a white wall in your hall; / At midnight it will fly away amidst the wind and clouds."[115]

While the worldly omnipresence of a transcendent divinity is expressed in the elaborate geometry of Islamic art, the Chinese notion of cosmic energy pervading the immanent order of the world gives birth to an aesthetic that values transformation, spontaneity, and potentiality. Contrast between the two traditions is epitomized by Islamic earthenware bearing interlocking units of rectangular Kufic script and Chinese porcelain decorated with cranes winging toward an imperceptible horizon.[116] China and Southwest Asia thus were committed to strikingly different views of reality that found expression in opposing aesthetic values and techniques.

Still, while they could obviously never be reconciled and they halted considerably short of fusion, the two traditions of design nonetheless learned from each other. When Egyptian, Syrian, and Persian potters began copying Chinese blue-and-white patterns in the early fifteenth century, they slowly adopted a more liberated sense of movement and space, opening up their designs to some of the vitality and spontaneity that characterized Chinese practice. For their part, Chinese craftsmen adopted elements of Islamic composition, such as banded ornamentation and stricter regulation of space. They became adept at rendering their own visual idiom in the context of Southwest Asian spatial organization. The result was an art unmistakably Chinese, yet with elements new to that culture, a coming together of traditions that proved irresistible in the Abode of Islam as well as around the world.

6

The Primacy of Chinese Porcelain

Korea, Japan, and Continental Southeast Asia, 1400–1700

Matteo Ricci recounted that when he showed some Chinese officials a European map of the world, they were puzzled to find the Middle Kingdom placed at its farthest eastern margin. When he later drafted a map for the Wanli emperor, he therefore so arranged it that "the empire of China occupied a more or less central position." Naturally, Ricci was concerned to respect the sensibilities of his hosts (and potential converts); yet he also believed that the exceptional nature of the Chinese domain justified the revision: "Considering its vast stretches and the boundaries of its lands, it would at present surpass all the kingdoms of the earth, taken as one, and as far as I am aware, it has surpassed them during all previous ages."[1] In any event, as Ricci realized, both Western and Chinese notions of cartographic propriety had more to do with cultural convictions than with strictly geographic considerations.

In the eyes of the Confucian ruling class, China graciously spread its culture to other countries while less civilized peoples expressed gratitude and ritual submission by presenting gifts to the imperial court, tribute to the Son of Heaven. Culture represented China's most significant export, establishing the image of the Middle Kingdom as a fountainhead of antique traditions, revered sages, and sacred texts. Naturally, Korea, Japan, and Vietnam felt the gravitational pull of China most powerfully. They constituted part of an interacting East Asian sphere of which China was the central domain, a realm that pursued a *mission civilisatrice* by sending abroad its ideographic script, state ceremonies, elite attire, law codes, bureaucratic methods, and Confucian canon along with its silks, paintings, and porcelains. Chinese served as the lingua franca of East Asia, Chinese writing its common medium for literate discourse. In some form, however attenuated and merely honorific at times, the Chinese examination system and Confucianism formed the basis of official ideology, administrative management, and scholastic knowledge.[2]

Chinese culture played the same role in East Asia that, in late antiquity, Hellenistic culture did in the eastern Mediterranean and Roman culture did north of the Alps: it provided a framework within which the elite traditions of other peoples found expression. The ruling classes of Korea, Japan, and Vietnam even adopted the Chinese concept of *zhongguo*, regarding their own respective polity as a political and cultural "Middle Kingdom," extending universal principles to people on the margins of high civilization. The cynosure of East Asia, China attracted many Buddhist monks from nearby kingdoms who thereafter carried its culture back to their own communities. After lengthy stays in Chinese monasteries, they returned home with sacred texts and relics, as well as with porcelains, paintings, and a passion for tea.

By the Song period, porcelain had become a central emblem of Chinese culture, an artifact peoples in other countries coveted and tried to emulate. They eagerly purchased the ceramic and incorporated it into their societies, sometimes in surprising ways. In general, the most creative response to Chinese porcelain by other ceramic traditions arose in regions with a sophisticated culture (as characterized by density of urban life, organized religion, and some measure of literacy), and peoples in relatively less developed areas sometimes abandoned their own pottery styles altogether. As far as ceramics were concerned, however, Korea, Japan, and continental Southeast Asia shared something more fundamental than a collective cultural context. The tectonic unit that underlies the geology of southern China also extends to parts of continental Southeast Asia (much of Vietnam, Cambodia, and Thailand) and, after a plunge under the Yellow Sea, reappears in the southern half of the Korean peninsula as well as in the southern islands of the Japanese archipelago.[3] Korea, Japan, and continental Southeast Asia thus possessed igneous-based materials, mainly clays, needed for making fine pottery. And as they were well situated to learn about Chinese kilns, glazes, and craftsmanship, they had the additional advantage of not having to develop a sophisticated ceramic technology entirely from scratch. Bringing a native genius and individuality to these twin advantages, material resources and geographic proximity, they not only mastered the attainments of China but also created pottery traditions with unique styles and repertoires.

In contrast, Sarawak in northwestern Borneo has excellent white clay that potters could have used for producing stoneware but which they exploited only for making rudimentary earthenware and crucibles in which to melt gold dust. Although Chinese craftsmen probably took certain glazing techniques to Borneo, Sarawak potters had no economic incentive to produce high-quality wares for a home market already abundantly supplied with jars, pots, and plates from China. As early as the Han period, Chinese merchants sailed to Borneo to trade ceramics and other commodities for pearls, gold, iron, camphor, hornbills, scented woods, and birds' nests (for gourmet soup).[4] Also, Sarawak potters did not work within a tradition of royal or aristocratic patronage, a powerful inducement for improving ceramics in Korea, Japan, and continental Southeast Asia.

"FIRST UNDER HEAVEN": THE CULTURE OF PORCELAIN IN KOREA

In a history of painting written around 1080, Kuo Ruoxi praised Korea as the only foreign nation "in which veneration for culture is taught, and which through the gradual infiltration of Chinese ways has attained at last a true refinement in artistic skill."[5] According to Chinese connoisseurs, pottery ranked among Korea's greatest accomplishments. Korean artisans replicated Chinese pottery from at least the Han period, though the principal influence on them came from southern China, not from the contiguous north. Sometime in the early centuries C.E., they began to produce stoneware, the earliest-known high-fired wares outside China. By the early ninth century, they made the transition from stoneware to porcelain with such success that the Song court greatly admired the green-glazed pottery Korean monarchs offered as tribute.

In his *Illustrated Account of Korea,* Xu Jing (1091–1153), a Song artist and calligrapher, wrote, "Koreans call the green color used in ceramics *kingfisher blue* [i.e., turquoise]. . . . The lion-shaped incense burners are most intricately made, and the color of the other pieces resembles the mysterious green of the vessels produced at the Yaozhou and Ruzhou kilns."[6] Located in the northern Chinese province of Shaanxi, the kilns of Yaozhou county in fact made only modest quantities of celadons, copies of vessels from Jiangsu and Zhejiang, the principal manufacturers of fashionable greenwares. Vessels made at the so-called Raozhou kilns actually came from Jingdezhen, best known in the Song for its jadelike, blue-tinted whitewares.[7] In short, Xu's statement indicates that Korean pottery owed a great deal to the techniques of potters from south of the Yangzi.

The western coast of Korea is highly irregular, sprinkled with thousands of inlets and islands. Unlike the inhospitable, smooth eastern shore, it presents a welcoming face to the sea and easy passage between the peninsula and southern Chinese ports. Kilns in Cholla province in southwestern Korea produced the best pottery, and the westernmost tip of Cholla points like an arrow to the delta of the Yangzi, five hundred kilometers across the Yellow Sea, a shallow (average depth, about forty-five meters), easily navigable waterway. Trade linked the two regions from the period of fragmentation, lasting from the third to the sixth century, that followed collapse of the Han dynasty. With the fall of the Han in 220, Korea broke free of Chinese control for the first time in four hundred years. The peninsula then entered the war-wracked era of the Three Kingdoms—Koguryo, Paekche, and Silla—all aggressive and expansionist, all tightly organized, each allying with powers in Mongolia, China, and Japan in the struggle for survival and supremacy. Division of Korea ended when the southeastern kingdom of Silla (with help from the Tang dynasty) consolidated control over its rivals, inaugurating the period of Unified Silla (675–935).

In battles of the Three Kingdoms, an economic strength of the southwestern kingdom of Paekche (18 B.C.E.–660 C.E.), which encompassed the future Cholla province, was that it dominated trade with southern China.[8] Disregarding the Chinese states of the north, it sent numerous embassies to those controlling the coasts of Zhejiang and Fujian. Many Buddhist monks from Paekche went on pilgrimage to southern China, and according to one account, first recorded in the Tang period, a Paekche monk beheld a vision of Avalokitesvara in a Zhejiang monastery. The Paekche king Muryong (r. 501–23) and his queen were buried in a brick-built, tunnel-shaped vault typical of those on the southern coast of China; objects entombed with them included gold crowns, bronze mirrors, a lacquered wooden coffin, and Chinese-style silver jewelry. A stone statue of a Chinese *qilin*-like animal guarded the entrance to the grave, clay bricks bore inscriptions of Chinese lotus motifs, and Chinese stoneware vases stood in niches along interior walls. Some of the southern Chinese pottery found in fourth-century tombs of the kingdom of Silla (57 B.C.E.–618 C.E.) entered the peninsula at Paekche ports.

Potters in Cholla province may have learned some techniques, such as the construction of modest versions of the dragon kiln, from southern Chinese potters who escaped local turmoil by moving to Korea. According to a Japanese monk who traveled to China in the Tang, migration went in both directions, for families from Paekche and Silla settled in coastal China, where they farmed, traded, and piloted riverboats. The ports of southern Korea and China also gave Korean polities a gateway to the rest of the ecumene. Cosmopolitan and refined, Paekche maintained close contacts with Japan, playing a key role in the transmission of Buddhist art and culture there during the formative Asuka era (552–710). Buddhist monks from Paekche also went on pilgrimage to India by ship, and a glass goblet from Alexandria in Egypt, excavated from a Silla tomb dating to 300 C.E., most likely came to Korea by sea. An Arab work some time later described Silla as a "gold-glittering nation."[9]

In the period of the Three Kingdoms, rulers established stringent control of their populations and resources by creating agencies to supervise collective labor, status relations, commerce, and craft production. On a national level, the kings of Unified Silla pursued similar authoritarian policies, a legacy carried forward for a millennium thereafter by the Koryo (918–1392) and Choson (1392–1910) dynasties. In the absence of extensive bureaucracies, the compact size of their realms aided centralization. Unlike China and Vietnam, the Korean state reached its natural limits, bounded by the Sea of Japan, the Yellow Sea, and the broad Yalu River, very early in its history, with territorial closure thereby giving it unparalleled stability.[10] Governance extended over just 220,500 square kilometers, a landmass equivalent to that of England, which came under effective centralized rule around the same time the Koryo regime supplanted Unified Silla. The Tang and Song empires possessed considerable space for many kiln centers, few of which received direction or supervision from the imperial court. But the royal houses of Korea and their ruling class of aristocratic literati *(yangban)* had the capacity to control pottery production in monolithic fashion, thereby ensuring striking consistency of aesthetic purpose for twelve centuries.

A government office managed pottery workshops that supplied vases, incense burners, and water pots for the court and aristocracy. Administrators organized potters into craft guilds and registered them on government lists. They took charge of the mining of clay and the provision of pigments, and they ensured that pottery painters correctly executed designs. An office for managing tile manufacture constituted part of the directorate that supervised construction of temples, palaces, and tombs. King Hyonjong (r. 1009–31) appointed officials to oversee dress codes and major handicrafts. King Sonjong (r. 1083–94) decreed that "craftsmen and merchants must serve the king in their given profession and skills and may not be appointed government officials."[11] The regime enforced strict segregation of ranks; the *yangban* elite of some fifteen hundred officials served the court, and their families lived in communities that excluded peasants. Government agents regulated pottery use according to class, with the privileged having access to fine vessels and the lower echelons making do with bamboo platters, terra-cotta, and crude stoneware. Masses of chattel slaves huddled at the bottom of the social ladder, constituting fully one-third of Korea's population by the Choson period, roughly the same percentage as in the antebellum American South.[12]

Rigorous control of handicrafts by the Korean state resulted in high-quality production serving an extremely restricted domestic clientele. Unlike the more or less self-sufficient entrepreneurs of Jingdezhen, with their eyes on the main chance and return on investment, Korean kiln owners had slight opportunity to sell their first-class pottery in foreign markets. Still, official supervision and patronage, combined with salutary influence from southern Chinese potters, paid aesthetic dividends. By the end of the first century of the Koryo dynasty, Koreans turned out celadon wares that even the Chinese admired for their excellence. Although Japanese potters generally refrained from imitating Song porcelains, perhaps because their copies would have fallen so short, Koreans rightly felt no such qualm. Their green-glazed vessels have graceful forms, unaffected style, and innovative decoration, comparable to the finest output of Song kilns.[13]

At the same time, the high value placed by the Koryo elite and their potters on spontaneity and vitality meant that Korean producers lacked the zeal for technical perfection and high gloss that characterized their Chinese counterparts. Koreans deemed slightly lopsided forms, unpredictable surface markings, and mottled glazes refreshing and attractive—a taste they passed on to the Japanese. Korean potters also turned out wares in natural, beguiling shapes, such as water pots in the form of a mandarin duck, pomegranate cluster, or dragon-headed tortoise; incense burners in the form of a lion, rooster, or fish-dragon; and water droppers in the form of a gourd, monkey, or lotus bud. Along with such wares, the Korean upper stratum also took up the Chinese habit of tea. According to Xu Jing, who visited the Koryo court in 1123:

> The people of Kao-li [Koryo] have become addicted to tea drinking, and many kinds of implement are made: a black tea bowl ornamented with gold, a small tea bowl of *kingfisher colour*, and a silver *ting* for heating water are modelled after Chinese wares. . . . Tea is set out there three times daily and is followed by hot water. The people of Kao-li regard hot water as medicinal.[14]

A distinctive Koryo technique known as *sanggam* (inlay), perhaps inspired by Chinese lacquerware with insets of mother-of-pearl, involved incising designs in the gray clay, filling the carved outlines with slip, a dilute black or white clay mixture, then applying glaze and firing. Requiring precise control of the furnace, the effect at its best achieved the look of delicate brush painting on a milky-green surface. Potters commonly decorated the porcelains with motifs of stylized cranes, clouds, willow trees, peony scrolls, and chrysanthemums.[15] The most illustrious Koryo ceramics, the inlaid celadons won the greatest praise from contemporaries. A Chinese envoy in Korea's capital of Songdo (present-day Kaesong) in the eleventh century praised the Koryo court for its sophisticated culture, all derived from his homeland—manuscripts, costumes, wines, inkstones, silk brocades—but he considered its pottery exceptional: "As for the 'secret color' of celadons, the 'secret color' of Koryo is first under Heaven. Although [potters of] other areas imitate them, none of them can achieve [the same qualities]."[16] King Euijong (r. 1146–70) impressed Chinese visitors with his summer pavilion outside Songdo, roofed with celadon tiles at scandalous expense. Yi Kyu-bo (1168–1241), a Korean poet and scholar, evoked the achievement of Korean potters in terms reminiscent of contemporary praise for Jingdezhen's *qingbai:*

> The felling of trees left Mount Namsan bare
> and the smoke from the fires obscured the sun.
> The wares produced were celadon bowls:
> out of every ten, one was selected—
> for it had the bluish-green lustre of jade.
> It was clear and brilliant as crystal, it was hard as a rock.
> With what skill did the potters work—
> it seemed as if they borrowed the secret from Heaven![17]

Centralized royal control and oversight by the *yangban* hierarchy explain the abrupt transitions in Korean pottery fashions. A change in dynastic power—a markedly rare event in Korean history after 675—meant a change in what potters had to produce for their masters. The great era of Koryo celadons came to a sudden end soon after 1231 as a consequence of repeated Mongol invasions of the peninsula. The Koryo court fled to an island off the west coast, bringing supervision and control of potters to a halt. After finally conquering Korea in 1259, the Mongols permitted the king to return to Songdo so long as he accepted his status as a docile vassal. Weighed against the great prize of China, Korea was a distracting sideshow to the Mongols, who wanted no trouble in the peninsula as they mounted their assault on the Southern Song. When the Korean court sent inlaid celadons painted with gold to Kublai Khan, he responded by berating it for the opulence of the vessels and forbidding further production of them.[18] After the Hongwu emperor forced the last Yuan ruler to flee to the steppes in 1368, the discredited Koryo royal house faced a bleak future.

Yi Song-gye (1335–1408), a Korean general, seized power in a coup in 1392 and became the first ruler of the Choson line. (Hongwu bestowed the name on the dynasty, taking it from an old Chinese designation for the region; in most Korean accounts, the dynastic designation is *Yi.*) He built a new capital at Seoul and began to change the kingdom's ideological orientation. Buddhism, which had come to Korea from China, had never suffered proscriptions such as those imposed on it by the Tang government after the An Lushan rebellion. During the reign of the Koryo, Buddhism became a virtual state religion, employed by the royal house to bolster its sacred aura and authority. Buddhism also reached down to the common people, among whom Avalokitesvara, in the protean figure of Guanyin, took on the role of the Korean mother-goddess of the Earth. Many of the inlaid celadons produced for the elite were used in Buddhist rites and for display on family shrines.[19]

Following the lead of the Ming, however, the Choson dynasty adopted Neo-Confucian teaching as its governing ideology and restricted the political and cultural sway of Buddhism, not least because the new rulers regarded Buddhist clergy as loyal to the deposed Koryo line. Indeed, Choson Korea developed into a more single-minded Confucian regime than China itself, though Confucianism remained a preoccupation of the *yangban* and did not filter down to commoners. The Choson dynasty spurned old-style celadons as lavish, corrupt, and pretentious, tokens of the allegedly pleasure-besotted Koryo sovereigns. In the service of Confucian devotion to frugality and simplicity, Choson royalty favored austere, practical ceramics, chiefly white porcelains with a blue-tinted glaze and rustic *punch'ong* (powder blue) vessels, stoneware coated with a slip of white clay. According to a census of the early fifteenth century, 324 kilns turned out those wares. Potters decorated both types with plain motifs, such as stamped or incised images of flowers, grass, birds, and fish. Natural and lively in character, the ornamented vessels did not aspire to high art or sophistication. The court reserved undecorated white porcelains for use in the royal palace and in official rituals.[20]

The Choson government suppressed the production and importation of colorfully ornamented pottery. Blue-and-white porcelain proved acceptable, though, inasmuch as the Ming court was Korea's political patron and cultural model. Choson rulers received gifts of the Jingdezhen ware from the emperor, and they responded by having their own potters imitate the style. Although the potters at first routinely copied Chinese blue-and-white, Korean individuality quickly asserted itself, with wares displaying the elegant simplicity typical of white porcelains and *punch'ong*. In addition, scarcity of fine cobalt pigment dictated spartan decoration and modest output. Although Koreans discovered their own source of cobalt ore in the 1460s, it produced only unsatisfactory grayish blue tones, so the Persian material still had to be imported at great expense; functionaries doled it out only to accomplished painters at select kilns. As with white porcelains, Choson rulers prohibited use of blue-and-white by commoners. Korean ambassadors presented blue-and-white vessels to the Ming court on tribute missions, but while attractive and admirably executed, they never received the acclaim from Chinese connoisseurs that inlaid celadons had centuries earlier.

In spring 1592 a foreign army once again invaded Korea. Obeying the commands of Toyotomi Hideyoshi (1536–98), the warlord-unifier of Japan, 150,000 troops stormed from Pusan in the southeast to Seoul, a distance of 355 kilometers, in only three weeks, their way cleared by devastating use of gunpowder weapons. Claiming a heavenly mandate for his ambition (but poorly informed about his adversaries), Hideyoshi aimed at nothing less than the conquest of China, after which he intended to march on Southeast Asia and India.[21] Unlike the Mongols almost three hundred years earlier, however, the Japanese ultimately failed in their attack, in part because the Ming sent tens of thousands of troops across the Yalu to support Korea.

Although a negligible matter amid the untold misery and hardship of the Korean people, the Japanese invasion derailed the ceramic history of the peninsula. Armies ravaged Cholla province and the countryside around Seoul, the main centers of pottery production, destroying kilns and forcing potters to flee for their lives. Moreover, when Japanese troops withdrew from Korea in winter 1598, after the death of Hideyoshi, they took with them at least sixty thousand prisoners, sometimes removing the inhabitants of entire villages. Hundreds of potters and their families numbered among those carried into captivity.

Korean pottery did not regain its vitality for two generations, for when the Choson court finally turned its attention to the industry, clay mines had to be reopened, kilns rebuilt, and new craftsmen trained. The wars spelled the end of production of *punch'ong* vessels, evidently because the Japanese killed or captured so many rustic potters. From the mid-seventeenth century, white and blue-and-white porcelains reigned supreme among the Korean elite, while commoners used rough, brown stoneware. With increased imports of cobalt pigment, potters produced vessels in blue-and-white, with a delicate silvery-blue tone and spontaneous brushwork, more extensively than before. (See figure 17.)

The Japanese assault, then, temporarily wrecked the Korean pottery industry but failed to destroy the high tradition of Korean ceramics. Well into the nineteenth century, the Choson royal house, the most enduring and stable dynasty in world history, the apex of a slave society and despotic hierarchy, encouraged the production of pottery that combined unaffected simplicity with warmth, refinement with humanity, subtle design with subdued shades. Partly as a consequence of the kidnapping of Korean potters, Japanese ceramic artistry also displayed those qualities as it came to maturity in the seventeenth century.

"CHINESE OBJECTS": THE CULTURE OF CHINA IN JAPAN

The invasion of Korea by Hideyoshi meant suffering for all involved. Japan experienced humiliating defeat and serious casualties among its lordly class of daimyo (lit., "great names") and their bands of samurai retainers. Japan would not return in force to Korea until its stunning victory in the Russo-Japanese war of 1905 enabled it to take over the peninsula without opposition. As a result of pouring men and money into aiding the Choson, the late Ming regime, already facing a financial emergency, failed to withstand the Manchu threat from the north a generation later. With their countryside and towns in ruins, Koreans naturally suffered the most. The toll on them could hardly be calculated, though the Japanese did their best. They kept track by cutting off the noses of Korean fatalities and packing them in salt tubs to be dispatched to Kyoto, where Hideyoshi's clerks counted tens of thousands in a given month. The contingent of warriors led by Nabeshima Katsushige (1580–1657), first lord of Hizen province in Kyushu (modern Nagasaki prefecture), claimed credit in 1597 for garnering 5,444 noses in just five weeks.[22]

Nabeshima, however, is not remembered for his battlefield heroics but rather for kidnapping scores of Korean potters and settling them on his estates. Along with the hostages of other daimyo, they never made it back home. Restricted to their villages and despised by neighboring Japanese, they could not move elsewhere. Daimyo kept guards at the kilns and required craftsmen to wear identification tags to monitor their movements. A local potentate placed one potter under house arrest for six years for begging to return home. A century later, a Japanese visitor to a so-called Korea Town on the coast of Kyushu, two hundred kilometers across the Straits of Tsushima from Korea, reported that the fifteen hundred residents in the area, mainly descendants of Nabeshima's prisoners, still gazed out to sea and longed for their ancestral soil.[23]

Nabeshima supervised foreign trade in Kyushu and developed the potteries under his control into the largest producers in Japan. He evidently had no aesthetic interest in pottery but rather hoped to augment his revenue and to present wares as gifts to other clan lords. By the opening years of the seventeenth century, his Koreans had unearthed kaolin at Mount Izumiyama near Arita, forty kilometers north of the town of Nagasaki; they discovered it near hot springs, evidence of ancient volcanic activity. The potters soon after introduced the Korean version of the dragon kiln, called the *noborigama* (rising kiln), to Japan, which not only increased production tenfold but also, for the first time, provided the high temperature needed to fire porcelain. It is because of such achievements that the Korean conflicts between 1592 and 1598 are called the Potters' Wars (or Teabowl Wars). The designation proposes that the merits of Japanese pottery derived from a virtual act of larceny inasmuch as Korean prisoners introduced new kiln technology and use of kaolin to the land of their captors. Korean potters, however, had worked in Kyushu long before Hideyoshi's assault, and their wares had been prized in Japan for generations. Karatsu, the port in Hizen from which Hideyoshi's warriors sailed for Pusan, sold plentiful Korean-style pottery to its neighbors.[24] Japanese kidnapping of Korean potters thus points to eagerness for acquiring more such expertise, the pitiless seizure of an opportunity arising from war, not the beginning of something new.

Japan knew only crude, red-colored terra-cotta before Korean potters arrived there in the early centuries of the common era, bringing with them knowledge of glazing, the potter's wheel, and high-firing techniques.[25] According to legend, a Paekche king sent a teacher to Japan in the early 300s to instruct a crown prince in the Confucian classics and Chinese poetry. Along with ceramics, other elements of Chinese culture to reach the Japanese islands by way of Korea during the same period included bronze technology, mounted warfare, ideographic script, the calendar, chronicle writing, and worship of Avalokitesvara/Guanyin. Shotoku Taishi (574–622), a crown prince and regent, granted official recognition to Buddhism, hired Korean architects to design Chinese-style temples, and brought a Korean monk to court to supervise the manufacture of paper, ink, and inkstones. The regent won posthumous elevation for his devotion when Buddhists declared him the incarnation of Kannon, the Japanese name for Avalokitesvara/Guanyin. During the time of the crown prince, Buddhists established pilgrimage routes dedicated to Kannon, lined with a string of thirty-three temples to memorialize the manifold forms taken by the bodhisattva.

In the late seventh century, elite refugees from the kingdoms of Koguryo and Paekche, fleeing the takeover of Korea by Unified Silla, brought Tang styles of painting, sculpture, and temple architecture to Japan. Soon after, some twenty Japanese diplomatic missions went to the Tang court and brought back extensive collections of Chinese texts. In the Asuka and Nara (710–94) periods and in the early Heian era (794–1185), Japan looked to the Tang regime as a model of imperial centralization. The Tang capital of Chang'an served as the ground plan for the Japanese capital at Heijokyo (modern Nara) in 670 and for that at Heiankyo (modern Kyoto) in 794. Japanese emperors *(tenno)* mandated Tang dress for men at court, and they imported Chinese green-glazed tiles at great expense to roof their palaces in Kyoto. When the emperor appeared enthroned during state ceremonies, large silk-covered panels depicting Confucian sages garlanded the wall behind him. Emperor Saga (r. 809–23), a skilled calligrapher and poet in Chinese-style *(kanshi)* verse, enormously admired Confucian moral teaching and political doctrine; he staged court ceremonies according to Chinese protocol and sponsored the compilation of three influential anthologies of *kanshi* poetry. He followed the counsel of his imperial predecessor, Cao Pi (r. 220–26), whose essay on Chinese literature declared that "writing is a great enterprise for governing the state," a phrase Saga appended as a title to one of his anthologies.[26] Substantially linked to the commerce of the mainland for the first time, Japan imported large quantities of Chinese merchandise. It also became the easternmost destination for goods carried on the Silk Road: Saga and other Heian monarchs stored Persian silver vessels and Persian gold-threaded brocades in the Shosoin Imperial Treasure-House at Nara.

Since indigenous high culture had yet to develop substantially, Japanese traditions came close to being overwhelmed by elite partiality for things Chinese in the Asuka and Nara periods. As much as they admired Chinese culture in its own right, however, Japanese monarchs also employed it to bolster their power and prestige, thereby setting a fashion emulated by the nobility. The court sent at least seventeen embassies to China, each comprising hundreds of men, before 779. Above all, the emperor's entourage aimed to control the import of *karamono* (Chinese objects), especially celadons, hanging scrolls, bronze incense burners, books, and writing utensils.[27] For the Japanese elite, situated at the outskirts of the ecumene, Chinese objects represented concrete signs of civilized life, verifications of cultural refinement. Around the same time that Chinese monks such as Xuanzang journeyed to India to acquire Buddhist texts and cult souvenirs, Japanese monks, licensed by the imperial court, visited Buddhist temples in Zhejiang and Fujian to collect scripture and *karamono.*

By the mid-eleventh century, however, demand for Chinese commodities outran the efficiency of government controls. As Japan shifted from an administrative system of trade to a private mercantile one, the court lost its monopoly on Chinese contacts. Concurrently, the Japanese economy expanded and a provincial military aristocracy gained power. The overpowering Chinese economy of the Song period, especially its exports of copper-cash, played a crucial role in stimulating these changes. In some years, cash imports to Japan came to 100 million pieces, the equivalent of the coinage minted annually by the Southern Song government. Since it sold in Japan for five times its worth in China, Korean and Chinese entrepreneurs gained fortunes trading the currency. Song cash spurred transition to a money economy, fostered growth of market towns, and enriched noble landholders. As the circulation of Chinese coins undermined the traditional, barter-based economy, contemporaries bewailed the plague of "money sickness," a grievance associated with the appearance of professional moneylenders in the mid-1250s. Japan did not mint its own copper coins until the sixteenth century, by which time it had begun exploiting its own copper and silver mines and exporting huge amounts of the two metals to China.[28]

In a long perspective, economic developments stemming from the flood of Chinese cash established a unique setting for the cultural role of porcelain. In Korea, the royal dynasty and *yangban* served as exclusive channels of Chinese culture and supervised production of restricted pottery forms. In Japan, the edgy collaboration of Buddhists in monasteries, merchant commoners in towns, daimyo in the provinces, the emperor and his court nobility, and a ruling generalissimo, or shogun, with his own entourage determined the value of Chinese things, as well as advances in Japanese ceramics. As central power weakened in the twelfth century and society became increasingly militarized, daimyo emerged as the focal group, the one linking all others. Even as late as the sixteenth century, the age of the three great unifiers of Japan—Oda Nobunaga (1534–82), Hideyoshi, and Tokugawa Ieyasu (1543–1616)—the hegemony of the *bakufu* (tent government, or shogunate) depended on the eminence of some 250 daimyo on their local estates.[29] Centuries before Hideyoshi's Korean adventure, provincial barons such as Nabeshima Katsushige maintained ties to monasteries and market towns, established pottery kilns, patronized eminent potters, and controlled the ports through which Chinese commodities and *karamono* entered the country.

Merchants imported Longquan celadons, Fujian whitewares, and Jingdezhen *qingbai* to Japan from early Song times. Archaeologists have dug up Chinese shards from a tenth-century Kyushu trading post, and antique fragments still wash up on beaches after storms on the Inland Sea, the major commercial highway in premodern Japan. Fujian and *qingbai* pottery was produced in white, the achromatic color favored by Buddhists; hence worshipers buried thousands of the vessels in sutra mounds in the late Heian and early Kamkurara (1185–1333) periods. The widely popular Pure Land Amida sect of Buddhism preached the messianic doctrine that mankind had crossed into the "Latter Days of the Law" (Mappo), a time of suffering, disorder, and darkness. From the late Heian, believers buried copies of Buddhist scripture in clay and metal cases to preserve them during the epoch of decline, which they thought had begun in 1052 (1,500 years after the Buddha Gautama's death) and expected to end with the Buddha Amida shepherding them into the Pure Land paradise. Most of the pottery cases came from Fujian kilns, made to order for the purpose.[30]

While the pessimism of the Latter Days of the Law represented a response to the tumult that accompanied the rise of samurai warriors from the eleventh century, it also expressed the Japanese sentiment of the fleeting nature of existence, typically epitomized in poetic imagery of impermanence—the scattering of autumn leaves, the fall of cherry blossoms, pine trees in winter, the moon at dawn.[31] Discussing Japanese taste in representations of nature, Rodrigues enumerates favorite painting scenes, such as those showing the moon reflected in water or snow descending at night: "All this is in keeping with their temperament, and makes them feel very nostalgic and quietly lonely."[32] That outlook had a lasting impact on Japanese taste in material culture, above all in ceramics. It appears in *The Tale of Genji* by Murasaki Shikibu (973–ca. 1013): acclaimed today as the world's first psychological novel, its plot was plundered for centuries by Japanese artists for illustrations on porcelains, silk hangings, and lacquer panels. Murasaki has her eponymous prince say, "I have been told that in China nothing is held to surpass the brocades of spring, but in the poetry of our own country the preference would seem to be the wistful notes of autumn."[33]

Sei Shonagon, Murasaki's contemporary, expressed the same view. In her *Pillow Book* (ca. 1000), a compendium of ruminations and complaints, she lists "Things That Give a Clean Feeling," all of them distinguished for being unadorned and colorless: "A new earthen cup. A new metal bowl. A rush mat. The play of the light on water as one pours a vessel. A new wooden chest." Yet after viewing a procession in the imperial palace, Sei, a consummate snob, recorded her approval at "how everything was done with proper ceremony in the Chinese style," thus alluding to fashions far removed from plain, homely things that evoke clean feeling. At a gathering in a palace chamber, she delighted in a brilliant Chinese screen painting of long-legged, monstrous creatures and "a large celadon vase, full of magnificent cherry branches; some of them were as much as five feet long, and their blossoms overflowed to the very foot of the railing."[34] So too, however much Genji waxed melancholy over the notes of autumn, he took pleasure in jaunts in imperial "dragon and phoenix boats . . . brilliantly decorated in the Chinese fashion" and in feasting with the emperor on Chinese tableware of silver and lapis lazuli.[35]

Style in Heian Japan was manifested in both sumptuous and subdued effects—vibrant paintings and inky drawings, Chinese crimson cloth and home-dyed pastels, gilded celadons and unadorned whitewares, *karamono* magnificence and native temperance, elegant simplicity *(kotan)* and splendor *(karei)*.[36] In rough measure, this came down to lavish colors for public occasions and spartan hues for private. The distinction persisted in Japanese tradition, for Rodrigues observes that in aristocratic households, "the rooms and apartments where guests are received are richly gilded with various paintings in colour," whereas rooms used only for private life are "excellently painted with black water-colour."[37]

Clearly, the reserved decoration and cool-toned shades of Song porcelains ideally matched Japanese taste in the private realm; hence, in the normal course of events, the immense prestige of China most likely would have impelled the Japanese elite to adopt the ceramic for tableware. Turbulent international relations, however, ruled out that prospect. With the Tang regime in chaos following the An Lushan rebellion, Japanese diplomatic and commercial relations with China deteriorated, finally to be severed in 894, the year of the last official embassy dispatched by Japan. Although regular commerce (including imports of pottery) began again around 1200 in the early Kamakura period, Japanese law permitted no more than five Chinese ships in its harbors at any time, so trade remained strictly limited. War between the Mongols and the Southern Song disrupted exchange between China and Japan for over a generation, and attempted Mongol invasions of Japan in 1274 and 1281 naturally brought it to a standstill.[38]

A hiatus of more than three centuries in intense relations with China proved crucial for the development of Japanese high culture inasmuch as it gave space for indigenous attitudes and values to come to the fore, curbing immoderate regard for the Middle Kingdom. In the realm of government, this meant that a Confucian bureaucracy effectively died stillborn. Great clans retained power at court and in the provinces, making Japanese politics akin to the superseded aristocratic system of the Tang. The civil examinations of the Nara and early Heian periods, which initially promised to create a Song-type bureaucratic polity, survived merely as hollow ritual. In China and Korea, Confucianism shaped official mentalities, though it had little impact on the mass of the populace. In Japan, the moral dimensions of Confucianism held sway, promoted by the court as a means to inculcate ethical behavior in all ranks, from the privileged to ordinary villagers, though the ideology rarely determined who had access to administrative office.[39]

In the realm of religion after 900, the government followed the Tang practice of supporting Buddhist temples; yet unlike in China, where Buddhism won converts largely within educated circles, Buddhist monks in Japan popularized the faith, laced with elements of Confucian ethics, among both commoners and elite. Emperor Go-Daigo (r. 1318–39) regarded himself as an avatar of the legendary Shotoku Taishi, backed Neo-Confucianism, surrounded himself with Zen (Chan) Buddhist advocates, and subsidized Zen monasteries. In addition, the sentiment of the Latter Days of the Law shifted the focus of religious behavior inasmuch as many believers regarded traditional Buddhism as having degenerated into empty formalism. In the Nara and early Heian periods, ritual, prayer, and textual study had been paramount; but in the Kamakura era, Zen, which emphasized meditation, spiritual experience, and sudden enlightenment, swept in from southern China, transmitted by hundreds of Japanese monks returning from extended pilgrimages. Numerous Chinese monks of like persuasion, fleeing the disorder of the late Tang, amplified the influence of their Japanese counterparts in Zen monasteries.[40]

In the realm of ceramics, as in government and religion, the hiatus after 900 had the effect of displacing Chinese forms in favor of Japanese. Cut off from supplies of clay and glazing materials from southern China, Japanese pottery fell into decline, with kilns abandoned and production reduced. Though Song porcelains reached Japan between the ninth and thirteenth centuries, they did so only in limited amounts, to be used for storage and display, buried in sutra mounds, or cherished as heirlooms (*densei*, "handed down") and *karamono*. The wares remained too uncommon and expensive to persuade the elite to adopt them as tableware. But by the time sizable porcelain imports resumed in the early Muromachi era (1336–1573), Jingdezhen had shifted from *qingbai* to blue-and-white, and Chinese fashion had recast Longquan celadons as outdated relics.

Between the reigns of the Heian and Muromachi, then, in the absence of plentiful Chinese monochrome pottery, the Japanese upper stratum adopted lacquered wood, usually tinted deep black, as their customary tableware. That remained the case until the late eighteenth century, when porcelain finally secured a dominant place on dining tables.[41] In short, if Chinese potters aimed to capture a share of the Japanese market in the Muromachi era, they would have to produce a ware that appealed to the distinctive taste of the Japanese. Their answer was the teabowl.

THE CULTURE OF PORCELAIN IN JAPAN

Tea first came to Japan in the early ninth century, introduced by the Buddhist monk Eichu, who had spent decades in China around the same time Lu Yu was writing *The Classic of Tea*. As with other aspects of Chinese culture, Emperor Saga became passionate about the beverage; he ordered tea bushes planted in the provinces and within the precincts of his palace. Although tea won many converts thereafter, Sugawara no Michizane (845–903), the great statesman and scholar-poet, suggested that whatever its alleged health benefits, it could not allay personal distress: "When one's heart is choked with grief and his innards knotted with anguish, not even a cup of tea will bring relief."[42] Perhaps it caused Michizane anguish to end official embassies to China in 894, seven years before the powerful Fujiwara clan defeated him in a power struggle. Exiled to faraway Kyushu, he wrote Chinese-style poems and drank bitter tea in his final wretched years. His vengeful spirit is said to have caused epidemics, bizarre weather, and the death of princes. To appease the wraith of Michizane, the Fujiwara in 947 built a shrine to him as the guardian deity of learning at Kitano, a pine forest north of Kyoto, which became a major pilgrimage site and eventually the setting for a remarkable tea fête convened by Hideyoshi. As with other things Chinese, however, Japanese interest in tea declined in the three hundred years after Michizane, though Buddhist shrines, temples, and monasteries kept tea drinking alive.

Along with a new focus on Zen principles, Buddhist monks returning home from China in the Kamakura period brought greater interest in secular Chinese customs than had earlier religious pilgrims, especially in calligraphy, garden design, and tea culture. In the arts, Zen valued spontaneity, irregularity, and the natural, precisely those aspects of Korean pottery that the Japanese most admired. Zen monks brought back quantities of teabowls produced by kilns in Zheijiang and Fujian. Known as *temmoku*—after Mount Tianmu in Zhejiang, a setting with many Zen temples—

the wares had a glossy black glaze with orange-brown streaks (or hare's fur decoration). The Huizong emperor, whom Kamakura and Muromachi dynasts regarded as a model ruler, had favored the somber teabowls, and Chinese traders shipped them in large quantities to Japan in the late Kamukura.[43] As excavations reveal, the trading junk that sank off Korea on its way to Japan in 1323 carried lots of *temmoku.*

Yosai, the Buddhist monk credited with reintroducing the tea plant to Japan in 1191, declared in his *Record of Curing Disease with Tea* that "in Japan we do not drink tea."[44] Under his influence, and with the decisive support of the imperial court, that quickly changed: tea culture took off as never before, even becoming the focus of renewed enthusiasm for Chinese things. For a while, *chawan,* the Chinese term for "teabowl," designated Chinese pottery in general.[45] The Southern Song technique of stewing tea leaves in a pot never caught on, however. Japanese tea preparation remained essentially the same as that developed during the Tang: the host put powdered tea into a heated cauldron filled with water, after which he scooped the liquid into a bowl, beat it with a bamboo whisk into a green lather, and served it to guests. Much more time-consuming than stewing and calling for about thirty utensils, the Tang technique became the centerpiece of *chanoyu* (lit., "hot water for tea"), the highly structured social and spiritual ceremony that emerged between the fourteenth and sixteenth centuries.

At the end of the Kamakura, a prominent politician wrote that "the popularity of *karamono* and tea grows ever greater" among propertied ranks.[46] Led by the military aristocracy, tea judging *(tocha)* sessions grew in popularity, often accompanied by gambling and drinking of rice wine *(sake).* Showy guest halls *(kaisho),* detached from the rest of the home, became the setting for flower viewings, poetry recitations, and incense competitions (in which participants identified various fragrances). They also served as venues for presentation of Chinese objects in an ostentatious decorative style that mandated use of ornamental lacquer screens, a triptych of scroll paintings, porcelain flower vases on stands, and bronze incense burners set on embroidered brocade. Under the Ashikaga shoguns (1336–1573) of the Muromachi period, tea gatherings shifted to special chambers *(shoin)* outfitted by daimyo (and their Zen advisers) with asymmetrical shelves, a decorative platform, an alcove, a built-in desk, and woven rush floor matting, or tatami. Like the guest halls, the *shoin* also functioned as a backdrop for exhibition of *karamono,* treasures shown off by the host while attendants served tea to guests. In time, the *shoin* became the basis for the main room of traditional Japanese residential architecture.[47]

In an attempt to appropriate the prestige of the emperor, Ashikaga shoguns collected Chinese art, validating their political ascendancy by acting as stewards of culture. Recently raised to supreme authority and sensitive to imputations of social inferiority from aristocrats with more illustrious pedigrees, the Ashikaga also used their Chinese connections and artworks to assert their authority over daimyo. Ashikaga Yoshimitsu (r. 1368–94) styled himself "King of Japan" in dispatches to the Chinese emperor, designated his gifts to the Chinese court "tribute," and, to some scandal, accepted status as a vassal of the Middle Kingdom. He promoted trade with China, which provided revenue to pay for his cultural largesse. He sponsored Noh drama, poetry competitions, and Neo-Confucian doctrine, and he probably ordered Japanese kilns to turn out imitations of Song celadons. He built the Temple of the Golden Pavilion (Kinkakuji), the most famous Japanese structure, in the outskirts of northern Kyoto. Eclectically combining styles from imperial and temple architecture, the building epitomizes Yoshimitsu's strategy of using his position as supreme champion of arts and culture to enhance his political power.[48]

Ashikaga Yoshinori (r. 1429–41) owned portraits of legendary Zhejiang monks, a thirteenth-century depiction of Avalokitesvara, and numerous other Chinese paintings. He put his Chinese possessions—porcelain vases, incense burners, writing utensils, calligraphic scrolls, and candleholders—on display when Buddhist and literati dignitaries came to visit from the continent, and, adhering to Chinese etiquette, his courtiers provided chairs when tea was served. When Emperor Go-Hanazono (1419–70) paid Yoshinori the honor of visiting the guest hall at his Muromachi Villa in 1437, the shogun set out more than a thousand items for viewing, including *Landscapes in the Four Seasons,* a paper-screen painting by the Huizong emperor. Certifying the authenticity and arranging the exhibition of such collections became a matter of professional expertise, giving rise to a class of art curators known as *doboshu* (companions). By the early fifteenth century, they also served as arbiters for shoguns and daimyo on the proper forms of the tea ceremony.

Ashikaga Yoshimasa (1436–90), the eighth shogun of his line, is renowned for his patronage of the arts, especially for building the Temple of the Silver Pavilion (Ginkakuji) at Higashiyama, a hilly retreat in Kyoto. He also is notorious for his political ineptitude and effective abdication of power. "The daimyo do as they please and do not follow orders," he moaned. "That means there can be no government."[49] During Yoshimasa's reign, 1449 to 1473, Japan plunged into the Onin War (1467–77), and the shogun, out of hopelessness or heartlessness, turned from anxiety about political stability to exclusively artistic diversions. The term *chanoyu* first appears soon after he took the title of shogun, and although his own taste in the tea ceremony generally remained old-fashioned, he employed Murata Shuko (1423–1502), a tea master *(chajin)* who encouraged turning the ceremony into a more restrained affair, purging it of the raucous conduct of the early Muromachi, such as gambling and sake drinking, as well as banishing shows of sumptuous *karamono.*

Cultural glamour increasingly took on reserved tones. Under Shuko's tutelage, servants vanished from the tea ceremony, which instead came to focus on the host himself displaying civility and skill in personally catering to select guests away from the madding crowd. The increasingly ritualistic drinking of the beverage moved from elaborate *shoin* to secluded tea rooms; the one built at the Silver Pavilion provided the standard layout that evolved into the distinctive tea hut of the sixteenth century. Preparing and sharing tea shifted to the tatami, for "etiquette on the mats" displaced use of the chair, one of the Chinese fashions introduced during the Kamakura period.[50]

Although Yoshimasa admired elegant Chinese pottery, and even sent his favorite celadon vessel to China to be repaired when it broke, he nonetheless heeded tea preceptors who directed him toward the ideals of the Way of Tea *(chado),* an approach to the ceremony that called for modest, plain wares and cultivation of an ethos of restraint and self-possession. Despite squandering funds on luxurious trappings, the shogun cultivated a reputation for ascetic taste. The new tea style had some of the appeal of monochrome ink painting in that it conveyed an aura of impoverished beauty, an intimation of the wistful notes of autumn, an aesthetics of the "cold and withered" *(hiekareru),* as practitioners called it.[51]

Heavily influenced by Zen Buddhism, the Way of Tea accentuated what became known as *wabi-chanoyu,* or *wabicha* for short (roughly, "austere tea"), a ceremonial mode characterized by lack of formality, affectation, and opulence. It embodied a spirit of detachment from the standards and responsibilities of conventional, everyday life, even a tacit critique of them. In the realm of elite taste, the rise of *wabicha* points to the same shift in values that took place in Korea almost a century earlier as reflected in the rejection of ornate celadons in favor of simple white porcelains and rustic *punch'ong.* Takeno Joo (1502–55), a tea master who carried forward the style endorsed by Shuko in Yoshimasa's circle, went so far as to seek out crude teabowls owned by common villagers for use in *wabicha.*

Chanoyu not only assumed a central position in elite social life, it also entailed employment of certain ceramics. Tea adepts placed astonishing value on specific types of teabowls from kilns in China, Korea, and Vietnam. They prized vessels characterized by coarse texture, patchy tonality, glaze drips and crackle, slight malformations, and inadvertent kiln markings. A studied rusticity, a sentimental idolization of country life by urban sophisticates, lay behind appreciation of such works. Indeed, the fashion bears a resemblance to the romantic cult of nature and folk culture that beguiled the intelligentsia of late-eighteenth-century Europe. The kind of botched wares that flea market hustlers sold at cut-rate prices in Jingdezhen became treasures beyond compare in Japan.

Merchants imported Jingdezhen blue-and-white porcelains to Japan from the fifteenth century, and tea practitioners often used them in the meal that came before the serving of tea; small water jars designed to look like wooden buckets were a popular item. But the only blue-and-white teabowls employed in the all-important tea ceremony itself were "old blue-and-white" *(ko-sometsuke),* a term coined in the nineteenth century. They typically bore unsophisticated decorations of bamboo clusters, rustic monks, and rough-hewn fishermen. Tea masters particularly fancied "worm-eaten" or "insect-nibbled" *(mushikui)* old blue-and-white, that is, pieces with glaze flaking away from the rim.[52] A late-seventeenth-century work instructed the naive practitioner on the proper, refined perspective: "Utensils used in the small tea room need not be entirely perfect. There are people who dislike even slightly damaged objects. This, however, is merely indicative of thinking that has not attained true understanding."[53]

In turning out *ko-sometsuke* and *temmoku* teabowls, Chinese potters faced the problematic task of mass-producing an emphatically idiosyncratic ware. For the Japanese enthusiast, every porcelain vessel in the tea ceremony had to be unique in the particulars of its shortcomings. The flaws of a bowl endowed it with an inimitable personality and thus with warrant for conferring a given name. Sen no Rikyu (1522–91), the most celebrated and influential tea master in Japanese history—in the late seventeenth century, his ideological heirs proclaimed him the guardian deity of tea—collaborated with Raku Chojiro (1516–92), a tile maker and perhaps the son of a Korean immigrant, in the design of teabowls. A famous one, with a warped rim, pockmarked black glaze, and repairs with black lacquer, received the sobriquet *Otogoze,* an allusion to a woman with homely features. Rikyu possessed a number of notable teabowls that he variously christened "Old Eggplant," "Snipe," "Bursting Bag," and "Swelled Bottom"; his colleagues dubbed other famous tea utensils "Turnip," "Hag's Mouth," "Flat Spider," and "Potato Head." In his quest for an ideal tea ceremony, Rikyu doggedly sought out imperfection, cultivating a sense of surprise as well as freedom from orthodoxy and hierarchy. He once created a minor sensation by displaying a flower without a porcelain vase in a tea hut, and he reportedly marred a porcelain teabowl that had too elegant a profile.[54]

PORCELAIN, POLITICS, AND THE JAPANESE TEA CEREMONY

Westerners found it hard to believe that the Japanese placed enormous value on blemished porcelain bowls and other mundane tea gear. The Portuguese Jesuit Luis de Almeida (1525–83) struck that note in the first European description of a tea ceremony:

> The way of drinking this [beverage] is to pour half a nutful of this powdered herb into a porcelain dish and then drink it mixed with very hot water. And for this purpose they have very old iron kettles, some porcelain dishes, a small receptacle into which they pour the water with which they rinse the porcelain dishes, and a small tripod on which they place the lid of the iron kettle. . . . The vessels into which they pour the *cha* powder, the spoon with which they pour it, the dipper with which they transfer the hot water from the kettle—all these utensils are regarded as the jewels of Japan, much in the same way as we value rings, gems, and necklaces made of many costly rubies and diamonds.[55]

That estimation also baffled Fra Alessandro Valignano (1539–1606), the Visitor, or inspector-general, of the Jesuit mission in the Indies:

> Quite often one of these vessels, tripods, bowls or caddies will fetch three, four or six thousand ducats and even more, although to our eyes they appear completely worthless. The king of Bungo [the daimyo Otomo Yoshishige of Kyushu, 1530–87] once showed me a small earthenware caddy for which, in all truth, we would have no other use than to put it in a bird's cage as a drinking-trough; nevertheless, he had paid 9,000 *taels* (or about 14,000 ducats) for it, although I would certainly not have given two farthings for it.[56]

Rodrigues observes that it is in keeping with the "naturally melancholy disposition" of the Japanese that "there are utensils, albeit of earthenware, that come to be worth ten, twenty, or thirty thousand crowns or even more; this is something that will appear as madness and barbarity to other nations that hear of it."[57]

Never before or since have humble objects played such a commanding role in conspicuous consumption. Even the Chinese silk brocade bags and boxes of paulownia wood used to protect the bowls had a place in the elaborate aesthetics of the tea ceremony, thereby becoming cherished and expensive objects in their own right. Rodrigues, however, went beyond the puzzlement of his Jesuit colleagues by recognizing that the apparent absurdity of the costly equipment made sense if seen from the perspective of competition for social esteem: "For what was bought and sold in such transactions was the artistic taste of both parties and not the item itself."[58]

One of the most remarkable transactions involved ruinously expensive tea caddies that Rodrigues describes as possessing "the special property of preserving the *cha* leaf from one year to another with such constancy that it always seems to be as fresh, even at the end of the year, as when it was poured in."[59] The brewed tea would be regarded as unpalatable if workers improperly handled the leaf or powdered tea, thereby ruining the spiritual harmony of the tea gathering along with the reputation of the host. At the most extreme, this meant that officials prohibited peasants who harvested tea leaves destined for the imperial court from eating fish for weeks ahead of time so that their breath would not pollute the tender plant. Called "Luzon jars" (Rusontsubo) by the Japanese, the most sought-after tea caddies were stoneware imported from Manila, where some fifteen thousand Japanese lived in the early seventeenth century.

According to Francesco Carletti, the stoneware actually came from Vietnam, Cambodia, and Thailand—a nice indication of the extent to which tea culture and ceramics linked Japan to the wider Asian commercial world. The Florentine merchant reports that officials boarded foreign ships in Nagasaki to search for the stoneware jars inasmuch as Hideyoshi aimed at monopolizing them, to the extent of threatening execution for anyone not handing them over. The Japanese elite, Carletti remarks, possess "an infinite number of these vases, which they regard as their principal treasure, esteeming them more than anything else of value. And out of vainglory and for grandeur they make a contest of who possesses the largest quantity of them, displaying them to one another with the greatest satisfaction."[60]

Rodrigues devoted twenty thousand words of his account of Japan to the social and spiritual significance of the tea ceremony. Its contradictions fascinated him—its retreat from the world and its use for political promotion, its function as a venue for solitary meditation and for camaraderie, its posture of material indifference combined with barefaced avarice in acquiring rare utensils. Above all, he highlighted the paradox that the ceremony purportedly embodied an ethos of temperance and poverty, yet it necessitated precious tea paraphernalia and furniture, Chinese ink paintings and ancient porcelain vessels, banquets of crane and the choicest fish, silk robes as bribes for cultivators providing the finest *cha,* and pretentiously humble tea huts made of rough cedar, wattle, and thatch, constructed by high-paid carpenters and expensively landscaped to appear entirely natural. "So this poverty is really very rich and wealthy . . . ," he concludes. "Indeed, it is so rich that it is beyond poor people, and it is so poor that even the rich and mighty can keep it up only with difficulty."[61]

Rodrigues believed that the culminating elaboration of the tea ceremony as a rich man's "imitation of poverty" stemmed from *chanoyu* being taken up by powerful merchants of the city of Sakai. Located on Osaka Bay, fifty kilometers south of Kyoto and facing the eastern shore of the Inland Sea, Sakai emerged as a political and economic force during the Sengoku period ("the age of Warring States"), which spanned from the outbreak of the Onin War in 1467 to Oda Nobunaga's consolidation of military authority in 1568. Like a handful of other cities, Sakai gained independence similar to that enjoyed by the ports of the Hansa on the Baltic and North Seas and by the thirty-odd German Imperial Free Cities *(Reichsfreistädte)* in the late medieval period.

Building on wealth from textile, iron, and lacquer production, Sakai emerged as an international entrepôt when Yoshimasa designated it in 1469 as the base for tribute embassies to the Ming court. That transformed the city into the most important Japanese gateway for Chinese cash and *karamono.* According to Luis Frois (1532–97), a Jesuit who lived there for several years, "No other place apart from Kyoto is as important as the city of Sakai. The Venice of Japan, it is not only large, wealthy, and full of commerce, but it is also like a central market for all the other provinces, and people of different regions are continually flocking there."[62] Allied with local Buddhist temples and daimyo, for whom it collected rents and acted as a financial broker, Sakai manufactured and imported munitions, indispensable commodities in the violent Sengoku period, especially after Nobunaga and Hideyoshi introduced gunpowder firearms to the battlefield. Rich and enterprising, Sakai rivaled Kyoto in its patronage of poets, painters, Noh drama, and temple construction.

As Rodrigues spells out, the conditions of their urban environment compelled the merchants of Sakai to develop a more restrained version of the tea ceremony. They shied away from Ashikaga-style guest halls and fancy *shoin* in favor of tea huts that accommodated only two tatami mats (3.3 square meters, adequate sitting room for about three persons). They made up for the absence of "lonely and refreshing places" in the hot, sandy plain outside Sakai by investing large sums in constructing rustic-style hermitages, often with arbors, springs, and rock gardens, in the congested city itself.[63] The tea huts provided locations, not otherwise available, for merchants to convene for frequent rounds of urbane conviviality. Rodrigues explains that in order that "the furnishings might be in keeping with the smaller hut, they did away with many of the utensils and items required by *chanoyu.*" Although still very expensive, this reserved approach brought the tea ceremony within financial reach of a broader privileged clientele and closer to the sober, luxury-spurning ideals of Zen. The new style was adopted by well-to-do merchants in Kyoto, Nara, and Hakata (in Kyushu), as well as by numerous daimyo throughout the realm.

However sincerely Sakai's merchant oligarchs espoused the tea ceremony, they also employed it to put themselves on a more secure social footing with daimyo, imperial courtiers, and Buddhist abbots. In principle, rigorous egalitarianism distinguished the tea ceremony, so it constituted a ritual space in which merchants could gather with their social superiors on terms of formal leveling.[64] Still, their status as commoners obliged Sakai's prominent entrepreneurs to be discreet. They steered clear of lavish exhibitions and the hiring of *doboshu* curators to oversee their *karamono* collections. Experts on the tea ceremony and Chinese objects perforce came from within their own ranks. Murata Shuko, founder of the *wabicha* style of tea, came from a merchant family in Sakai, as did Takeno Joo, who sold lacquered-leather armor to samurai warriors. Sen no Rikyu, the most exalted tea master of all, enjoyed financial independence as a member of a wealthy Sakai clan of fish wholesalers. He also sometimes dealt in munitions: Nobunaga once sent him a thank-you note for supplying one thousand musket balls.

Nobunaga gained control of Sakai in 1569 with the support of Imai Sokyu (1520–93), a manufacturer of armaments and a distinguished tea master. From having only a passing interest in the tea ceremony, the warlord converted to enthusiasm for it virtually overnight, making a show of attending tea gatherings with leading merchants. He bullied the governing tea men of Sakai into surrendering treasured stoneware caddies. He bestowed famed tea utensils on his chief lieutenants; he gave his general Hideyoshi a dozen pieces as a reward after a crucial, hard-fought campaign. Nobunaga even reserved to himself the right to determine whether military aristocrats could hold formal tea ceremonies and acquire valuable tea equipment.[65] After he came to power, Hideyoshi emphasized the significance of tea culture for both himself and his predecessor: "The tea ceremony was [Nobunaga's] Way of Politics. He gave me permission to perform the tea ceremony, and I was most honored. I will never forget it."[66]

The practice of awarding porcelains to followers and designating who could take part in tea ceremonies expressed and reinforced hierarchical relations. Nobunaga thus integrated tea culture into his conception of the Way of Politics, a violation of the cocoon of spirituality and fellowship that ostensibly exemplified the tea ceremony. Since the fourteenth century, *chanoyu* had developed into the epitome of aesthetic discrimination, the ideal incarnation of the "cold and withered," while also becoming a key social rite of privileged ranks, an elite vehicle of communication and sociability. In the late sixteenth century, however, *chanoyu* effectively became a casualty of its astounding success as warlord-unifiers struggled to bring an end to the turbulent Sengoku era.

In the eyes of Nobunaga (and his successors, Hideyoshi and Tokugawa Ieyasu), the Way of Tea had become too significant to be allowed even nominally to withdraw into its own sheltered enclave and to stand for a counterculture isolated from and implicitly critical of the Way of Politics. In addition, extending control over *wabicha* proved expedient for Nobunaga since it demonstrated to political and mercantile leaders—at least from his perspective—that decorum and sophistication graced his mandate to rule. He followed the tradition established by powerful rulers from the Asuka to the Muromachi of using symbolic conduct and cultural distinction to shore up political dominance.[67]

Significantly, Nobunaga took Rikyu into his entourage soon after he negotiated Sakai's capitulation. The tea master advised Nobunaga in his hunt for famous tea utensils (*meibutsu*, "pieces of renown") held by daimyo and the last of the impotent Ashikaga shoguns.[68] When Nobunaga died by assassination in 1582, Hideyoshi proved even more zealous than his predecessor regarding matters of tea. As Japan's most powerful warlord, he laid claim to Nobunaga's outstanding collection of tea implements, and he retained the services of Rikyu, who became a close political adviser, using *wabicha* ceremonies as venues to reconcile powerful daimyo, such as Otomo Yoshishige, to the ruler's policy of national unification.

In 1585 Rikyu assisted Hideyoshi in serving tea to Emperor Ogimachi (r. 1560–86) in Kyoto. The event coincided with Hideyoshi's accession to the office of imperial regent (Kampaku) and marked Rikyu's designation as "Tea Master of Japan" *(tenka gosado)*. Of course, Hideyoshi never contemplated performing the unprecedented ceremony in a wattle-daubed tea hut erected on the grounds of the imperial palace. Instead, he held it in a gold-plated tea room with a luxurious brocaded carpet on the floor, a portable apparatus he subsequently took with him on military campaigns; the emperor, regent, and tea master sipped frothy tea from solid gold teabowls.[69] There is no indication that these opulent circumstances offended Rikyu's aesthetic sensibility.

Two years later, in the most spectacular demonstration of tea culture of the era, Hideyoshi, once again seconded by Rikyu, put on a tea party in the Kitano pine forest of Kyoto, at the pilgrimage shrine dedicated to Michizane. Hideyoshi invited all well-known tea practitioners to attend, even poor men, and he prohibited those who failed to show up to engage in *chanoyu* thereafter. Participants built hundreds of tea stalls, and Hideyoshi shared tea with countless guests; he set out his gold-plated tea room and famous utensils for them to gawk at. With his status as autocrat of tea culture and unifier of the realm—the one corroborating the other—unreservedly conceded in the public festival, Hideyoshi brusquely dismissed the assembly days ahead of time.[70]

Four years later, to the shock of everyone, Hideyoshi ordered Rikyu to commit suicide by disembowelment, or seppuku. In 1591, at the age of sixty-nine, in a Kyoto house surrounded by a regiment of guards, the Tea Master of Japan took his own life. His death poem intimates that his fate recapitulated that of an earlier defeated statesman and that his reputation eventually would reach similar godlike heights as a consequence of his pitiful end:

> That fellow Rikyu,
> What great good fortune:
> To think that he will turn out to be
> A second Michizane![71]

Rikyu's suicide remains the murkiest event in the history of *chanoyu.* It most likely stemmed from the disquiet of Hideyoshi at Rikyu exercising an authority in the Way of Tea superior to his in the Way of Politics, an incongruity that contravened the ruler's drive for supremacy in all spheres. According to a contemporary chronicler, Rikyu's eminence endowed him with unchecked aesthetic sway: "In [tea] objects he liked, [Rikyu] declared good points bad and bought them for mean prices. In vessels he disdained, [Rikyu] declared bad points good and bought them at high prices. He called new old and old new. No he made yes, false he made genuine."[72]

The calculated eccentricity of tea culture introduced disturbing notes of capriciousness and contradiction into public affairs, a situation intolerable to Hideyoshi, a notoriously prickly and haughty ruler. As Rodrigues recognized, the art of tea hinged on ineffable "artistic taste," not on the intrinsic value of homely porcelain bowls. Tea culture fostered a cultlike exclusivity, informed by secret teachings *(mitsuden)* and oral traditions *(kuden)* known only to the elite, dependent on a mentality that had attained "true understanding."[73] Nothing like *wabicha* existed in China, where tea culture remained a subsidiary feature of literati connoisseurship, drinking tea lacked egalitarian connotations, the Confucian literati and imperial court set aesthetic standards, merchants aped the taste of their social superiors, and nobody conferred transcendent value on mottled porcelains. In Japan, however, the radical aesthetics of Zen, through its manifestation in the Way of Tea, posed a challenge to the existing order, uniting the tea ceremony—one of the most extraordinary subcultures in world history—with notions of social leveling, spiritual reformation, and exemplary austerity. A vital expression of elite culture, *chanoyu* nonetheless endorsed principles subversive of authority and hierarchy. As such, it could not be allowed to stand. Oda Nobunaga and his successors successfully domesticated the tea ceremony, marginalizing its troubling aspects and placing its performance under official scrutiny.

With Rikyu's suicide as an object lesson, urban tea men surrendered their lead to the warrior elite. Furata Oribe (1544–1615), a samurai in origin and heir apparent to Rikyu, emerged as the most talented and influential tea master serving Tokugawa Ieyasu, the first shogun of the Tokugawa dynasty of the Edo period (1615–1868). He made Oribe a daimyo, situating the tea ceremony firmly within the social milieu of the provincial aristocracy. This made the ceremony more appealing to imperial courtiers, who until then had shown slight interest in it, perhaps because merchant commoners so thoroughly dominated tea culture. Oribe instructed his fellow daimyo in *chanoyu,* worked closely with daimyo-sponsored kilns, and produced tea wares attractive to courtiers; he organized tea gatherings that separated participants by social status and provided room for daimyo to bring their samurai retainers.[74] Endlessly inventive in designing pottery, a champion of both Rikyu-style flawed tea wares and elegant porcelains with dazzling glaze, Oribe, like Rikyu, proved too eccentric and forceful a tea master for his authoritarian master. Ieyasu ordered him to commit seppuku, apparently because the shogun regarded Oribe as a "defiler of treasures," given to cutting up valuable Chinese calligraphic scrolls to fit his tea hut and breaking shapely porcelains so that he could glue them back together to suit his ideal of imperfection.[75]

The fates of Rikyu and Oribe reflected the constraints of Japanese public order in the closing stages of the age of Warring States. They also marked the point at which the tea ceremony, shorn of subversive elements, entered the mainstream of Japanese popular culture. It gained considerably in popularity under the Tokugawa, though as a diverting pastime, comparable to calligraphic expertise and competitive incense sniffing, rather than as an expression of Zen mysticism or as a vehicle for a radical social message. It took its place alongside Kabuki drama, puppet theater, and other leisure pursuits of the middling classes in the emerging urban pleasure quarters, the well-known "floating worlds" (*ukiyoe*) of the Edo period.[76]

No tea master after Rikyu and Oribe served in such a high capacity or with such official approbation. In Edo Castle, headquarters of the Tokugawa, in-house functionaries supervised the tea ceremony after Oribe's suicide; by the mid-seventeenth century, they occupied a respected niche in government administration. They also looked after the shogun's porcelain collection, which included over one thousand blue-and-white bowls. Among other responsibilities, they organized the annual procession of hundreds of peasants bearing stoneware caddies to Edo Castle, a ritual that obliged onlookers to bow deeply as the tea jars passed by. For the wider society being introduced to tea, printed manuals spelled out instruction on all phases of the ceremony in stultifying detail. A strict code of etiquette stipulated proper demeanor between social groups, as when a manual instructed a low-echelon tea novice that if an aristocrat "urges you to join him, you must sit cross-legged on the bench [in the arbor near the tea hut] without letting your legs dangle in front of him."[77]

National unification and formation of the Tokugawa regime witnessed the comeback of the imperial court as a center of culture after two centuries of relative obscurity. Once again it became an arbiter of taste, especially in the fields of landscape gardening, flower arranging, architecture, and *chanoyu.* Extravagant accouterments returned to favor in the tea ceremony. As the latter won imperial patronage and reached a larger urban audience, manufacture of Japanese porcelain, made possible by the innovations of kidnapped Koreans, increased in response to demand for fine wares. Kilns near Arita in Kyushu began turning out blue-and-white porcelains around 1620, often with motifs quite different from those on Jingdezhen wares, including spacious, abstract designs strikingly modern in appearance.[78] At the same time, Chinese culture retained its high reputation, so decoration in Chinese-style brushwork and with traditional Chinese motifs, such as mandarins at repose in a pavilion set in idealized natural surroundings, remained popular. (See figure 18.)

Under the impact of court and middling-class taste, the color range of pottery expanded several decades later, with decoration in translucent, jewel-like enamels on the white porcelain surface, sometimes drawn from patterns on Noh costumes. Potters also took motifs from Chinese manuals, such as the *Illustrated Compendium of Eight Styles,* printed in Anhui province around 1616 and widely used in Japan. The Nabeshima kilns took the lead in making the novel enameled vessels. Kakiemon, a much-admired pottery produced by several Arita kilns, commonly displayed the bright orange-red shade characteristic of the persimmon *(kaki).* A wider palette of enamels, including vivid shades of gold, turquoise, cobalt blue, and iron red, appeared on Imari pottery, named after a port near Arita from which the VOC shipped the wares.

Jingdezhen took advantage of the growing Japanese market in the early seventeenth century. Since the Tokugawa regime cut off most trade with Europeans in 1639 as part of its policy of stamping out Christianity, the VOC served as commercial middleman from its base at Deshima, a dreary, cramped island in Nagasaki harbor, not much more than a mudflat, where authorities confined the Dutch merchants. For its part, Jingdezhen aggressively expanded overseas markets because of disruptions in demand at home. It had to find new customers, in part because the court halted its huge orders when the Wanli emperor died in 1620. Some 25 percent of Chinese ceramic exports went to Japan in the early decades of the seventeenth century; the VOC delivered 800,000 porcelains in both 1635 and 1637. But the struggle between the Ming and Manchu brought the marketing surge to a halt, cutting Jingdezhen's exports in half. Even after the Qing dynasty came to power in 1644, warfare and rebellion plagued Jiangxi province, destroying Jingdezhen kilns in the late seventeenth century. In response, seeking alternative supplies, the VOC turned to Japan. Dutch merchants placed their first large order with Kyushu kilns in 1658, and they sent almost one million Japanese pieces abroad in subsequent years. Loads of wares went to kingdoms in continental Southeast Asia, with the result that beer mugs originally made in Japan for the Dutch market ended up gracing the tables of Thai aristocrats and Buddhist abbots. Many Japanese porcelains carried a forged Chinese reign mark on the bottom since customers in Europe regarded it as a guarantee of quality.[79] (See figure 19.)

Japanese porcelain, the tardy offspring of Chinese and Korean traditions, barely two generations old, seemed on the verge of winning dominance over Jingdezhen in the first new marketplace to open up for Chinese ceramics in almost a thousand years. Dutch troubles at home played a role in preventing that from happening.[80] The United Provinces fought wars with England in 1665–66 and 1672–74, while war with France in 1672–78 resulted in French troops overrunning much of the republic's territory, halted only by the expedient of breaching the dikes and flooding the approaches to Amsterdam. By the time the Dutch recovered from this disaster, Chinese pottery production was rebounding as well. In 1683 the Kangxi emperor ordered Jingdezhen's kilns rebuilt, and when production came back on line, the volume and expertise of Jingdezhen eventually demolished the Japanese competition. Overseas trade revived, with more than nine thousand Chinese merchants on board 193 Chinese ships going to Japan in a single year.

Japanese competition also faced obstacles in the Tokugawa regime itself inasmuch as it adhered to a strict version of Neo-Confucianism that drove it to discount commercial profit as a revenue source and instead rely on agrarian taxation. In addition, the Tokugawa adhered to the mercantilist view that export of precious metal harmed the economy; but whereas Western nations generally tried to boost stocks of specie by expanding commercial exchange, Japan disdained foreign trade and focused exclusively on product innovation to eliminate the need for spending silver on imported merchandise. Shoguns could enforce such a policy inasmuch as they held a virtual monopoly over the country's silver mines.[81]

As a consequence of Dutch commercial setbacks, Tokugawa policy, and Jingdezhen initiative, Chinese pottery in Japan in the early eighteenth century cost a quarter the price of Japanese wares. Jingdezhen's finest blue-and-white became popular in the tea ceremony since "insect-nibbled" wares held scant charm for most new practitioners. The porcelain city also cut into the Japanese export market by expertly imitating colorful Kakiemon and Imari vessels for Europeans. Not only that, Jingdezhen copied Japanese copies of Dutch blue-and-white earthenware—which themselves copied Chinese blue-and-white from the Wanli era—and sold them to the VOC for its Western customers. Faced with this onslaught, Japanese kilns fell into decline, and many folded by 1720, including those producing Kakiemon wares.[82]

Clearly, Jingdezhen's centuries of experience dealing with large overseas markets could not be gainsaid. In the long run, however, Western competition turned out to be far deadlier than that from Japan. Harbingers of the threat first loomed in the early eighteenth century, as Chinese and Japanese porcelains reached Europe in unprecedented amounts, and European entrepreneurs scrambled to cash in on the rage for Asian ceramics. Dutch enamelers in Delft converted Chinese blue-and-white into pseudo-Japanese wares by "overdecorating" them with their own versions of Chinese and Japanese designs in red and gold enamels, then firing them at low temperatures. They took their motifs from embellishments on Chinese blue-and-white, Kakiemon, and Imari, as well as from travel books such as Nieuhoff's *Embassy from the East India Company.*[83] Another serious omen for Jingdezhen emerged at the Meissen manufactory, the trendsetter in the early days of European porcelain. By turning out table services imitating Kakiemon and Imari, ornamented with Chinese dragons in red and gold, Japanese tigers dappled in black and yellow, and Indian flowers in glowing shades, the potters of August II appealed to affluent customers who wanted colorful pottery in fashionable Asian styles. Both Delft and Meissen produced a truly hybrid cultural commodity, an amalgamation of Chinese pottery, Japanese enamel colors, and European chinoiserie ornament, the creation of an ecumenical style, the end result of bringing together motifs, patterns, and shades from around the world. (See figure 20.)

Jingdezhen manufacturers knew about the European innovations because merchants of the VOC and the EIC, extending the round robin of Eurasian cultural exchange, sent commissions to the porcelain city to copy the Japanese-style wares of Delft and Meissen. Inasmuch as Jingdezhen lacked central direction, however, there was no way its many kiln owners could perceive, much less respond to, the threat implicit in the ingenious European replications. In any event, Jingdezhen had succeeded since time immemorial in overpowering all competition by its capacity to produce high-quality copies in volume, so there was no reason to think that Western rivalry would be any more troublesome than that already encountered from the Japanese. By the end of the eighteenth century, however, Wedgwood and his Staffordshire colleagues would show that was not the case.

"EARTHS RIGHT FOR POTS AND CERAMICS": VIETNAM AND CHINA

Archaeologists in recent years have investigated a number of fourteenth- and fifteenth-century shipwrecks in the waters of Southeast Asia, a minute sample of the countless vessels that came to grief there during that time.[84] Excavation of the Pandanan wreck near Palawan Island, a 400-kilometer sliver of land stretching southwest between Mindoro Island in the Philippines and the great island of Borneo, yielded more than 5,000 Chinese and Vietnamese ceramics. A Chinese ship, the *Turiang,* sank 160 kilometers off the Malaysian coast, perhaps as a result of being top-heavy with a sizable cargo of Thai pottery, supplemented with wares from China and Vietnam. The Hoi An wreck, a Thai ship found by fishermen off central Vietnam, went down with some 250,000 pieces of pottery, mostly from kilns in the Hong River delta, including cobalt-decorated stoneware and a large number of *kendi* jars. A Chinese junk, discovered in the waters south of Mindoro, may have been delivering an order commissioned by a wealthy aristocrat: its cargo included 5,000 ceramics, mainly Jingdezhen blue-and-white but also pottery from Vietnam and Thailand, along with bronze cannons, copper basins, writing boxes, lacquerware furniture, and elephant tusks. An Indonesian wreck discovered beneath the Java Sea carried around 100,000 pieces of pottery from Fujian, as well as Jingdezhen *qingbai,* Thai *kendi,* bronze weights, and 190 tons of iron bars.

Besides their pottery cargoes and regrettable fates, the ships had other things in common. They all sank on the way to ports in continental and maritime Southeast Asia, locations where merchants found the best customers for relatively unsophisticated jars and plates. They all carried pottery from different regions of East Asia, evidence of the cosmopolitan nature of the trade by the fourteenth century. The Chinese, Vietnamese, and Thai wares are similar in shape and decoration, suggesting a common pool of knowledge about markets thousands of kilometers apart, probably as a consequence of clients in the Philippines and Indonesia ordering desired pottery by sending drawings and wooden models to various kiln centers.

A few of the Vietnamese teabowls and Thai stoneware caddies on the wrecked ships might have ended up in Japan, imported from the Philippines to share shelf space with Jingdezhen *ko-sometsuke,* Zhejiang *temmoku,* and Korean *punch'ong* in the collections of merchants and daimyo. On the other hand, some of the best Vietnamese blue-and-white may have been manufactured for the Southwest Asian market inasmuch as Arab and Persian merchants living in Thang-long (sixteenth-century Tonkin, present-day Hanoi) imported cobalt from Persia and commissioned pottery for their homelands, just as their compatriots did in Quanzhou and Canton. The stoneware potters of Vietnam so skillfully copied Jingdezhen blue-and-white that Southwest Asian customers reportedly could not tell the difference.

Ceramics of continental Southeast Asia, like those of Korea and Japan, were greatly influenced by China. As Vietnam bordered on China, however, its pottery owed a much greater debt to the technology and aesthetic traditions of the Middle Kingdom than did that made in Cambodia and Thailand. According to legend, a Chinese potter came to the Hong River valley in the second century C.E., where he built the first kiln and passed on his skills to the natives. The story may contain a residue of truth, for potters in northern Vietnam during the Neolithic period did in fact produce wares similar to those made south of the Yangzi, an area with close ties, based on common linguistic and ethnic features, to coastal Vietnam.[85]

Both southern China and coastal Vietnam are archipelagic in nature, made up of deltas, shallow waters, and inshore islands that the Chinese referred to collectively as "the inner sea" *(neihai)* in contrast to "the outer sea" *(nanhai),* that is, the deep reaches of the South China Sea.[86] Hemmed in by hills and linked by river networks, both areas comprise islandlike enclaves oriented to saltwater commerce and linked historically to similar societies in the Archipelago. A different historical trajectory could have led the twin littorals of southern China and Vietnam to coalesce into an enduring "blue China," an independent state shielded from both steppe warrior invasion and Beijing domination.

From the last centuries before the common era, however, both southern China and northern Vietnam became incorporated into the Middle Kingdom. Having extended his rule south of the Yangzi, the emperor Qin Shihuang ordered invasion of northern Vietnam in 214 B.C.E.[87] After the fall of the Qin dynasty, the Han maintained control there, attracted by valuable products such as coral, tortoiseshell, cinnabar, malachite, and woods of teak and camphor. According to a Tang poet banished to the area, the natural resources of the Chinese protectorate of Vietnam, named *Annan* (the pacified south) by the Tang, included "Forests rich in cassia and juniper; / Earths right for pots and ceramics."[88] Yet though Chinese immigrants built Han-style tombs and Buddhist monasteries in the fertile, triangular-shaped plain of the Hong River, Chinese of high rank spurned northern Vietnam as a miserable backwoods, a land of droning heat, pestilent vapors, and perfidious peoples. The Viet (or Yue) ruling class, however, adopted fundamental aspects of Chinese culture, such as ideographic script and Confucian education, and they sought enrollment within the hierarchy of imperial administration.

In the tenth century, after the collapse of the Tang dynasty and a millennium of Chinese rule, the Viet elite rebelled against foreign domination. Following a period of disorder, the Ly dynasty (1009–1225) came to power, establishing the first independent state in the region. It made Buddhism a virtual state religion and encouraged expansion of the ceramic industry to serve the construction needs of the new capital of Thang-long, as well as the building of many temples and monasteries. Under the Ly and and Tran (1225–1400) dynasties, Annan expressed its Chinese heritage by regarding itself as a Middle Kingdom, spreading the gifts of high civilization to the Champa region of central Vietnam (now Binh Dinh province), a territory of loosely associated communities with close ties to Thailand and Java. The claim bred deep resentment among China's leaders, who saw the onetime protectorate as fittingly part of their own imperial realm.[89] Chinese culture thus represented a double-edged sword for Annan, giving it confidence in its own identity while simultaneously persuading the Yuan and Ming that the region must be brought back under the bona fide Middle Kingdom.

Chinese culture continued to hold large sway in elite circles of Annan, not only because of the outlook of Viet aristocrats, but also as a consequence of turmoil to the north impelling well-to-do Chinese to seek safety in the Hong River area. When the Southern Song fell to the Mongols, thirty ships filled with refugees escaped to Hanoi, and when Mongol troops invaded Annan in 1284, they discovered more than four hundred senior Song bureaucrats and court officials living there. At least some reminders of home surrounded the exiles, for indigenous potters turned out excellent stoneware facsimiles of Jingdezhen *qingbai* and Longquan celadons. Acknowledging the skill of the potters, Kublai Khan demanded white pottery in tribute from Annan, along with pearls and rhinoceros horn, and some Chinese potters of the Yuan copied Hong Valley wares for export to Southeast Asia.[90]

In the early Ming period, the Yongle emperor ordered armies totaling 215,000 into Vietnam, yet another effort to regain the territorial rights held until the fall of the Tang. Although the Mongol invasion lasted only four years, the Ming endeavor persisted for almost a generation, from 1407 to 1427, causing great loss of life, destruction of Buddhist temples, and the fall of the Buddhist-oriented regime of the Tran. In the end, however, the Ming could not afford the huge expense of occupation, and they found the lines of communication between China and its would-be protectorate too extended for secure control of the country.[91] Within a few years of the death of Yongle in 1424, Chinese troops, by order of the Xuande emperor, finally withdrew from the unpacified south. They left behind a changed country, for the years of invasion and upheaval represented as great a transition for Vietnam as the contemporaneous shift from Koryo to Choson did for Korea.

Le Loi (r. 1428–33) founded the Le dynasty (1428–1527), a regime ideologically based on Neo-Confucianism, devoted to a Chinese-style bureaucratic government, and focused on imperialist expansion. The superpower of East Asia, China had used gunpowder weapons to defeat Vietnamese resistance, but during the decades of Ming occupation, the Vietnamese adopted the military technology and turned it against their oppressors. In 1471 they employed firearms to conquer the Champa region to the south, taking a giant step toward rounding out the borders of present-day Vietnam. The Cham people had made stoneware for several generations, but after conquest by northern Vietnam, pottery imports from there took precedence, and local kilns went out of business. When they resumed again in the early fifteenth century, the Cham kilns produced blue-and-white stoneware, copies of wares from the Hong River that were themselves modeled on Jingdezhen porcelain.[92]

Chinese and Muslim merchants imported Jingdezhen blue-and-white to Vietnam from the early fourteenth century inasmuch as it was on the route from Quanzhou to the Persian Gulf and the Indian Ocean. Hong River potters had followed Chinese traditions for centuries, adding individual touches to their own wares, such as spirited brush painting and distinctive shades on celadons, which expressed an independent cultural identity. In all likelihood, however, they shifted entirely to production of blue-and-white stoneware only in the aftermath of the Ming invasion, when conditions of peace and prosperity made such industry practicable. They developed their own vigorous styles for blue-and-white, shunning many of the standard designs used on Jingdezhen wares. Given the intensity of Buddhism in Vietnam, the lotus naturally remained a highly popular decoration; but idiosyncratic motifs included a creature with a dragon's tail and elephant's head, jungle scenes with hunters, panels of overlapping petals and peonies, serrated bands with pineapple-like designs, images of peacocks and parakeets, and dragons with large central fins rising in clouds. Vietnamese potters also adopted motifs from India, including the *garuda,* an unusual avian incarnation of the Hindu deity Vishnu, and the *makara,* a beast combining features of a fish, crocodile, and elephant.[93]

In the fifteenth century, during Vietnam's most fruitful period of pottery manufacture, the Le dynasty encouraged the export of ceramics to increase the kingdom's revenue. The Ming government restricted private overseas trade much of that time, thereby providing Vietnamese potters (as well as those of Thailand) with an opportunity for selling their wares in foreign markets. Merchants in southern Chinese ports also may have relocated to Vietnam to avoid the Ming restrictions. Blue-and-white pottery was exported from the Hong Valley and Champa to the Archipelago and Southwest Asia. At least one Vietnamese *kendi* in polychrome was buried in a sixteenth-century Thai stupa monument, suggesting that Buddhists there valued the ware. Some kilns filled orders from Japanese connoisseurs for plain blue-and-white stoneware to use in the tea ceremony.[94]

Most important, Vietnamese potters made ceramics for the court and mosques of the kingdom of Majapahit in Java. Court officials in the capital of Trowulan specially commissioned blue-and-white tiles decorated with images of cranes, deer, lotuses, and chrysanthemum sprigs. Royal dignitaries had a monopoly on use of tableware in red, yellow, black, and gold, and they controlled circulation of the finest ceramics. Muslim and Vietnamese merchants supplied blue-and-white and polychrome tiles for the gates and walls of mosques at Demak, the leading port on the north coast of Java. Many of the pieces have decoration resembling that on Southwest Asian pieces, indicating that the Islamic community in Java, which traded with both China and countries on the Indian Ocean, ordered the tiles from Vietnam when China temporarily closed its harbors to foreign trade.[95] Javanese artisans, however, did not cover buildings with tiles in the all-encompassing, interlocking fashion of Southwest Asia but rather set them on walls independently as architectural adornment, as if each piece were to be appreciated for its own sake.

Exports of pottery from mainland Southeast Asia fluctuated according to the status of Chinese shipping, expanding when China went into a period of withdrawal, losing ground when it reemerged. By the mid-sixteenth century, Jingdezhen and Fujian kilns began sending their wares to the Archipelago once again, putting an end to the bonanza enjoyed by manufacturers outside China. But mainland Southeast Asia had a second opportunity in the late seventeenth century, during the Ming-Manchu dynastic turnover. As with Japanese kilns, those in Vietnam and Thailand benefited from China's time of troubles for more than two generations. They also used the VOC as middleman in the trade: between 1663 and 1682, the Dutch exported around 1.5 million pieces from Hanoi to Batavia; from there, the wares went off to the southern Philippines and the Sulu Archipelago, a cluster of islands immediately southwest of Mindanao. But when Jingdezhen recovered its footing after 1680, Vietnamese potters had to make do with selling their products only in local communities.[96]

By the late seventeenth century, then, as a consequence of Jingdezhen's retreat from and return to overseas trade, something like a common market for pottery and a common pottery style had been created throughout East Asia.[97] Japanese, Vietnamese, and Thai kilns modeled their wares on Chinese porcelain, and Jingdezhen copied them all when it came back into competition. By that time, Westerners had joined the circuit of ceramic exchange, and designs from Dutch pottery were making their way onto wares in Persia, Vietnam, Thailand, Japan, and China. (See figure 21.)

"GOODS FROM CHINA": KINGDOMS OF THE KHMER AND THAI

Khmer and Thai kingdoms on the lower Mekong River experienced a cultural environment very different from that to the east on the Hong River. According to a legend recorded by Chinese officials, an Indian Brahman named Kaudinya came to Cambodia in the first century C.E.; he married a dragon-princess *(nagi)* and spread the teachings of the Buddha. The myth accounts for the early transplanting of Sanskrit culture into mainland Southeast Asia. In fact, from the fourth century, Indian art and coinage, Sanskrit terminology, and Hindu-Buddhist religious traditions influenced continental Southeast Asia from present-day Burma to the Archipelago. Converts to the new faith identified indigenous deities and spirits with Hindu gods and Buddhist bodhisattvas. As usual, Avalokitesvara occupied a central place in the pantheon. Early in the common era, Indian merchants, especially from the coast of Coromandel, established trading colonies in settlements on the Gulf of Siam. Powerful Indian states, most notably the Gupta Empire (320–ca. 550), the first long-lived Indian political entity, provided the region with models of religious and political organization.[98] Since kingdoms of the subcontinent communicated with Mekong River polities by sea, however, those states had little fear of conquest by Indian powers. On the other hand, Chinese invasion of Vietnam seemed to confirm that the Middle Kingdom represented a lively threat, thereby making Indian political models an appealing alternative to Khmer and Thai communities. Although numerous Chinese merchants settled in the region, they came from the ports of southern China, remote from (and often hostile to) centers of authority in the north. Chinese traders never promoted Chinese political institutions abroad, while Chinese officialdom regarded the seafarers with suspicion as both merchants and expatriates, consumed by self-seeking profit and treacherously self-exiled from the Middle Kingdom.

Khmer and Thai kingdoms developed modes of governance remote from both China and the Chinese-style state of Vietnam, where Confucian ideology and strong administrative traditions circumscribed the sacral role of monarchy.[99] Those kingdoms comprised shifting federations of communities and lacked clearly articulated state structures of finance, law, and military command. Kings maintained loose unity by distinguishing themselves as enshrining cosmic notions of political space. The ruler, conceived of as a "God-king" (Devaraja) or "Universal Monarch" (Cakravartin), nominally established a realm in harmony with the Hindu-Buddhist universe, built monuments and ritual centers reflecting cosmological structures, and figured as the linchpin of religious ceremonies. Court protocols dramatically exhibited the monarch's preeminence over those not part of the god-sanctioned dynasty of royal blood, with aristocrats obliged to prostrate themselves before the king, who presented himself clad in gold cloth, seated on a silver throne, and elevated above all spectators.

A principal obligation of the monarch was the building of temples, veritable sacred mountains of stone, to honor his dynasty and guardian deity. The "Lord of the Heavens" protected the king, one of whose titles was "Lord of the Mountains."[100] In the seventeenth century, a Dutch merchant in the kingdom of Ayutthaya (1351–1767) in Thailand (called *Siam* by Westerners) described the king performing a pious ritual that confirmed his legitimacy: "Once every year . . . the king of Siam shews himself by water and land in state to his people, going to the principal Temple of the Gods, to offer there for the welfare of his Person and Kingdom."[101] Generally weak in administrative authority, Khmer and Thai monarchs achieved exalted status by virtue of their embodiment of religious principles, compelling political loyalty by being seen as avatars of Hindu gods and Buddhist bodhisattvas. This endowed their kingdoms with considerable resilience, helping to explain the failure of Islam (as in the Hindu-Buddhist kingdom of Bali) to make great advances in continental Southeast Asia.

Building magnificent temples and royal palaces placed great demands on ceramic technology, especially for tiles and sculptural relief. Buddhists and Hindus built at least one hundred temples over 250 square kilometers in the Khmer-speaking region of what is now northwestern Cambodia. The base of the twelfth-century Shwedagon pagoda at Angkor Wat includes a frieze of rectangular ceramic panels in yellow and green depicting the life of the Buddha; the Sulamani pagoda of the same period boasts glazed cornice tiles decorated with a motif of lotuses. In addition, large ceramic statues topped many temples and stood in niches along their walls.[102]

The source of the pottery technology ultimately was China, most likely by way of Vietnam and Champa. Glazed pottery and efficient kilns developed incrementally in China and Vietnam, yet they made a sudden appearance in Khmer-speaking regions in the lower Mekong River basin around the ninth century, suggesting they arose from outside influence. Khmer stonewares of the late ninth century are shaped like Chinese porcelains, and Khmer tiles copy Chinese ones. Artisans used both Khmer and Chinese ceramics in constructing Angkor Wat, the core of the most important early state of the region, centered on the Tonle Sap (Great Lake), a body of water that gave the area efficient links to the outside world by way of the Mekong River.[103]

Built in the reign of Suryavarman II (r. 1113–ca. 1150) and devoted to the cult of Vishnu, Angkor Wat attracted pilgrims from India and (as a palace inscription declared) "goods from China."[104] Zhou Daguan, a Chinese ambassador who visited the Khmer kingdom in the late thirteenth century, noted the large number of Chinese living there, as well as the strong demand for Chinese commodities, saltpeter and porcelain in particular. The palace-temple complex of Angkor Wat housed over ten thousand persons, prompting a European visitor to the ruins in 1609 to conclude that it must have been built by the Roman emperor Trajan (r. 98–117), renowned in Europe for his grand construction projects.

As was true in many other countries, privileged ranks in Khmer kingdoms monopolized glazed ceramics, whether indigenous stoneware, Vietnamese pottery, or porcelain. Chinese merchants may have been responsible for introducing pottery production at Sisatchanalai (also known as Sawankhalok, "the Place of Heaven") in northern Thailand, a center that eventually grew to include one thousand kilns spread over an area of more than six kilometers. Chinese incising techniques and Chinese embroidered silks served as major sources of inspiration for embellishments on Khmer pottery. Above all, Khmer potters copied metal vessels from India; a sharp profile and angled base on their vessels are among their most distinctive features. Chinese merchants evidently sent samples of Indian-style Khmer stoneware to Chinese kilns, where potters made facsimiles to be shipped back for sale in Cambodia. Suryavarman II maintained diplomatic relations with the Southern Song court, though those tailed off with the rise of internal conflict after his death. With the sacking of Angkor Wat by invaders from Champa in 1177, Khmer ceramic art largely disappeared, displaced by imports of Chinese porcelain.[105]

Legend claims that King Ramkamhaeng (d. 1317) of the kingdom of Sukotai (ca. 1238–1419), the first polity established by Thai-speaking people, paid state visits to Kublai Khan in the late 1290s, supposedly returning with a Chinese bride and hundreds of potters from Hebei province. Presumably this fable stems from the migration of Chinese potters to the Thai region during the death throes of the Southern Song regime. They established kilns at Sukotai in north-central Thailand and introduced celadon production to the older kilns at Sisatchanalai. Wares from those pottery centers carried on the Khmer practice of reproducing Indian shapes and ornament, including decorative patterns associated with Hindu gods. The most popular motif on dishes was a fish, one of the Buddhist emblems of good fortune as well as a reference to the god Vishnu, who was reincarnated as a fish and rescued mankind from a flood.[106] Potters turned out many *kendi* jars, which priests used at wedding ceremonies for decanting purifying water onto the hands of participants. After the founding of these productive kilns, Thai wares for the first time began joining Chinese and Vietnamese ceramics in the extensive maritime marketplace of the Archipelago. (See figure 24.)

Located in the Chao Phraya River basin, ninety kilometers upriver from the Gulf of Siam, the Thai kingdom of Ayutthaya, after incorporating Sukothai in 1438, maintained the overseas pottery trade as a royal monopoly, customarily working through resident Chinese merchants. Indeed, expansion of Chinese commerce encouraged the foundation of Ayutthaya, the capital city. But merchants there could not sustain exports of Thai ceramics when Chinese seaborne trade revived after its fifteenth- and late-seventeenth-century disruptions. Thai pottery thus suffered the same rapid decline in international markets as did Vietnamese wares. Nonetheless, Ayutthaya continued to prosper as one of the busiest cosmopolitan centers of East Asia, even recovering rapidly from a destructive Burmese invasion between 1563 and 1569. Chinese merchants played an important role in this. They invested in pepper and sugar plantations, and making use of good-quality native teak and low labor costs, they built ships in Ayutthaya, saving as much as 90 percent of the expense of doing so in China.[107]

In its tribute trade with Ming and Qing China, Ayutthaya exported forty-four different kinds of goods, including tons of pepper, aromatic woods, and deerskins. It also showed itself more adept than other states at taking advantage of that trade: while Ayutthaya's diplomatic emissaries made the time-consuming overland journey from southern Chinese ports to Beijing to pay their respects to the Son of Heaven, the kingdom's tribute ships, commanded by Chinese mariners, sailed home with merchandise and then speedily returned for a second (illicit) load.[108] Chinese formed the largest expatriate group in the capital city, accounting for as much as a third of the population of ten thousand. Jeremias van Vliet (1602–63), an agent of the VOC in Ayutthaya, wrote in 1638 that "in the kingdom of Siam many Chinese are still living who enjoy reasonable freedom in trade throughout the whole country and are well respected by the previous as well as the present king, so that some of them have been appointed to high positions and offices and others are considered the best factors, traders, and sailors."[109]

In addition to the Chinese and Dutch, Indian and Japanese entrepreneurs kept commercial agents in Ayutthaya. These outsiders made it possible for the Universal Monarch to profit from being a "Great Merchant," as a French diplomat said of

King Phra Narai (r. 1658–88), one of the last powerful monarchs of his dynasty.[110] Joost Schouten, a director of the VOC in Ayutthaya in 1636, wrote that King Prasat-Thong (r. 1629–56), founder of the dynasty and father of Phra Narai, "himself is also a Merchant, and hath his own Ships and Factors, trading to Choromandel and China, being for that cause more favored and privileged than any other Prince."[111] Nonetheless, the king took care to appear aloft from disreputable trade. As Schouten describes him, he was aided in this artifice by elaborate ritual, backed by wielding absolute power over priests, grandees, and aristocrats:

> When he gives them Audience, he is richly clothed and crowned, sitting upon a golden Throne, at whose feet his Gentlemen and Attendants reverently kneel. . . . All, as well Strangers as Subjects, who have audience of his Majesty, whilest they are in his presence, must continually kneel, with folded hands and heads hanging down; when they speak to him, it must be in this humble posture, loading him with titles and praises; his answers are esteemed Oracles, and his commands unchangeable.[112]

"AN IMITATOR OF THE KINGS OF ASIA": LOUIS XIV IN 1686

Chinese traders delivered large quantities of porcelain to Ayutthaya, both humdrum and high-quality pieces. Muhammad Rabi' bin Muhammad Ibrahim, a Persian diplomat at Phra Narai's court in 1685, attended a royal banquet in which the king, styled the "Lord of the White Elephant," dressed in Persian costume and servants laid out more than fifty dishes in porcelains topped with silver lids. In the same year, the Jesuit Guy Tachard (1648–1712) arrived in Ayutthaya on an ambassadorial mission for Louis XIV. He was accompanied by Joachim Bouvet, who continued on to China to serve at the court of the Kangxi emperor. Tachard recorded seeing splendid porcelains everywhere in the king's palace, especially set in niches along walls. He and his colleagues purchased fifteen hundred Chinese vessels for Louis. They also negotiated with Phra Narai for dispatch of the first Asian diplomatic legation to Europe. French officials and Jesuits had high hopes of converting the king to Roman Christianity, thus gaining Ayutthaya as an Asian beachhead that would smooth the way for trade with China and thereby counter the VOC, an enterprise that functioned as an arm of the Dutch state, France's foremost opponent.[113]

In 1686 three envoys from Ayutthaya arrived in France, an occasion soon commemorated in *The Siamese Ambassadors in Paris,* a tapestry produced by the royal Beauvais workshop. Crowds gawked at the exotic visitors everywhere they went; the envoys toured all the high spots, including Parisian palaces, the tapestry works, and the Trianon de Porcelaine (soon to be demolished). They presented Louis with a host of gifts from Phra Narai, most notably fourteen hundred pieces of Kangxi porcelain, described by one observer as "the most remarkable ever found in the Indies."[114] An aristocrat told the envoys that with the porcelains on exhibit in a royal apartment, it seemed that "the Indies are more present in this *cabinet* than in the Indies themselves, since it contained the best of all the beautiful things ever produced in the East."[115] The porcelains ignited the French rage for chinaware that would burn even brighter at the turn of the century with the destruction of aristocratic silver plate in the War of the Spanish Succession.

In the most extraordinary reception ever granted by Louis XIV, the Siamese ambassadors processed to their royal audience down the glittering Hall of Mirrors at Versailles.[116] Having ordered a meticulous study of the protocols of Phra Narai's court, the French king flaunted himself as a Siamese despot, seated on a silver throne on a high platform, dressed in a suit of cloth of gold set with huge diamonds, flanked by silver urns and candelabra, and, in a marked break with established protocol, surrounded only by individuals (including bastards) sharing the royal blood. Eyes ritually averted from this awesome spectacle, the Siamese diplomats prostrated themselves before the throne, their noses grazing the floor. Looking beyond the custom-bound, constrained sovereigns of the West, the Sun King finally had discovered a monarchical style that measured up to his lofty standards for splendor and extravagance. Adam Smith's appraisal of Louis captures the singular nature of the moment: "He had a step and deportment which could suit only him and his rank, and which would have been ridiculous in any other person."[117]

From his post in Beijing, Bouvet later applauded the unprecedented performance in the Hall of Mirrors, just as he commended the autocratic modes of the Kangxi emperor to "the most Glorious Monarch upon Earth," as he fawningly described Louis XIV in 1697.[118] Regalia and riches, however, did not stop the enemies of Phra Narai from imprisoning the sick, elderly monarch in his own palace after a revolution in 1688, a bloody response to his unpopular strategy of accommodation to France and the Christian religion. Ministers who had promoted that policy lost their heads, and the king died an ostensibly natural death within a few months.

The reception of the Siamese ambassadors was not forgotten in France, for memory of it lingered well into the next century, helping to solidify the image of a profligate, despotic, and therefore perhaps superfluous monarchy. Immediately after the death of Louis, the marquis de La Fare (1644–1712), in *Mémoires et réflexions de Louis XIV* (1715), wrote that the king in 1686 had revealed his ambition to be "an imitator of the kings of Asia, whom slavery alone pleased."[119] The marquis d'Argenson (1694–1757) attacked Louis for his indulgence in "Asiatic luxury." Montesquieu worried in *The Spirit of the Laws* that the French monarchy appeared to be degenerating into an "oriental despotism," with the king "referring everything to himself exclusively[,] . . . the state to its capital, the capital to the court, and the court to his person alone."[120] In his *Essaie de le despotisme* (1775), the comte de

Mirabeau (1749–91), a leader in the opening years of the French Revolution, criticized the Bourbon monarchy and likened it to an Asian despotism.

In some small measure, then, the Sun King's infatuation with Asian majesty came home to roost. Perhaps the context and connotations of Phra Narai's gift of Kangxi porcelains should be set alongside the destiny of a pair of cannon also presented to Louis by the Siamese envoys: almost two meters long, "inlaid with silver, mounted on carriages also inlaid with silver," they were seized by a mob from the royal palace in Paris on 14 July 1789 and used to storm the Bastille.[121]

7

The Triumph of Chinese Porcelain

Maritime Southeast Asia, the Indian Ocean, and Southwest Asia, 1400–1700

According to Matteo Ricci, the Confucian elite regarded peoples beyond their empire with scorn, differing "but little from the beasts of the fields and the forest," because they lacked the social and political virtues characteristic of the Middle Kingdom. As he explained, "The few kingdoms contiguous to their state, of which they had any knowledge before they learned of the existence of Europe, were, in their estimation, hardly worthy of consideration." In summer 1598, when Ricci at last achieved his ambition to enter Beijing, he discovered that the Chinese there made so little distinction between foreigners that he faced the danger of being taken for Japanese at the very moment when Hideyoshi's warriors in Korea seemed poised to cross the Yalu River and invade China.[1]

Inasmuch as Ricci's writings circulated widely among the Western intelligentsia, they helped establish an enduring image of imperial China as implacably arrogant and aloof, a hermetically sealed realm disdainful of the wider world, foreign contact, and overseas trade. Ricci also contributed to the European perception that the Chinese were inherently hypocritical. He pointed out that the vaunted tribute system of the empire, the paramount international expression of China's cultural superiority, functioned as an elaborate charade:

> The term tribute is more of a name than a reality. . . . When the time arrives [to depart for China], the so-called ambassadors forge letters in the names of various kings who are supposed to be sending them, and the letters are filled with the highest praise of the Emperor of China. . . . The Chinese people realize that the whole thing is an imposture, but they make nothing of the deception. Rather, by way of adulation to their king, they permit him to believe that the whole world is paying tribute to the

> Chinese Kingdom, whereas, in fact, China is really paying tribute to the other countries [by buying their commodities at inflated prices].[2]

TRADE AND TRIBUTE IN CHINA

Although Ricci was correct that the way the Chinese dealt with tribute pandered to imperial self-satisfaction and fostered official duplicity, there nevertheless was a great deal more to tribute than that. In fact, the practice survived for many centuries because it represented an effective, flexible, and usefully muddled blend of cultural propaganda, reasoned diplomacy, and economic pragmatism. From as early as the Han period, tribute payments to the imperial court developed as an institutional expression in foreign relations of the centrality of the Middle Kingdom, as an instrument whereby dealings with alien peoples could be translated into traditional terms of respect and deference from inferior to superior.

In principle, Chinese cultural superiority found expression through ritual and aesthetics: "barbarian" ambassadors performed the kowtow *(koutou)* ceremony, which included three genuflections and nine prostrations, before the Son of Heaven (or sometimes, as in the late Ming, even before his empty throne) and presented tribute gifts in the form of commodities from their own countries. For his part, the emperor gave his tribute clients ritual offerings, including official calendars, umbrellas, formal costumes, and gold seals of office. In reality, acknowledging the preeminence of the distant emperor was a cheap price for a ruler to pay in order to gain the opportunity to ship large amounts of merchandise to China under the rubric of tribute. And not only did a ruler gain an economic benefit from his nominal submission, but being recognized as a tribute client of the emperor could be parlayed into an advantage at home in competition with both internal political rivals and powerful neighboring enemies.

Emperors generally recognized that they served their own interests by encouraging overseas trade or, at least, not obstructing it. A Northern Song emperor commanded maritime trade commissioners "to attract foreign tribute [from the Archipelago], and exchange [Chinese goods] for spices, raw medicine, rhinoceros horn, ivory, pearls, and borneol [a camphor-based drug]."[3] Southern Song and Yuan monarchs promoted foreign commerce even more vigorously, with both regimes keeping huge quantities of silk in stock to bestow on ambassadors in return for merchandise presented under the guise of tribute payments. In 1385 Korea sent 5,000 horses, 225 kilograms of gold, 14,000 kilograms of silver, and 50,000 bolts of cotton. A few years later, a tribute mission from the kingdom of Ayutthaya arrived with more than 77,000 kilograms of aromatic wood, only a fraction of which was bestowed on the Hongwu emperor; the rest was put on sale. A Japanese mission numbered twelve hundred persons and over half a million kilograms of goods in nine ships, including 12,000 kilograms of sulfur, though only 4,500 was the established tribute, and 37,000 swords, only 100 of which were presented to the court. Moreover, gifts bestowed on tribute envoys invariably turned up for sale in foreign markets. In 1393 representatives of Thailand were awarded 19,000 pieces of porcelain, which they sold at a handsome profit when they returned home. When a ruler in the Ryukyu Islands ordered his envoy to purchase huge amounts of porcelain in 1404, the Yongle emperor permitted it on the grounds that benighted foreigners did not comprehend Chinese tribute-trade prohibitions.[4] In truth, this was little more than a face-saving justification, for overseas clients of the Middle Kingdom had a clear, even ruthless appreciation for how to manipulate the tribute system for their own advantage.

During the early Ming, however, the link between trade and tribute came under concerted assault, with long-term, damaging effects on Chinese merchants. The Hongwu emperor regarded the customary, generally informal practices of tribute-trade as fraudulent and pernicious, an affront to Confucian ethics and a corrupting influence on his bureaucracy. He detested the Mongols as alien conquerors of the Middle Kingdom and saw himself as the divinely ordained restorer of traditional values and institutions. Basing himself on the supposed principles of Han and Tang monarchs, he held to the revolutionary belief that China should conform to Confucian standards in all areas of state policy.[5]

In regard to foreign relations, that meant a break with the liberal, expansive policies toward foreigners and maritime commerce that had been promoted by emperors since at least the tenth century. For the first time, tribute relations took on a truly systematic character, transformed from ceremonies and mutual, tacit accords into a set of bureaucratic regulations that spelled out operations and obligations down to the smallest detail, such as which foods were to be served to tribute envoys and which to their servants.[6] The system was still nominally in place when Ricci arrived in China, and he mistakenly assumed it was ancient in character, reflecting China's immemorial perspective on the outside world.

In 1374 the Hongwu emperor abolished the office of maritime trade commissioner because not only did the bureaucrats fail to supervise trade, but they often opposed the very idea of supervision itself. He later decreed that "severe punishment will be inflicted on those who desire to do business privately with foreigners."[7] He prohibited his subjects from using many foreign products, such as spices, incense, and tortoiseshell, and he required all ships sailing overseas to purchase an imperial seal in advance. He also commanded all Chinese living overseas to return home on pain of death. Perhaps to assert the legitimacy of his new dynasty, Hongwu took the unprecedented step of dispatching envoys to invite states into a tribute relationship with the Ming court. Equally unparalleled, he coupled those summons with restrictions on tribute-trade, permitting only those countries paying tribute to engage in commerce, though under draconian regulations regarding the size, timing, and number of tribute missions. He aimed to severely reduce the commercial dimensions of the tribute system and to reorganize it as an exclusively hierarchical relationship between imperial center and subservient client. He believed his extreme measures would cauterize the wound inflicted on the Chinese moral and political order by Mongol influence and foreign exchange. He apparently never recognized (or cared) that by imposing new tribute rules he was disrupting trading customs and networks that had existed for hundreds of years.[8]

An immediate result of Hongwu's prohibition of private maritime commerce and his sweeping cutbacks in tribute-trade was to imperil the livelihood of numerous coastal inhabitants and foreign merchants. Some overseas business continued under different auspices, however, for by a stroke of the pen, the emperor had transformed ordinary traders into legally proscribed smugglers. In an attempt to stop illicit commerce, he placed rigorous controls on the coastal provinces, in some prefectures conscripting as much as a third of the adult male population into the military forces. For all practical purposes, the imperial administration had embarked on a campaign against the most enterprising ranks of its society.

After Hongwu's son, the Yongle emperor, seized the throne from his nephew in 1402, he had to deal with the tumultuous situation created by the Ming founder. Yongle shared his father's authoritarianism but not his stiff-necked devotion to Confucian fundamentalism. He sought to extend imperial control by a forward policy of military force, most notably by launching an invasion of Vietnam and by leading armies into the steppes against the Mongols in five grueling campaigns. In the southern coastal provinces, he continued Hongwu's policy of outlawing overseas trade, even ordering the people of Fujian to convert their oceangoing vessels into "flat-nosed ships" *(pingtouchuan)* suitable only for river traffic.[9] Faced with widespread lawlessness and economic hardship among the coastal population, however, Yongle took the action for which he is now most famous: he dispatched to the southern seas the largest maritime expeditions the world had ever seen.

THE TREASURE SHIPS OF ZHENG HE, 1405–1433

The Yongle emperor appointed Zheng He (1371–1435), a senior military officer, to command a fleet of 317 ships carrying some 28,000 men that left China in 1405. Six more such armadas would be sent out over the next generation, each voyage taking about two years. By the time the last of them returned home in 1433 (in the reign of the Xuande emperor), they had taken imperial emissaries and Chinese merchandise to Southeast Asia, the Indian Ocean, and Southwest Asia. Yongle did not merely invite tribute clients to China: figuratively speaking, he went out soliciting them. His employment of maritime power represented an extraordinary break with traditional imperial policy, the most audacious effort by a Chinese sovereign to restructure the relationship between the Middle Kingdom and the wider world. For the first and last time, yellow China attempted to encompass and direct the seaborne enterprise of blue China. This is reflected in the production at Jingdezhen of blue-and-white porcelains with sea animal motifs—flying shrimp, winged dragon-fish, mollusks, turtles, and wave-cresting dragons—a fashion unique to the early Ming period.[10]

Scores of treasure ships *(baochuan)* formed the nucleus of Zheng He's fleets. The greatest vessels ever built, each was some 2,000 tons in capacity, carried around 600 men, boasted nine masts, and measured perhaps 76 meters in length.[11] In contrast, Vasco da Gama in 1498 commanded about 170 men in four ships ranging from 70 to 300 tons in capacity, each with a maximum of three masts and with no vessel more than 25 meters long. There were as many medical personnel in the Ming armada as there were mariners in da Gama's. The largest fleet ever assembled by the Portuguese in Asia was the forty-three ships sent to raise a Muslim siege of Melaka in 1600, about the same number as the junks in the Ming flotilla that supplied Zheng He's men with rice and water. The total European and Eurasian population of *Ásia portuguesa* at any time in the sixteenth century—including soldiers, bureaucrats, clerics, women, and servants—was never more than 10,000 strong, about one-third the complement of men in a single Ming expedition.[12]

Zheng He, the agent of Yongle's maritime imperialism, was a Muslim from the landlocked southwestern province of Yunnan.[13] Born into the Ma family, he proudly traced his lineage back six generations to Sayyid Ajall Shams al-Din, the Turkic-Muslim governor of Yunnan appointed by Kublai Khan. Ma He's grandfather and father both claimed the title "hajji," a designation that signified they had fulfilled the obligation to go on pilgrimage to Mecca. After Ming armies subjugated Yunnan in 1381, Ma was among many Yunnan boys rounded up, castrated, and enrolled in the burgeoning eunuch establishment that served members of the new dynasty. By employing eunuchs, who were altogether separate from the formal imperial bureaucracy, Ming rulers gained a measure of independence and initiative from the Confucian elite.[14] That was to be crucial for the maritime expeditions, for Yongle put his eunuch entourage in charge of them against the opposition of his own civil administration.

Loyal and competent, Ma commanded troops under the future Yongle emperor in his rebellion against Jianwen, Hongwu's successor. As a reward for Ma's valor in a decisive encounter, Yongle in 1404 rechristened him *Zheng,* the name of the battle site. The emperor later awarded him the prestigious title "San Bao Eunuch" (San Bao taijian, the Triple-Jeweled Eunuch), a reference to the *triratna* (triple gem), the heart of Buddhist veneration—the Buddha himself, his teaching *(dharma),* and his community *(sangha).* As commander of the treasure ships, Zheng He's title was "Admiral and Representative of the Emperor."

As the Ming armadas would be visiting many ports in Southwest Asia, Zheng He's Muslim background undoubtedly played a role in his appointment. Several of his eunuch captains (though not Zheng himself) went on pilgrimage to Mecca on the seventh voyage, discovering (as they reported) that the walls of "the Heavenly Hall" (the Ka'aba) were "formed of clay mixed with rosewater and ambergris, exhaling a perpetual fragrance."[15] Word of arrival of Chinese ships on the Arabian coast came to the attention of Sultan al-Ashraf Barsbay (r. 1422–38) of Mamluk Egypt. As the historian Ibn Taghri Birdi (1411–69) records, "A report came from Mecca the Honored [to Cairo] that a number of junks had come from China to the seaports of India, and two of them anchored in the port of Aden."[16] Eager for trade, the sultan permitted the Chinese ships to enter Jedda, the port on the Red Sea nearest Mecca. Merchants of Venice, the sultan's most important partners in the spice trade, apparently never learned that official representatives of Cathay, the virtual dreamland of their very own Marco Polo, had anchored such a comparatively short distance from their trading outposts in the eastern Mediterranean.

The failure of Zheng He to go to Mecca, despite anchoring on the Malabar coast, suggests he was not a deeply pious Muslim. In fact, he displayed the same religious forbearance for which Sayyid Ajall had been known. On a plaque Zheng He affixed to a mosque in Shaanxi province, he associated himself with that thirteenth-century ancestor; but the admiral also established a Daoist shrine in the Nanjing boatyards, where the treasure ships were built, and he worshiped at a Buddhist temple in the city before he sailed for the Indian Ocean. As military commander of Nanjing, he supervised construction of the Buddhist Baoen Temple, or Porcelain Pagoda, a task that stretched over nineteen years. In Quanzhou, where he recruited members of the Muslim Pu family as commanders and translators for his fleet, he prayed for a blessing on his voyage at the Holy Tomb of Muhammad's presumed disciples.

At Changle harbor in Fujian province, a staging post for the fleet, Zheng He built a temple to the Queen of Heaven, the Buddhist patron-goddess of mariners, and he erected an altar to her in a two-story cabin at the stern of his flagship. In ports of Southeast Asia, he bestowed gifts on shrines dedicated to Guanyin, another holy guardian of seafarers.[17] In *Voyage of the San Bao Eunuch to the Western Ocean* (San Bao taijian xia xiyang), a 1597 novel by Luo Maodeng, when Zheng He prays to the Heavenly Queen during a storm, she appears holding a red lantern to pacify the wind and waves. She also proclaims her presence by the spectral appearance of Saint Elmo's Fire, an electrical discharge at the peak of the mast during a storm. Luo drew some of his material from a chronicler on the voyages who recorded that when a gale once endangered the junks, the Queen appeared astride a giant turtle and guided the fleet to safety. During another tempest, Zheng He spurned a proposal to sacrifice crewmen to propitiate angry spirits and instead threw overboard dummy figures filled with the blood and intestines of shipboard geese.[18]

On the whole, the admiral's assorted devotions are a case in point of Ricci's charge that the Chinese believed "the more different ways there are of talking about religious questions, the more beneficial it will be for the common good." Appropriately, Zheng He received the same posthumous reward as the Jesuit—elevation to celestial status. In northern Javanese ports, descendants of Fujianese and Guangdong emigrants set up cultic altars, usually decorated with blue-and-white porcelain bowls, to Sam Po, that is, the Triple-Jeweled Eunuch, the viceroy of the Son of Heaven. In legends of the Archipelago, Zheng He appears in the guise of "the Master of the Sky" (Dampu Awang), the commander of "fully laden ships" *(jong sarat),* bearer of spectacular riches from the Middle Kingdom.[19]

Zheng He's treasure ships made a stunning impression when they coasted into foreign harbors, their sails dyed red with henna, their railings festooned with yellow banners, their hulls painted with giant images of white seabirds, their masts towering over everything in sight. When thousands of troops marched off the junks and built fortified warehouses *(chinakotta),* it surely inclined their hosts to consider that a client relationship with the Ming emperor was an offer they could not refuse. This was reinforced by the consideration that the number of soldiers in every expedition was greater than the population of any port between Canton and Calicut. According to Ma Huan (ca. 1380–ca. 1460), a Muslim translator on Zheng He's staff and author of *The Overall Survey of the Ocean's Shores* (Ying-Yai Sheng-Lan), the principal source of information on the voyages, "chieftains and heads of the barbarous tribes all vied to give [the emperor's representative] greeting."[20]

In the course of the seven Ming expeditions, representatives of some seventy states became tribute clients of the Yongle emperor, many formally acknowledging Chinese suzerainty for the first time. The Middle Kingdom became the arbiter of the rise and fall of powers throughout the Archipelago. A cluster of harbors on the northern coast of Java asserted their independence from the kingdom of Majapahit after the Ming fleets appeared on the scene. Under Chinese protection, Brunei and Palembang renounced allegiance to Majapahit and the kingdom of Ayutthaya. Most significantly, the Chinese armada provided an umbrella of protection for Melaka against its powerful enemies in Java and Thailand. Yongle acknowledged Melaka as a tribute client, declaring in an official inscription that it "wished to be better than barbarian and wanted to be permanently part of the imperial domain."[21] Zheng He made the port his central depot and command center in the Archipelago. His role as godfather of Melaka, the key figure behind its swift rise to autonomy and prosperity, was recognized in the seventeenth century when a shrine to Sam Po was set up there. His protection and sponsorship proved of utmost significance inasmuch as the city-state immediately became a leading commercial center, an achievement that eventually enabled it to become a hub for the spread of Islam in the region.

At Melaka and elsewhere, "chieftains and heads of the barbarous tribes" hailed the ships of Zheng He because of desire for the cargo in their holds. As Francesco Carletti wrote later about fleets of Chinese junks delivering goods to Spanish Manila, "They have enough so that they could furnish the whole world."[22] A Portuguese observer explained around the same time, "When the Chinese smell silver, they will bring mountains of merchandise."[23] The armada of Zheng He functioned as an emporium offering a wealth of products, including nails, cast-iron kettles, axes and hoes, copper basins, bronze jewelry, lead and zinc ingots, vermilion, saltpeter, lacquerware furniture, bed furnishings, fans, umbrellas, embroidered velvet and taffeta, carpets, tapestries, thread and needles, clothing, dyes, glass beads, paper and ink, candles, pickled plums and lychees, raisins, sugar, dried rhubarb, chicken and geese, wheat flour, salted meats, and preserves of ginger, orange, pear, and peach.[24]

Zheng He returned to China with at least 180 kinds of tribute goods, most prominently silver, spices, sandalwood, precious stones, ivory, ebony, camphor, tin, deer hides, coral, kingfisher feathers, tortoiseshell, gums and resins, rhinoceros horn, sapanwood and safflower (for dyes and drugs), Indian cotton cloth, and ambergris—which the Chinese knew as "dragon's spittle" and used for making perfume. The Ming fleets brought back such a huge amount of Persian cobalt ore that Jingdezhen could draw on it for decades after the conclusion of the voyages. Exotic animals, especially a few ostriches, elephants, and giraffes, also figured in the consignments.[25]

Although the Chinese government realized a profit from its monopoly on the sale of silk and merchants had to pay duties both on commodities sold overseas and on those imported from there, the fleets of Zheng He represented a serious drain on the treasury. An estimate of 1424 put the price tag of a single voyage at one million taels of silver (or a billion pieces of copper-cash).[26] In addition, the obligation to produce enormous quantities of silk fabric and porcelains for the voyages put considerable strain on weavers and potters, especially as they were not paid the full market price for their products. Along with opposition of the Confucian establishment to eunuch administrators, the financial burden of the maritime expeditions explains why there was just one more after Yongle died in 1424, a fleet dispatched by the Xuande emperor to return hundreds of stranded tribute envoys to more than a dozen countries.

Perhaps those sorriest to see the retreat of the treasure ships were natives of Fujian and Guangdong who had settled in the Archipelago, many during times of turmoil such as the collapse of the Southern Song and Yuan regimes. In some respects, those coastal provinces had more in common with maritime Southeast Asia than with the heart of the Middle Kingdom: it was easier to sail to the Philippines or Borneo than to journey overland to Hunan or Hubei. Coastal inhabitants thrived on maritime traffic and regarded the sea as a profitable thoroughfare, not a hindrance. As a Ming functionary declared in 1639, "The sea is the paddy field of the Fujianese."[27] All but a few of the Chinese loanwords in the principal Malay and Indonesian languages derive from dialects of Fujian, and the word *junk*, commonly regarded as Chinese in origin, most likely derives from the Javanese *jong*.[28]

Overseas Chinese generally maintained contact with their homeland, forming a trading diaspora that overlapped with and complemented the Muslim network. Although Chinese tradition in Java maintains that Zheng He founded the major trading centers on the northern coast, he in fact relied on the collaboration of Javanese Muslims and Chinese (many of whom were Muslim) already resident in long-established ports.[29] According to Ma Huan, Huihui (Muslims) from Southwest Asia, as well as emigrants from Fujian and Guangdong, dominated the principal Javanese ports of Tuban, Demak, Gresik, and Surabaja. He reported that all the residents there were "very fond of the blue-patterned porcelain-ware" of the Middle Kingdom, which they purchased with Chinese copper-cash, the chief currency of the region.[30]

Chinese literature commemorates the association of porcelain with the adventures of Zheng He. In *Voyage to the Western Ocean* (Xia Xiyang), an anonymous play written for the Ming court in the early fifteenth century, Zheng He lures a hostile Sulu chief on board his flagship with the promise of showing him a magical tree that bears porcelains instead of fruit. On another occasion, the admiral lulls hostile kings into complacency with an offering of porcelains in the form of a Chinese junk; he also hands out porcelains fashioned like a tree, accompanied by the menacing suggestion that through magic he can spy on enemies from a great height. Enchanted by the gifts, the rulers board Zheng He's flagship, where the admiral browbeats them into submission. In *Voyage of the San Bao Eunuch,* the surprise is on Zheng He. Luo Maodeng portrays him meeting in Arabia with envoys from Mecca, who present him with tribute gifts of porcelain. Astonished to be presented with merchandise from his own land, Zheng He asks how the Meccans obtained the chinaware. They reply that it came by way of a "thousand-league camel," that is, by caravan on the Silk Road. The San Bao Eunuch, however, thinks to himself that only the magic of the Heavenly Queen could have brought the porcelains to so distant a place.[31]

Along with silk cloth, porcelain was the most sought after commodity in places visited by Zheng He. For just one of the voyages, the Ming court commanded Jingdezhen to make 443,500 pieces. If an equivalent amount were carried in all the expeditions, a total of 3,104,500 pieces would have been exported to the Archipelago and the Indian Ocean between 1405 and 1433. That fell a great deal short of satisfying demand, however—which explains why markets opened up at the time for Vietnamese and Thai imitations of chinaware. By comparison, in the late seventeenth century, Chinese and VOC merchants shipped at least 800,000 porcelains to Batavia annually for regional consumption, which would amount to 21,600,000 pieces over the same number of years covered by Zheng He's fleets. And not only did the latter venture abroad just seven times in the course of a generation, they called only at harbors of the most important tribute clients, bypassing many trading centers that had long relied on exports from China. As a result, some localities, such as Borneo and the Philippines, experienced something of a porcelain famine in the early Ming, a sharp reduction in the quantities of the ceramic that had been received in the Song and Yuan eras.[32] However fully laden, even Zheng He's treasure ships could not satisfy the prodigious demand for porcelain in overseas communities.

PORCELAIN TRADE IN MARITIME SOUTHEAST ASIA

Between 1611 and 1615, the Dutchman Peter Floris journeyed through Asia as chief factor of the *Globe,* a frigate of the East India Company. It was the seventh voyage of an EIC ship to Asia and the first to trade in the Bay of Bengal and the Gulf of Siam, making layovers in the ports of Ayutthaya, Burma, and the Coromandel coast. In the isolated Andaman Islands in the Bay of Bengal, Floris noted an intriguing phenomenon: "Heere uppon a little ile wee founde a greate percell of broken porseleyn of all manner of sortes. . . . From whence it was come wee coulde not knowe, for wee sawe no signe at all of any junckes or shipps which might there have bene caste awaye."[33]

Shattered porcelain on the shorelines of Asia represents the flotsam and jetsam of myriad catastrophes. The sea passage from Canton to the Persian Gulf was the lengthiest in regular use in global commerce until Spanish inauguration of the route from Manila to Acapulco in the early 1570s. In the course of centuries, ships were destroyed on coral reefs or in storms, their crews lost, their cargoes often cast up at the water's edge. As a consequence, porcelain shards are everywhere on the beaches of Southeast Asia and the Indian Ocean, often turned up even today by casual digging in the sand. Two tons of porcelain rubble have been collected from the coast of Cebu, an island of the central Philippines, just north of Mindanao. Hundreds of thousands of pieces piled up in the river deltas of northwestern Borneo, some dating to the ninth century. Shards are scattered on the shores of the Malay Peninsula, the eastern coast of Sumatra, the littoral of the Gulf of Siam, and approaches to the southern Burmese port of Martaban. They also are found in abundance on the treacherous, reef-edged passage between India and Sri Lanka, as well as on the shallow seaboards of the Red Sea, southern Arabia, and the Persian Gulf. Every lagoon, island, and headland on the East African coast likewise is littered with fragments of chinaware.

At all times the quantity of porcelain arriving in maritime Southeast Asia represented the best gauge of prosperity for both buyer and seller, the clearest indication of whether Chinese merchants were surging overseas or in government-enforced retreat. The Archipelago responded unfailingly to the rhythms of Chinese seaborne commerce, with autonomous ports, ambitious chiefs, and royal potentates competing for trading advantage by sending tribute emissaries to the Middle Kingdom and by enticing Chinese merchants to patronize their markets. Even more than continental Southeast Asia, the Archipelago served as China's undeveloped but resource-rich hinterland, a cheap, easily accessible source of forest produce and a sprawling marketplace avid for manufactured goods. It was Jingdezhen's favorite dumping ground for its coarsest products, and the kilns of Fujian and Guangdong found some of their best customers there among natives of those provinces who had migrated to the harbors of Borneo, Java, and Sumatra. Moreover, Chinese ceramics often functioned as the "small change" of commerce in areas without a dominant cash economy. A ship wrecked off Java in the tenth century did not carry enough Javanese and Chinese coins for the purpose of trade, but the great variety of ceramics on board, from the finest to the most routine, certainly smoothed the progress of commerce.[34]

From sometime during the Tang, not long after the creation of porcelain, the ceramic began entering the life of maritime Southeast Asia. In the early thirteenth century, Zhao Rugua (1170–1231), a superintendent of trade and minor member of the Song imperial family, recorded in his *Description of Barbarian Peoples* (Zhufan zhi [ca. 1225]) that merchants long had traded Chinese pottery and other goods for Southeast Asian pearls, tortoiseshell, hemp, beeswax, and incense woods. In the Philippine Islands, Zhao relates, outsiders greatly fear natives who skulk in the jungle and shoot arrows at passersby, yet "when the trader takes them a porcelain jar, they bow and take it, and then uttering cries of joy, run away with it."[35]

However patronizing the tone, Zhao's anecdote indicates that many peoples in maritime Southeast Asia responded to porcelain in a fashion dramatically different from those in Korea, Japan, and Vietnam. Since potters in the Archipelago made only rough, unglazed earthenware in bonfires and small, crude kilns, the peoples of the region enormously admired the Chinese imports and came to discount their own terra-cotta. In some cases, having employed mainly seashells, woven holders, or bamboo for eating their meals and storing their foodstuffs, they made the transition to porcelain with only slight experience of common crockery. A Chinese observer in 1618 declared that the inhabitants of Banjarmasin in southeastern Borneo stopped using banana leaves for dishes only when chinaware came into their hands.[36]

Sarawak in northwestern Borneo, an iron-smelting area, imported such huge amounts of chinaware in the twelfth century that vessels of the ceramic supplemented copper-cash as a trade currency. Maintaining close commercial ties with China from at least the Song period, the port of Santubong on the Sarawak River imported porcelains and provided over fifty tons of iron for a pagoda in Hebei province. On the swampy coast south of the river, the Melanau people traded for chinaware from Santubong, including vessels of Longquan celadon and Jingdezhen blue-and-white, which they used for storing oils, medicines, cosmetics, and magic charms. The traditional dowry given by the father of the bridegroom to the family of a high-status bride comprised a sword, gold bracelets, and blue-and-white plates inscribed with Chinese script. Trudging up steep footpaths into the remote, mountainous interior, Melanau traders brought porcelains to hunter-gatherers who never laid eyes on the sea.[37]

In Borneo and elsewhere, coastal peoples represented the major consumers of porcelain at all times, though inland areas received increasing amounts from the time of the late Song. By the Ming period, the Kelabit of Borneo, living twelve hundred meters above sea level, had access to Chinese, Vietnamese, and Thai pottery.

They had a penchant for offbeat and colorful items, such as porcelain pots shaped like ducks, crayfish, and parrots. Coastal residents preferred an assortment of small wares—bowls, plates, jarlets, cups, saucers—but hefty jars and other large pieces were the principal articles in interior communities.[38]

Many martaban jars, also called dragon jars or *pesaka* (heirloom jars), came from Burma, Thailand, and China. Tall, conical-shaped, and brown-glazed, the stoneware jars stood in the longhouses of warring clans in the Kalimantan region deep in Borneo, sometimes with porcelain plates suspended from vines above them. (With vast tracts of Borneo rain forest currently being hacked down, hundreds of martabans have come onto European auction blocks, where they fetch high prices.) Thomas Forrest (ca. 1729–1802), an English traveler, visited the longhouse of a chief on the island of Luzon that had "much the appearance of a china shop," with some thirty porcelain jars, each holding at least seventy-five liters, displayed on shelves; another chief treated Forrest to a feast on fifty-three porcelain dishes, including several sizable tureens. The Englishman also noted that humble villagers of New Guinea and small neighboring islands often possessed "a china plate or basin."[39]

Sturdy and handsome, with a thick wall, solid foot, and interior glazing, stoneware martaban jars had been shipped to the Archipelago since the Han dynasty, after which they were joined by those made of porcelain. They were costly items: the VOC paid as much as twenty-one guilders (or about five taels of silver) for a martaban at a time when a blue-and-white vase went for ten guilders and a humble jug for a small fraction of the coin. Ibn Battuta reports seeing martabans in Malabar ports that stood a meter high and held as much as two hundred kilos. On voyages from the Coromandel coast to the Hindu-Buddhist kingdoms of Southeast Asia, traders sometimes carried them filled with holy water from the Ganges; in the early eighteenth century, the dragon jars proved handy for shipping balls of opium, each as big as a person's head, from Bengal to China. Because of the robust construction of the jar, merchants also employed it for taking metallic mercury—a liquid almost fourteen times heavier than water and used for making red ink, red lacquer, and alchemical compounds—to China. Chinese traders packed martabans with a range of products, including salt pork, ginger, wine, rice, honey, and sugared citrons. Moreover, since a goodly number of small porcelains fit inside the massive jars, they made for excellent ballast.[40]

Peoples of the Archipelago used their own terra-cotta jugs for holding water as the porosity of the vessels caused evaporation on the surface, removing heat from the interior and thus cooling the liquid. Martabans, however, proved superior for containing alcoholic liquors made from rice, honey, or sugarcane, as unglazed earthenware could not keep fermented beverages in potable condition for more than a brief time. The jars also kept insects and other vermin from invading large stores of fish sauce, pickled bamboo shoots, rice, dried meat, and lime paste (necessary for the common practice of betel-nut chewing).

Sawn in half and fitted around a corpse in a fetal position, martabans in the Philippines and Borneo also commonly served as coffins for individuals of prominent status, cherished possessions whose use supposedly placated the spirits haunting burial grounds. Jars bearing the design of a dragon, regarded as a symbol of recreation and of rebirth after death, were especially valued for prestigious funerals. People treated the jars as womblike receptacles for processes that seemed to renew and recycle life, such as the fermentation and decomposition of natural things. Jar burial was an ancient custom in the Philippines, perhaps introduced from China soon after the Neolithic; but during the Song, martabans supplanted native earthenware jars in elaborate burial ceremonies. Sometimes the funerary jar with its remains would be set overlooking a cliff or river so that the soul of the deceased would find trouble-free passage to the hereafter. In secluded mountain regions of the southern Philippines, martabans occasionally still serve as ossuaries today.

Chinese pottery imports devastated the ceramic traditions of maritime Southeast Asia. Joined in due course by Chinese-style stoneware from Vietnam and Thailand, porcelain displaced indigenous terra-cotta for all important cultural functions, such as rituals of harvest, feasting, matrimony, and interment. In the Philippines, the switchover began in the Song period as Chinese merchants established links with trading port chiefs in Luzon, Mindoro, the Sulu Archipelago, and the western littoral of the Mindanao Sea. A multitude of small-scale maritime bosses (routinely doubling as pirates) dominated the sea lanes to the Moluccas, some eight hundred kilometers south of Mindanao. Ginger and cinnamon were grown in China, but the Moluccas were the exclusive source of cloves, mace, and nutmeg, spices keenly sought by Chinese merchants.

Some prosperous and strong rulers of the region even made tribute missions to China. Two chiefs from Sulu paid obeisance to the Yongle emperor in 1417, escorted by a retinue of 340 persons and presenting tribute gifts of deer hides, spices, and a pearl reported to weigh 213 grams (7.5 ounces). Yongle bestowed on them the title "king," as well as court robes of gold silk embroidered with dragons, 100 taels of gold, 200 rolls of silk, 2,000 strings of copper-cash, and numerous fine porcelains. By the late Ming period, Chinese merchants routinely shipped a special miscellany of pottery to the islands, catering to the demand of chiefs for certain vessels and fashions. The flood of chinaware dealt a deathblow to customary styles: although Philippine craftsmen continued to turn out huge amounts of terra-cotta, the unique design traditions that informed their production rapidly withered away as trade wares usurped first place in indigenous culture. At Calatagan, one hundred kilometers southwest of Manila, archaeologists have exhumed fifteenth-century graves whose bodies had blue-and-white porcelain dishes lying under the head and covering the pubic region. The second-rate status of Philippine earthenware is indicated by its being positioned at a remove from the corpse.[41]

Porcelains and other prestige trade goods had powerful economic and social effects in the Philippines and elsewhere, stimulating the rise of trading networks and more specialized craft production. Above all, warfare increased as chiefs sought to expand their sway by winning official recognition from China, purchasing Chinese weapons, and controlling the entry and distribution of Chinese merchandise. By the fifteenth century, porcelain constituted up to 20 percent of all ceramics in the Philippines, even at ports and islands off the beaten track; at some coastal settlements frequently visited by foreign merchants, Chinese, Vietnamese, and Thai wares amounted to as much as 40 percent of the pottery in use.[42] Archaeologists reckon that the Iron Age of the Philippines stretched from 500 B.C.E. to 1000 C.E.; but given the quantities of chinaware imported from the Song period onward, as shown by thousands of tons of shards excavated in the islands, they have coined the term *Porcelain Age* to designate the period of Philippine history from the late Tang to the arrival of the Spanish at Cebu in 1565.

Java had an experience similar to the Philippines, though its commercial contacts with China went back as far as the late Han inasmuch as its ports served as obligatory way-stations on the sailing route between China and the Indian Ocean. Potters in the villages of central and eastern Java produced unglazed, red-colored wares since the early common era, often modeled on Indian forms such as the *kendi*. Terra-cotta sculptural art flourished from the eighth to the tenth century, with numerous kilns on the Brantas River in central Java producing tiles, figurines, and building ornaments. Chinese pottery began to arrive in large amounts during the same time, when central Java experienced a construction boom; shards of Guangdong wares have been excavated from around the gigantic stepped stupa of Borobudur on the Kedu Plain, built in the heyday of the Sailendra dynasty (ca. 775–860) of the Srivijayan kingdom (670–1025) of central Java and southern Sumatra.

By the time Chinese overseas trade mushroomed in the Song, the kingdom of Majapahit in eastern Java had supplanted Srivijaya and extended its sway through much of the island. Majapahit sovereigns adopted Chinese copper-cash as their common currency and as the standard for government levies; Javanese metalworkers replicated the coin in tin and lead, complete with Chinese ideographic inscriptions. The kingdom's rulers launched a program of building Hindu-Buddhist temples *(candi)*, multistory monuments dedicated to the worship of deceased rulers, new deities whose ashes reposed in porcelain jars. As with similar temples being built at the same time by Khmer and Thai royal dynasties, the construction placed great demands on native ceramic technology and encouraged imports of Chinese pottery.

From the fourteenth century, Chinese and Muslim traders came to Java for spices. That was a consequence of the policy of Gaja Mada (d. 1351), a powerful Majapahit minister, who promoted bringing spices from the Moluccas to his kingdom, a distance of some sixteen hundred kilometers (or one month's sail), from where they could be transshipped to China or the Indian Ocean. In enforcing this strategy, he relied in part on Chinese-descent mariners in northern Javanese ports who had trading contacts with the Spice Islands. To obtain the riches available in Java, Chinese merchants carried enormous amounts of porcelain there. To compete with the imports, and surely out of admiration for them, Javanese potters abandoned their own artistic styles and manufacturing techniques for Chinese ones. They eventually did little more than produce earthenware copies of a wide range of Chinese vessels. In particular, they replicated martaban jars down to the last detail of decoration.[43]

THE CULTURE OF PORCELAIN IN MARITIME SOUTHEAST ASIA

Porcelains and other Chinese merchandise did not spread evenly in the Archipelago, for there was a marked divergence between two cultural environments, a difference reflected in uses and perceptions of porcelain. In terms of high culture—as characterized by urban life, long-distance trade, organized religion, and some measure of literacy—towns on coasts and in fertile river valleys provided a distinctive setting. Wet-rice *(sawah)* agriculture supported the relatively dense populations of small principalities, such as the Brunei Sultanate and the city-state of Melaka, as well as good-sized kingdoms, such as Hindu-Buddhist Majapahit and Muslim Mataram (ca. 1600–1755) in Java. Depending on seaborne exchange for customs revenue and mercantile earnings, harbormasters, administrative officials, and urban businessmen controlled imports of porcelain, which naturally accumulated primarily in their own localities. They also employed the chinaware as a commercial resource with which to extract desirable forest commodities from peoples unfamiliar with structured governments and complex economies.

Small, fiercely independent groups dominated most of the Archipelago, especially on innumerable islets and in the interior of the Philippines, Borneo, Sumatra, and the Malay Peninsula. Living in isolated valleys and impenetrable forest, they belonged to narrowly defined bands or lineage groups, combined hunting with slash-and-burn agriculture, regarded harbor states with hostility, and communicated little with the outside world. They were the sort of people typified by Zhao Rugua as lurking in the jungle and harassing interlopers. They remained entirely illiterate and only lightly touched (if at all) by the religions of India and Southwest Asia that had spread through the Archipelago since the early centuries C.E.. Raiding and feasting figured among them as central, related activities; sometimes the heads of enemies hung from funerary poles as warriors held a celebratory meal after a victorious clash. Porcelains filtered into their districts in restricted amounts, brought by agents of the distant entrepôt and traded for stuff such as tortoiseshell, rhinoceros horn, kingfisher feathers, and exotic birds.

Although there was some overlap in perceptions, porcelain generally signified different things in the two environments. Peoples in the trading towns of Borneo, Java, and Sumatra, often living alongside and marrying expatriate Chinese, treated porcelain as a well-known, prized commodity, signifying both civilized life and continuing association with the Middle Kingdom. They regarded possession of the pottery as a voucher of social reputation, economic standing, and political prestige while also valuing its patent domestic utility. They undoubtedly shared some of the Chinese notions about the magical features of the ceramic, the fantastic consequences of furnace transformation, and it accordingly had a central role in their ritual life.

The status of porcelain in comparatively sophisticated communities in the Archipelago is nicely indicated by the reception accorded Ferdinand Magellan (ca. 1470–1521) and his mariners. Commanding five ships and 237 men, Magellan left Seville in August 1519 for what turned out to be the first circumnavigation of the world. Almost two years later, on 27 April 1521, he and forty of his men lost their lives on Mactan, a pocket-sized island close to Cebu, in a futile skirmish with local warriors. After seventeen months, only one ship, the *Victoria*, limped back to home port, captained by Juan Sebastián de Elcano (ca. 1476–1526), crewed by eighteen scurvy-wracked sailors, and loaded with enough cloves to make a modest profit for investors in the voyage. In the end, the closest Magellan himself came to the treasures of Asia took place just before his death. According to Antonio Pigafetta (ca. 1480–1534), the Italian chronicler of the voyage, Magellan and a local chief on Cebu ate tortoise eggs from a porcelain dish, and at a feast given by a headman on the coast of Mindanao, the commander and his men banqueted from "large porcelain platters full of rice, and others full of pork."

After Magellan's death, the *Victoria* sailed east to Brunei, where the Muslim ruler, Raja Siripada, greeted the mariners with "silk-draped elephants . . . [accompanied by] twelve men, each with a porcelain tray covered with silk in order to carry their presents."[44] His attendants laid out a banquet of fish, capon, and peacock in porcelain bowls, along with rice wine in porcelain cups, each "big as an egg." Siripada's extravagant welcome of the Europeans, however, involved more than just bringing out the best tableware for distinguished guests: he employed the apotropaic force of his porcelains to establish a harmonious footing with unsettling strangers, perhaps as a way to avoid the sort of carnage that had marked the encounter on Mactan.

Among peoples of the Archipelago distant from maritime commerce and urban communities, porcelain took on even greater cultural and spiritual significance than it had in Siripada's domain. In regions where porcelains (and stonewares) circulated sparingly, gaining control over them meant gaining control over people. Chinaware and other foreign trade goods conferred wealth, prestige, and power on those who acquired and distributed them.[45] In effect, a chief (or proverbial "big man"), the mediator between alien vendors and the native population, severed porcelains from economic exchange and turned them into political capital for his own profit. This was the natural outcome of the economy itself not having an independent existence as such in the local community. Rather, considerations regarding material goods and resources were embedded in the culture as a whole, in the assemblage of households and kinship groups that subsumed all aspects of production and consumption. In such an environment, the value of porcelain jars and bowls had virtually nothing to do with their pecuniary cost. Rather, when people traded chinaware, they did so to settle social debts between individuals or groups, not to gain a strictly economic advantage.

Prestige merchandise from outside the community became a bankable cultural asset with which the big man rewarded followers, cemented alliances, and boosted his status and reputation. The chief who owned the porcelain jars that Forrest saw in the Philippines employed them as exhibitions of power and as markers of exclusivity, not unlike Japanese daimyo showing off their stoneware Luzon jars (as Carletti said) "out of vainglory and for grandeur," with no regard for their economic value outside their own self-validating subculture. This estimation ran down the social scale in the Archipelago, for the ordinary villager who possessed a single porcelain plate or basin also regarded it as a species of social currency, treasured it as an instrument of sumptuary distinction, and therefore endowed it with unique significance.

To people who possessed only terra-cotta, or perhaps did not even have that, the ethereal qualities of porcelains—radiant color, entrancing design, silky-smooth surface, resonant tone—made the ceramics seem not merely functional commodities for the domestic economy but also talismanic objects to be comprehended in exalted terms. Porcelains inspired the same sort of veneration that enveloped bronze cauldrons in ancient China. As with the metal vessels in the Shang and Zhou, monopolized by the powerful and vital for religious ceremony and ritual feasting, porcelains in much of the Archipelago were perceived as communal entities imbued with cosmological power.

This perspective was immeasurably enhanced by virtue of the porcelains (unlike the bronzes) being novel exotic articles; something impossible for native artisans to produce, scarcely possible for them to imagine how it could be done.[46] Porcelain thereby took on remarkable cultural and symbolic significance, gaining spiritual weight and energy by having journeyed an unfathomable extent, as if from the wraithlike realm of revered ancestors, capricious daemons, and star-dwelling deities. Since porcelains entered the community as foreign objects, people related them not to familiar containers, such as rattan baskets, bamboo mugs, and vegetable gourds, but rather assimilated them to otherworldly contexts. Earnest piety and distance conflated earthly and celestial geography. Porcelains trailed the emotive power and status of the enigmatic craftsman-god shaping his wares in an alien milieu, himself a numinous aspect of the omnipotent, primordial spirits who made all things. Regarded as enchanted and magical, porcelain vessels had extraordinary status as tokens of anticipated rank in the afterlife, as chambers inhabited by spirits, and as a means of communication with divinities. Chinaware mediated between the human and animal worlds, the heavens and the earth, the quick and the dead, the present and the future.

A Borneo legend related that the deity Sanjumang cured the illness of a Majapahit princess and founded a royal dynasty, after which he shaped an infinite number of porcelains from lunar clay and gold, filled the caverns of seven mountains with the items, and then flew back to his home in the sky. In the Kalimantan region, the Jar Deity, closely associated with martabans, lived in the heavens and had to be ritually consulted whenever an exceptional jar was traded or intentionally broken. A sacrifice to the Deity was required if the vessel shattered accidentally, else the spirit who had occupied the jar would be outraged and vengeful. The Pal'awan of Palawan Island in the Philippines believed that shooting stars striking earth caused porcelain jars to be formed. The Tinguian from the present-day region of Abra in the Philippines and the Ifugao from the province of that name regarded porcelains as bequests from the same gods who gave them rice and sugarcane and who instructed them on ritual performances that guided their lives and protected their communities.

Blowing across the mouths of martabans produced vibrating moans that shamans interpreted as celestial warnings against calamity; cords strung across the top of the jars created a percussion instrument that enabled shamans to speak with spirits of the dead. The Tagbanuwa on Palawan Island tapped their porcelain vessels to give a musical ring in order to summon gods and spirits to feasts, just as the Chinese did with pottery and bronze bells in the Shang era.[47] A Moluccan chief in the late seventeenth century alluded to the inimitable timbre of chinaware when he told a VOC commander that he felt "like a fine porcelain dish upon which rap both the Dutch [his allies] and those people of Ternate in the Moluccas [his enemies]. . . . The Lord God have mercy on my brain pan."[48]

Island peoples conceived of dragon jars as virtuosi of transmutation, possessing mortal characteristics and sharing in the sociability and misfortunes of human lives, as when a jar owned by a Borneo chief reportedly wailed pitiably when his wife died. Incorporated into families and clan networks, jars were given names, married to each other, buried with ceremony, and passed down through generations; they were seen to converse, play with one another, turn into animals and forest spirits, assume human form, heal the sick, tell fortunes, and deliver prophecies.

People in a remote district of southern Sumatra credited a martaban jar with bringing Islam to them. A Philippine legend said that a male Luzon jar, famed for its chatter, made audacious journeys around the Archipelago but eventually settled down on a secluded island with a jar of the opposite sex. Gender was determined by shape: a male jar typically had a narrow waist and high, wide shoulders, while a female one had sloping shoulders and a round body. The Iban Dayaks of Sarawak believed that porcelain jars could turn into men and men into jars. The Melanau of Borneo held to the fable that a wild boar turned into a jar, a magical change that warranted draping the vessel in costly yellow cloth and keeping it in a protective basket. Another boar, speared to death, instantly came back to life as a jar and scampered away from the hunters. A host of natural things—hooked fish, dead python, trapped turtle, plucked fruit—had the capacity to metamorphose into chinaware jars or plates, the transformation making the wondrous vessels efficacious for curative ceremonies, childbirth rites, and supernatural communication. Shamans and chiefs vigilantly cared for such miracle-working objects, bringing them out in public only on special ritual occasions.[49]

The Bataks of northern Sumatra used Ming jars for holding medicines and magic potions. In Borneo, celadon plates were believed to reveal the presence of poison when food was served on them—a superstition that probably came from Southwest Asia—and porcelain jars were regarded as enhancing the potency of the medicines they held. On the island of Sulawesi (east of Borneo), people purified water by casting celadon shards into it. The Iban of Borneo used chinaware jars for collecting spirits required for bodily health, and their ritual experts made medicine by grinding chips from dragon jars into a fine powder. Among the Tinguian people, a spirit would release a child from illness or possession when a shaman drank rice wine from a porcelain cup. On Java and Sulawesi, women placed newborns, with their placentas, on porcelain dishes, and nine days after giving birth, the mother ritually bathed herself with water from a Chinese pot. An infant's welfare was safeguarded by keeping the water in which it was washed in a Chinese jar, often in tandem with a piece of copper-cash. Severe infirmities, people in Borneo supposed, called for the application of oil kept in chinaware jars, containers seen as so potent that their very presence at the doorway of a longhouse kept disease and demons at bay.[50]

Among the Dyaks of Kalimantan, if a martaban ornamented with two dragons stood at the entrance of a home, it indicated that a female virgin dwelled within. In a fertility rite after nuptials, Dyaks danced and chanted around a pole before which was placed a porcelain jar festooned with fruit and leaves. As part of a Melanau wedding ceremony, women bathed the bride and groom in water from a Chinese jar, and the liquid also was poured over the right foot of the bride when she entered her husband's home for the first time. Women rubbed coconut oil from a celadon jar on the temples, knees, and elbows of the couple to guarantee fruitful reproduction. For the same reason, the jar remained under the bed of the newlyweds during their first nights together.[51]

The Kelabit of Borneo suspended Chinese pots along with the heads of their enemies from the rafters of their longhouses so that the spiritual force of the latter would be siphoned off into the containers. Sealed with a wooden stopper to retain the magical energy, the pots had such substantial value that an owner could exchange one for a slave or sacrificial victim. Warriors used Kangxi porcelain pots in the form of ducks and crayfish to offer libations to enemy heads, always careful that women and outsiders did not pollute the precious vessels by touching them. Among the Iban of Sarawak, the possessions of a dead warrior were tied to the top of a dragon jar, and only a comrade who had taken a head in combat could remove them. The Iban placed Chinese pottery at the base of ornamented poles from which the heads of enemies dangled. The Punan Ba numbered among their foes, and they too adorned their funerary columns with porcelains to commemorate the loss of valiant warriors. At the foot of the poles, they set out blue-and-white dishes with offerings of food for the spirits of the men they mourned.[52]

In the shrouded green jungle of the Archipelago, gleaming porcelains radiated a spellbinding aura. The vessels seemed to provide access to a remote, mysterious world, a sense that one could get in touch with, and perhaps have an effect on, awe-inspiring forces. Chinaware played a vital role in the whole cycle of existence—birth, sickness, marriage, feasting, worship, war, and death. Naturally, beliefs and practices taken for granted by the Kelabit and Punan Ba were less in evidence in the trading towns and urbane courts of maritime Southeast Asia. Traces of them lingered, however. Despite their formal censure of idolatry and heathen custom, Islam and Christianity in fact won converts by freely assimilating a host of local folk spirits, rituals, and customs. Old and new seldom had trouble reaching accommodation. The salvation religions of Southwest Asia clearly failed to stamp out the notion that porcelain could be employed as an avenue of contact with the supernatural; the ceramic continued to serve as a gesture of reverence for transcendent power, an adornment linked with longstanding devotion. Spanish priests in Manila poured holy water into blue-and-white bowls set in baptismal founts, and they had blue-and-white porcelains cemented into the walls of their churches.[53] Similarly, Muslims in Batavia, Demak, and Melaka built mosques with walls of brick inlaid with blue-and-white plates, and they decked out the gates of shrines and mosques with blue-and-white tiles from China and Vietnam. And Raja Siripada staged a pageant with his porcelains and elephants to intimidate and pacify his worrisome Spanish visitors.

THE CULTURE OF PORCELAIN ON THE SWAHILI COAST

In the four voyages between 1414 and 1433, Zheng He's treasure ships sailed beyond the Archipelago. After steering north through the Straits of Melaka, they turned west for the Indian Ocean. At an average speed of 2.5 knots (111 kilometers) a day, the fleet covered the 1,750 kilometers to Sri Lanka in about a month. From there, it moved on to the Maldive Islands, where merchants loaded up on ambergris, coconut-fiber rope, and tons of cowry shells. The islands served as a useful stopover for the fleet, most of which would then proceed north to Calicut on the Malabar coast while a contingent sailed on to East Africa. It took about a week to reach Somalia, which in the late fifteenth century boasted a "Zheng He village" (Zheng He tuan) on the coast, as well as a midden of porcelain shards more than a meter deep.[54]

The monsoon system gave the ports of East Africa connections to Arabia, some 5,000 kilometers to the north, as well as to the coast of Malabar, 4,100 kilometers to the northeast (measured from the equator). From April to October, the summer monsoon impelled ships from Africa to India; from November to March, the winter monsoon blew them back to the Swahili coast. By Zheng He's time, about thirty-five Swahili ports used the monsoon winds to take part in Indian Ocean commerce, the most prominent being Mogadishu (now the capital of Somalia), Pate, Malindi, Mombasa, Kilwa, and Sofala. Ruled by merchant oligarchies and titular kings, they preserved their political autonomy inasmuch as the lowland coastal belt was too shallow and extensive—1,750 kilometers from Mogadishu to Kilwa, just north of Mozambique's Cape Delgado—to permit formation of a centralized state. A north-south coastal axis thus emerged rather than an east-west axis linking the seaboard and the interior.

Historically, the Swahili coast looked outward to Arabia, India, and Southeast Asia more than to the vast African heartland. The latter begins some sixteen kilometers from the coast, where an escarpment rises twelve hundred meters above sea level, the rim of a plateau that extends to the great lakes that are the ultimate source of the Nile River. The name *Swahili,* "Lands of the Coast," is Arabic in origin, deriving from *sahil* (pl. *sawahil*), "margin" or "boundary." Africans founded the coastal towns, however, not (as legend has it) Arabian, Mesopotamian, or Persian merchants.[55] Ibn Battuta, who visited the coast in 1331, observed that the "majority of its inhabitants are Zanj [i.e., Bantu], jet-black in colour, and with tattoo-marks on their faces."[56] He commended them for their Muslim piety and well-built wooden mosques.

Although Africans made up the majority of the Swahili people, significant numbers of Arabians, Persians, and Indians added to the mix of population. In terms of culture, coastal inhabitants belonged to the cosmopolitan mercantile world of Islam. Their port cities acted as intermediaries between the polities of the Indian Ocean and the African interior, exporting ivory, ambergris, iron ingots, timber (especially mangrove poles), leopard skins, gold, and slaves. Gold came mainly from the stone-built settlements clustered in the area of Great Zimbabwe on the southern margin of the Zimbabwe plateau, a variety of locations that date from the eleventh to the sixteenth century. Archaeologists have excavated Indian glass beads, Southwest Asian pottery, and porcelain there. Slaves too came from the interior, usually men captured in war, and merchants sold them throughout West and East Asia. Zhao Rugua reports that "many families [in China] buy black people to make gatekeepers of," referring to them as "devil slaves" *(kui-nu).*[57] According to Ibn Battuta, Africans also served as elite warriors for the king of Sri Lanka and as men-at-arms on great Chinese junks, such as the one in Calicut on which he booked passage for Canton in a private stateroom for himself and his ever-present slave girls.

Merchants imported rock crystal, glass beads, fine metalwork, cotton cloth, spices, weapons, and porcelain to the Swahili coast. Chinese copper-cash first arrived there during the late Tang period, around the same time Islam began spreading down the littoral from Arabia and Iraq, winning over the indigenous population. The economic demands of the Fatimid and Ayyubid dynasties of Egypt and of the Abbasids of Mesopotamia—especially for timber, iron, and slave labor for swamp drainage and construction projects—provided a powerful stimulus for the establishment of trading towns. By the twelfth century, worshipers had built mosques as far south as Madagascar, and Islamic forms of burial, with the head of the body facing east toward Mecca, appeared around the same time.

Porcelain first reached the coast in the late Song, but it did not become plentiful until Ming blue-and-white began to arrive not long before the expeditions of Zheng He. Swahili merchants never took significant amounts of it inland, however, just as they never attempted to spread Islam there. Their intense social individuality, stemming from their character as tight-knit oligarchies presiding over small, self-governing communities, stymied cultural interaction between the seaboard and the Bantu heartland. This exclusivity was intensified by the seaports not developing durable institutions for sustaining and promulgating scriptural tradition. Significantly, potters on the coast and those in the interior preserved entirely distinct customs. In any case, places associated with Great Zimbabwe—settlements primarily for relatively penurious warriors—stimulated little demand for foreign pottery and other Asian luxury goods. Almost all inhabitants of the arid plateau relied on hardscrabble farming or followed an itinerant, cattle-herding way of life, neither of which lent itself to purchasing exotic commodities such as Persian and Chinese pottery. When chinaware did make its way inland, however, it predictably took on the magical quality it enjoyed in the Archipelago: a sixteenth-century Portuguese traveler recounted that natives rubbed their bodies with blue-and-white vessels to assuage a range of infirmities.[58]

The Swahili elite, known in the Kiswahili language as *Waungwana,* "persons of gentility," regarded porcelain solely as an exotic article to be displayed as a status symbol. They kept trade in foreign merchandise in their own homes and storehouses, away from the public markets of commoners. As described by Ibn Battuta, the customs of Mogadishu were typical: "Each merchant on disembarking goes only to the house of the young man who is his host. . . . The host then sells his goods for him and buys for him."[59] Local businessmen extended trading concessions to foreign merchants by incorporating them ritually into blood brotherhoods; the putative kinsman generally acknowledged the honor by presenting his hosts with gifts of porcelain.

"Persons of gentility" restricted the access of commoners to exotic commodities, hereditary offices, fertile farmland, fishing rights, and weapons. Nor were those outside privileged circles allowed to have homes as grand. Whereas most dwellings on the coast were one-story, mud-and-wattle huts, the elite built mansions from coral blocks quarried from reefs and dressed in lime plaster. With windowless walls rebuffing the outside world, the rooms were perfumed with incense and hung with silk cloth; special niches in galleries displayed the owner's finest chinaware and could be viewed only by those of suitable rank. Since porcelain had so central a role in validating the status of the *Waungwana,* they made certain it never spread appreciably up the escarpment or down the social ladder. Unlike what happened in the Philippines and Java, then, indigenous terra-cotta traditions remained intact. As revealed by archaeological excavation, only 5 percent of pottery shards excavated on the Swahili coast derive from trade imports.[60]

Wealthy merchants and royal dignitaries paid for public structures such as tombs, mosques, and palaces. They decorated the buildings with intricate coral carvings, as well as with insets of porcelain and Chinese-influenced Southwest Asian pottery. At many sacred sites associated with ancestors and honored dead, such as rock shelters and headlands, Muslims left porcelain bowls as offerings to propitiate spirits, believing the dead would drink from them when the living departed the site. Porcelain plates were shattered over the tomb of the deceased; gifts of porcelain were placed in caves for spirits who watched over a lagoon or stretch of coast. Worshipers employed porcelain as decoration in the vaults of mosques and in the spandrels of mihrabs, the niches indicating the direction of Mecca. Muslims decorated the facade of a large domed tomb on Pate Island in the Lamu Archipelago (in present-day Kenya) with pieces of Chinese and Southwest Asian blue-and-white. Since blue dyes were unknown in East Africa (until the Portuguese planted indigo in the region), blue was a compelling color there, leading Swahili inhabitants to regard imported blue cloth as exceptionally valuable and alluring. The same consideration enhanced the status of blue-painted pottery.[61]

Kilwa (in present-day Tanzania) emerged in the twelfth century as the leading commercial power on the coast, chiefly because it controlled the port of Sofala, located over one thousand kilometers south of Cape Delgado. Merchants from Sofala ventured far inland to trade for gold, perhaps all the way to Great Zimbabwe. In exchange for the yellow metal, Kilwa supplied glass beads, metalwork, textiles, and pottery. It held sway as the wealthiest city on the coast for several hundred years before the arrival of the Portuguese. The gold trade financed the building of the Great Mosque, a fourteenth-century building with eighteen cupolas and barrel vaults that the Portuguese compared favorably with the Great Mosque of Córdoba in Spain. A portico of the Kilwa mosque had mid-fifteenth-century blue-and-white porcelain bowls embedded in its roof, some with Islamic geometric motifs, others with the usual Chinese thematic repertory. The Kilwa citadel possessed a similar grandeur: perhaps built by craftsmen from Egypt or Mesopotamia, it was a double-story structure with a large bathing pool, storage rooms for merchandise, and walls ornamented with insets of porcelain and Persian blue-and-white. Much of the latter was custom-made for East Africa, copied from porcelains unloaded in Siraf and Hormuz.[62]

Like caves and other sacred shelters, pillars were regarded as dwellings for honored spirits. Tall and tapered, the structures rose near tombs and small mosques, usually as memorials for a holy man or an illustrious ancestor. Pieces of blue-and-white porcelain encircled the base of the Great Pillar of Malindi (10 meters high) on the Galana River (Kenya), and a frieze of Ming plates and bowls crowned the Great Pillar of Mambrui (8 meters high), just north of Malindi. In Mombasa (Kenya) broken seventeenth-century Chinese pots adorned the base of the Mbaraki Pillar (14 meters high) and the walls of a nearby mosque. In the sixteenth century, Muslims set Portuguese imitations of Chinese blue-and-white into the walls of the Small Mosque at Mnarani, a town between Malindi and Mombasa.[63]

From the Horn of Africa to Cape Delgado, Swahili merchants for centuries measured their good fortune by the porcelains arrayed in the gallery niches of their elegant coral mansions. Their prosperity and power came to an abrupt end with the arrival of the Portuguese, however. The newcomers established fortresses on the coast to support their annual voyages between Lisbon and Goa. They drove the Swahili out of the gold and ivory trade with the hinterland; they took over Kilwa, pillaged Mombasa, and monopolized the routes to the Red Sea and the Malabar coast. The ports could not muster the unity needed to resist the invaders; nor could the Swahili elite call upon broad social support to salvage their political eminence. Impotent and insolvent, the prey of outside forces, the Swahili ruling class ceased to play a major role in the commerce of the Indian Ocean. Yet the Portuguese experienced their own collapse little more than a century later: by the early 1600s the Ya'aruba sultans of Oman had seized control of the coast, and European competitors, mainly the Dutch and English, had dislodged the Portuguese from supremacy in the trade between Africa and India.[64]

In *The Soul's Awakening* (Al-Inkishafi [ca. 1810]), Sayyid Abdalla bin 'Ali bin Nasir contemplates the passing of the "persons of gentility" who ruled the city-state of Pate for so long. Both mordant and melancholy, his verse depicts Providence wreaking vengeance on the Swahili for the sins and corruption of its resplendent merchants, men besotted by chattels and profit at the cost of their eternal salvation. The poet suggests they should have feared false times when they did feast:

> And in their halls dwelt Beauty everywhere
> And Veneration stalked them all their days.
> Their homes set with Chinese porcelain
> And every cup and goblet was engraved
> While, placed amidst the glittering ornaments,
> Great crystal pitchers gleamed luminous. . . .
> Their lighted mansions [now] echo emptily;
> High in the rafters flutter bats.
> There are no murmurings, no happy shouts,
> And on the carved bedsteads spiders spin their webs.
> Where once in the wall-niches the porcelain stood
> Are now the ragged nests of wild birds.[65]

"I THOUGHT ALL INDIA A CHINA SHOP"

For the fleet of Zheng He, Calicut was "the great country of the Western Ocean," its ambassadors taking precedence over all others.[66] The admiral's traders certainly found their largest and most profitable market there. After the treasure ships docked, weeks of haggling followed between Zheng He's agents and those of the Hindu king over prices to be set for a mountain of things. The merchants of Calicut naturally were familiar with porcelain and other Chinese products since they had been coming to the port in Indian and Chinese ships for centuries before the voyages of Zheng He. After the Ming expeditions ended, Muslims of southwestern India took the lead in the seaborne trade, usually making a round-trip from their home ports to Melaka.

An EIC factor in India in the seventeenth century reported that "all sorts of China ware are heere both in great esteeme and use which beinge sorted of all sizes, pryces and fashions will vend here at least 100 tonns per annun."[67] In Goa in 1578, Francesco Pasio (1544–1612), an Italian Jesuit on his way to China with Ricci, expressed astonishment at the chinaware on sale in the Portuguese trading center: "There is so much porcelain and at such an attractive price that I saw nobody there making . . . similar things in clay, for porcelain is sold so cheaply that they could not sell clay pots for any price without wasting their time and losing money. And they even have porcelain under the bed to be used at night."[68] When Sir Thomas Roe (ca. 1581–1644), ambassador of James VI (r. 1603–1625) to the Mughal emperor Jahangir (r. 1605–27), arrived in India, he looked forward to some gratifying purchases: "I thought all India a China shop, and that I should furnish all my Frendes with rarietyes."[69]

Roe and other customers, however, had to guard against being palmed off with Persian copies of Jingdezhen blue-and-white, a fiddle exploited by the VOC and EIC in India and Europe at times when the Chinese government embargoed overseas trade. According to John Chardin (1643–1713), who traveled in Persia in the 1670s, potters there made earthenware that is "pure Enamel, both within and without, like the China Ware. The Grain of it is as fine and transparent as that is, whereby it happens that one is often deceived in that Earthen Ware, and that one cannot distinguish it from China Ware, so beautiful and lively is the Varnish of it. . . . They say that the Dutch mix that Persian Ware with the China Ware, and import it into Holland."[70] The sophisticated consumer looked for the equivalent of a trademark. In dispatching a shipping order from Coromandel to China, a VOC agent stipulated that "it should be borne in mind that the aforesaid porcelains are most desired when they have a blue mark drawn like a character on the bottom."[71]

On the face of it, India would seem to be a lucrative market for porcelain: its population reached 150 million in the late sixteenth century, and its Hindu and Muslim aristocrats were renowned for their coffers stuffed with gemstones and precious metal. Yet the subcontinent was the only region of the ecumene where porcelain effectively met defeat. Goa aside, as Roe soon discovered, all India proved not to be "the China shop" he had imagined. And only in that port, amply supplied by Portuguese carracks, were porcelains so abundant and cheap that Europeans could employ them for chamber pots. To be sure, Muslim merchants and nobles valued the ceramic, and Muslim sovereigns of the Delhi Sultanate and Mughal Empire zealously collected it, putting their marks of ownership on vessels in the form of drilled holes and engravings. And they were knowledgeable enough about porcelain to discriminate between wares from various kiln centers.[72] But Muslims numbered less than 20 percent of the population, and the vast majority of Indians adhered to Hindu religious and cultural practices. As a corollary of their devotion, Brahmin priests and Rajput warriors recoiled from using porcelain, and Indian potters turned out only unadorned terra-cotta.

An English agent for the VOC observed in 1614 that Persian merchants in India "eat from porcelain, while the Gentus [i.e., "Gentiles" or Hindus] do not. . . . Most of them observe the heathen rule."[73] That so-called rule stemmed from Hindu anxieties about spiritual pollution, which dictate that utensils made of terra-cotta, which are porous and difficult to clean, must be discarded after use. From ancient times, Indian artisans displayed impressive skill in making large terra-cotta sculptures for cultic ceremonies; but apprehension about pollution reduced conventional ceramics to a stagnant art from at least the Gupta period of the early common era, leading to the production of incalculable quantities of undistinguished, eminently disposable pottery.[74]

As a result of clay vessels being identified with pollution-fraught terra-cotta, scrupulous Hindus shunned all ceramic wares and instead used metal dishes or improvised receptacles. (Lacking a well-developed caste system, Hindu-Buddhists of Southeast Asia did not adhere to such exacting strictures.) Ibn Battuta remarked that in Malabar and Sri Lanka, Hindus often served food on banana leaves to their guests. In Hindu households even today, pottery is imperative for various ceremonies; but since many believers presume that evil spirits may be absorbed by cooking pots and vessels when perturbing events take place—such as illness, death, and menstruation—wares are smashed immediately after use, with a new set of crockery purchased when the occasion for ritual pollution has run its course.

Not surprisingly, porcelain always cost too much for such ready destruction, and even glazed earthenware seemed a pointless investment. Indian craftsmen, perforce, halted on the terra-cotta rung of the pottery ladder. Eager to embellish their tombs, mosques, and forts, Mughal dynasts were forced to import brilliantly colored tiles from Kashan, the Persian center of tile production, and when enemies of the Mughals, Muslim kings and nobles of the sultanate of Golconda (1518–1637) in southern India, wanted to decorate their buildings with similar tiles, they found that local potters lacked the facility to make them; huge quantities had to be imported from distant Kashan.[75]

For their own good reasons, then, Indian potters never adopted the innovations of Southwest Asian artisans, such as luster glazing and tin glazing, and they had no inducement to turn out imitations of porcelain. In addition, for more than a millennium, Hindu caste restrictions, which aimed to minimize the peril of spiritual pollution, contributed to the immobility of Indian earthenware. In the huge caste of potters, called *kumbhakars* (makers of *kumbha,* round-bottomed water jars), members of a given subcaste *(nyat)* abjured social relations with those belonging to another, even when a number of such subcastes worked in the same town.[76] Under the impact of such weighty conventions, there could be no long-term pooling of expertise regarding clays and kilns; hence pottery skills, even within the humble ambit of terra-cotta, remained undeveloped. In striking contrast to everywhere else in the ecumene, potters in India not only had no interest in stealing secrets from one another, but they had no secrets to steal.

THE CULTURE OF PORCELAIN IN TIMURID PERSIA

A detachment of Zheng He's fleet steered northwest from Calicut to the island of Hormuz, just off the coast of Persia. With a fair wind in the right monsoon season, the voyage took about twenty-five days. Located at the choke point where the Arabian Sea flows into the Persian Gulf, Hormuz flourished by transshipping merchandise from India and China to the Tigris-Euphrates, inland Persia, and Transoxiana. Ma Huan recorded that the markets of Hormuz offered commodities of every description, especially pearls, rubies, topaz, jade utensils, and brocaded velvet. The island's sultan paid tribute to the Yongle emperor with such articles, as well as with consignments of lions, lynx, and leopards.[77]

As always, Zheng He's stevedores carried quantities of Jingdezhen blue-and-white off the treasure ships. The pottery was not entirely new to Hormuz, however, since Chinese junks had taken some there in the mid-fourteenth century, shortly before China fell into the civil war that led to the overthrow of the Yuan dynasty. Persian merchants in the port who sold cobalt ore to Zheng He's factors knew that potters used it to produce the Jingdezhen vessels, though they surely did not realize that Persian tin-glazed, blue-decorated earthenware had stimulated creation of the Chinese blue-and-white style the better part of a century earlier.

In a relay of influence typical of ceramic history, after Jingdezhen blue-and-white disembarked in the land of its inspiration, it in turn decisively influenced Southwest Asian pottery. Royal courts and artisans developed a new ceramic style, modeled on Ming porcelain designs, which became an integral part of a broad cultural renaissance in the Islamic world. The territories most affected were those ruled by the emperors of Safavid Persia (characterized by Westerners as the "Great Sofi"), Mughal India (the "Great Mughal"), and the Ottoman dominion of the Balkans, Anatolia, Syria, and Egypt (the "Great Turk"). Beyond subscribing to Islam, the three regimes shared a great deal: they had links to Mongol tradition, derived from Turkic horse warriors of Central Asia, came to power by conquest after 1450, relied on gunpowder weapons to reinforce imperial authority, and sought cultural legitimacy to secure and promote their rule.[78] The three emperors who presided over the greatest cultural flowering of their respective realms all lived around the same time and served as model rulers for their successors: Shah ʻAbbas (1587–1629) of the Safavids, Akbar (1556–1605) of the Mughals, and Süleyman (1520–66) of the Ottomans.

The Muslim sovereigns looked back to cultural traditions that first emerged at the courts of the Timurid dynasty (1378–1506) of Persia, Iraq, and Afghanistan. One of the most famous conquerors in history, Timur (ca. 1330–1405) is best known in the West as Tamerlane, from the Persian version of his name, *Temur-i lang*, "Temur the lame." He pursued his ambitions in Central Asia in the context of the legacy established by the Mongols. Although a protracted struggle over election of the Great Khan led to rupturing of the "Great Mongol Empire" in the mid-thirteenth century, it remained a settled principle that political leadership must be in the hands of provincial khans belonging to the house of Genghis. The power of Timur, however, stemmed solely from his own astonishing energy and ability. After sweeping aside the minor Mongol lords of Transoxiana, he tried to make up for his discomforting lack of political legitimacy in various ways. He married a woman from the dynasty of the khans, concocted a spurious genealogy linking himself to the family of the Prophet, erected shrines for Sufi holy men, sponsored religious endowments, and patronized Persian poets and painters.[79] Because it was identified with Mongol China and regarded as a token of high culture, he also favored blue-and-white porcelain.

At least in a minor key, Timur, like Korean kings, Ashikaga shoguns, and Philippine chieftains, called on the prestige of China to burnish his cultural credentials. Collecting porcelain along the way in his career of conquest represented an obvious means of doing so. When in 1398 he pillaged Delhi, capital of the Tughlaq dynasty (1321–1450) of the Delhi Sultanate, he acquired quantities of Ming blue-and-white. Two years later, when he sacked Damascus in Mamluk Syria, porcelain once again figured among the spoils prisoners carted to his capital, Samarqand. Potters captured in Damascus, many of them adept at copying Chinese blue-and-white, trained Timur's potters to do the same.[80] Their expertise accounts for mausoleums in Samarqand being faced with white tiles painted with cobalt to resemble chinaware and for a Samarqand mosque, built in honor of Timur's niece, being decorated with medallions of stylized blue lotuses taken from porcelain. Miniature manuscript paintings by Junayd, created in Baghdad around 1396 to accompany a collection of poems, show numerous examples of blue-and-white porcelain being used in convivial gatherings of Timurid notables. Junayd's work proved influential in establishing an association in later Persian art between princely entertainment and Chinese blue-and-white.[81]

The Castilian ambassador Clavijo recorded that Timur habitually used porcelain at his court. Aside from the blue-and-white looted from Delhi and Damascus, the ceramic also came to Samarqand on the Silk Road, transported in enormous caravans of camels and carts, such as the one Clavijo saw in 1405. For safety, tribute missions journeyed with the caravans as well. Timur presented the Hongwu emperor with steel swords, armor, precious stones, and thousands of horses, and he reciprocated with the standard gifts, mainly silk and porcelains. But Hongwu also sent along imperious messages insinuating that Timur held title as a vassal of the Middle Kingdom. In response, Timur in 1405 began marching his army to conquer China, only to die a few hundred kilometers north of Samarqand. The same year marked the departure from China of the first Ming maritime expedition.[82]

When the treasure ships of Zheng He reached Hormuz about a decade later, some of the porcelains unloaded certainly ended up in the hands of Timur's successors. They lacked Timur's zeal for conquest and devoted their energies to political consolidation and lavish cultural patronage. Sons and grandsons of the conqueror, they sponsored a coherent style in the arts, tempering Turkic-Mongol aggression by circumscribing it within the framework of Persian literary and aesthetic traditions, a strategy pioneered by the Ilkhanids.[83] The result was a remarkable revival of art and culture that spread through the Islamic world at the same time the Italian Renaissance was transforming intellectual life in Europe.

The Timurids tirelessly cultivated China, sending envoys there every few years, usually with letters specifically requesting porcelains. Shah Rukh (r. 1405–47), Timur's successor in Persia, several times welcomed Chinese ambassadors bearing gifts of chinaware. *Procession Scene,* a Timurid manuscript painting of the early fifteenth century, depicts a fantasy version of such an embassy: in a rocky desert landscape, nine men in colorful Chinese dress escort a donkey cart weighed down with enormous blue-and-white jars.[84] In fact, blue-and-white is a common feature in manuscripts sponsored by the Timurids, such as the vessels depicted in texts of Rashid al-Din's history of the world, a work originally sponsored by the Ilkhanids.

The most influential illustrations were those created for the famed *Book of Kings* (Shahnama), an epic poem by Firdausi (ca. 934–ca. 1026) that recounts adventures from the mythic and historical past of Persia. The Ilkhanids had made the *Book of Kings* a major focus of their painting academies, and, in like fashion, the Timurids incorporated it into their patronage, thereby affirming their ties with traditions as far back as the Sassanids. Porcelain is a recurring decorative figure and symbolic motif in Timurid productions of the *Book of Kings.* A 1444 text opens with a double-page frontispiece, *A Royal Feast in the Garden,* which portrays a royal Timurid couple being presented with a dozen pieces of blue-and-white by three Chinese ambassadors in black headgear.[85]

Iskander Sultan (1409–35), Timur's grandson, encouraged manuscript painting at Shiraz in Persia, developing forms and standards that prevailed for the next two centuries in the Persian-speaking world. Ulugh Beg (1394–1449), another grandson, built a *chinikhana* (porcelain house) in Samarqand: the pavilion had vessel-shaped niches for pottery, and porcelain tiles, specially ordered for the building, embellished the walls. In 1411 Ulugh Beg ordered the release of all craftsmen in Samarqand who had been enslaved by Timur. Damascus potters soon found employment throughout the Timurid realm, in particular in the cities of Tabriz, Shiraz, Kashan, and Isfahan, thereby helping to disseminate Timurid design techniques based on blue-and-white porcelain.[86] By the end of the fifteenth century, the popularity of Timurid pottery throughout Southwest Asia meant that artisans skilled in the traditional geometrical style of the Islamic world were increasingly adopting elements of free-floating Chinese design.

Timurid modes of Chinese styles also spread in the Islamic world as a result of the increasing use from the thirteenth century of high-quality paper.[87] Southwest Asian potters no longer had to learn their brushwork by working with unfired ceramics but could practice on sheets of the new material. Even more important, they no longer had to rely on chancy memories or vessels at hand for their decorative repertoire but could take their motifs from pattern books. This new practice also facilitated the exchange of designs among diverse media—textiles, architecture, calligraphy, metalwork, illuminated manuscripts, and pottery.

The common aesthetic programs that sprang up in the Islamic world after 1400 stemmed not only from the undeniable magnetism of the Timurid cultural renaissance but also from pattern books bringing an appealing, coherent style to widely scattered artists and courts. Thus at the same time that a long-term convergence of the arts began in Ming China as a consequence of the introduction of blue-and-white porcelain to a centralized imperial procurement system, the arts of the Safavid, Mughal, and Ottoman Empires experienced a comparable, ongoing fusion. Something similar also began in the West from the 1450s: invention of the printing press set up conditions for cross-cultural exchange as book workshops for the first time brought together painters, manuscript illuminators, goldsmiths, engravers, metalworkers, and scholars—all engaged in fashioning a product that in time would create a self-styled Commonwealth of Learning throughout Europe.[88] This development intersected with Asian ones, for Western entrepreneurs in the late seventeenth century started to send pattern books to India and China to be used for designs on textiles and pottery intended for European consumers. As a result, motifs, designs, and symbols from diverse cultures began to combine promiscuously on a global scale.

Potters, paper, and publications circulated through Southwest Asia and South Asia. After the fall of the Persian Timurid dynasty in 1506, all three of the great Muslim realms claimed the heritage of Timur and his successors; they inhabited a common cultural space with common cultural references.[89] Shah Isma'il (r. 1501–24) brought the Safavids to power in Persia, the first truly native dynasty there in eight hundred years. As a war leader and religious zealot in his youth and a listless drunk in his later years, he paid little attention to cultural endeavors. His son Shah Ṭahmasp (r. 1524–76), like Timur's heirs, turned to gentler pursuits, becoming a painter of some skill and patronizing art on a grand scale. His many commissions included an enormous text of the *Book of Kings*—760 folios in length with 256 miniatures, many including porcelains—that he presented to the Ottoman sultan Selim II (r. 1566–74). Fernão Mendes Pinto, the Portuguese traveler, claims that only Tahmasp's merchants were licensed to purchase the finest imperial porcelain in China.[90]

Tahmasp was far surpassed, however, by his successor, Shah 'Abbas, known as "the Great," the most splendid figure of the dynasty. He rebuilt the capital of Isfahan, embellishing it with parks, palaces, and public squares. He encouraged use of colored tile, so the city reportedly blazed with vivid hues in the sunlight. Ambrosio Bembo (1652–1705), a Venetian traveler in 1671, described a marble bridge with twenty-seven arches, with "a very beautiful cornice of porcelain of various designs" above every arch; the royal palace made a majestic sight, "decorated with ornaments in porcelain and gold," and across from the palace stood a mosque with a "dome and façade in fine polychrome porcelain." Although homes of the upper stratum appeared drab on the outside, Bembo found that many had interior walls decked out with porcelains. In the palace apartments of Shah Sulaiman (1666–94), known as the Nightingale's Gate (Bab-i Bulbul), he noted that all the rooms had gilt mirrors, European paintings, and "some cabinets full of fine Chinese porcelain."[91]

In 1611 Shah 'Abbas bequeathed 1,162 porcelains to the shrine of Sheik Safi (1252–1334) at Ardabil, near the southern shore of the Caspian Sea. The pottery formed part of a huge endowment that established 'Abbas as the foremost champion of the dynasty. The gift, which included more than four hundred blue-and-white vessels, was put on display in a "porcelain house," a large octagonal building decorated in blue and gold. With so many pieces of blue-and-white concentrated in one location, the collection became a source of inspiration for Persian potters. Kashan, a center of pottery manufacture three hundred kilometers south of the Ardabil *chinikhana*, experienced a flowering under 'Abbas, producing high-quality wares in tints ranging from powder blue to rich lapis lazuli. Working from the originals, potters as a rule faithfully copied the patterns and shapes of the porcelains; but they also painted vessels with Chinese designs from different periods, inventively fashioning a hybrid style that owed its particulars to China while at the same time conveying an overall Persian sensibility.[92] (See figure 22.)

The support Shah 'Abbas gave to the Persian pottery industry was part of his program for the economic prosperity of the empire. Adhering to the same mercantilist views as his royal contemporaries in Europe, he aimed to stop the drain of precious metal to India, that notorious abyss for gold and silver.[93] He established royal workshops for the manufacture of silk and cotton in many cities. In Isfahan, tens of thousands of weavers produced damask, satin, and brocade, sometimes decorated with patterns of roses, peonies, and flowering trees taken from porcelains. A

Safavid panel of thirty-two blue-and-white tiles from a palace in Isfahan portrays a merchant in Western costume presenting an indigo-dyed textile to a woman while a pilgrim flask decorated with a flowering tree perches on the nearby grass.[94] Most remarkably, 'Abbas had three hundred Chinese potters and their families brought from China and settled near Isfahan—something that must have been done without the permission of Chinese officials. The artisans helped improve the quality of Persian pottery, though they naturally were limited in what they could achieve by the clays of the region.

PORCELAIN AND THE MUGHALS: INDIA FROM BABUR TO AURANGZEB, 1526–1707

Shah 'Abbas encouraged merchants to export blue-and-white earthenware to India. Huge supplies from Persia, along with the low status of clay vessels in India, account for the Mughals not developing a ceramic industry to support their mammoth architectural projects. Moreover, despite frequent political clashes with the Safavids, the Mughals always looked to the art, literature, and pottery of Persia as the highest expression of their own Timurid heritage. So too did Muslim sultans of the Deccan (in south-central India), who viewed the Mughals as mortal enemies but remained culturally close to Safavid Persia and the Ottoman Empire.[95]

In the early eighteenth century, the Indian artist Bhawani Das produced a painting now titled *The Mughal Dynasty from Timur to Aurangzeb*. The sovereigns sit on a marble-balustraded terrace overlooking a cypress garden, a blue floral carpet cushioning their feet. Timur takes center stage, flanked by the Mughal founder Babur (1483–1530), Humayan (r. 1530–40, 1555–56), Akbar (1556–1605), Jahangir (r. 1605–27), Shah Jahan (r. 1628–58), and Aurangzeb (r. 1658–1707). The artwork embodies a genealogy of Timurid majesty, and as if porcelain figures as a signature emblem of the family tree, two elegant blue-and-white vases filled with flowers rest in splendor on a jeweled gold tray in the middle of the semicircle of sovereigns. In the same artist's *The Sons of Shah Jahan* (ca. 1680), two blue-and-white porcelains are once again placed prominently in the foreground.[96]

In 1526 Zhahir al-Din Muhammad Babur invaded India from Kabul in Afghanistan and established what became the dynasty of the Mughals, a name derived from an Indo-Persian form of *Mongol*. He had impeccable credentials, for he descended from Genghis Khan on his mother's side and from Timur on his father's. "Shake loose your Turkish locks," wrote a Timurid poet, "for in your ascendant are royal fortune and Genghis Khan's position."[97] A Mughal manuscript painting of 1630, *Timur Handing the Imperial Crown to Babur*, encapsulates the Timurid connection. The regime Babur established became the largest and most powerful in Indian history, nominally lasting until the British exiled the twenty-third of his line from Delhi in 1858. His engaging memoir, the *Baburnama*, became a preferred text for his successors to copy and illustrate. He himself, however, did not live long enough after his Indian conquest to do more than plan magnificent gardens. Like his fellow Timurids, he was enthusiastic about porcelain: he always ate meals from porcelain plates and traveled with a favorite porcelain cup. In the *Baburnama,* he fondly recalls that when he lost his chinaware in a river mishap, a chieftain gave him "a porcelain cup exactly like the one that had disappeared into the water."[98]

In an illustrated manuscript of the *Baburnama* commissioned by Akbar in 1590, Mughal courtiers are portrayed dining on blue-and-white porcelain. Similar scenes are pictured in the *Timurnama,* a chronicle of the life of Timur, also ordered by Akbar. In fact, these were accurate renditions of his own practice, however idealized and formulaic. When the emperor had himself depicted in a watercolor painting, *Akbar Receiving the Iranian Ambassador* (1590–95), porcelains were included in the work, probably copied from vessels he kept in the *chinikhana* display rooms in his palace.[99]

As with his great contemporary Shah 'Abbas, Akbar's high regard for porcelain played a decidedly secondary role in his grand scheme of things. Above all, Akbar aspired to mold the Hindu and Muslim elites into a cohesive ruling class whose loyalty to himself and the Timurids rose above religious differences. Rajput warriors were the core of his military force. He abolished the poll tax levied on all Hindus, restricted the power of the Muslim religious establishment *(ulema),* ordered the translation of Sanskrit classics into Persian, and prohibited forcible conversion to Islam. He presided over formal religious disputations by spokesmen for various beliefs—Hindu Brahmans, Portuguese Jesuits, Persian Zoroastrians, *ulema* scholars, and Sufi holy men. The *Akbarnama,* a 2,500-page chronicle by the Persian scholar Abu'l-Fazl (1551–1602), includes a painting by Narsingh, *Akbar Discourses with two Jesuit Priests* (ca. 1605), that testifies to the emperor's catholic interests. The chronicle goes on at length about Akbar's role as an instrument of Providence appointed to bring stability and harmony to mankind, an inspired guide whose message and spiritual charisma transcend the limits not only of orthodox Islam but of all outward religious forms as well.[100]

Naturally, Jesuits waxed enthusiastic about Akbar, including Ricci, who spent five years in India and anticipated that the emperor would convert to Christianity. He was the most powerful Asian ruler ever to give their faith a sympathetic hearing. In the realm of art, he exploited Christian iconography as it served his distinctive conception of sacral kingship. For their part, Jesuits applauded Akbar's employment of Christian imagery while remaining courteously opaque about the extent to which it pointed to common ground with Muslim and Hindus belief. That was the same policy Dentrecolles pursued with his parishioners as he cultivated a fruitful confusion between images of Guanyin and the Virgin Mary. Angels allowed for an equivalent mystification in India. Abu'l-Fazl, who maintained that Akbar descended from the Prophet Muhammad, said that the angel Gabriel had conveyed an emanation of God's light to the emperor. In *The Nativity* (ca. 1600), a manuscript painting by Manohar (d. 1620), who specialized in Christian themes for Akbar, an angel descends from the sky toward the Holy Family bearing a large blue-and-white porcelain bowl with a gold lid, seemingly fluttering from a *chinikhana* in Paradise.[101]

Akbar's heady mix of religious syncretism and messianic fervor proved too idiosyncratic and heterodox for his successors. As a 1618 letter by a Jesuit makes clear, however, Jahangir at least shared his father's interest in Western art and distant cultures: "The rich and curious pieces from the world are all in the hands of this king, and it seems that all Europe is involved in making pieces for him."[102] In *Jahangir Entertains Shah 'Abbas* (ca. 1618), among the items on display are a German silver case, a table probably made in Venice, a Venetian glass bottle shaped like a pilgrim flask, and a blue-and-white porcelain bowl. Given the sums Jahangir spent on porcelain and Western objects, Sir Thomas Roe thought "the best trade for him were to remove the China Shops and Pawn [the Royal Exchange of London] into Agra," his imperial city.[103]

The courts of Akbar and Jahangir gave birth to an international hybrid art, mixing together elements of Timurid fashion, European engraving, Indo-Portuguese craftsmanship, and Chinese design. In *Jahangir and Prince Khurram Feasted by Nur Jahan* (1617), a painting commissioned by Nur Jahan (ca. 1581–1645), Jahangir's powerful wife, the eponymous subjects—Prince Khurram became the next emperor, Shah Jahan—are portrayed sitting in a room in which porcelain pilgrim flasks and an icon of the Virgin Mary are installed in marble niches. Nur Jahan also had the tomb in Agra of her Persian father (and Jahangir's chief minister) decorated in Safavid style, its walls covered with images of wine cups, pomegranates, and long-necked porcelain vases.[104] Roe commented that Jahangir prized chinaware and crystal "more than gold and silver, horses and jewels," and he reputedly once almost beat a man to death for shattering a beloved piece of porcelain.[105] Mumtaz Mahal (1593–1631), wife of Shah Jahan, is depicted in a portrait holding a rose in one hand, a blue-and-white demitasse in the other.

According to the French doctor François Bernier, the emperor Aurangzeb used chinaware along with gold and pewter tableware. Yet he had little else in common with either his great-grandfather, the visionary, open-minded Akbar, or with his grandfather Jahangir, a ruler addicted to opium, solely devoted to courtly culture, and dominated by his wife.[106] Jahangir's son Shah Jahan and then (especially) Aurangzeb shifted the Mughal regime to a policy of rigorous Muslim orthodoxy by enforcing application of Qur'an-based law, levying discriminatory taxes on Hindus, demolishing Hindu temples, dismissing Rajput commanders from the army, and abolishing patronage of official chronicles and manuscript illuminations.[107] There would be no more imperial commissions for portraits of Mughal courtiers dining on porcelain, no more imperial *chinikhana*.

Yet the change of course stemmed from more than the grim fanaticism of Aurangzeb: although the Timurid renaissance sank deep roots in the Mughal ruling class, it remained very much a thinly spread, minority culture in a vast realm dominated by disaffected Hindus. Porcelain is representative of this—created in an alien land, cloistered in palace collections, depicted in closely held manuscripts, and scorned by Hindus. It never assumed the sort of public role it had in Korea, the Archipelago, and, later, in Europe. The Timurid renaissance in India was a hothouse phenomenon, even more isolated from the general populace than were austere forms of the tea ceremony from everyday life in Tokugawa Japan.

Shah Jahan's most enduring accomplishment remains the building of the Taj Mahal in Agra, now counted among the most exquisite monuments in the world. A tomb for Mumtaz Mal, its very structure in fact embodies a hard-edged, orthodox message inasmuch as its dome is a visual replica of the throne of God on Judgment Day, when eternal damnation awaits all nonbelievers.[108] The legacy of Aurangzeb was considerably less constructive. A contemporary of Louis XIV, he bears a resemblance to that monarch in that he too plunged his country into never-ending wars and drained his kingdom's resources to finance them. Like the French king, he resorted to melting down his household silver when his coffers dried up. His most disastrous conflict took place when he led eighty thousand troops in a virtual jihad in southern India for the last twenty-five years of his reign. Yet since the Sun King never forsook his slippered ease in Versailles to face danger and distress in the field, the last powerful Mughal despot seems more like a throwback to Timur himself, the fearsome begetter of the lineage. By the time Aurangzeb died at the age of ninety in 1707, Mughal India lay in shambles. Although the dynasty nominally hung on for another 150 years, the breakdown of the imperial system gave the British an opportunity to establish themselves among vulnerable regional powers and ultimately to seize control of the subcontinent.

THE CULTURE OF PORCELAIN IN THE OTTOMAN EMPIRE

Like the Safavids and the Mughals, Ottoman sultans saw themselves as heirs to Timurid culture, though they lacked the family ties of their competing dynasts. In the early fifteenth century, artists from Tabriz relocated to Turkish potteries at Iznik (ancient Nicaea), ninety-six kilometers east of Istanbul on the road to Damascus, where they introduced new techniques and Timurid chinoiserie styles. In the 1470s, responding to court demands, Iznik began to make blue-and-white pottery, copying Jingdezhen wares that Safavid monarchs presented to the sultan. Perhaps those gifts inspired Iznik craftsmen to paint mythical *qilin* and Buddhist lions onto their own pottery.[109]

At the western edge of the Timurid culture sphere, the Ottomans received only modest amounts of porcelain until they expanded their realm. In 1514 Selim I (r. 1512–20), using gunpowder weapons, defeated Shah Isma'il in the critical battle of Chaldiran. After sacking Tabriz, he took a thousand camel loads of plunder back to Istanbul, including chinaware. He made a captive Safavid designer head of the Ottoman court workshop *(nakkashane)* in Istanbul, from which Timurid-flavored blue-and-white designs poured forth in subsequent decades. The captive also brought along a number of fifteenth-century Timurid illuminated manuscripts, which served as a source of designs.[110]

Having dealt with the Safavids, Selim turned on the Mamluks in 1516 and 1517, conquering Syria and then Egypt. This made the Ottomans a world power, with geopolitical interests in the Indian Ocean. In their fight against the Portuguese, who tried to blockade shipment of spices to the Red Sea, the Ottomans sent aid to Muslims in southern India, to Atjeh in northern Sumatra, and to Java. The seizure of Egypt also gave the Ottomans a great deal of chinaware and the forced labor of many Egyptian potters, who in previous years had abandoned their own ceramic traditions to copy Ming porcelain.[111]

With new expertise available, as well as stocks of porcelain as models, Iznik artisans began to turn out large quantities of tin-glazed copies of Jingdezhen blue-and-white. Potters termed stylized lotus and peony blossoms *hatayi*, designs "from China." By the mid-sixteenth century, however, their own native traditions, combined with the influence of Italian maiolica, led to a reduction of emphasis on blue-and-white and an expansion of the color palette to red, emerald green, and turquoise. Aiming to create the atmosphere of gardens indoors, Iznik and Syrian potters painted exuberant, intricate designs emphasizing the natural world, including tulips, carnations, hyacinths, and flowering trees. They also copied portrait medallions from Italian pottery, which they encircled with a spiral pattern derived from the sultan's elaborate imperial monogram *(tughra)*. Ottoman potters were so skilled at replicating porcelain that Italian consumers often could not tell the wares apart. Italian potters returned the compliment by decorating their wares in *porcellana* style, with a winding leaf-stem design taken from Iznik vessels.[112] (See figure 23.)

Some porcelains reached Europe from Southwest Asia in the half century before Vasco da Gama brought the first ones by sea from India. Between 1442 and 1498 the Mamluk sultan gave fifty-eight porcelains to the Venetian government as well as pieces to Charles VII (r. 1403–61) of France and to Lorenzo de' Medici (1449–92), the effective ruler of Florence. In the same half century, Venice and Faenza began to make replicas of Chinese blue-and-white, decorating them with arabesques, flowers, and animals from the Islamic bestiary.[113] As always, however, Europeans aspired to produce their own porcelain. In 1518 a Venetian sought the privilege "in this glorious city of Venice to make good and excellent works of porcelain of every kind, like that called Levantine, which is transparent."[114]

In 1479 the Venetian artist Gentile Bellini (ca. 1429–1507) went to Istanbul, where he had a commission to paint the portrait of Memhed II (r. 1451–81), "the Conqueror." Bellini may have returned to Venice with the three blue-and-white porcelains that his younger brother Giovanni (1430–1516) included in *The Feast of the Gods* (1514), a painting later retouched by Titian (1488–1576); or he may have used porcelains already in the hands of the Venetian government. Depicting a legend recounted by Ovid, Greek gods engage in drunken debauchery: a satyr balances a blue-and-white dish on his head, gazed at enticingly by a maiden hugging a blue-and-white bowl to her bosom; Neptune leans close to Cybele, a fertility goddess, his hand caressing her thigh, a blue-and-white bowl overflowing with fruit next to the couple.[115] These are the first certain representations of porcelain in Western art. Europe was about to catch up to the Safavids, Mughals, and Ottomans.

"THE MAKING OF EARTHEN POTS": THE HISPANO-MORESQUE LEGACY

Ottoman and Italian potters of the sixteenth century, influencing each other and both replicating Chinese porcelain, unwittingly shared a heritage that derived from the artisans of ninth-century Iraq. Indeed, the Southwest Asian innovation of cobalt decoration on tin-glazed earthenware revolutionized ceramic traditions at both poles of the ecumene. In China, it led to the creation of blue-and-white porcelain, and in the West, it transformed the coarse brown earthenware that had been the dominant pottery made there since the fall of the Roman Empire. From the thirteenth century, tin-glazing and painted ornament inspired European potters to reevaluate their craft, to turn from producing mundane, utilitarian crockery to making vessels with vivid colors and imaginative patterns.

The Southwest Asian techniques reached Málaga in Andalucia in the early thirteenth century, probably carried there by potters migrating from Fatimid Egypt in the troubled last days of the regime. Ibn Battuta declared that merchants exported gilded Málaga pottery everywhere. The grandest vessels were in the famous Alhambra palace outside Granada, built in the late fourteenth century by the Islamic Nasrid dynasty (1230–1492). The "Alhambra vases" (as they were called from the eighteenth century) were the largest lusterware jars ever made: they were 1.25 meters high, with flaring, winglike handles and adornment in blue, white, and gold. Rudolf II (r. 1576–1612) of the Holy Roman Empire later purchased one, secure in the knowledge that the Christian god had used it to turn water into wine at the Wedding of Cana.[116]

In the 1350s, spurred by political upheaval in Granada, potters carried knowledge of tin-glazing north from Andalucia to Manises in Valencia. Tin-glazing transformed everyday life there, for the number of ceramic forms used in the area increased from about ten to more than forty, many with glazed interiors. By the early

1400s merchants were exporting large quantities of vessels in what is now termed Hispano-Moresque style to Italy, though some tin-glazed imports had already arrived there from Southwest Asia and North Africa. In a custom reminiscent of Muslim shrines in Java, headhunter poles in Borneo, Spanish cathedrals in Manila, and funerary pillars on the Swahili coast, Italians decorated the walls of churches and civic monuments with lustrous ceramics.[117]

Hispano-Moresque imports stimulated widespread adoption of tin-glazing in Italy, especially from the late fourteenth century. In Tuscany and Umbria, potters added rich colors—copper green, manganese purple, egg yellow, and rusty orange—when they copied the floral patterns and arabesques of the Hispano-Moresque vessels. In the sixteenth century, they often employed a pseudo-Chinese style, taken from Iznik pottery, in painting stylized plants and animals on their wares. More important, they also used Renaissance pictorial conventions taken from paintings and engravings for decorative scenes.[118] The new *istoriato,* "story-painted," pottery elevated the respectability of earthenware in the eyes of sophisticated patrons; they came to regard clay as a noble substance, worthy of admittance to the realm of art and thus to the households of the privileged. In 1557 Piccolpasso assured his readers that "the making of earthen pots . . . will not diminish the greatness and worth" of princes.[119] The Florentine sculptor Luca della Robbia (1400–82) developed a technique for making polychrome pottery figures and established an influential workshop that was continued by his nephew Andrea della Robbia (1435–1525). Some Italian potters in the sixteenth century even started to put their names on the bottom of their wares, a practice begun by Yixing teapot makers in Jiangsu province around the same time. Italian potters also copied bronzes from the Roman classical period, to be used (as in Song China) to adorn the studios of scholars.

Tin-glazing techniques reached Antwerp in the Netherlands in the early sixteenth century, carried by Guido di Savino (d. 1541), identified there as a "Venetian potter" *(veneetsienpotbacker).*[120] He was part of a significant talent drain from the Mediterranean to northern Europe of glassmakers, soap producers, weavers, mirror makers, and printers who sought employment in the up-and-coming region. The migration of artisans signaled that Atlantic-oriented cities of the north, first Antwerp and then Amsterdam, were soon to displace Venice and the Mediterranean in commercial and industrial supremacy.

Northern Europeans previously had known only drab earthenware, so they readily embraced the vibrant colors made possible by tin-glazing. Venetian *porcellana* designs soon appeared on tile pavement and pottery in Antwerp. When Philip II's unpaid troops ravaged the city in the infamous Spanish Fury in the first days of November 1576, three of Guido's sons fled to England, where they introduced tin-glazing to potteries in London and Norwich; another son went to Spain, his craftsmanship enriched with Italian and Flemish variations in design and color.

After roughly 150 years, then, the Hispano-Moresque legacy had come full circle. Things had changed in the meantime, however. The coming of chinaware to Europe since the voyage of Vasco da Gama meant that the art of tin-glazed earthenware, for all its vitality and creativity, was destined for obliteration. The potters of Lisbon who greeted Philip III in 1619 with their "pilgrim art," their resourceful imitations of porcelain, were on the verge of having their craft supplanted by the real thing.

8

The Decline and Fall of Chinese Porcelain

The West and the World, 1500–1850

In March 1602 two VOC ships from the province of Zeeland captured the Portuguese *San Jago* off the island of St. Helena in the South Atlantic, a convenient layover for carracks on the way home from Goa. The auction of its cargo of porcelain in Middleburg attracted considerable attention, assisted by the VOC, which ceremonially presented packages of dishes and bowls to many town councils and dignitaries. A year later, Dutch captains took an even more spectacular prize. In early February 1603 the Portuguese carrack *Santa Catarina* sailed from China for India with a cargo of silk, colored damask, lacquer furniture, spices, and seventy tons of unrefined gold. It also carried sixty tons of porcelain (about 100,000 pieces), probably intended for sale in Goa. Near the Straits of Melaka at the end of the month, two VOC ships commanded by Admiral Jacob van Heemskerk (1567–1607) attacked the carrack at anchor.[1] The daylong battle resulted in half the cargo being destroyed by fire, though what remained was impressive enough. Auction of the porcelain and other merchandise in Amsterdam yielded the VOC about 3.5 million guilders, or 35,000 kilograms of silver, a staggering sum given that a laborer earned no more than 250 guilders a year. At 5,000 guilders for the price of a first-rate house in Amsterdam, the auction netted enough to buy some 750 houses in the most exclusive district of the city. At 100,000 guilders for a carrack, it was sufficient to build a fleet of thirty-five, more than the total number of ships that had departed Amsterdam for Asia in the ten years after 1592. When the East India Company of the United Provinces was formed in 1602, its start-up capital from share subscriptions totaled just short of 6.5 million guilders, so plundering the *Santa Catarina* brought in fully 54 percent of the value of the VOC's entire stock.[2] Shareholders must have been pleased with Admiral Heemskerk.

Large quantities of porcelain thus first began to arrive in the United Province of the Netherlands as a by-product of the global conflict between the Spanish-Portuguese empire of Philip III and his erstwhile Dutch subjects. A year after the *Santa Catarina* auction, Hugo Grotius (1583–1645), a scholar-politician whose major work, *Of the Law of War and Peace* (1625), is considered the first comprehensive treatise on international law, argued for the righteousness of the Dutch piracy and enthused over its bounty: "When the prize from the *[Santa Caterina]* was recently put up for sale, who did not marvel at the wealth revealed? Who was not struck with amazement? Who did not feel that the auction in progress was practically the sale of a royal property, rather than that of a fortune privately owned?"[3] The auction caused an international sensation and convinced the VOC to start purchasing porcelain in China. With a bow to Portuguese precedence in the trade, the Dutch coined the term *kraakporselein,* "carrack-porcelain," for the imports. The only Asian pottery the Dutch had been acquainted with up to 1603 was blue-and-white earthenware imported from Persia; but things changed so rapidly by 1614 that Johannes Pontanus (1571–1639), a Danish resident in Holland, wrote, "The East India traffic has brought a large amount of porcelains to the Netherlands. . . . [T]hat is why one must conclude about the porcelains, the abundance of which grows daily, that only because of these navigations, they come to be with us in nearly daily use with the common people."[4] The Western mania for porcelain had begun.

"SO EMPIRES ARE EXCHANGED": THE PORTUGUESE AND THE DUTCH IN EAST ASIA

Despite the market for porcelain in Europe, massive quantities were not imported until the Dutch entered the trade after the *Santa Catarina* bonanza, one hundred years after Vasco da Gama first presented Manuel I with a dozen pieces. Neither the Portuguese crown nor its merchants had obtained exceptionally large amounts of chinaware or spread it much beyond Portugal itself. A fair number of pieces reached neighboring Spain, including some embellished with the monogram of Charles V (1500–58), king of Spain and Holy Roman Emperor, the father of Philip II. The presence of a small Portuguese colony in Burgundy, arising from the marriage of Isabella of Portugal (1397–1471) to Duke Philip the Good (r. 1419–67), accounts for modest amounts of porcelain reaching there. In 1521 a Portuguese merchant in Antwerp gave the German artist Albrecht Dürer (1471–1528) three pieces of *porcelana* in payment for a portrait of himself.[5] Naturally, some pieces filtered into Europe from the Islamic world, though usually they went unrecorded. The historian Paolo Giovio (1483–1553) boasts that he received a superb porcelain escutcheon *(uno scudellone di porcellanissima)* as a souvenir from an agent of Charles V after that monarch conquered the Muslim stronghold of Tunis in 1535.[6]

Yet, except for a number of porcelains owned by royal dignitaries, the ceramic remained generally uncommon outside the Iberia Peninsula. Aiming for a quick return on their investment, Portuguese monarchs kept prices high, thereby reducing the market for the pottery (and other goods) and providing no incentive for increasing supplies; the crown regarded the Asia trade as a royal monopoly, not as a merchant enterprise. It classed porcelain as one of the "trifles" *(miudezas)* on the *carreira da Índia,* along with items such as amber beads, seed pearls, and gilded lacquer boxes.[7] For their part, Portuguese merchants in Asia never devoted themselves fully to sending goods to their home country. Antonie van Diemen (1593–1645), governor-general of Batavia for the VOC, observed, with a trace of contempt, "Most of the Portuguese in Asia look upon this region as their fatherland. They think no more about Portugal. They drive little or no trade thither, but content themselves with the interport trade of Asia, just as if they were natives thereof and had no other country."[8]

Large-scale imports of porcelain to Europe also had to wait on Dutch initiative because the Portuguese had the bad luck to reach China at a time when the imperial government regarded foreign merchants and maritime trade with hostility. The policy established by the Hongwu emperor more than a century earlier still determined conditions of seaborne exchange. The Portuguese, however, compounded their ill fortune with deplorable behavior. The first Portuguese ships, eight vessels commanded by Fernão Pires de Andrade, entered Canton harbor in 1517. Tomé Pires was with them: having just finished writing his *Suma Oriental* in Melaka, he had reluctantly agreed to serve as Portugal's first ambassador to China. Portuguese sailors killed some local villagers, but Chinese trade superintendents, probably as a result of judicious bribes, still allowed Andrade to buy silk and porcelain. Things went downhill with the arrival of Simão Andrade, Fernão's younger brother, with four ships in 1519: he angered Chinese officials by violating Chinese law, including constructing a fort, erecting a gallows, and executing a sailor. Finally, after more outrages, Chinese naval vessels attacked the Portuguese mariners, who fled Canton, leaving behind a reputation, not entirely unjustified, as kidnappers, slavers, and cannibals.[9]

When Pires reached Beijing for an audience with the emperor in 1521, officials turned him away, pointing to the 1511 Portuguese conquest of Melaka, a Chinese tribute client since the voyages of Zheng He, as an offense to the Son of Heaven. Thrown into prison, Pires vanished from sight, perhaps beheaded shortly after, perhaps to die years later under house arrest in a remote province. The Portuguese spent the next thirty years engaged in smuggling on the Zhejiang and Fujian coasts, finding accomplices among countless Chinese violators of the imperial trade interdict. Other collaborators came from among Japanese merchant clans in Nagasaki, Sakai, and other port cities: they could breach Japanese laws with impunity because feeble Ashikaga shoguns, in the midst of the Sengoku era, lacked the power to keep them in check. The Portuguese had to wait until 1557 to receive permission from the imperial government to occupy a permanent settlement close to Canton, the port that became renowned as Macao.[10]

The golden age of Portuguese trade followed, for Macao became the commercial linchpin of China, Japan, and India. Japan experienced a mining boom in silver in the fifteenth century as a result of improved techniques for processing ore introduced from Korea. The Tokugawa invested silver in the silk industry, importing the raw material from China and setting up weaving workshops.[11] The combination of Japanese silver, Chinese silk, and imperial restrictions on Chinese merchants led to the Portuguese playing the role of middlemen. Ralph Fitch (ca. 1550–1611), an English traveler in the late 1580s, described the most powerful economies in East Asia as linked by triangular trade:

> When the Portugales goe from Macao in China to Japan, they carrie much white silke, Gold, Muske and Porcelanes; and they bring from thence nothing but Silver. They have a great Carake which goeth thither every yeare, and shee bringeth from thence every yeare above 600,000 crusadoes [1.2 million guilders]; and all this silver of Japan, and 200,000 crusadoes [400,000 guilders] more in Silver which they bring yearly out of India, they employ to their great advantage in China: and they bring from thence [for trade in India] Gold, Muske, Silke, Copper, Porcelanes, and many other things very costly and gilded.[12]

During the years when the Portuguese were bringing "nothing but Silver" from Japan, fleets of Chinese junks sailed to the Philippines to trade goods for the same commodity. According to a governor of the Philippines in 1576, the islands represented nothing more than "an archipelago of China."[13] The West came to exert a powerful influence on the Chinese economy in the late sixteenth century as a consequence of Spanish seizure of the Philippines in the late 1560s, followed a decade later by the start of enormous output from Peru's silver mines. About fifty metric tons (176,370 taels) of the precious metal went across the Pacific annually between 1571 and the 1640s. Spanish America, the Philippines, and China thus became connected by a stream of silver: once a year, a galleon packed with it left Acapulco for Manila, where it paid for silk, porcelain, and other merchandise. In the late sixteenth century, porcelain was more commonly found among the Spanish in the New World than in Europe, and silk cloth was cheaper in Mexico City, where fourteen thousand laborers wove fabric from silk yarn, than in Seville or Madrid. In China, however, the influx of silver caused severe price inflation, and, coupled with crop failures and the expense of aiding Korea against Hideyoshi's invasion, the imperial government by the 1630s faced bankruptcy and lacked the financial resources to defend itself against Manchu enemies.[14]

Facing intense competition from the Spanish and other Westerners, the Portuguese remained on top of the pyramid of Asian trade for little more than half a century. Ironically, Philip II, their sovereign since 1580, inadvertently set their fall in motion. In 1598 he imposed an embargo on Dutch trade with Spain, and with flotillas of redundant carracks in their home ports, the Dutch for the first time launched into trading ventures around the world—to West Africa for slaves, Venezuela for salt, Taiwan and Brazil for sugar, Persia for silk, and the East Indies for spices.[15] In the decades after plundering the *Santa Catarina*, the Dutch ousted the Portuguese from Western leadership in Asian trade.

As early as 1605, the Dutch conquered the Moluccas from the Portuguese, thereby achieving almost total control of the world's supply of nutmeg, mace, and cloves. A Dutch fleet attacked Manila in 1617, though without success. By 1623, from its central headquarters in Batavia, the VOC commanded ninety ships in Asia, backed by two thousand troops in twenty forts. Governor-general van Diemen had legal warrant from his home government to act with sovereign authority. He drove the Portuguese from Sri Lanka and blockaded Goa and Macao; his Dutch troops and Southeast Asian auxiliaries conquered Melaka. In 1635 the Tokugawa shogun threw out the Portuguese "Southern Barbarians," leaving the Dutch the sole Westerners on the Japanese scene, the only ones with access to the Japanese silver so important for fueling Asian trade. In the early 1660s the Dutch won control of the coast of Malabar. When in 1662 the Portuguese ceded Bombay to England, as the dowry of the Portuguese princess Catherine of Braganza (1638–1705) on her marriage to Charles II (r. 1660–85), only Goa remained in the possession of the Portuguese crown in India.

In contrast to the Portuguese, Dutch merchants, under management of the VOC, put their energy into dispatching commodities to their homeland. Directors, captains, and factors maintained close coordination between what consumers wanted at home and what the VOC purchased abroad. And unlike the Portuguese crown's neglect of investment in Asian commerce, the governors of the United Provinces and especially the burghers who controlled Amsterdam, made sure that the VOC, as a joint-stock company engaged in operations half a world away, had the financial and institutional resources to take commercial risks and deliver merchandise to Dutch markets.[16] As a consequence, the Dutch dominated world trade in the seventeenth century, and Amsterdam emerged as the world's first central, dominating entrepôt. Striking a note of triumph in a poem of 1639, Joost van Vondel (1587–1679) celebrated the global reach of the Dutch:

> Great Java yields to us her treasures rich,
> China her porcelain. We of Amsterdam
> Sail to all coasts and seas where profit calls,
> Even to the Ganges meeting with the waves.
> For love of gain no port is strange to us.
> With Portugal we share the sea and land,
> And rest from strife with one who's ceded much.
> Who doubts, may cast his eye on town and fortress:

I'll show him other towns, another Fatherland
With different stars. So Empires are exchanged:
Our harvests garnered from such widespread fields
That one vast Indian warehouse holds the East.[17]

"WHERE PROFIT CALLS": CHINA, THE DUTCH, AND THE INTERNATIONAL POTTERY MARKET

From the beginning of the seventeenth century to the end of the eighteenth, the VOC imported at least 43 million pieces of porcelain to Europe. That sum represents only the official total, however: a great deal more, impossible to determine, came in as private shipments by VOC employees. Of course, the primary market at first was in the United Provinces, the most dynamic economic region in Europe. With the most ample diet, most luxuriously appointed homes, and most highly developed market for the decorative and applied arts in Europe, the Dutch Republic (as it was from 1648) led the way in the demand for new tableware.[18] By 1614 ordinary porcelain bowls and plates were already "in nearly daily use with the common people." The VOC quickly became particular about importing special items, what one trader called "the curious pieces, without which the profits can only be small."[19] Van Diemen wrote to Amsterdam in 1638, "As Europe begins to get crammed with common porcelain, we shall after this curtail the assortments of the same and order according to the demand more exquisite and fine assortments."[20]

In response to the flood of porcelain, Dutch potters turned out millions of pieces of tin-glazed earthenware, with imitations of blue-and-white chinaware displacing those with Italo-Flemish polychrome decoration. In fact, potters in Rotterdam and other urban centers had been taking designs from Italian copybooks and engravings for several generations, thereby effectively preparing themselves for the task of responding to the visual culture of the Chinese. Stimulated first by the migration of tin glazing from Antwerp and Venice and then by porcelain imports, a typical Dutch pottery manufacturer increased its inventory from about 1,500 objects in the 1570s to over 10,000 by the 1650s; the size of the largest workshop in Delft grew from ten employees in 1570 to about sixty a couple of generations later.[21]

Pottery production began in Delft in 1614 when an entrepreneur received permission from the States-General, the governing assembly of the United Provinces, to open a manufactory to make vessels like those of "the Indies." Yet the industry did not really take off until two calamities took place around midcentury at opposite sides of the globe. The first arose from an extraordinary political decision in China. Struggle between the Ming and the Manchu disrupted export of Jingdezhen's ceramics, and even when the Qing dynasty came to power in 1644, turmoil in Jiangxi province and on the southern coast continued to curtail pottery output. After expelling the Dutch from Taiwan, Zheng Chenggong (1624–62), the Ming loyalist and warlord known to Westerners as "Koxinga," financed his military resistance to the Qing by seaborne trade and domination of the Fujian coast. To counter this, the Qing government resorted in 1660 to the draconian expedient of ordering removal of the entire coastal population from a 32-kilometer corridor stretching from the province of Guangdong to Jiangsu, a distance of 2,400 kilometers.[22]

No other state in the world could have aspired to enforce such a breathtaking directive. Nor would a government with a forceful maritime perspective have retreated from its shoreline to resolve its military predicament: as always in China, domestic security trumped seaborne interests. In any event, the Qing lacked the resources to carry out the edict with complete success, not to mention humanely: perhaps millions of farmers, mariners, and fishermen fell into poverty, wandered as refugees, and died miserably. In 1662 the Qing government issued another extraordinary command: "All oceangoing junks are to be burned; not an inch of wood is allowed to be in the water."[23] While this can stand comparison with the most far-reaching diktats of the Hongwu emperor, the motivation for the policy was not as radical: the Qing did not aim to erect permanent barriers between China and overseas countries but rather to enforce emergency measures to regain control of its maritime frontier. When the empire resolved the threat from Ming loyalists in 1684, the government reopened ports and resettled coastal districts.

The second midcentury calamity took place in Holland. On 12 October 1654, 41,000 kilograms of gunpowder stored underground near the Delft city center exploded, killing at least 500 residents and leveling 200 homes, as well as a number of breweries. Potteries moved into the devastated area and greatly expanded production of what came to be known as delftware. Using high-quality clay discovered a short distance to the south to make their vessels, Delft potters covered the inside of them with a transparent lead glaze and the outside with decoration painted on tin glaze. The odd names of their workshops—the Metal Pot (De Metalen Pot), the Greek A (Der Grieksche A), the Double Jar (De Dobbelde Schenckan)—derive from the breweries that blew up in what Delft still remembers as the fearsome *Donderslag,* the "Thunderclap."[24]

Before the *Donderslag,* earthenware from Italy, Spain, Portugal, France, England, Denmark, China, and Japan competed in the markets of Amsterdam and other Dutch towns; but afterward, Delft replaced Rotterdam as the center of Dutch pottery manufacture, and production of delftware ended the bulk of European pottery imports. Of course, turmoil in the late Ming and early Qing cut off most imports of porcelain. This called forth the resilience and enterprise of the VOC, which found alternative sources of pottery in Japan, Vietnam, Ayutthaya, and Persia. In Delft, potters acted in the same fashion as their counterparts in Vietnam and Thailand: they stepped into the breach opened by disruptions in Chinese exports and turned out their own blue-and-white wares as substitutes. With porcelain imports reduced and uncertain from midcentury, Delft opened new potteries almost every year for decades. Drawing on an established community of engravers, painters, and printers for labor and inspiration, potters used decoration from chinaware, the Bible, classical mythology, still lifes, proverb books, emblems, and escutcheons. By 1680 the industry employed 1,600 craftsmen (or 20 percent of Delft's workforce), had a fixed capital of 300,000 guilders, and turned out at least 300,000 pieces annually.[25]

Delft blue-and-white, often called *porcelyn* because of its glossy appearance, entered the international pottery market: the VOC took it to Safavid Persia, Mughal India, the Swahili coast, the Archipelago, and the Americas. The Dutch company also shipped Persian blue-and-white to Sri Lanka, Bengal, Ayutthaya, Batavia, and Amsterdam. Dutch pottery going to the Indian Ocean and Safavid pottery taken to Holland often boasted a Chinese reign mark on the bottom, the perennial sign of high quality. When the Qing government lifted its ban on overseas trade in 1684, however, Jingdezhen's exports drove Delft and Safavid imitations out of China's traditional markets as thoroughly as it expelled blue-and-white pottery made in Vietnam and Thailand. Faced with Chinese competition, the Kakiemon kilns of Japan stopped production in the early eighteenth century. In Holland, Delft lost its quasi-monopoly in the home market as porcelain imports flowed in again, eventually constituting as much as 40 percent of the total consumption of ceramics in Amsterdam. (See figure 22.)

Between the 1620s, when the Wanli emperor cut short his enormous orders for pottery, and the 1680s, when peace was restored on the Chinese coast—a period labeled "Transitional" in Chinese porcelain history—Jingdezhen's production was transformed. Enriched by silver profits, Quanzhou merchants formed commercial syndicates that bought up many privately owned kilns and made the porcelain city more attuned than ever to the demands of the market, responsive to both middling ranks in Chinese cities and an expanding number of overseas customers. The Kangxi emperor and his successors took a vigorous interest in Jingdezhen, appointing officials to oversee development of new wares and promoting novel decorative techniques, especially enamel colors brought to imperial workshops by Jesuit craftsmen.[26] These developments made Jingdezhen an effective collaborator with European East Indies companies, especially the VOC and the EIC.

The resurgence of Jingdezhen in the international pottery market after 1684 was just one aspect of China's renewed commercial dominance of East Asia: after its prolonged time of troubles, Chinese trade expanded as never before, achieving a position of supremacy it would retain into the early nineteenth century. The collapse of Ming resistance in China and the end of the Sengoku era in Japan destroyed the raison d'être for permitting Westerners, whether Portuguese or Dutch, to have a commanding role in East Asian exchange. After being driven from Taiwan by the Ming and having its trading privileges curtailed by the Tokugawa shogun, the VOC gave way before the Chinese, ceding precedence to them in direct trade with continental and maritime Southeast Asia. The Dutch rested content with Chinese junks delivering porcelain to Batavia rather than fetching it themselves.[27] China did not "yield" its porcelain, as Vondel's poem boasted, but instead conducted business on its own terms. Although the VOC continued to control the spice traffic, it was never able to cobble together the sort of lucrative triangular trade—China, Japan, and India—that had proved such a dependable moneymaker for the Portuguese.

Samuel Pepys (1633–1703), the well-known diarist and high naval bureaucrat, wrote in 1664 that the Dutch in the Indies scorn the English, "saying that whatever their masters do or say at home, they will do what they list and will be masters of all the world there, and have so proclaimed themselves Soveraigne of all the South Seas."[28] Reality, however, routed their ambitions: within a few decades, Dutch merchants in Asia made a living in the same fashion as their longtime Portuguese antagonists, that is, focusing on intra-Asian exchange and leaving the most rewarding seaborne commerce in Chinese hands. The strength of the Chinese position was that merchants in the Middle Kingdom could rely on the long-established Chinese diaspora throughout East Asia, including Chinese residents in the new, booming entrepôts of Manila and Batavia, a vast and experienced trade network that considerably outmatched Dutch fleets and forts. In the meantime, the English East India Company, which for much of the seventeenth century took second place to its Dutch competitor, established itself in Bengal, a power base from which it eventually expanded in the subcontinent and thereby surpassed the achievements of the VOC.

TABLEWARE OF EUROPE: EARTHENWARE, PEWTER, AND SILVER

Westerners in China had available a wide assortment of porcelain wares as a consequence of the versatility of Chinese potters and the industrial capacity of Jingdezhen. Importing huge amounts of the pottery, the East India companies thereby transformed the everyday life of a large part of the European population between 1600 and 1800. Chinaware, however, was not simply a neutral object to be used and admired, for it had a significant effect on Western society: it played a leading role in a consumer revolution, as great a change as had occurred in everyday life since the revival of sophisticated urban culture in the eleventh century.[29]

Most consumer wealth in the average household before 1500 added up to little more than a bed, stools, and a few pieces of terra-cotta; but from the inauguration of global trade, Europeans purchased a multitude of new goods.[30] To be sure, the revolution in consumption fully reached into all ranks of society only in the late nineteenth century, when household purchasing power had increased and the industrial revolution had made its effects fully apparent; but the first manifestations of that transformation took place two centuries earlier. Indeed, a heightened demand for consumer products preceded and stimulated movement toward machine production in Great Britain.[31] Porcelain and its imitations ranked foremost among the commodities newly available, and from the sixteenth century, they proved central to altering viewpoints and practices in elite taste, daily life, and social conventions.

With a population of some 100 million in 1700, Europe offered a receptive and lucrative market for good tableware. Before 1400 most people used only trenchers, some formed from loaves of bread cut horizontally and some made from wood trays with a depression in the middle to hold liquid.[32] Drinking cups often were made of horn and ash wood. Peasants valued their few ceramics and considered it worthwhile to repair terra-cotta cooking pots with a plug of lead. Stoneware from the Rhineland enlarged the ceramic repertoire in Germany, England, and the Netherlands during the fifteenth century, but standards for tableware still remained abysmally low. In England, unglazed earthenware mugs replaced "black jacks," tankards made of leather coated with pitch, only in the late sixteenth century. James I (r. 1603–25) quaffed his beer from a wooden mug, a choice perhaps conditioned by the indulgence he extended to his gay courtiers, who boisterously smashed costly glass vessels at banquets.

Genre paintings of the seventeenth century indicate the scarce quantities and wretched quality of the crockery on most tables: almost all vessels are terra-cotta or coarse glazed earthenware. Although the canvases cannot be viewed as snapshots of the past, there is every reason to believe they more or less reflect everyday reality.[33] In *Tavern Scene with Two Men and a Boy at Table* by Diego Velázquez (1599–1660), a single crude bowl, drinking glass, pitcher, and knife are on the table. In *Prayer before the Meal* (ca. 1667) by Jan Steen (1626–79), a peasant family of four eats a meal of porridge with a single spoon and an earthenware bowl; a dog licks out an earthenware cooking pot tipped over on the dirt floor. *The King Drinks* (ca. 1634) by Jan Miense Molenaer (ca. 1610–68) portrays thirteen persons around a table: a glutton guzzles from a storage jar while the rest of the company pass around a pewter tankard, a Rhenish stoneware jug, and a glass goblet. Molenaer's *Tavern of the Crescent Moon* (ca. 1637) depicts fifteen persons with just three flagons on the table. In all such renderings, the individuals drinking and eating invariably outnumber the pieces of crockery on hand.[34] In contrast, Dutch still lifes, which naturally aspired to a different aesthetic effect, commonly feature a range of vessels, from terra-cotta to silver, that never appear together in genre tableaus.

While low-status groups habitually used wood and terra-cotta for table vessels, middling ranks favored pewter. Pepys recorded his annoyance at a tavern meal in 1663: it was "very unpleasing that we had no napkins nor change of trenchers, and drunk out of earthen pitchers and wooden dishes."[35] As a top official moving in highly regarded London circles, Pepys dined on pewter most of the time. But while individuals who had only wooden trenchers surely would covet his tableware, pewter had drawbacks that help explain its eventual replacement by glazed pottery. Metalsmiths made common pewter from low-grade tin alloyed with 15 percent lead; hence scratches occurred easily, requiring hours of scouring by sand to efface. (A marginal advantage of scratches, however, was that they made it more difficult for food to slip off the plate.)[36] In addition, the nature of pewter manufacture meant that designs on the wares were hardly possible, thereby isolating pewter as a medium from rapidly changing decorative fashions in the late seventeenth century.

Beer drinkers preferred pewter mugs and would continue to do so into the early 1800s; but as tea drinking soared in Britain after 1700, per capita beer consumption fell dramatically. So too did use of pewter mugs, which tea drinkers naturally did not employ because the brew made the vessels too hot to handle. The price of tin therefore declined, provoking Cornwall tin miners to denounce pottery manufacturers for their loss of income and employment. In 1776 a mob of miners stormed a building in Exeter that held stores of Staffordshire wares. A friend cautioned Wedgwood "not to trust [himself] among the miners of Cornwall, the tin trade being then so low."[37]

Silver plate ranked infinitely higher than pewter. It was the most prestigious tableware, fashioned of material perceived as noble, an artifact only the very wealthy could afford. In sixteenth-century England, diners were served "in silver vessels, if they be of the degree of barons, bishops and upwards," whereas lesser ranks at the table had to settle for pewter.[38] A complete silver service in Britain in the early eighteenth century cost about £600 (roughly $103,000 in present-day terms). The annual income of an earthenware-using artisan was about £25 ($4,300) and that of a pewter-using clergyman or naval officer no more than £60 ($10,320); polite society took for granted that a gentleman required a minimum income of £300 ($51,500) a year.

Given its enormous cachet, silver plate validated the social status of a household, its very patina, a sign of age on its surface, supposedly declaring the refinement, high standing, and venerable lineage of the owner.[39] It served as an impressive decorative item, but when it was used on the table in the early seventeenth century, other sorts of vessels generally mingled with it helter-skelter. The dining equipment of Charles I (r. 1625–49) included silver plate, crystal cruets and saltcellars, cups of agate and heliotrope (a bloodstone), forty-five pewter dishes, a few wooden platters and porcelain plates, and (as an inventory listed) "one great Purselaine Bason sett in a Foote & Frame of Silver and gilt, with two handles to it in a leather case."[40] This eye-catching article was valued at £42 ($7,200). That pales in comparison to the £3,600 ($619,200) that William III (r. 1689–1702) paid for a silver dining table engraved with his coat of arms, trophies of war, and the emblems of England, Scotland, Ireland, and France.

Vessels of silver plate functioned as a practical investment and repository of value. Before the development of deposit banking in the late seventeenth century, silver utensils represented a frozen asset, a resource to cash in when necessary, usually by pawning. Still, merely putting the precious metal on show in a sideboard or on a table made that déclassé expedient less likely. Debts owed to irksome tradesmen, the bugbear of milord, could more easily be waved away so long as the defaulter girded himself with massy silver. "Money spent on strong substantial plate," a young English nobleman was counseled in 1630, "will doe you more service and credit than in your purse."[41]

Silverwares encountered hard times in the sixteenth and seventeenth centuries, however. Trying to promote the sale of Spanish tin-glazed tableware, Philip III endorsed sumptuary laws in 1600 barring the use and manufacture of vessels of precious metal. A Spanish comic drama some years later has a queen declare that "if from plates of dirt I must now eat," she nonetheless still will drink from her remaining silver goblet.[42] In the War of the Spanish Succession, Louis XIV bullied his nobles into melting down their plate in the name of patriotism. English aristocrats went through a similar experience in the Civil War (1642–49), when they handed over their silver plate to support the cause of Charles I and replaced it with tin-glazed tableware from London, Bristol, and Delft. Despite their sacrifice, the king lost his head, and it was not until restoration of the Stuart monarchy in 1660 that silver services made a comeback. Still, Pepys complained that at a Lord Mayor's banquet in 1663, diners had to put up with wooden plates because the Civil Corporation's silver service had been melted down during the Civil War.[43]

Price inflation, generated by population increase and imports of American bullion, also put European owners of silver services under siege. With prices rising by about 400 percent for a century after 1550 and with the supply of coins remaining the same, cashing in silver plate and switching to pottery often proved irresistible. The increasing distinction and respectability of pottery eased the pain of divorce. Though still expensive, a porcelain armorial dinner service, priced at £100 ($17,200) in the early eighteenth century, came to only 17 percent the cost of a silver service.

In the decades after 1600, the Western elite recapitulated the experience of Chinese gentry and literati of the Song era in turning from precious metal tableware to high-quality ceramic, a transition that was part of a general move to tone down portentous display in favor of elegant living. In addition, European potteries turned out such huge quantities of wares and the East Indies companies imported so much porcelain, that the price of good ceramic vessels in due course came within reach of all social groups. By the end of the eighteenth century, porcelain and its imitations—above all, the wares of Wedgwood—came to occupy all the spaces in the gamut from trenchers and terra-cotta to pewter and silver.[44]

"TOGETHER AT TABLE": FROM COMMUNAL TO INDIVIDUAL DINING

Matteo Ricci devoted several pages to the highly structured etiquette that accompanied dinner in a Chinese gentleman's home: a sequence of invitations sent out beforehand, an opening round of tea, pouring of wine, service of dishes, presentation of toasts, and bowing ceremonies at departure. He emphasized that his hosts "do not use forks or spoons or knives for eating, but rather polished sticks, about a palm and a half long, with which they are very adept in lifting any kind of food to their mouths, without touching it with their fingers."[45] João Rodrigues observed that at a Japanese dinner, chopsticks are used just once:

> They are changed many times while they are eating, and so there is no need for napkins for they do not touch anything with their hands and all the food arrives at the table already cut up. . . . They are greatly astonished by eating with the hands and wiping them on napkins which then remain covered with food stains, and this causes both nausea and disgust.[46]

Christianity would have won few souls in China or Japan if missionaries had not learned some new table manners, for Westerners routinely appalled their hosts by their lack of decorum and hygiene. Fernão Mendes Pinto recounts that in Japan, when "the king of Bungo" (a powerful daimyo) invited him and his Portuguese companions to dinner,

> he had the table set with generous portions of very clean, well-prepared delicacies, served by very beautiful women, and we all very heartily threw ourselves upon what they placed in front of us. However, the charming, courtly remarks of the ladies, and the way they made fun of us eating with our hands, provided the king and queen with far greater pleasure than any kind of farce that could have been presented for their amusement, since all these people are accustomed to eating with two sticks. . . . [T]hey consider it a very dirty thing to eat with our hands, as is our custom.[47]

In fact, the mockery and repugnance of Pinto's hosts say less about a contrast between sophisticated Easterners and uncouth Westerners than it does about different modes of etiquette arising from contrasting technologies of consumption. For Europeans, wiping one's hands on napkins, the tablecloth, or pieces of bread represented mannerly behavior if they were greasy from handling food. In China and Japan, carving meat was a menial task done out of sight of the diners, and it came to the table in bits small enough to be manipulated with chopsticks. In Europe, carving was a singular public honor, and touching meat with hands remained common practice well into the seventeenth century. In *Banquet of the Officers of the St. George Militia* (1616) by Frans Hals (1580–1656), a sumptuous portrait of upper-class burghers costumed in the finest lace and velvet, the captain stands at the head of the company, steadying a haunch of roast beef with his left hand as he prepares to carve it with a knife daintily held in his right. As prescribed by etiquette manuals, he touches the meat with just three fingers. His skill and courtly performance establish a befitting tone for the polite gathering.[48]

When spoons and knives were the only utensils available, grasping the beef or fowl by hand in order to slice it scarcely could be helped. Perhaps introduced to Europe from the eastern Mediterranean in the eleventh century, the fork first came into significant use in fifteenth-century Italy. Seen as an Italian affectation, and difficult to manipulate for those first taking it up, it gained acceptance in Europe quite slowly—not least because Louis XIV, the model for courtly culture, persisted in eating with his fingers—but by the early eighteenth century, it made almost as much difference in reformed conduct at the dining table as increased amounts of tableware.[49]

Before the seventeenth century, a person rarely ate alone. With a dearth of spoons, mugs, and plates, meals necessarily were social acts of consumption involving a degree of intimacy, even between strangers, that is alien to the modern world. As genre paintings indicate, communal-style dining was routine: people drank from collective cups, shared bowls, and ate from the same trencher and spoon. An etiquette manual counseled, "When ye shall drink, wipe your mouth clean with a cloth, and your hands also, so that you shall not in any way soil the cup, for then shall none of your companions be loth to drink with you."[50] Traveling through Germany in 1580, Montaigne noted with revulsion that when soup came to the table in taverns, "everyone fishes together, for there is no individual serving." He remarked that it was all the same to him whether the dishes were of wood, pewter, or silver; but, he said, "I no more like drinking out of a common cup than I would like eating out of common fingers."[51]

A sixteenth-century Italian writer cautioned that the beaker and bowl in the center of the table are there "for lord and cleric without distinction and nobody dreams of an extra cup."[52] As an etiquette manual pointed out, "Some are so nice, they will not eat Pottage, or any thing of that nature, in which you have put your Spoon unwip'd, after you have put it into your own mouth."[53] Share-and-share-alike habits died very slowly, however. As late as 1763, the English novelist Tobias Smollett (1721–71) remarked that the French, however beastly otherwise at the table, "will not drink out of a tankard, in which, perhaps, a dozen of filthy mouths have flabbered, as is the custom in England."[54]

Communal meals began to retreat among the upper stratum as pottery became common on the table, with commensurate changes in notions of hygiene, self-discipline, and social propriety. The ultimate in individualized consumption of food was the dinner service as it provided diners with a circumscribed social space, thereby promoting self-restraint in interaction.[55] When dinner services and the fork came into wide use among the well-to-do, questions of etiquette shifted from consideration of how to share a bowl or grasp the roast beef to the matter of how properly to employ one's exclusive set of cups, plates, and cutlery.

Wealthy Italians used matching appointments in tin-glazed earthenware in the early sixteenth century, but until the seventeenth century, the term *service (servizio)* referred only to the ceramic equipment in which food was served; vessels of earthenware, pewter, glass, and silver clustered together on the table without a formal designation.[56] The first complete dinner services, with settings for every person, came into fashion in China around the time Westerners first sailed there. According to Zhu Yan, early Ming emperors began to use vessels with matching designs and color, with twenty-seven pieces in each service, on their banquet tables. The Jiajing emperor (r. 1522–66) ordered Jingdezhen to produce 1,340 dinner sets decorated with dragons, including 26,350 bowls and 50,500 plates.[57] Though it is possible that Western merchants or Jesuits carried word of the fashion for matching wares to Europe, the notion most likely was suggested by the intrinsically pleasing ensemble look of different shapes of blue-and-white porcelain on the table.

The VOC and the EIC started to import porcelain dinner services in the early eighteenth century, and, in response, Meissen began to manufacture large quantities at about the same time. The standard service comprised about 130 pieces, including 60 dinner plates, 24 soup bowls, 21 serving platters, 4 sauceboats, 1 fish platter, 6 tureens, 6 saltcellars, and 6 salad bowls. Items such as cruets, oval dishes, wine coolers, and candlesticks often supplemented this collection. In the course of the eighteenth century, British families that could afford the extra expense—some ten times the price of ordinary settings—commissioned over four thousand services with coats of arms.[58] In fact, Chinese dinner services had an impact that extends to the present day. The West was imprinted on blue-and-white pottery in the early modern period, shortly before the introduction of sets of tableware; hence the configuration of the modern dinner plate, essentially a flat surface embellished with a border design and a central pictorial emblem, is a legacy of the porcelain of the Ming dynasty.

The more exalted the rank, the more extravagant the dinner service. In 1737 Count Heinrich von Brühl declared his wealth and power as the chief minister of Saxony by commissioning the Swan Service in rococo style: 3,000 pieces with painted decoration, including serving dishes masquerading as birds, shells, and flowers. The ornate Snowballflower *(Schneeballblüten)* coffee service represented another triumph of rococo design: made by Meissen in 1739 for Maria Josepha, daughter of Augustus III of Saxony, it took the blossom of the Japanese snowball tree as its decorative fantasy theme.[59] In 1763 Horace Walpole (1717–97), the English writer and politician, viewed a dinner set commissioned by George III (r. 1760–80) and Queen Charlotte (1744–1820) for the latter's brother: "I saw yesterday a magnificent service of Chelsea China which the King and Queen are sending to the Duke of Mecklenburg. There are dishes and plates without number, an epergne [a table centerpiece with holders for flowers and fruit], candlesticks, salt-sellers, sauce-boats, tea and coffee equipages, in short, it is complete and cost twelve hundred pounds!"[60]

The neoclassical Cameo Service produced by Sèvres in the late 1770s for Empress Catherine II (the Great, r. 1762–96) of Russia included 797 pieces and cost 331,317 livres, the annual wages of over a thousand French workers. In 1783 Louis XVI ordered an even costlier service from Sèvres: it comprised about 800 pieces with over a thousand decorative scenes and cost 480 livres a plate. The managers of Sèvres estimated it would take twenty-three years to complete the set. Incredibly, the monarch continued to receive reports on its progress until a few weeks before going to the guillotine in January 1793.[61]

Dinner sets as grand as the Swan and Cameo Services were housed in massive palaces. Farther down the social ladder, room had to be made for more modest collections. From the seventeenth century, homes became larger, first in the Dutch Republic and then in England and France; specialized rooms emerged for sleeping, cooking, dining, and the domestic servants.[62] Builders provided storage space to accommodate the increasing number of possessions in well-to-do households, such as chests, cupboards, mirrors, paintings, clavichords, tea tables, chandeliers, upholstered furniture, textile wall hangings, and tableware. A 1678 Dutch book on etiquette advised that a well-furnished home must have Venetian mirrors, Amsterdam gilt leather, a silver service, and Indian *Kraakporselein*. Indeed, some people may have sought out products to fill the new spaces available. One VOC bureaucrat bought an Amsterdam mansion and stocked it with a sizable painting collection, ten porcelain dinner services, and forty-one porcelain tea sets.

Smokeless stoves, made of glazed earthenware or porcelain tiles, were adopted in privileged circles, making it possible to cook standing up instead of crouched before a hearth. Stoves also advanced the separation of cooking from presentation, lackeys from masters. Increasingly excluded from the common domestic table and intimate family circle, servants were given additional tasks. They were trained in what became known as dinner *á la française,* that is, the sequence and technique of presenting food and drink, with menus detailing stages of the repast and table plans dictating placement of cutlery, wine glasses, and dinner services.

By the end of the eighteenth century, the taste for grandiose banquets had faded as private dinners in a separate dining room took precedence over formal and ceremonial public dining. Invention of the dumbwaiter meant that servants need not enter the *salle à manger* to deliver courses of the meal. Intimate dinners, involving cultivated conversation within a like-minded circle, superseded the ritual display and extravagant feasts that had been a standard elite entertainment for centuries. This reflected the values of privacy and intimacy that were keynotes in enlightened circles. Restaurants emerged in the late seventeenth century in major urban centers, and dining in one naturally entailed being part of a crowd; but diners were provided with separate tables for their individual group and an exclusive set of utensils for every individual at the table.

Dining etiquette, a hallmark of modernity, was comprehensively reformed in the three centuries after 1500. The exhortations of fourteenth-century *Tischzuchten* (table discipline) manuals could have little effect before the advent of both the printed book and a colossal amount of pottery.[63] Treating one's dining companions courteously became significantly more feasible once an abundance of vessels on the table eliminated need for sharing; many soup bowls make good neighbors. Dinner services made dining both individualized and standardized, with each set of utensils the same around the table and the social code for appropriate behavior known to everyone. In the late seventeenth century, a French aristocrat noted the change that was still largely under way:

> Formerly one ate soup from the dish, without ceremony, and wiped one's spoon often on the boiled fowl. . . . Now each person eats his soup from his own plate. One must make polite use of spoon and fork. . . . Never use the same plate for different courses, but change the plate frequently. That is what plates are made for, just as you are given napkins to wipe your mouth. At table as elsewhere, after all, one should think of one's neighbors.[64]

A century or so later, etiquette manuals no longer had to hammer home such elementary reminders, for instruction in table manners in polite society was devolving into a matter of adults coaching youngsters how to eat without making a mess. An English writer's judgment in 1797 that sitting "together at table is perhaps one of the strongest characteristics of civilization and refinement" indicates how radically dining conventions had altered since Pinto and his companions so heartily threw themselves upon what their Japanese hosts placed in front of them.[65]

"THE LAST STAGE OF POLITE ENTERTAINMENTS": CUISINE AND TABLE DECORATION

High-quality pottery on the table reached a critical mass for the first time in Renaissance Italy, which, not surprisingly, also began to define new standards for dining decorum. In the fifteenth century, tin-glazed earthenware became fashionable among the increasingly prosperous middling rank of people who could not afford silver plate or pewter but who desired something better than terra-cotta and wood. In addition, new modes of private expenditure took hold in which those who did not have the wherewithal for patronage of sculpture and painting could express their elite status by purchasing cameos, bibelots, and tableware.[66] Displaying maiolica as an ornament in the home served as the precedent for later displays of Chinese and Japanese porcelain. Increased use of ceramics altered the look of the table; napkins, tablecloths, and scented water for washing hands became essential items. Although silver plate still ruled the high table, pottery steadily gained ground. At the Este court of Ferrara in 1565, a wedding party banqueted on twelve thousand tin-glazed plates decorated with the coat of arms of the ducal house.[67]

While Renaissance Italians led the way in defining elegant living and judicious spending, the Dutch pioneered new standards of convenience and variety in gastronomic culture. From the early eighteenth century, when the VOC opened an office in Canton, its agents met with members of Chinese merchants' associations in contact with brokerage firms in Jingdezhen. The Dutch handed over drawings, engravings, and wooden models of earthenware for potters to copy. As best they could, they relayed the detailed, often prolix instructions dispatched by company directors in Batavia and Amsterdam:

> We want a good lot of fifteen to twenty thousand fine butter dishes, and also eight to ten thousand fruit-dishes to be sent, and all kinds of ordinary and small bowls, caudle cups [used at the lying-in to drink to the health of a new mother], half, third and quarter sizes and other kinds, but those cups must be with straight or sheer sides, and not flaring and flat-lipped like common caudle-cups generally are, because the kind with sheer sides are worth a quarter more than those with flaring sides, and even in large lots they will find buyers. When the same are available, of a kind of eight-sided, medium-sized porcelain dishes to which on one side can be added other smallish dishes, so that standing on a table and joined to each other they have the shape of one dish, we should gladly be supplied with one lot, to wit, of eight or more dishes and members, because some curious people have demanded them expressly, having seen them several times.[68]

Mercifully, however, sometimes instructions from the Dutch were succinct: "a herring must be painted on the herring dishes."[69] (See figure 24.)

Tableware custom-ordered from Jingdezhen or made at Delft included a variety of utensils that crowded into Dutch cupboards from the seventeenth century: soup tureens, beer tankards, radish saucers, pickle shells, cruets, artichoke cups, ice pails, salad bowls, saltcellars, stem goblets, melon and strawberry dishes, sauceboats, butter tubs, butter coolers, monteiths, cheese cradles, chestnut baskets, juice pourers, tea canisters, custard cups, cream pitchers, mustard pots, dessert plates, oval saucepans, and ornamental nobby bottles *(Knobbelflesen).*

New foods and modes of preparation called for a range of tableware and kitchen paraphernalia. Printed recipe books, an innovation of the Italian Renaissance, included greater varieties of meat, vegetables, and fresh fruit. Platters of peacock and heron abandoned the table while dishes of mutton and veal moved in. Rice, maize, shallots, peas, mushrooms, artichokes, cucumbers, squash, pears, and figs entered the diet of the privileged ranks and very slowly trickled down the social ladder. Pineapple, however, remained a costly delicacy out of the reach of all but the wealthy. From the late sixteenth century, sugar imports from Brazil and the West Indies gave cooks the means to produce unprecedented quantities of puddings, sweet pies, sorbets, jellies, tarts, and marmalades—each calling for its own distinctive pottery.[70]

Butter became a foundation of sauces, a triumph of dairy-breeding northern Europe over the olive-growing Mediterranean as decisive as that which took place in industrial leadership during the same time. Plentiful ceramics were essential for dairy activities such as making butter and cheese: milk pans, jars, pitchers, and crocks. In the 1770s Wedgwood won a monopoly on the production of pots and tiles for English dairies, providing wares with patterns simulating the grains of marble, jasper, and porphyry. When dairying became a chic aristocratic diversion in

France, Louis XVI commissioned a pleasure dairy for Marie Antoinette, equipped with porcelain utensils in the fashionable neoclassical "Etruscan" style.[71] The queen and her ladies played at milking cows and churning butter, enacting a fantasy of rustic delight akin to that which impelled affluent merchants in Japan to build pretentiously humble tea huts.

During the Middle Ages, the high cost of spices naturally restricted their use to the wealthy. Ginger, cinnamon, and nutmeg functioned primarily as markers of social discrimination, not as condiments to disguise tainted meat; they set off the aristocracy from lower social ranks as surely as Pantagruelian feasts of roast swan and stewed venison.[72] According to a thirteenth-century English account, one royal feast included

> the head of a boar, larded, with the snout well garlanded. . . . And then there were a great variety of cranes, peacocks, and swans, kids, pigs, and hens. Then they had rabbits in gravy, all covered with sugar, Viaunde de Cypre [ground meat blended with minced dates and marinated in spiced wine] and Mawmenny [chicken pudding with almond milk or spiced wine] . . . and then quite a different multitude of roasts, each of them set next to another: pheasants, woodcocks, and partridges, fieldfares [thrushes], larks, and roasted plovers, blackbirds, woodcock, and song-thrushes . . . and fried meat, crisps and fritters, with sugar mixed with rosewater. And when the table was taken away, sweet spice powder with large dragees [sugar-coated fruits or nuts], maces, cubebs [spicy Javanese berries], and enough spicerie, and plenty of wafers.[73]

Medieval aristocrats favored dishes with diverse flavors, accounting for menus with offerings such as spit-roasted pigeons coated with saffron and stuffed with spiced mutton suet, geese stuffed with sugared oysters, and capon slathered with almonds, cinnamon, and quince. As spices poured into Europe in the sixteenth century, privileged ranks abandoned them and adopted new social signifiers. The gastronomic culture of the rich and well-born became characterized by what were regarded as "natural" foods, such as vegetable soups and herb-seasoned meats. Nicolas Boileau (1636–1711), the French satirist and critic (and a favorite of Louis XIV), mocked a diner besotted with spices: "You like nutmeg, it's put on everything!"[74] Dishes calling for a clash of flavors—sweet with sour, bitter with sweet, savory with sour—were spurned in favor of straightforward broths and meat cooked in its own juices. The founding text for French cooking in the age of the Sun King, *Le Cuisinier français* (1651) by François Pierre de la Varenne (1618–78), led the way in establishing *nouvelle cuisine.* The cookbook sparked numerous follow-ups, most notably *Le Cuisinier royal et bourgeois* (1691), by François Massialot (1660–1733), which continued to be edited and brought up to date into the late eighteenth century.[75]

Modern dining protocols emerged, with the elite distinguished from their inferiors by correct employment of *fourchettes* and *serviettes.* Displays of expensive ceramic tableware served the same social function, as well as being essential to the new commensal sensibility and genteel protocols. Ceremonial feasting in the Middle Ages took the form of showy gatherings, with a lavish carcass on the table, redressed in its skin or festooned with feathers, tinted with cinnamon or spinach juice, exhibited with theatrical panache, and consumed before a crowd of spectators. When the festive occasion ended, an almoner doled out the juice-soaked bread trenchers to the poor at the kitchen door. By the mid-seventeenth century, however, aristocratic fashion in dining shifted to the vessels that accompanied the relatively unadorned food, to table services whose refinement testified to the lofty status and urbane taste of the host.[76] Pyramids of spiced songbirds and grilled quail were exiled from the center of the table in favor of new forms of display.

Naturally, the rich merely changed their modes of affectation, not their ostentatious bent. Table decoration at the court of the dukes of Burgundy in the fifteenth century included a gold dragoir (a platter for ceremonial serving of spices) weighing nine kilograms and embellished with pearls, diamonds, and rubies; supposed unicorn horns, elephant tusks, and fossilized shark's teeth set in gilt mounts also decorated the table. Shiploads of sugar opened up new possibilities for table exhibits, though sugar sculptures *(trionfi)* were as ephemeral as mayflies compared to tusks and teeth. Specialty cooks *(chefs de l'Office)* sculpted vases of flowers from painted and gilded almond paste; they built toy palaces and temples from sugar bonded with gum tragacanth (resin from a Mediterranean shrub), with the finished works sometimes weighing over one hundred kilograms. At a feast in Brussels in 1565 celebrating the nuptials of Alessandro Farnese (1545–92), a cousin of Philip II, and Maria of Portugal (1538–77), all the plates, glasses, and candlesticks were fashioned from sugar, as was the three-thousand-piece mock-up of palaces, ports, galleons, carriages, and sea monsters illustrating the stages of the bride's voyage from Lisbon, with each of the sugar-cities en route measuring two meters long and one high.[77] At a 1667 Amsterdam dinner, a delegation of Portuguese Jews presented Cosimo de' Medici, Grand Duke of Tuscany, with "a triumph of sugar, representing a ship, finely worked with its decks and inner rooms in fullest detail."[78] August II held a dinner in 1719 in which sugar mountains shaped in the letter *A*, a tribute to his sponsorship of Saxony's mining industry, towered over the banquet tables. In 1757, when the vogue for sugar decorations had petered out nearly everywhere, the wedding of an Amsterdam burgomaster featured a sugar temple—one meter wide, two meters high, and three meters long—with eight gilded sugar columns, a floor of colored sugar, and a roof of spun sugar, all decorated with images of flowers, fruits, musical instruments, and river gods. The temple was set on a plateau of mirrored glass, with candlelight illuminating the saccharine extravaganza from below.[79]

Trionfi quickly fell from fashion, replaced by prancing porcelain figurines, Lilliputian models of splendor and privilege. Gerrit van den Brenk, an Amsterdam chef who wrote *A Dialogue between a Lady and a Pastry Cook* (ca. 1750), told his readers, "Nowadays many statues and figurines are made of Saxon porcelain, whereas sugar statues are hardly used any more."[80] Walpole felt the new fashion demeaned his beloved porcelain: "Jellies, biscuits, sugar plums and creams had long since given way to Turks, Chinese and Shepherdesses of Saxon china. . . . By degrees, whole meadows of cattle, of the same brittle materials spread themselves over the table[,] . . . confectioners found their trade moulder away, while toy-men and china shops were the only fashionable purveyors of the last stage of polite entertainments."[81] Calling upon the expertise of his Meissen manufactory, August III launched a cavalcade of china dolls onto his table, including Harlequin, Pantaloon, mandarins, the muses, and Olympian gods—pint-sized versions of the actual masked promenades staged a few years earlier by August II and his courtiers. In Dresden in 1748 the British ambassador reported being overwhelmed by a flamboyant porcelain spectacle: "When dinner was brought on, I thought it the most wonderful thing I ever beheld. . . . In the middle of the table was the Fountain of the Piazza Navona at Rome, at least eight foot high, which ran all the while with Rose-Water, and 'tis said that the piece alone cost six thousand dollars *[Thalers]*."[82]

In 1770 Sèvres produced an even grander ceramic monument for a banquet at Versailles celebrating Marie Antoinette's marriage to the future Louis XVI. Measuring ten by four meters, the centerpiece included a series of terraces with cascading fountains; a frieze of royal emblems on fifty-six Doric columns encircled an arcade topped by a large pottery statue of Louis XV. Wedding guests gaped at the 6,576 vibrantly colored porcelain blossoms.[83]

"THE CONTAGION OF CHINA FANCY": COLLECTING AND CREATING PORCELAIN

Horace Walpole, one of England's foremost porcelain connoisseurs, kept his collection in a China Room at Strawberry Hill, his estate in Middlesex. As a companion recorded, on a trip to Paris in 1765 Walpole purchased, for the extraordinary sum of £100 ($17,200), a tea cup and saucer by Sèvres, "perfect jewels that deserve to be set in gold frames to be admired and looked at, but never to use for fear of breaking them."[84] Lacking Walpole's fortune but sharing his passion for the Chinese drink, Dr. Samuel Johnson (1709–84), the English critic and lexicographer, described himself as a "hardened and shameless tea-drinker . . . who with tea amuses the evening; with tea solaces the midnight; and with tea welcomes the morning."[85] As he wrote to a friend, however, he was "not yet so infected with the contagion of China fancy" that he cared to spend money on ruinously expensive, gilded porcelain that could shatter in an instant.[86]

By the last half of the eighteenth century, costs fell considerably as the EIC imported massive quantities of chinaware for serving the new hot liquors that had entered Europe along with other foodstuffs. According to a London newspaper in

1659, "Theire ware also att this time a Turkish drink to be sould, almost in evry street, called Coffee, and another kind of drink called Tee, and also a drink called Chacolate, which was a very hearty drink."[87] The extraordinary popularity of coffee and tea—hot chocolate remained chiefly a Spanish taste—both customarily taken with sugar, prompted Voltaire to remark that "Martinique, Mocha and China furnish the breakfast of a servant girl." Indeed, that was scarcely an exaggeration: inventories of eighteenth-century Amsterdam pawnshops reveal that those living in poor households routinely consumed coffee and tea and that three-fourths possessed a few pieces of porcelain, though often chipped and broken. The Dutch minister François Valentijn (1666–1727) wrote in 1726 that coffee had "broken through so generally in our land that maids and seamstresses now had to have their coffee in the morning or they could not put their thread through the eye of their needle."[88]

When large amounts of porcelain began to enter Europe after 1600, however, it first caught on only among those who could best afford it. The crowned heads of Europe, from the kings of Portugal to the tsars of Russia, came down with *la maladie de porcelaine.* Like palaces and ermine robes, massed displays of the ceramic functioned as assertions of power and magnificence. Porcelain became the currency of social emulation among the royalty of every country and spread down the social ladder to aristocrats, country gentry, and prosperous burghers.

Henry IV bought a chinaware dinner service, and at his marriage to Marie de' Medici, an observer noted that "tables at the banquet were decorated with gold, silver and porcelain vases." Their son, the future Louis XIII (r. 1610–43), "daily drank his broth" from a porcelain bowl.[89] Sir William Cecil (1520–98), the first Lord Burghley and the queen's Lord Treasurer, gave Elizabeth I, Henry's contemporary, a "Porrynger [a small bowl with a handle] garnissed with golde, the cover of gold with a lyon on the toppe thereof."[90] In 1604, a year after James I, Elizabeth's successor, came to the throne, William Shakespeare (1564–1616) made his sole reference to porcelain: "a dish of some three pence," says garrulous Pompey: "your honours have seen such dishes; they are not China dishes, but very good dishes" (*Measure for Measure,* act 2, scene 1). James received a gift of porcelain plates from Sultan Iskandar Muda (r. 1606–36) of Atjeh in Sumatra, an attempt to enlist the king's support against the sultan's European enemies. Charles II had access to a much larger collection after he married Catherine Braganza of Portugal, whose dowry included lots of chinaware.

After the Glorious Revolution of 1688, William III and Mary II brought some eight hundred pieces of porcelain and delftware from the Dutch Republic to England, along with a new continental fashion. The popularity of porcelain throughout Europe stemmed not only from its use in dining but also from its incorporation in the new consumer vogue for interior decoration, a trend that burgeoned as the elite built increasingly spacious homes. Princess Mary went so far as to design rooms and buildings to fit her collection. She installed a porcelain chamber in her country home outside Amsterdam, and she employed Daniel Marot (1661–1752), a French Hugenot architect and designer, in the court at the Hague. He introduced the popular fashion of massing porcelains on mantelpieces and placing them in front of mirrors on shelves and sideboards. According to Defoe, at Hampton Court (begun in 1689) Queen Mary introduced the English to the "fatal excess" of "piling their china upon the tops of cabinets, scrutores, and every chymney-piece, to the tops of the ceilings, and even setting up shelves for their china-ware, where they wanted such places, till it became a grievance in the expence of it, and even injurious to their families and estates."[91] The furnishings at Hampton Court were intended to evoke Louis XIV's Trianon de Porcelaine (torn down two years earlier): the porcelain collection was displayed in rooms filled with pseudo-lacquer furniture in blue and white and with swathes of blue-and-white silk hanging from wall fixtures.

As the great aristocratic dynasty of the Dutch Republic and an investor in the VOC, the House of Orange naturally did much to spread the fashion for porcelain throughout Europe. William III's aunt, Louise Henriette (1627–67), married Frederick William II (the Great Elector, r. 1640–88) of Brandenburg-Prussia, who built the Oranienburg Palace to house porcelain. The Great Elector claimed that Louise Henriette was obsessed with her collection: "My wife says she is frightened that if there were a fire, the servants would remove the furniture, etc., whereas she is more concerned about the porcelain."[92] Another of William III's aunts, Albertina Agnes (1634–96), married William Frederick of Nassau-Dietz (d. 1664) and constructed the Oranienstein Palace in 1683 for her porcelain collection. Frederick William I of Prussia in 1702 built a mirrored *Porzellankabinett* for his collection of four hundred Chinese pieces in his palace at Charlottenburg near Berlin. Influenced by his visit to Holland, Tsar Peter (the Great) of Russia (r. 1682–1725) set up a porcelain chamber in his palace, Monplaisir, near Peterhof. Of course, William III's great enemy, Louis XIV, also promoted the fashion for porcelain. He inherited several hundred pieces from both Cardinal Mazarin (1602–61) and Marie de' Medici, his grandmother.[93] After he received Kangxi porcelains from the Siamese ambassadors in 1686, he often dined *en porcelaine* at Versailles. In the next decade, dozens of shops dealing in Chinese wares opened in Paris.

European princes, however, aspired to manufacture their own porcelain rather than purchase it from China. In the 1450s Venetian glassworkers produced an opaque glass, called *lattimo* because of its milky appearance, in an attempt to simulate porcelain. Maestro Antonio, a Venetian alchemist, tried to create the ceramic in the late fifteenth century, perhaps inspired by some vessels in the treasury of the Basilica of San Marco, but he produced only a glasslike material. Francesco I de' Medici, Grand Duke of Tuscany and an avid patron of alchemy, poured a fortune into trying to create porcelain by combining ground glass, powdered crystal, and Vicenza clay. Although his product turned out as white as porcelain, it proved virtually impossible to fire, carve, or shape; his potters managed to turn out just a few dozen pieces. Researchers at Saint-Cloud near Paris in 1698 apparently produced something similar, but no vessels or documents survive from the experiments. After visiting Saint-Cloud in 1701, Tschirnaus told August II that its product clearly was inferior to Chinese porcelain.[94]

Also in 1701, reports circulated around central Europe, even coming to the attention of Leibniz, that a journeyman pharmacist and self-proclaimed alchemist in Berlin named Johann Friedrich Böttger had succeeded in transmuting silver coins into gold. Sought after by Frederick William I, who desired a treasure-house of gold, the nineteen-year-old charlatan fled for nearby Saxony, where he fell into the hands of August the Strong. Failing to transmute base metal into gold for the king—August demanded sixty million ducats as soon as possible—Böttger instead was forced to work with Tschirnhaus on a formula for porcelain, the new "white gold." Penned up in a laboratory in the fortifications of the Jungfernbastei in Dresden, Böttger bitterly wrote over the door, "a goldmaker has been turned into a potmaker."[95] Tschirnhaus, Böttger, and their associates arguably formed the first research-and-development enterprise in history, driven by visions of enormous profit and haunted by fears of industrial espionage.

Clever and industrious, Böttger also was lucky: Tschirnaus's death, just when the final breakthrough was made in creating a version of porcelain, allowed the putative alchemist to claim all the credit for the achievement. The first wares produced in early 1709 actually consisted of red stoneware, a material similar to that used for the much-admired Yixing teapots, a score of which August held in his mammoth collection. Within a short time, beds of kaolin were discovered in Saxony, which improved the product and made for increased production. In 1709 August opened the Royal Saxon Porcelain Manufactory at Meissen, immediately northeast of Dresden. Yet despite severe methods of guarding against spies and betrayal, such as locking up his own craftsmen, August failed to keep the secret of porcelain manufacture for himself. Within a few years, workers knowledgeable about the ceramic recipe and the creation of high-temperature kilns escaped from Meissen and peddled the precious information to other princes. Problems with constructing effective kilns and locating sources of kaolin delayed Meissen's rivals, but by 1760 some thirty porcelain manufactories dotted the map of Europe, about half of them in the states of Germany.[96] In the mid-eighteenth century Duke Karl Eugen (r. 1737–93) of Württemberg expressed a commonplace when he declared that a porcelain manufactory was "an indispensable accompaniment to splendor and magnificence."[97]

Of course, the Chinese eventually suffered a great deal more than August the Strong from the spread of the secret. They lost their monopoly on porcelain, their longest, dearest-held treasure. There is a certain irony in the consideration that they did so as a direct consequence of the triumphal expansion of their ancient trade in pottery into the new, predatory marketplace of Europe.

WESTERN AND CHINESE PORCELAIN IN THE EIGHTEENTH CENTURY

The craze for Chinese porcelain declined swiftly in the West in the last decades of the eighteenth century. It was all the fashion in 1700, with huge quantities arriving every year. The trade reached its height around the mid-1750s, with some thirty ships loaded with porcelain leaving Canton every sailing season; but in the last years of the century, the VOC and the EIC suspended pottery shipments, bringing the commerce to an abrupt end.[98]

This was not simply a consequence of commercial competition from Western porcelain, since the Chinese imports remained somewhat cheaper throughout the century, and they were regarded as of higher quality than their European counterparts. Rather, the fall in popularity resulted mainly from changes in elite fashion, aesthetic sensibility, and intellectual perspective. In 1807 Sir Joseph Banks of the Royal Society wrote a treatise in which he paid tribute to Chinese porcelain collected by his wife for her dairy at their Spring Grove estate at Isleworth in Kent. He maintained:

> At the latter end of the 17th and the beginning of the 18th century, China ware was much in vogue, it was at that time a necessary ornament of the Table of the Opulent & was consequently imported in great quantities & sold at high prices; Fashion however began about 1740 to introduce European Porcelain in its stead, and as the dearer ornaments will always be preferred to the cheaper by those who are able to purchase them, however preferable the cheaper may be, Oriental China has at length fallen into total neglect.[99]

In addition, while admiring "the pure white, the beautiful colors & the semitransparent brilliancy of the glaze of China Ware," Banks applauded English manufacturers, especially his friend Josiah Wedgwood, for making vessels that combined those qualities with the aesthetic standards of classical antiquity.[100] The decline of Chinese porcelain, then, followed in part from the success of European potters and patrons in situating their wares in the context of Western cultural traditions.

Such a definition necessarily meant a rejection of the aesthetic ideals of China. This was a striking volte-face, for no other culture ever struck Europe with such concentrated force: Chinese philosophy, government, art, architecture, and landscape design seized the imagination of Europe's elite, especially from the late seventeenth to the middle of the eighteenth century. French publication at the turn of the century of Jesuit reports on China and Jesuit translations into Latin of Confucian texts established the image of a society uniquely governed by natural law, secular values, and paternal benevolence.[101] Voltaire proclaimed that "their empire was already governing like a family . . . while we in small numbers were erring in the forest of the Ardennes."[102] In his preface to du Halde's *Description* in 1738, the English translator held up the Chinese sovereign as a model for George II (r. 1683–1760):

> The Chinese emperor raises taxes only for the public good; hears all complaints; redresses grievances on all matters; lets no one oppress the people; pursues all advantages for the public good; encourages manufactures and trade; never does anything contrary to the "Inclinations of the People"; and invites his ministers to examine his conduct and correct his mistakes. . . . This is the glorious Light in which the Chinese History presents their Monarch to our View; and this is the Light in which the Inhabitants of these Islands figure to themselves Your Future Reign.[103]

Samuel Johnson, who would later turn sour on China, recommended du Halde's volumes, declaring that readers "will be amazed to find that there is a Country where Nobility and Knowledge are the same, where Men advance in rank as they advance in Learning, and Promotion is the Effect of virtuous Industry."[104] Striving to capture the essence of the idyllic empire, the elite constructed pagodas and Chinese-style meandering gardens everywhere in Europe. In his *Tratados históricos,* Domingo Navarette had highlighted the emperor's ritual plowing before the Altar of Agriculture at the vernal equinox, prompting Voltaire to ask, "What are our European monarchs going to do when they hear about such examples? Admire and blush, yet above all: copy."[105] In France and Austria, royal heads dutifully took to plowing ceremonially on the first day of spring.

During a time of increasing secularism and remarkable economic change, the West defined its identity and relationship with other powerful societies by measuring itself against China. Europe soon entered its own age of empire, however, with a consequent decline in esteem for all things Chinese. Around the middle of the eighteenth century, with Western power growing in Asia, European merchants familiar with China began to counter the idealized Jesuit portrait of that country, detailing the presumed corruption and debility of its administration; intellectuals soon turned against China as an exemplar for their own society. In like fashion, Europeans stopped looking to Chinese porcelain as an ideal. The craze for Chinese ceramics had heralded a period of China worship, and when the Chinese model toppled from its pedestal, Chinese porcelain fell along with it.

During the first half of the eighteenth century, however, Chinese porcelain still enjoyed enormous prestige in the West. Indeed, potters in both Europe and China attained a level of artistic excellence and technical skill that has never been surpassed. Yet while China's achievement was based on centuries of tradition, Europe's was founded on astonishingly recent progress: in less than two generations, Western potters located sources of kaolin and china stone, constructed high-firing kilns, developed new ceramic formulas, and so completely mastered their craft that the Chinese copied their wares and adopted their decorative techniques. In fact, the rapid development of European pottery stemmed from the same consideration that accounted for the exceptional achievements of Chinese craftsmen in the Song period: a large number of competing pottery centers stimulated experiment and innovation, thereby increasing the pool of industrial expertise as well as the variety and quality of production. Blue-and-white porcelain, however, was solely Jingdezhen's creation, and its triumph in China in the Ming period effectively crushed other ceramic centers, especially those manufacturing celadons. By the sixteenth century, a combination of natural resources, capital investment, and imperial patronage transformed Jingdezhen into the colossus of Chinese potteries. Having nothing to learn from competitors at home, the entrepreneurs and craftsmen of the porcelain city responded instead to their Western rivals. In the end, however, Jingdezhen collectively failed to respond to the innovations of Staffordshire that radically changed the nature of pottery production.

Control and support of Jingdezhen by the Qing dynasty greatly facilitated the Chinese response to Europe, for the rulers employed Jesuits in the imperial palace who were well informed about Western artistic innovations. They also forged closer links between the pottery center and Beijing than had earlier regimes. The Kangxi emperor appointed a resident official to supervise reconstruction. As director of Jingdezhen from 1683 to 1688, Zang Yingxuan oversaw rebuilding of the kilns and ensured that the imperial household received the huge amounts of porcelain needed for its daily life and routine rituals. Equally energetic supervisors followed him: Lang Tingji (1705–12), Nian Xiyao (1726–28), and Tang Ying (1729–56). Because of his long term in office and his technical expertise, Tang was particularly important; he looked back to the period of the early Ming as a model of excellence, and his innovations in glaze colors helped bring porcelain decoration closer to the styles employed for painting. The importance of the office of supervisor became clear in 1786 when its abolition to cut bureaucratic costs hindered Jingdezhen's ability to compete in world markets.[106]

The Kangxi emperor established workshops in his Beijing palace for porcelain, lacquerware, jade carving, painting, and calligraphy. He also promoted the brilliant new "foreign colors" *(yangcai* or *falang)* that Westerners had devised for cloisonné and glass working. As had happened four centuries earlier with Persian cobalt, new color came to Chinese porcelain from the outside world. The pigments involved the use of lead arsenate to create opaque white enamel to which colorants could be added. Tints in which green predominated gave an adamantine sheen, known as *famille verte* (green family) in the West and *yingcai* (hard colors) in China. Pastels derived from employing colloidal gold chloride to create lilac, and ruby-magenta tones were called *famille rose* (rose family) and *fencai* (powdery colors). Mixed with other pigments, the new hard and powdery colors produced a wide range of hues.[107] The European marketing of Prussian blue after 1704 and of Rose Pompadour after 1756 rounded out a palette that gave an intense, new experience of color to the world. In the West as well as in China, porcelain graduated from blue and white to all the shades of the rainbow.[108]

Eager to acquire European color innovations, the Kangxi emperor assigned two Italian Jesuits, Giuseppe Castiglione (1688–1766) and Matteo Ripa (1682–1746), to teach enameling techniques to Chinese artisans, and he requested the Jesuit general in Rome to send more handicraft experts. Under the Yongzheng emperor, the new colors and designs became prominent on porcelain. Castiglione was a friend of Nian Xiyao, resident director of Jingdezhen, and he persuaded him to translate Andrea Pozzo's influential *Perspectiva Pictorum* (1693–98) under the title *Shi Xue* (Visual Learning). Nian Xiyao wrote that Castiglione had taught him "the way to give three-dimensional effect to the drawing of an object with its particular light and shade aspects."[109] A porcelain painter in the imperial workshop even reproduced one of Castiglione's perspective paintings on a dish.

Western colors and design techniques also greatly appealed to the Qianlong emperor, the foremost connoisseur and art collector of the Qing dynasty. He wanted the "oceanic styles" *(yang shi)*, as the innovations were called in China, put in the service of "copying the ancients," the convention that had a powerful impact on the decorative arts. Of course, under Qianlong, there was renewed interest in manufacturing vessels modeled on Song wares, using illustrated catalogues of antiquities as guides to forms and decoration. The West had long followed a tradition of copying its own ancients, most notably during the Renaissance. Painters, sculptors, and architects from Giotto to Michelangelo adhered to notions of human form and coherent space that they regarded as faithful to classical Greece and Rome. In their *istoriato* vessels, Italian potters extended Renaissance precepts of classical composition and subject matter to the newly respectable medium of clay. But the popularity of Chinese design in the West, especially its asymmetry and lack of European-style perspective, represented a rejection of the classical-Renaissance tradition, even liberation from it. Significantly, designs on blue-and-white porcelain and their chinoiserie variations never made as great an impact in Italy as elsewhere, no doubt because classical and Renaissance standards were so deeply rooted there. Italian potters even drew their fantasy decoration from ancient Roman motifs—dolphins, satyrs, grinning skulls, grotesque beasts—thereby filling the arena that the exotic Chinese menagerie occupied in northern Europe.[110]

Westerners regarded Chinese pictorial space with some ambivalence, not unlike the way the Islamic world had viewed it much earlier. On the one hand, they cherished its feeling of freedom, its sense of space as fluid and unlimited. Charles Lamb (1775–1834), who spent thirty years working at Leadenhall Street for the East India Company, wrote that as a youth he loved those "lawless, azure-tinctured grotesques, that, under the notion of men and women, float about uncircumscribed by any element, in that world before perspective—a china tea-cup."[111] On the other hand, the legacy of the Renaissance maintained its force: Western chinoiserie designs routinely imposed the regularity of perspective on Chinese compositions, bringing the license of the exotic scenes under the dominion of Western principles of design.[112] This suggests that, despite its passion, the Western love affair with Chinese art was destined to cool in time.

As millions of pieces of blue-and-white porcelain poured into Europe beginning in the early seventeenth century, the decoration on them appealed to educated elites disenchanted with classical-Renaissance design and drawn to the fashionable eccentricities of mannerism. Later in the century, chinoiserie patterns reflected a Baroque taste for serpentine curves and dynamic rhythms. After the death of Louis XIV in 1715, with the heavy hand of the Sun King removed, Meissen and other porcelain manufactories embraced the rioting flamboyance of rococo style. The plasticity of porcelain allowed realization of the most extravagant decorative fantasies in material form, while the exoticism of Chinese patterns and themes was custom-made for sophisticated circles weary of both the balanced certainties of Renaissance design and the heroic sweep of Baroque ornament. In a work on garden projects in 1683, Sir William Temple (1628–99) employed the term *sharawadgi*—perhaps from the Chinese *sarowaichi,* signifying elegant disorder—to characterize the asymmetries of Chinese art. In landscape gardening and elsewhere, he wrote, the Chinese imagination is

> employed in contriving figures, where the beauty shall be great, and strike the eye, but without any order or disposition of parts, that shall commonly or easily be observ'd. And though we have hardly any notion of this sort of beauty, yet they have a particular word to express it; and where they find it hit the eye at first sight, they say the *Sharawadgi* is fine or is admirable, or any such expression of esteem. And whoever observes the work upon the best Indian gowns, or the painting upon their best screens or purcellans, will find their beauty is all of this kind, [that is,] without order.[113]

Walpole and others later applied *sharawadgi* to pleasing, irregular patterns as they appeared on porcelain, wallpaper, lacquerware, and furniture.

Treatises on Chinese style provided hundreds of engravings that professional and amateur artists plundered for decoration on pottery, wallpaper, silverware, textiles, and furniture, as well as for models of gardens, gazebos, and pagodas. Among the most important texts were Nieuhoff's *An Embassy from the East India Company* (1665) and *Treatise on Japanning and Varnishing* (1688) by John Stalker and George Parker. Robert Sayer's *The Ladies Amusement* (1760), drawing on these works, decreed that Chinese subjects, since they typically display luxuriant fancy and carefree coloring, sanction unrestrained liberty with traditional forms.[114]

The most influential treatise on Chinese architecture and landscape gardening was *Designs of Chinese Buildings* (1757) by Sir William Chambers (1723–96). The author, who had visited Canton, was commissioned by the Dowager Princess Augusta (1719–72) to build a Chinese-style aviary and a "House of Confucius" in the royal gardens at Kew in London. His most famous work, however, was the Great Pagoda of Kew, constructed in 1761 and inspired by depictions of the Porcelain Pagoda of Nanjing. Other buildings had ostensibly been based on Chinese models—most outstandingly, the Trianon de Porcelaine—but the Kew pagoda was the most accurate copy of a Chinese structure yet erected in Europe. Its balconies were painted in blue and red, and eighty gilded dragons projected from its ten octagonal stories. From Prussia to Russia, the Great Pagoda inspired scores of imitations.[115] Unlike the Porcelain Pagoda, it is still standing.

"PORCELAIN ELEPHANTS AND CHINA GODS": THE DECLINE OF CHINESE PORCELAIN IN THE WEST

Blue-and-white porcelain had kindled enthusiasm for Chinese design in arts and crafts, and when reaction set in around the mid-eighteenth century, pottery took a leading role in the revival of classical style. Ironically, neoclassicism in ceramics sprang from the pursuits of one of the last great aficionados of chinoiserie, Charles IV (1716–88), King of Naples and the Two Sicilies and great-grandson of Louis XIV. In 1738 Charles married Princess Marie Amalia (1724–60), daughter of August III, thereby uniting the Bourbon dynasty of France with the Saxon line of August the Strong. The bride brought seventeen Meissen dinner services to her Neapolitan palace. Ambitious for the splendor and magnificence of a modern prince, Charles built the Capodimonte porcelain manufactory near Naples with workmen sent from Meissen. When he succeeded to the Spanish throne as Charles III in 1759, he packed up the establishment (including tons of porcelain paste) in three ships and reassembled it in the gardens of the Buon Retiro, a palace outside Madrid. There he built the Saleta de la China, a fairy-tale room entirely sheathed in panels of blue-and-white porcelain embellished with rococo tracery.[116]

Ten years after his wedding, pursuing the newly fashionable aristocratic hobby of archaeology, Charles started excavations on the ancient Roman site of Herculaneum, which had been destroyed (along with Pompeii) by the eruption of Vesuvius in 79 C.E. The pottery unearthed there, mistakenly identified as Etruscan, caused a sensation. Naples soon became a center for collectors of antique vessels, and August III commissioned the art authority Johann Winckelmann (1717–68) to report on the discoveries. Clambering around the volcano looking for new finds, the German savant dined on pigeons cooked in the smoldering lava. His Vesuvius-inspired books on Greco-Roman painting and sculpture became fundamental texts for neoclassicism, benchmarks for pottery connoisseurs such as Sir William Hamilton and pottery manufacturers such as Wedgwood. As Winckelmann decreed, "Most porcelain is in the shape of ridiculous dolls. From it originated the childish taste which has become so widespread. In its place we should strive to emulate the eternal works of classical art."[117] Excoriating porcelain figurines adopted from the Italian *Commedia dell'arte,* such as Harlequin and Pantaloon, he encouraged manufacturers to reproduce ceramic replicas of antique statues, miniatures of noble pieces such as the *Apollo Belevedere* and the *Dying Gladiator.* In effect, Winckelmann agreed with the venerable view of Chinese literati: "the only way to become great . . . is by imitation of the ancients."[118]

Immediately after Hamilton arrived in Naples to take up his diplomatic post in 1764, he began to collect antique pottery. The four-volume catalogue of his antiquities, published at a cost of £6,000 (or $1,032,000 in present-day terms), with colored illustrations and a text by the self-styled Baron Pierre d'Hancarville (1719–1805), a connoisseur of classical art, a con man, and a pornographer, was a publishing masterwork of the century. It became the indispensable guide for the decoration and forms of neoclassical ceramics throughout Europe; its color plates were incorporated into pattern books and used in producing countless imitations of antique objects. The catalogue helped persuade the British Parliament in 1772 to purchase Hamilton's collection for £8,400 ($1,444,800), with the pottery forming the nucleus of what evolved into the British Museum.[119]

Rococo style faded quickly from ceramics and other art forms after the discoveries at Herculaneum reached a wide audience. European wars had an impact as well. Meissen production, which led the field in rococo design and chinoiserie, suffered ruinous setbacks in the War of the Austrian Succession (1740–48). King Frederick II (the Great, r. 1740–86) of Prussia invaded Saxony—the Porcelain Regiment marched in his ranks—and looted precious materials from the pottery manufactory. In the Seven Years' War (1756–63), Frederick occupied Meissen once again, halted production for seven years, and plotted to move the entire works to Berlin. To celebrate victory in the war, he held a musical concert at Meissen and then withdrew from Saxony with one hundred crates of porcelain, the last of the inventory in the stockroom of the manufactory. When Meissen finally reopened, potters and artists had to be retrained. Some practiced their craft by reproducing plaster models of antique statues; managers dispatched a few painters to Paris to learn up-to-date designs. In order to regain its share of a market no longer enamored of rococo style, Meissen turned to copying the neoclassical wares of Sèvres.[120]

Always restrained in its chinoiserie, Sèvres emerged as the dominant force in European porcelain when Meissen was beset by its midcentury troubles. The French manufactory began specializing in wares in *à la etrusque* style. The Cameo Service that Sèvres made for Catherine of Russia in the 1770s highlighted neoclassical fashion and set a new standard for elegant tableware. Catherine suffered from what she called "cameo fever," a symptom of her devotion to the classical world; Minerva, goddess of wisdom and the arts, with whom the queen identified, served as the chief symbol of the service. Following the trend set by the Cameo Service, the Sèvres dinner ensemble commissioned by Louis XVI in 1783 was decorated with scenes from classical mythology and Roman history. When Marie Antoinette later wanted a dinner service for her faux dairy at Versailles, Sèvres modeled it on the antique vessels in Hamilton's catalogue.[121]

With the triumph of neoclassicism, the dazzling porcelain statuettes of Meissen, an artistic legacy of the Italian Renaissance and the most innovative genre of rococo, fell out of fashion. Exiled from banquet tables to cabinet shelves, they became the first knickknacks, eventually to suffer a debased twentieth-century reincarnation as Hummel figurines. Porcelain lost its function as a manifestation of aristocratic self-esteem and magnificence; no longer an emblem of princely splendor, it evolved into a token of middle-class gentility.[122] Pushed from the *Porzellanzimmer* and reception hall into the boudoir and kitchen, it entered the increasingly separate sphere of females, now leading patrons of the booming consumer economy. John Gay (1685–1732), in *To a Lady on her Passion for Old China* (1725), satirizes their supposed infatuation with porcelain:

What ecstasies her bosom fire!
How her eyes languish with desire!
How blest, how happy should I be,
Were that fond glance bestow'd on me!
New doubts and fears within me war:
What rival's near? a *China* Jar.

China's the passion of her soul;
A cup, a plate, a dish, a bowl
Can kindle wishes in her breast,
Inflame with joy, or break her rest.[123]

In *The Metamorphoses of the Town* (1730), Elizabeth Thomas (1675–1731) depicts Olympian gods so incensed at English gentlewomen engaging in "Scandal o'er a Dish of Tea" that they destroy the tea sets of the gossips—"Each Cup ('tis said) was broke to Shatters, / Now as it broke, the Liquor scatters"—and then reconstitute the utensils by transforming the women into lovely but speechless "China-Cups."[124] A generation later, in *The Ladies Amusement,* Sayer portrays frivolous upper-class women visiting newfangled shops in London, filling time by browsing the goods on display:

Some brittle wares, we now must see,
  Delft, china, glass and stone;
We'll say they're cracked, we'll say they're dear,
  And a shopping we will go.[125]

Given prevailing gender stereotypes, porcelain became identified with domesticity, coarse taste, slack morals, and incessant shopping. Critics perceived both porcelain and women as fragile, decorative, and, in the end, merely clay in a stylish wrapping. As an English writer remarked in 1751, "the finest and most beautiful bodies are but earthen vessels as well as chamber-pots."[126] A taste for Chinese porcelain thus became synonymous with effeminacy, and (as a journalist put it in

1755) a soft spot for porcelain and chinoiserie by a man smacked suspiciously of a "delicate make and silky disposition."[127]

By the 1750s the image of porcelain, both Chinese and European, fused with a host of antifeminine clichés and convictions that had been gathering momentum for two generations. There was a widespread sense among intellectuals, publicists, and politicians—articulated with greatest force in England but present as well in Holland and France—that the supremacy of fashion, exotic commodities, and troubling financial innovations worked to destabilize traditional society and its values. Defoe stood out around 1700 for his hostility toward China, porcelain, and luxury trade; but by the outbreak of the Seven Years' War, such views had become commonplace. "Effeminacy" evolved into a codeword for the degeneration of masculinity that supposedly had resulted from opulent fashions, self-indulgence, and Asian commodities.

Critics perceived social hierarchy as under attack on all fronts. Government-sponsored banks, a national debt, joint-stock companies, lotteries, volatile markets, and foreign exchange eroded the stature and power of landholding elites. Defoe denounced moves "to stock-jobb the Nation, Couzen the Parliament, ruffle the Bank, run up and down the Stocks, and put the Dice upon the whole Town."[128] He wrote spirited articles on the menace of "Lady Credit," a goddess of reckless appetite and debauched imagination. In the course of the century, such attacks became ever more strident as critics portrayed other termagants joining forces with Lady Credit in plays, novels, and pamphlets—Lady Luxury, Lady Fortune, Lady of the South Sea, Lady of the Bank. The assorted iconographic females invariably appeared as hysterical, emasculating, and bewitched by material possessions. The antifeminine rhetoric stemmed from anxiety and resentment at the commanding role women played in the emerging consumer society, their enhanced influence in establishing fashions, and (in Great Britain) their increasing share of stock ownership in the Bank of England.[129]

The novel force of fashion swept through social classes, seemingly obliterating material distinctions that had fixed social identity since time immemorial. The vogue for low-cost Indian cottons appeared to open the door to social anarchy, permitting the servant girl to vie publicly with her mistress, "the Tradesman's Wife to out-do the Gentleman's Wife."[130] According to Jonas Hanway in his *Essay on Tea,* the hoi polloi mimic their betters by guzzling tea, "an epidemical disease," as a consequence of which "the different ranks of people are too much confounded."[131] When both great aristocrats and parvenu merchants purchased relatively inexpensive porcelain, privileged status could no longer be determined simply by the flaunting of silver plate. Time-honored artistic standards apparently meant little amid the craze for alien ceramics bedecked with peculiar adornment, inscrutable calligraphy, and outlandish deities. Elizabeth Wortley Montagu (1720–1800), a writer on aesthetics and leader of the group known as "bluestockings," supported

British manufactures (such as Wedgwood's pottery) and protested that "we must all seek the barbarous gaudy [taste] of the Chinese . . . and Apollo and Venus must give way to a fat idol with a sconce on his head."[132] Her contemporary, James Cawthorn (1719–61), a poet and critic, attacked the fad for placing "porcelain elephants and china gods" on every mantelpiece.[133] In *The Citizen of the World,* Oliver Goldsmith (ca. 1730–74), the novelist and playwright, describes a fashionable lady's apartment decorated "in the Chinese manner; sprawling dragons, squatting pagodas, and clumsy mandarins, were stuck on every shelf."[134]

War with France intensified such perceptions, fueled by resentment about imports from there of allegedly effeminate commodities such as perfume, toilet water, pomades, and umbrellas.[135] During the war, Smollett asserted that women in France were markedly subject to the "caprice of fashion," as a consequence of which "France is the general reservoir from which all the absurdities of false taste, luxury, and extravagance have overflowed the different kingdoms and states of Europe."[136] In his novel, *The New Héloïse* (1761), the Swiss Jean-Jacques Rousseau (1712–78) has his French spokesman (who has visited China) condemn the Chinese in terms Robinson Crusoe would have applauded: they are "learned, craven, hypocritical, and devious; speaking much without saying anything, full of wit without a bit of genius, abounding in signs and sterile in ideas; polite, fawning, clever, sly, and knavish." In the eyes of the leading apostle for the values of Nature, the worst abomination of the Chinese was to sin against that sacrosanct realm, for they were responsible for the European vogue for gardens littered with "porcelain flowers, magots [porcelain figurines] . . . and fine vases full of nothing."[137]

In English eyes, both France and China were metaphorically feminized and thereby devalued: the former as England's leading imperial rival and spawning ground of fashion, the latter as the great stumbling block to English domination of global commerce. Just as the French allegedly were obsessed with "fripperies, fancies, & Chinese trash," so too China was seen as dwelling in a fantasy world isolated from economic reality—that is, the presumed necessity for free trade—every bit as infantile and capricious as its ubiquitous porcelain.[138] Critics lumped together French and Chinese taste as degenerate, ostentatious, and epicene. Around the time of the Seven Years' War, the English began to stereotype the French as "frogs," a slur they formerly reserved for the Dutch. In *The Tryal of Lady Alluria Luxury,* a satire published anonymously in the middle of the war, Sir Oliver Roastbeef commands a jury that convicts Lady Luxury of transforming the honest "English hospitable Table" into "nothing but Frenchified disguised dishes," servings of "high-seasoned Ragouts and masqueraded Poisons."[139] During the war, Hanway patriotically lamented, "As the modern phrase expresses it, *We live in hot water,*" enervated financially and spiritually by consuming tea from China and lace, brandy, and baubles from France.[140]

An English newspaper in 1759 bemoaned the nation's lack of heroes and condemned commanders who dress "like an ape, stinking with perfumes and ointments," preferring affable dissipation to confronting danger on the field.[141] The backdrop to the denunciation was the notorious case of Admiral George Byng (1704–57), commander of the Mediterranean fleet in the Seven Years' War and an eminent devotee of porcelain. He came from a family with several expensive armorial services, and he filled his estate in Hertfordshire with a large chinaware collection.[142] In May 1757, after he ordered the retreat of his armada before the French, the latter captured the strategic stronghold of Minorca in the Balearic Islands. As a consequence, he was court-martialed for cowardice and shot on the quarterdeck of HMS *Monarch,* his own flagship. The controversial execution provoked Voltaire's well-known gibe in *Candide* (1759): "In this country it is a good thing to kill an admiral from time to time to encourage the others *[pour encourager les autres]*."[143] The English press lampooned Byng with a broadsheet depicting him sailing away from the enemy: his allegedly spineless, decadent character is revealed by the colorful Chinese porcelain arrayed on the cabin shelves of the *Monarch.*

England had heroes enough, however. In the same year that Byng retreated at Minorca, Colonel Robert Clive (1725–74) of the East India Company defeated a huge Indian army at the battle of Plassey, some 120 kilometers northwest of Calcutta. As it turned out, this gave the British their first significant foothold in the province of Bengal, the base from which the EIC soon would seize additional territories and go on within a few decades to win effective control of the subcontinent. A year before the end of the Seven Years' War, just after British fleets captured the ports of Manila and Havana, Walpole said, "I wish we had conquered the world and had done! I think we were full as happy when we were a peaceable, quiet set of tradesfolks as now we are heir-apparent to the Romans and over-running the East and West Indies."[144]

By the terms of the Treaty of Paris (1763) at the end of the war, France ceded Canada, Senegal, and a number of islands in the Caribbean to Britain; the latter also laid claim to Louisiana east of the Mississippi and to Spanish Florida. The island nation thus emerged as the greatest naval and commercial power in the world, possessing the largest European colonial empire and the most profitable network of seaborne commerce. In *London and Westminster Improved* (1766), the architect John Gwynn (1713–86) noted the epochal change with gratification:

> The English are now what the Romans were of old, distinguished like them by power and opulence, and excelling all other nations in commerce and navigation. Our wisdom is respected, our laws are envied, and our dominions are spread over a large part of the world. Let us therefore, no longer neglect to enjoy our superiority; let us employ our riches in the encouragement of ingenious labour, by promoting the advancement of grandeur and elegance.[145]

Contemplating a newly redrawn map of the world, one of Gwynn's compatriots had a haughtier perspective: "I shall burn my Greek and Latin books. They are the histories of little people. We subdue the globe in these campaigns, and a globe as big again as it was in their day."[146]

Although imperial China and Chinese porcelain enjoyed an excellent reputation at the beginning of the eighteenth century, by the end of it, both had fallen precipitously in esteem. The more Westerners learned about China—that is, the more they looked beyond the superficial, idealized image set forth in du Halde and the Jesuit relations—they less they liked what they saw; the more porcelain and chinoiserie were seen to clash with neoclassical norms, the more contempt was heaped on Chinese aesthetic standards. In 1778, forty years after Samuel Johnson praised du Halde's rosy portrait of China, James Boswell (1740–95), his companion and biographer, noted that Johnson "expressed a particular enthusiasm with respect to visiting the wall of China." Yet the learned doctor had otherwise come to scorn the Chinese, who (he said) had not even been able to invent an alphabet. Boswell recorded their exchange when Johnson castigated "East Indians" as "barbarians":[147]

Boswell: "You will except the Chinese, Sir?"
Johnson: "No, Sir."
Boswell: "Have they not arts?"
Johnson: "They have pottery."

JOSIAH WEDGWOOD, "VASE MAKER GENERAL OF THE UNIVERSE"

Wedgwood disdained the styles of rococo and Chinese porcelain as corrupt, an affront to proper proportion and respectable taste. He regarded Hamilton's catalogue as the bible of neoclassicism, and when he made a Greek-style ceramic portrait of Hamilton, he told the diplomat that antique pottery "may be the means, not only of improving and refining the public taste, but of keeping alive that sacred fire, which [your] collection of inestimable models has happily kindled in Great Britain."[148] Some enthusiasts even made the suggestion that modern Etruria would rise superior to the ancient world inasmuch as production of high-quality Etruscan pottery collapsed when Romans succumbed to vulgar silver plate; but, as an admirer told Wedgwood, "when English luxury seems at the height, your elegant taste has put to flight Gold and Silver vessels, & banished them from our tables."[149]

Neoclassicism embodied all Wedgwood's ideals—rationality, republican virtue, discipline, sobriety, capitalism, and Whig supremacy. He agreed with Winckelmann that antique masterpieces ennobled the human spirit. He saw first-rate, well-designed pottery as part of a cultural crusade, with the properly designed teacup an emblem of civilized life, a force for progress in the new Age of Improvement. He issued pottery medallions supporting abolition of slavery—in a best-selling item, a kneeling slave pleads, "Am I not a man and a brother?"—and moral rehabilitation of reprobates in the British convict colony of Australia. As he wrote to his partner, Thomas Bentley, in 1769, they should dedicate their business to "the pursuit of *Fortune, Fame, & the Public Good.*"[150]

On an earthier note, Wedgwood rejoiced that the public was "randy for the antique."[151] With waning of "the contagion for China fancy," Wedgwood profited from what he called an epidemic of "violent *Vase madness.*"[152] Classically minded connoisseurs and collectors favored the vase because its circular form offered a fine medium for friezelike compositions of figures and repeating patterns taken from antiquity. When Wedgwood ceremonially opened the Etruria manufactory in 1769, he ordered the first vessels decorated with figures from Hamilton's catalogue, which he also pillaged for designs for cameos, plaques, intaglios, and bas-reliefs.

Most famously, Wedgwood produced a replication of a work in cobalt-blue glass with sculptural relief in white cameo, believed to have been made during the reign of the Roman emperor Augustus (r. 31 B.C.E.–14 C.E.) and most likely modeled after a vessel made of onyx or agate. After painstaking experiment and sizable investment, Wedgwood at last produced a remarkable ceramic reproduction in blue and white known as the Portland Vase—in fact, a skeuomorph that traced its lineage back to an antique skeuomorph. It was a singular triumph, making him a leader of neoclassical fashion in Europe. He arranged a private showing for influential patrons, sent it on a promotional tour of Holland and Germany, and later produced a limited edition of some thirty-five copies.[153]

Another celebrated creation was the Green Frog Service, produced in 1773–74 for Catherine of Russia; each piece bore the emblem of a frog in reference to the monarch's palace, La Grenouillère ("The Froggery"), in St. Petersburg. Comprising fifty-two dinner settings (and 952 pieces) it represented, Wedgwood told Bentley, "the noblest plan ever yet laid down or undertaken by any Manufacturer in Great Britain."[154] Although Wedgwood barely broke even making the Green Frog Service for the tsarina, he employed it (like the Portland Vase) to enhance his reputation and publicize his main line of wares; titled dignitaries were given private showings of the service before it went off to Russia. Decorated with 1,244 views of English landscape, garden scenes, and notable residences, including Etruria Hall, the Green Frog represented an iconographic program of commercial prosperity, national purpose, and political freedom—ideals the Russian Minerva ceaselessly extolled and selectively pursued. Decoration on the service was almost entirely in monochrome, for in Wedgwood's view gilding and vibrant color evoked rococo decadence, unrefined taste, and aristocratic profligacy. Just as he believed that English civilization would redeem societies mired in Asian despotism, he also sought to wean high-class pottery from its associations with political absolutism. As d'Hancarville declared in his text for Hamilton's catalogue, "Porcelain is nearly always made into idiotic puppets."[155]

Wedgwood spent substantial amounts of money and energy producing his ceramics. He purchased kaolin clay from Cornwall, France, Germany, and (through Joseph Banks) the convict colony in New South Wales. In 1768 his agent bought five tons of first-rate clay from Cherokee Indians in western North Carolina and shipped it to England at the cost of £615 (the contemporary equivalent of $106,000); using it sparingly for his most expensive items, it lasted Wedgwood for two decades. He also arranged for a Canton agent of the EIC to send him samples of clay from Jingdezhen. In the end, however, he did not rely on kaolin to create creamware, his most lucrative product and the end result of lengthy research. (After winning the patronage of Queen Charlotte in 1765, Wedgwood promoted it as *Queen's ware.*) Superlative earthenware made from white-firing clay and ground flint, it is still the basis for many contemporary dinner services.[156] It provided a hard, white surface and fired more reliably than the various materials used for making European porcelain. Wedgwood used it for an extensive range of wares, including imitations of classical antiquities by coloring glazes to make vessels resemble porphyry and agate. He asked Bentley, "Don't you think we shall have some Chinese Missionaries come here soon to learn the art of making Creamcolour?"[157] Wedgwood also achieved fame among potters by creating jasper, an entirely new ceramic formula that he considered "my porcelain."[158] A white stoneware, jasper made an ideal material for neoclassical ornament, especially when colored with cobalt.

In using blue as a principal color and painting Chinese reign marks on the bottom of some of his wares, Wedgwood, like all other European potters, traded on the reputation of Jingdezhen. At the same time, unlike his continental competitors, he was not devoted to creating a ceramic recipe like that for Chinese porcelain. Rather, he aimed to turn out an incomparable product with the best raw materials, whether earthenware or stoneware. And like the entrepreneurs of Jingdezhen, he needed a commercially successful commodity to survive and flourish. In contrast, continental manufactories functioned primarily as vanity enterprises, with rulers such as Louis XV and the Electors of Saxony treating their potteries like private stockrooms and toy boxes, sustaining them with state subsidies and authoritarian directives. Sèvres never made a profit, and one-half its annual income came from sales forced on cowed aristocrats. The marquis d'Argenson quoted Madame de Pompadour as telling courtiers that "not to buy as much of the porcelain as one could afford is simply not to be a good citizen."[159]

The son of a hapless potter from the town of Burslem in Staffordshire, and determined to make his own way, Wedgwood aspired to be "Vase Maker General of the Universe," as he told Bentley in 1769.[160] He was an organizational and entrepreneurial genius, not only adopting the elaborate division of labor used at Jingdezhen (as described by Dentrecolles), but also introducing other innovations to Etruria, such as apprentice training, foreman management, clocking-in, and female employment. He introduced strict, military-style discipline to Etruria to ensure punctuality, cleanliness, thrifty use of materials, and (above all) an alcohol-free workplace.[161] One of his declarations forecasts the rigors of the emerging industrial age: he told a correspondent that he aimed "to make such *machines* of the *Men* as cannot Err."[162]

Wedgwood's introduction of machine power to production made Etruria a great deal more efficient than Jingdezhen, which remained organized on a labor-intensive, workshop basis. Wedgwood was the first Staffordshire potter to exploit steam power, perhaps with some advice from his friend James Watt. He used the steam engine to ground flint, prepare enamel colors, and mix clay. He led the way in employing the engine lathe in pottery manufacture (for making patterns on clay). He invented a kiln thermometer, or pyrometer, an enormous help in controlling the firing process and thus reducing wastage—an achievement that won him election as a Fellow of the Royal Society in 1783.[163]

Wedgwood substantially speeded up production by adopting the new transfer-printing process, which involved inking a copper-plate engraving, printing it on paper, and then transferring the image to the ceramic piece. By the last part of the century, Jingdezhen craftsmen faced the difficulty of keeping pace, by hand-copying, with English manufactures decorated with mechanical devices. At the same time, Wedgwood promoted the kind of cost-effective transportation that linked Jingdezhen with the wider world. He sponsored turnpike construction, which enabled coal to be delivered to the Staffordshire region at reduced rates, and he figured as a leading champion of the Trent and Mersey Canal, cutting the first sod for it in 1766 and making sure it ran past the front gate of Etruria Hall.[164] As he wrote to Bentley in 1765, "I scarcely know without a good deal of recollection whether I am a Landed Gentleman, an Engineer or a Potter, for indeed I am all three and many other characters by turns."[165]

Wedgwood supplemented Etruria's industrial innovations with new marketing methods that made its production highly responsive to consumer demand and stylistic changes. He pioneered techniques that became fundamental to modern commerce, such as market research, stock inventory, traveling salesmen, money-back guarantees, pattern books, sales catalogues, newspaper advertising, glamorous showrooms, and product endorsement by the rich and famous.[166] About a year's voyage from European markets, Jingdezhen could not match the coordination with rapidly shifting fashions made possible by these novel entrepreneurial tactics.

In 1786 Joseph Banks wrote that Wedgwood's "genius & ingenuity has put the business of Pottery in England so much above its original mechanical rank[,] . . . excelling both as an Art & a Science."[167] In close collaboration with other Staffordshire potters, Wedgwood made English pottery universally renowned. A magazine cartoon of the British icon John Bull depicted him with Wedgwood's name stamped on his hat and with Etruria's pottery serving as his face. Having inherited £20 from his father, Wedgwood died with a fortune of £300,000 ($51,600,000 in present-day terms), making him one of the two dozen wealthiest men in Great Britain, perhaps the only one not born to the aristocracy. His epitaph declared that he had "converted a rude and inconsiderable manufactory into an elegant art and an important part of National Commerce."[168] Indeed, those were Wedgwood's ambitions from the time he began his career as a small-town potter. As he said in 1775 when preparing a show of his finest wares, he aimed to "Astonish the world all at once, for I hate piddling you know."[169]

"NO MORE FROM CHINA, CHINA BRING": THE DECLINE OF CHINESE PORCELAIN IN THE WORLD

Wedgwood's aspiration to be "Vase Maker General of the Universe" seemed on the verge of realization by the time he died in 1795. Led by Etruria, Staffordshire potteries exported 84 percent of their annual production by the late 1780s. Bentley often referred to Wedgwood as "Generalissimo," for he approached foreign markets as if waging a military campaign, with Etruria as his *machine de guerre.*[170]

He lobbied ambassadors and English travelers to talk up his products in Spain, Denmark, Turkey, the Netherlands, Naples, and even China; he employed native speakers in several European languages to write letters urging foreign luminaries to buy his pottery. As a Swiss traveler noted in 1797, Wedgwood's unrelenting, decades-long campaign created

> a commerce so active and universal, that in Travelling from Paris to St. Petersberg, from Amsterdam to the farthest point of Sweden, from Dunkirk to the southern extremity of France, one is served at every inn from English earthenware. The same fine article adorns the tables of Spain, Portugal & Italy, and it provides the cargoes of ships to the East Indies, the West Indies and America.[171]

Wedgwood's pottery was purchased and copied everywhere from Portugal to Russia. Instead of complaining about Chinese porcelain draining silver bullion from their treasuries, European rulers turned their wrath on English tableware. The director of Meissen complained in 1774 that "the incredible number of English stoneware" entering Saxony had ruined his manufactory and damaged the economy.[172] A few years later, Stanislaus Augustus of Poland (r. 1764–95) established a pottery near his Belvedere palace in a vain attempt to stop the flow of precious metal from his kingdom. Both the Buon Retiro manufactory of the Spanish crown and the Portuguese manufactory of Miragaia began to copy blue-and-white Wedgwood pottery instead of Chinese porcelain.[173] Wedgwood expressed confidence that an appealing assortment of vases and an effective agent "will insure us success in the Conquest of our sister Kingdom [of Ireland]."[174] In short order, a nascent pottery industry in Ireland closed down as a result of imports of Wedgwood creamware. Huge amounts of the same commodity going to North America ensured that English colonists (including some potters who had absconded from Etruria) would not employ Cherokee clay to compete with Staffordshire, as Wedgwood long had feared.

In 1769 Wedgwood asked Bentley, "And do you really think we may make a *complete conquest* of France?"[175] When French import duties dropped in 1786, the answer delighted him: inexpensive, hard-wearing, and attractive, Staffordshire products with neoclassical designs claimed the lion's share of the market and all but destroyed the French tradition of tin-glazed earthenware. French workers purchased Wedgwood tea services; Sèvres copied Etruria's forms and decoration. Meissen likewise produced what it termed *Wedgwoodarbeit,* replicas of Etruria's manufactures. Delft suffered the price of not developing designs independent of Chinese style: Dutch customers turned against delftware, and sales of both delftware and Chinese porcelain gave way before the onslaught of Wedgwood creamware. The potteries of Delft shut down in the late eighteenth century, after which the place appeared to visitors like a ghost town, *une ville morte.* In like fashion, Wedgwood also drove Italian tin-glazing kilns out of business.[176]

Creamware supplanted Delft exports to America. The Puebla potters of Mexico, who had copied blue-and-white chinaware for almost two centuries, turned to duplicating Staffordshire wares. Etruria produced customized wares for the Ottoman market, painted (Wedgwood bragged) "with proper subjects for the Faithfull amongst the Musslemen."[177] Pottery from Staffordshire flooded the markets of Southwest Asia, and Persian craftsmen painted English transfer-printed decoration onto their blue-and-white earthenware. Wedgwood produced six-gallon oval basins in various colors "to please the fancy of a black King in Africa," keeping the price low so that additional sales would follow.[178] English ceramics reached East Africa in the early nineteenth century and dominated it after the great Taiping Rebellion (1850–64) destroyed the kilns of Jingdezhen. Wedgwood creamware joined chinaware on the great pillars of the Swahili coast, and it displaced both Chinese and Spanish-style pottery in the Philippines. A song in a play staged at Covent Garden in London in 1788 nicely sums up the worldwide fate of Chinese porcelain: "No more from China, china bring, / Here's English china ware!"[179]

"I shall be glad to give you joy on the Conquest of Peking," Wedgwood wrote to Bentley in 1792, just as the McCartney ambassadorial mission to open trade with China set off from England.[180] George Lord McCartney (1737–1806) presented the Qianlong emperor with gifts of English manufactures. According to the British record of the embassy, Chinese officials and courtiers feigned indifference to the merchandise, yet in fact "all eyes were fixed . . . on the vases, which were amongst the finest productions of the late Mr. Wedgwood's art. Of porcelain every Chinese is a judge. The specimens of the beauty of European manufacture were universally acknowledged and extolled."[181] Wedgwood's copy of the Portland Vase attracted special attention, and Jingdezhen at once tried to copy it, though without success.[182] Since McCartney's mission failed to open China to British trade, Wedgwood himself never cracked that market. Chinese resistance eventually collapsed, however, and within a generation, the Wedgwood firm was one of many Staffordshire concerns shipping pottery to the homeland of porcelain.

The decline of Chinese porcelain in Europe and the triumph of English pottery in international markets reflected a great reversal in the relationship between the West and Asia. Europe rejected an idealized image of China, as well as Chinese porcelain and chinoiserie, at the same time it began to exercise greater commercial and political dominion on the far side of the world. Until the eighteenth century, Westerners had flourished in the East by occupying enclaves that gave them entry to commerce but little political power. Their possessions were of negligible territorial significance, merely modest harbors on the margin of powerful empires (Goa and Macao), isolated strategic ports (Melaka and Batavia), and outposts valuable for their access to crops (the Spice Islands) and Chinese trade (Manila).[183]

Everything changed after midcentury, however. From 1757, the English East India Company carved out a state in Bengal and then used sales of Indian opium to transform its financial dealings with China. Defoe had a glimpse of that lucrative future: in *The Farther Adventures,* Robinson Crusoe takes opium to China to trade for porcelain.[184] Opium reversed the balance of payments with Europe: after 1814 China for the first time experienced a net outflow of silver, as much as 13 percent of its supply by 1850. In disputes with the Qing government, British commanders employed the military aggression that had worked so effectively in India. As John Gwynn had advised at the conclusion of the Seven Years' War, Westerners did not neglect to enjoy their superiority. In the First Opium War (1839–42), the British used steam-driven gunboats to destroy Chinese ships, bombard Canton, and force the Qing to capitulate to their terms, including possession of Hong Kong, payment of huge indemnities, and extensive trade privileges. A British observer of the conflict derided the Chinese navy as "a monstrous burlesque."[185]

At the end of the war, British seamen swarmed over the Porcelain Pagoda of Nanjing, using hammers and pickaxes to rip off white tiles as souvenirs.[186] The vandalism reveals how far Chinese power and Western admiration of China had fallen, for the action would have been inconceivable several generations earlier, when European trading companies still kowtowed to Chinese authority and when Europeans regarded the Porcelain Pagoda as an eighth wonder of the world, a symbol of the brilliance of Chinese culture. Chinese humiliation in the Opium War impelled Wei Yuan (1794–1856) to write the *Illustrated Gazetteer of Maritime Nations* (Haikuo tuzhi [1847]), in which he urges the Qing to lead the empire's ancient tribute states in expelling Westerners from Asia.[187] He invokes the memory of Zheng He, the towering figure of Chinese maritime adventure, who, as it happens, had supervised construction of the Porcelain Pagoda more than four hundred years earlier. In Wei's perspective, the Middle Kingdom paid a momentous price for ending the Ming expeditions and turning its back on the sea.

The coming of porcelain to the West after 1500 inaugurated more than two centuries in which Europeans abandoned their own varieties of tin-glazed earthenware in an attempt to match the achievements of the Chinese. "Porcelain disease" represented the earliest manifestation of high regard for the culture of the world's most ancient empire at a time when the West first was moving onto the global stage. The high opinion in which Chinese porcelain was held, as well as the prices it commanded, impelled European princes, potters, scientists, and alchemists to replicate it. Creation of European porcelain in the opening years of the eighteenth century, however, signaled the beginning of the end of an extended tutelage to the distant culture. Half a century later, when the utopian image of China itself was being radically recast, the rise of neoclassicism marked the end of the supremacy of Chinese porcelain and chinoiserie design. Fatefully, the reaction against an idealized China, Chinese porcelain, and Chinese aesthetics took place around the same time English potters pioneered industrial and entrepreneurial techniques that soon made their products ascendant not only in Europe but around the world as well.

After centuries of supremacy, Jingdezhen finally had encountered an adversary it could not defeat. Representing the climax of handicraft industry before the age of the industrial revolution, its labor-intensive methods and large-scale, decentralized structure served it excellently when its only challenges arose from relatively small kiln centers in Japan and continental Southeast Asia, potteries that did little more than copy the methods and products of the porcelain city. A behemoth that predictably rolled over the competition, Jingdezhen had developed to its fullest, most proficient extent by the time Dentrecolles arrived to discover its secrets. It was caught, however, in the trap of its own technical superiority: precisely because it had reached such a peak of efficiency, and therefore experienced no impetus for far-reaching reform, it proved unable to respond to a vigorous new opponent, one that created innovative products and techniques by experimental means and employed mechanized production to turn out its wares.[188]

Moreover, the backing given to Jingdezhen by the Qing court proved halfhearted and haphazard, perhaps reflecting certain drawbacks emerging in the central government itself as the century drew to a close. At the very moment British potteries were pushing into foreign markets customarily dominated by China, the imperial administration abolished the all-important office of director of the Jingdezhen kilns. When the porcelain city most needed central direction and renovation, nobody took charge. Things were different several generations earlier, when the Kangxi emperor restored the kilns, promoted innovations, and thus helped bring Chinese pottery back into international markets. Significantly, Delft was beaten back by Jingdezhen only a century before Wedgwood's international triumph. By 1800, however, both Delft and Jingdezhen represented the past while Staffordshire stood for the future.

In fact, in its handicraft technique and conventional organization (though not in its scale), Delft resembled Jingdezhen—as well as the potteries of Korea, Japan, Vietnam, Thailand, Persia, the Ottomans, Meissen, and Sèvres—more than it did the new style of factory production in the English Midlands. And on top of the advantages that Wedgwood and his Staffordshire colleagues won by enterprise and ingenuity, their invasion of global markets was eased immeasurably by virtue of being able to ship their pottery anywhere in the world inasmuch as Western nations dominated the seas.

In a long perspective, the influence of Chinese porcelain on the West was a brief episode in the millennium-long history of the ceramic's ecumenical impact. The encounter with the West proved decisive, however, for it was by way of European seaborne commerce that Chinese porcelain expanded its cultural influence in the new global ecumene. And it was that encounter which propelled the West into competition with Chinese porcelain and ultimately to victory over it.

The Chinese ruling elite since before the common era had regarded their culture as a model for the rest of the world, a way to impart virtue and civility to those whom they considered barbarians. Chinese porcelain had carried that culture vast distances, reshaping ceramic traditions, circulating in societies in manifold ways, and compelling the wonder of peoples everywhere. As Chinese porcelain lost its global markets, however, the "barbarians" of the West forcefully came to China, convinced of their superiority and bearing their own redeeming cultural messages. The awe of Qing officials before Wedgwood's Portland Vase, the sale of British pottery in Chinese markets, and British sailors clambering onto the Porcelain Pagoda were events of considerable symbolic significance. They represented a turning point in the history of the world. For the West, they signaled the rise to global dominance; for China, they marked the end of an epoch.

Epilogue

The Pilgrim Art

In the late eighteenth century, Louis-Sébastien Mercier expressed astonishment at the exhilarating, cosmopolitan life of Paris. The people thronging the streets, he said, included Japanese, Indians, Persians, Laplanders, Hottentots, and Quakers. He noted that his contemporaries took up novelties in clothing and tableware with enthusiasm, akin to "electricity passing from one to another." The commodities available in the city gave him a powerful sense of connection with the wider world:

> If one likes to travel, one can voyage a long way in imagination even while dining in a good house. China and Japan have furnished the porcelain in which aromatic tea boils; with a spoon made from the ore of Peruvian mines, one takes the sugar that unfortunate Negroes, transplanted from Africa, have raised in America; one sits on brilliant Indian fabrics, from that land over which three great powers have fought a long and cruel war.[1]

Mercier singled out the most significant trades in the new global economy: porcelain, tea, silver, sugar, slavery, and Indian painted cottons. In terms of commercial importance and political impact, chinaware counts as the least of these. Its role in the circuit of cross-cultural exchange, however, was uniquely important not only in the eighteenth century, but long before. The history of the lotus design exemplifies this. Persian potters in the Ilkhanid period copied undulating lotus scrolls from the borders of Chinese plates, unaware that the foliage was a mutation of acanthus patterns and vine scrolls carved on classical temples of Southwest Asia in the Hellenistic period. The acanthus and vine motifs had been transferred in Sassanian Persia from Greek temples to silver vessels, which merchants then traded eastward along the Silk Road.

In their long journey, the Hellenistic designs merged with Buddhist artistic themes that came from India. By the sixth century C.E., the distinctive lotus design appeared as sculpted decoration in Buddhist cave temples in northern China. As a boddhisattva, Avalokitesvara, soon to be transformed into Guanyin, had a special affinity for the lotus blossom. Sometimes the motif appeared on Nestorian Christian stone crosses in the Tang, the lotus emerging from the cross just as the Buddha ascends from the pristine bloom. Chinese craftsmen adopted the lotus design for ornamentation of vessels in silver and then ceramics. When merchants exported porcelains with lotus decoration to Ilkhanid Persia in the fourteenth century, potters there copied the foliage onto their earthenware. Two hundred years later, the lotus appeared so often on Chinese blue-and-white in Europe that botanists referred to it as the "porcelain flower." Those unfamiliar with the stylized plant termed it "the artichoke pattern."[2] (See figure 14.)

Porcelain thus played a central role in the transmission of a prominent artistic theme from region to region—from Southwest Asia to China and back—and from medium to medium—from architecture to silverwork to sculpture to silverwork to porcelain to earthenware. Nor did the motif's pilgrimage stop there. Once Persian potters had painted the lotus design onto a plate or bowl, craftsmen adopted it as ornamentation on textiles and buildings. Because ceramic workers in Persia (and other places) invariably made tiles, transfer of a design from utensils to architecture was commonplace. Thus potters decorated tiles with medallions of stylized lotus flowers on the mosque dedicated to Timur's niece in Samarqand, and patterns of lotus blossoms and Chinese plum branches appear on the Masjid-i-Jami, a sixteenth-century mosque in Kirman.[3]

Pottery adopted and transmitted symbols, themes, and shapes from all other media, including jade, lacquer, sculpture, metalwork, coins, textiles, engravings, woodblock prints, and painting. Decorative patterns and shapes journeyed from medium to medium, from country to country, were adapted to different cultures, took on innovative readings, and promiscuously mingled cultural referents. While China and its porcelain had a dominant role in this far-reaching exchange, the ecumene as a whole collaborated in the creation of a ceramic culture that in significant measure transcended territories and peoples.

Two kinds of flasks exemplify the remarkable fusion of influences typical of this process. *Kendi* is a Malay word deriving from the Sanskrit *kundika* (water pot) and denoting a metallic vessel used for ritual ablutions and drinking. With a round body and a mammiform spout set at a sharp angle to the shoulder, the *kendi* is designed so that it will not touch the drinker's lips, thereby avoiding pollution; it is filled from a large opening, much like that of a tea kettle, on top of the bulbous body. The earliest known *kendi*, found in northwestern India, dates from the second millennium B.C.E.[4] Along with other ritual paraphernalia associated with Buddhism and Hinduism, it spread into continental Southeast Asia by the early centuries of the common era. The Khmers considered silver or copper *kendi* essential for Hindu rituals of pouring sacred waters over the king at his installation, though they later also used *kendi* made of pottery. The Khmer and Thai favored *kendi* in zoomorphic shapes, such as ducks and geese; the Vietnamese made some shaped like elephants and storks. A *kendi* is sculpted in relief on a thirteenth-century temple of the Khmer royal city, Angkor Thom, and a real one was deposited in a chamber of a stupa in central Thailand. Relief sculptures of *kendi* also appear on the stupa of Borobudur in central Java. (See figure 4.)

Potters first copied *kendi* into porcelain in China in the Tang period, though the vessel never held a major place in the Chinese ceramic repertoire. Confucians used the vessels as water droppers in calligraphy, with some shaped like phoenixes and dragons with outstretched wings. Koreans sometimes replicated *kendi* in celadon and used them as water sprinklers in Buddhist rituals, a practice also followed by the Japanese. Merchants sold miniature Chinese versions of the ware in the Philippines and Java, where indigenous peoples incorporated them in burial rites, marriage ceremonies, and folk divination. In Bali, a bride signified submission to her husband by pouring water over his feet with a *kendi.* A sixteenth-century Chinese *kendi* found in Malaysia carried decoration of Buddhist auspicious symbols and Islamic inscriptions. Jingdezhen made a blue-and-white porcelain *kendi* shaped like a crescent moon, the symbol of Islam, to appeal to buyers in Southwest Asia.

China exported many *kendi,* often in fanciful animal forms, to Southwest Asia, where artisans copied them in earthenware in the seventeenth century. The porcelain collection of the Ottoman sultans included seven Ming elephant *kendi* fitted out with silver-gilt mounts. Shah ʿAbbas collected several kinds of *kendi,* including one in the form of an elephant and decorated with carnation motifs. In the seventeenth century, Persian potters translated the vessel into a hookah *(kendi-qalian).* Some *kendi* were imported to Europe, where German potters replicated the zoomorphic types, sometimes embellished with Dutch versions of Chinese designs; they featured in paintings by Jan Brueghel the Elder (1568–1625) and Willem Kalf, the latter depicting a *kendi* accompanied by oysters and a lobster. In the course of this remarkable peregrination, the *kendi* shed all connections with Indian religious ritual and became instead a scholarly utensil, a magic instrument, and an appealing curio for peoples of various cultures.

The pilgrim flask (*bianhu,* "flattened flask," in Chinese) traced a similarly wide circuit. An early version appeared in Nabatean pottery in Petra (now in Jordan) in the second century B.C.E., and the shape—a tapering neck and compressed moon-shaped body—resurfaced in Mesopotamian and Roman pottery, often decorated with images of Eros, Pan, and the Medusa. At the turn of the first millennium C.E., potters made such clay flasks at several locations on the Mediterranean coast. Early

Christian pilgrims employed small replicas for carrying holy water and sanctuary oil, objects classed among "blessings" *(eulogiae)* by the faithful. From the fifth century, flasks stamped with the image of Saint Menas were turned out at Abu Mina, a popular Egyptian pilgrimage center, forty-five kilometers southwest of Alexandria. Cheaply manufactured and of slight practical utility, they had value for pilgrims mainly as prestige items and pious souvenirs, affirmation of having made a taxing journey to a sacred site or shrine. Many flasks were decorated with representations of travel: Saint Isidore, a protector of seafarers, in a boat; the Virgin Mary fleeing into Egypt; the Magi on the road to Bethlehem; and Jesus entering Jerusalem on a donkey.[5]

Silk Road merchants took pilgrim flasks to Central Asia, where they became conflated with leather saddle flasks since the shapes are much alike. Persian earthenware and metallic flasks entered China in the Tang period, often bearing Hellenistic decoration, including acanthus patterns, dancing girls, and piping boys. Chinese craftsmen simulated the flask in porcelain, and, embellished with designs from Greece and Persia, they became prestige items as funerary goods in the Song period. In the Yuan and Ming periods, pilgrim flasks were made for export to Southwest Asia, often with Islamic-style floral decoration in the center. Comparable flasks made in the reigns of the Yongle and Xuande emperors are decorated on both sides with brocade patterns, floral scrolls, and Southwest Asian geometric patterns.[6] (See figure 13.)

In the sixteenth century Jingdezhen potters decorated a pilgrim flask, probably made to order for Portuguese Christians in Goa, with a choir of angels at Bethlehem. Persian potters produced a blue-and-white flask in 1523, copied from a Chinese porcelain example from a century earlier; they adorned it with a nightingale on a rosebush, perhaps an adaptation of a Chinese phoenix on a plum branch. Merchants traded Southwest Asian copies of Chinese flasks by way of Ottoman Anatolia to central Italy, where early-sixteenth-century potters in Umbria reproduced them in blue-and-white earthenware. Venetian artisans, who called the vessel *una inghistera fracada*, "a flattened bottle," reproduced them in glass and earthenware, decorated with floral and vegetal patterns in enamel paints. Around the same time, Flemish artisans made a large facsimile of the pilgrim flask by linking two nautiluses with a gilt hinge; the handle was in the shape of an elegantly coiling snake, and pearls and garnets set off the body.[7]

Philip II's agents in Manila and Macao commissioned a number of blue-and-white pilgrim flasks for the king. A Bavarian count, one of Philip's Protestant opponents, had a flask made for him in 1581, decorated with his coat of arms and anti-Catholic caricatures, including monkeys in ecclesiastical vestments. A German replica of a Chinese flask from about 1600, which ended up in August II's Japanese Palace, is so embellished with silver chains, gilt mounts, mother-of-pearl, and emeralds that it is scarcely recognizable. Finally, in the early eighteenth century, Meissen produced *Pilgerflaschen* decorated with Chinese floral motifs and with landscape scenes copied from Japanese lacquerware.[8] (See figure 1.)

The history of *kendi* and pilgrim flasks suggests the way in which ceramic vessels lent themselves to a bewildering cross-fertilization of influences. Styles, shapes, and decorative themes wandered as freely and as far as any Buddhist or Muslim pilgrim, though the concepts informing them generally remained behind. In the Chinese tradition, multivalent meanings clustered around symbols, such as the lotus, peony, and winter-flowering plum, evoking, respectively, ideas of spiritual attainment, female sexuality, and dynastic loyalty. In the West, severed from their roots, these plant forms were reduced to quaint embellishment.

Dutch treatment of Chinese compositions was typical. Delft potters painted their wares with fanciful versions of Chinese themes, such as the lotus and peony, so that their so-called *porcelyn* mimicked the appeal of the foreign product. They thus developed a make-believe Chinese style that proved so popular and attractive that Jingdezhen potters later imitated the novel chinoiserie patterns when their porcelain once again entered Southwest Asian and European markets after the time of troubles in the early Qing.

In effect, the end result of this cultural encounter between East and West was a creative imagining of China, a way of assimilating and domesticating it. Chinoiserie designs, especially those with figural and landscape elements, acted as filters that reduced the complexity of Chinese visual culture to stereotyped constituents, thereby rendering it picturesque and accessible rather than potent and enigmatic. Painted in pseudo-Chinese style and planted in a European perspective, the lotus retreated from Buddhism, the peony lost its sexual charge, and the flowering plum shed its links to political alienation. In like fashion, imperious Confucians turned into quaint mandarins, solitary Daoists became affable gentlemen, and the bodhisattva Budai, the chubby bringer of prosperity whose symbol is a bag of cash, changed into a gluttonous Christian friar. Sometimes pottery painters employed the exotic imagery for mere comic relief: a Meissen saucer of the early eighteenth century depicts huge mosquitoes hovering over a Chinese mandarin attempting to drive his wheel-less chair harnessed to two tortoises. Chinese potters reinforced this benign, reductive image of Chinese culture when they replicated the Western fantasy of China on their own porcelain exports. (See figure 20.)

In the end, something was gained, as well as lost, for the circuit of cross-cultural exchange promoted both innovation and misreading. Immense simplification inevitably took place when foreign craftsmen and artists reformulated artistic representations of Chinese culture; yet that very simplification resulted in creation of new decorative patterns with international currency. The world grew closer together through mutual misunderstanding.

Porcelain and its imitations played a primary role in cross-cultural exchange because potteries linked to remote markets pursued the strategy of replicating foreign artistic forms to win customers. From the seventh century to the sixteenth, the principal circuit of exchange ran between China and Southwest Asia, with blue-and-white becoming the standard color scheme in all cultures. The circuit thereafter expanded to encompass the West, including European settlements in the Americas. Chinese potters copied lotus scrolls and Persian silverwork in the eighth century, Arabic calligraphy and Islamic basins in the fourteenth, biblical illustrations and Dutch flagons in the eighteenth. Just as enterprising, Southwest Asian and European potters reproduced the Chinese decorative repertoire and devised their own versions of it to compete with the Chinese imports. Completing the circuit, the chinoiserie patterns migrated to China, where craftsmen copied them for the marketplaces of Southwest Asia and Europe. Global patterns of trade thus fostered the recycling of cultural fantasies, the creation of hybrid wares, and the emergence of a common visual language. (See figure 21.)

Turkish potters at Iznik in the early sixteenth century combined Chinese ornament, Ottoman court design, and European silverwork shapes to produce wares that appealed to customers throughout the Mediterranean. Venetian potters adapted floral patterns from Iznik pottery and designs *alla porcellana* from Chinese potters, while Chinese craftsmen replicated Venetian glass ewers in porcelain. Florentine potters decorated a pilgrim flask with a grotesque mask from ancient Rome statuary and tulips from Iznik ware. At the same time, Italian earthenware, itself shaped by pottery from Islamic Spain, influenced Iznik pottery, which sometimes combined medallion portraits in Renaissance *istoriato* style with spiral scrolls derived from the sultan's imperial monogram *(tughra)*.

From the late sixteenth century, Spanish galleons carried hundreds of thousands of porcelains from Manila to Acapulco, after which mules transported it 450 kilometers over the mountainous *camino de China* (China Road) to Mexico City. The town of Puebla, the chief center of Mexican pottery production, lay along the route, and the porcelains inspired potters to produce their own distinctive blue-and-white earthenware in order (as an eighteenth-century priest boasted) "to emulate and equal the beauty of the wares of China."[9] Chinese designs on Mexican pottery, such as chrysanthemums and cranes, mixed with images of the prickly pear cactus and the quetzal, icons of the shattered Aztec culture. Many porcelains also reached Peru, where they influenced native potters, who turned out wares combining traditional Inca motifs (such as a bird in flight with extended wings) with patterns taken from Chinese porcelain and Chinese-influenced Puebla vessels.

Japanese potters in the seventeenth century decorated blue-and-white beer mugs for the Amsterdam market with Dutch versions of Chinese lotus scrolls and with human figures drawn from Japanese traveling puppet shows, while Japanese officials placed orders with Delft and Jingdezhen, providing wooden models of the wares they desired. Agents of the Dutch East India Company ordered plates from

Japan with the monogram "VOC," and potters bordered that central motif with emblems of phoenixes, pomegranates, and a flowering camelia; panel decorations—a standard device on *kraakporselein*—consisted of bamboo plants and peonies. The coat of arms of Horatio Walpole (cousin of Horace) appears on a Kakiemon plate, the reverse side of which bears a traditional Japanese design of a tiger prowling in bamboo. Decoration on a Meissen service, probably made for August II, includes the armorial shield of the Elector of Saxony and King of Poland surrounded by Kakiemon-style flowers and sheaves of grain. The Porcelain Pagoda is featured on a Japanese teapot, copied from an illustration in a Dutch account of travel in China. Potters in Jingdezhen copied a square Dutch gin bottle in blue and white, after which the Chinese object was replicated in Japan, Persia, and Italy. In the late seventeenth century, a French earthenware ewer from a Nevers pottery had a shape that ultimately derived from Southwest Asian metalwork, a color scheme (blue and white) from China, and ornamentation from diverse sources, including classical antiquity (a coiled serpent), early Christianity (a winged angel), and China (exotic birds). (See figures 5 and 19.)

Jingdezhen artisans in 1690 made a plate decorated with a scene of a tax riot in Rotterdam, copied from a Dutch commemorative coin, with representations of the Eight Precious Objects of Buddhism around the central medallion. Chinese plates of 1722, commissioned by Dutch merchants, satirized the financial debacle of the South Sea Bubble, accompanied by advertisements for delftware and with human figures derived from the *Commedia dell'arte*, a fashion the Dutch copied from Meissen. A Chinese tureen and platter from the same time was copied from a French faience model, which was most likely based on pieces of Parisian silver, and when the Chinese objects reached France, potters in Rouen soon reproduced them.

Pottery made at Chelsea in the early eighteenth century imitated Meissen porcelain wares, which imitated Japanese imitations of Chinese vessels. A Worcester manufactory produced a teapot based on a contemporary European form that derived from the stoneware teapots of Yixing; it was ornamented with Asian-inspired motifs, especially Japanese Kakiemon designs in vivid enamel colors. Chinese copies of Dutch wares included marks on the bottom signifying Delft manufacture, while a Chinese copy of a Meissen copy of a Japanese dish came with a forged Meissen mark. A Chinese plate of 1750 decorated with a painting of a knight and his squire, surrounded by Chinese scenery and birds, represented a four-step adaptation: it derived from a pattern on a Meissen porcelain service of 1742 that replicated a Dutch engraving, which in turn copied a woodprint in a French translation of Miguel de Cervantes's *Don Quixote* (1605–15). A Chinese tureen of the late eighteenth century had a similarly tangled genealogy: its precursor was a Staffordshire creamware variation on a Sèvres porcelain bowl, which itself stemmed from a piece of French silverwork whose shape originated from an engraving in Hamilton's catalogue of pottery excavated at Herculaneum. (See figure 11.)

To be sure, tracing such connections induces a certain vertigo; but the exercise conveys the extent to which artists, craftsmen, and entrepreneurs around the world during the early modern period were relaying, integrating, and generating cultural forms. The history of the flowering tree design, a counterpart to the earlier creation of the lotus motif, exemplifies the extraordinary circuit of exchange at work. Sometime after the fifteenth century, Persian painters, in adapting themes from blue-and-white porcelain, translated the blossoming plum into a flowering tree, a theme that soon appeared on Persian earthenware and Indian cotton fabric. In the seventeenth century, European potters and textile producers adopted the Persian chinoiserie design, after which it mutated into the most characteristic of the patterns known as chintz (from Hindi *chint,* "many-colored"). Sending the design back to its source, English merchants commissioned Indian weavers to produce cottons with chintz patterns. In the eighteenth century, the merchants even sent drawings of the Indo-European chinoiserie fantasy to China as instructions for porcelain artists. The flowering-tree motif also surfaced on Japanese and Meissen porcelain of the eighteenth century and as an element in the famous blue-and-white Willow pattern, still the most popular ornamentation ever devised for tableware.[10]

The flowering-tree decoration is representative of ceramic development in that it sprang from the intersection of art and commerce as well from the long-distance collaboration of anonymous craftsmen in diverse media. Most significantly, it is impossible to say which culture was responsible for creating the celebrated design, since China, India, Southwest Asia, and Europe all played significant roles in its thematic development and geographic expansion.

Porcelain artistry of the seventeenth and eighteenth centuries suggests that the various regions of the global ecumene, across the countless boundaries dividing them, joined forces in the formation of a common cultural tradition. Although the tastes of elites played a crucial role in shaping it, the tradition stemmed far more from the ingenuity and enterprise of the potters themselves, in China, Japan, Southwest Asia, Europe, and the Americas. By the end of the eighteenth century, craftsmen around the world had created a collective visual language, a *koinē* of ceramic art. The potters of Lisbon acclaimed their copies of Chinese porcelain as examples of "the pilgrim art," products of a globally integrated circuit of aesthetic and commercial exchange. Given the peripatetic and imitative nature of ceramics, extended and intensified by the exemplary status of porcelain, potteries active in long-distance trade shared a common legacy, however provisional and indiscriminate.

The lotus scroll and the flowering tree, the *kendi* and the pilgrim flask, tureens and platters did not represent either high art or monumental achievement. At their best, they embodied a novel and creative cultural synthesis, enhanced by the charm of surprising associations; at their worst, they epitomized a sort of international kitsch, a harbinger of the tourist art of a later century. In neither case did they engage the attention of intellectuals striving to comprehend the emergence of a new global consciousness at the beginning of the modern era. Nevertheless, porcelain and its imitations provide the first and most widespread material evidence for sustained cultural encounter on an ecumenical scale, perhaps even for intimations of truly global culture.

NOTES

INTRODUCTION

1. Cited in Boxer 1986: 12; see also 52.

2. Cited in Parker 1998: 4; see also 165–67.

3. Cited in Padfield 2000: 2. On commercial networks, see Newitt 2005: 169; Flynn and Giráldez 1995.

4. Von der Porten 1972.

5. Cited in Haller 1967: 221.

6. Lane 1973: 293; Morga 1971: 19–20; Hess 1973.

7. Blair and Robertson 1915: 6:197.

8. The medal and related iconography are discussed in Parker 1998: 4.

9. Blair and Robertson 1915: 5:254; see Headley 1995: 641–45.

10. Cited in Schurz 1939: 27.

11. Brown 1995: 105–07; Shulsky 1998; Ray 1991: 300.

12. Pilgrim flasks from Philip's collection are now held by the Metropolitan Museum of Art, New York; the British Museum, London; the Gemeentemuseum, The Hague, the Netherlands; and the Peabody Essex Museum, Salem, Massachusetts.

13. Graça 1977: 45–47; Scheurleer 1974: 47; Loureiro 1999: 33.

14. Cited in Mudge 1985: 44.

15. P. Rawson 1984: 6, 100–103.

16. Blair and Robertson 1915: 1:78.

17. Collett 1993: 504–7.

18. On development of industrial ceramics, see Kerr and Wood 2004: 781–88.

19. Ricci 1953: 6.

20. Vickers and Gill 1994: 54–76.

21. Carswell 1985a: 22.

22. Mercier 1929: 125.
23. The quotation is taken from the reproduction of a handbill in Farrington 2002: 81.
24. Kelly 2004: 15; Zacks 2002: 6.
25. Dryden 1958: 1:57.
26. See Glassie 1999; Agnew 1993; Appadurai 1986.
27. Stevens 1982: 76.
28. Piccolpasso 1980: 2:61. Kerr and Wood 2004 provides specialized detail on clay, glazes, and kilns.
29. See McNeill 1963: 296–97. The concept of the ecumene is set forth in Hodgson 1974: 1:109–10.
30. Cited in Spence 1998: 18, which emphasizes the significance of this promise for Columbus; see Fernández-Armesto 1992: 41, 43.
31. Cited in Loureiro 1999: 33.
32. See Daniels 1996: 412, 479.
33. Chaudhuri 1985: 15; McNeill 1982: 24–25; Adshead 2004: 68–100. Some of the extensive literature on the theory of world systems is collected in Frank and Gills 1993.
34. Smith 1976: 2:1976.
35. Cited in Parker 1998: 3.
36. Montesquieu 1989: 393 (bk. 19, chap. 21).
37. See McNeill and McNeill 2003: 178, 201–2; Christian 2004: 390–91.

CHAPTER 1. THE PORCELAIN CITY

1. Unless otherwise noted, all translations from Dentrecolles's letters are from the original French text as provided in Bushell 1910: 81–222. English translations are available in Tichane 1983: 51–128; Burton 1906: 84–122 (but incomplete); and du Halde 1738–41 (also incomplete). All citations from Tang Ying's *Description of the Twenty Illustrations of the Manufacture of Porcelain* (Taoyetushuo) (1743) come from the translation in Bushell 1910: 7–30. Dentrecolles and Tang consulted some of the same documents compiled around 1795 by Lan Pu (and edited some fifteen years later by Zheng Tinggui) in *Jingdezhen taolu* (Potteries of Jingdezhen), edited and translated in Sayer 1951. *Tao Ya* (Pottery Refinements), published in 1906 under the pen name Ji Yuansou (edited and translated in Sayer 1959), comprises a collection of observations on Chinese ceramics, mainly from the eighteenth century, including some made by Tang and Lan. On Dentrecolles, see Dehergne 1973: 73–74, 351; Thomaz de Bossierre 1982; Tichane 1983; Rowbotham 1966: 255–56.
2. Cited in Thomaz de Bossierre 1982: 77.
3. Cited in Harrison-Hall 1997: 195.
4. Cited in Hochstrasser 2007: 142.
5. Sayer 1951: 37–38.
6. Cited in Dillon 1992: 278.
7. Cited in Gerritsen 2009: 119.
8. Cited in Elvin 1973: 285; see Dillon 1992: 278.
9. Cited in Gerritsen 2009: 139.
10. Cited in Brook 1981: 170.

11. Sayer 1951: 87.

12. Ledderose 2000: 85–101; Deng 1999: 81–82.

13. Smith 1976: 1: 31–36.

14. Cited in Lightbown 1969: 240.

15. Information on porcelain exports is taken from Ho 1994: 37; Deng 1997a: 276 and 1999: 60; Young 1999: 74; Godden 1982: 57, 60–62; Jörg 1982: 93, 149; Volker 1954: 226–28; Wästfelt, Gyllensvärd, and Weibull1990: 27; Clunas 1987: 16.

16. Cited in Sung 1966: 146.

17. Cited in Foust 1992: 82.

18. Jörg 1986: 59; Sheaf and Kilburn 1988.

19. Medley 1966; Yuan 1978.

20. Sayer 1959: 54–55.

21. Foster 1965; Caiger-Smith 1993–94.

22. Coomaraswamy and Kershaw 1928–29.

23. Sayer 1951: 17; 24; see Macintosh 2001: 45–46; Jörg 2002–3: 25–26.

24. The significance of skeuomorphs in pottery production is emphasized in Vickers and Gill 1994: 106–7.

25. Sayer 1951: 49; see Lam 1998–99.

26. Bushell 1910: 6; see Sayer 1951: 82.

27. Dillon 1976 is the most thorough survey of Jingdezhen. Staehelin 1966 describes porcelain production in annotations to eighteenth-century watercolor illustrations of it, a format also followed by Tang Ying.

28. Smith 1976: 1: 15.

29. Bushell 1910: 73.

30. Bai 1995; Howard 1994: 14–15 and 1997: 127.

31. Godden 1979: 17; Howard 1974: 84–85; Whitman 1978: 1:225; Mueller 2000: 19–20; Hallberg and Koninckx 1996; Kee 1999: 95.

32. Esten, Wahlund, and Fischell 1987: 86; see Stuart 1993: 56.

33. Cited in Scheurleer 1974: 146, 162.

34. Cited in Beurdeley and Beurdeley 1971: 147.

35. Cited in Ward 2001: 379; see Sayer 1959: 12–13.

36. Cooper 2001: 317.

37. Cited in Lightbown 1969: 263.

38. Cited in Hochstrasser 2007: 137.

39. Sayer 1951: 45–47.

40. Sayer 1951: 33.

41. Sayer 1951: 34–35.

42. Sayer 1951: 33.

43. Sung 1966: 154.

44. Sayer 1951: 105.

45. Bushell 1910: 38.

46. Sayer 1951: 32.

47. Names of colors are provided in Sayer 1951: 55; Sayer 1959: 16; Bushell 1910: 49–50; Kerr 1993: 152–53.

48. Sayer 1951: 49.
49. Groeneveldt 1880: 87; Peng 1994: 1:xxiv.
50. Carletti 1964: 149–50.
51. Dillon 1976: 30, 32, 35, 38, 43.
52. Cited in Volker 1954: 50.
53. Cited in Emerson, Chen, and Gates 2000: 244.
54. Cited in Kerr and Wood 2004: 19.
55. Cited in Kerr and Wood 2004: 211.
56. Cited in Gernet 1985: 88, 102.
57. Gernet 1985: 64–72.
58. Ricci 1953: 105.
59. Ricci 1953: 105.
60. Ricci 1953: 267.
61. Gernet 1985: 83, 92–93; Pagani 1995: 76.
62. On the Rites Controversy, see the essays in Mungello 1994.
63. Ricci 1953: 113, 98.
64. Guy 1963: 120.
65. Sung 1966: 155. On Chinese pottery deities before the Ming period, see Kerr and Wood 2004: 206, 243–44.
66. Bushell 1910: 38, 47–48, 63, 127.
67. Sayer 1951: 81, 83.
68. Sayer 1951: 119–20; see Sung 1966: 155.
69. Sayer 1951: 103–4; 78, 85.
70. Bushell 1910: 63.
71. Cited in Kerr and Wood 2004: 166.
72. For the following, see Hayden 2003: 134, 138–39; Amiran 1965; Moore 1995: 47–48; Bellwood 2005: 114, 158. Rice 1999 provides a survey on research into the origins of pottery.
73. Simpson 1997; Bottéro 2001: 85, 207–8; David, Sterner, and Gavua 1988: 365–66; Bellwood 2005: 54–55.
74. For the following, see Weinberg 1965; Hay 1986: 84; Matson 1989: 15; Cauvin 2000: 44; Chang 1999: 50–53.
75. Cited in Bottéro 2001: 99.
76. Mitchell 2004: 74.
77. Berzock 2005: 73, 100, 136; Gilbert 1989: 220.
78. Cited in Barley 1994: 53.
79. Beckwith 1970: 43; Miller 1985: 122–23; Huyler 1996: 19–20.
80. On widespread images of potters in the Americas, see Lévi-Strauss 1988.
81. Tedlock 1985: 347–48; Salles-Reese 1997: 53–54.
82. Cited in Shoemaker 1997: 635.
83. Rosenthal 1989: 1:263, 257–66.
84. Cited in Ritter 2003: 43.
85. Origen 1998: 193.
86. Piccolpasso 1980: 2:109, 68–69.

87. For the following, see Clunas 1997: 128–29; Godden 1982: 63–64, 118; Hansen 1990: 133, 139, 145–46; Watson 1985; Jörg 1995: 112.

88. Boxer 1953: 213.

89. Cited in Thomaz de Bossierre 1982: 27.

90. Cited in Hansen 1990: 146.

91. Little 1990; Brook 1981; Dillon 1992: 285.

92. Dillon 1976: 125–26; Hsu 1988: 147–48; Sayer 1951: 36.

93. Smith 1976: 1:32, 35; Ricci 1953: 12.

94. Cited in Little 1983: 16.

95. Du Halde 1738–41: 1:325; see Staehelin 1966: 68.

96. Pinto 1989: 170.

97. Cited in Schafer 1963: 17.

98. Ricci 1953: 261.

99. Atwell 1982: 68–69, 79. The significance of silver in world trade is emphasized in Flynn and Giráldez 2002. On collection of customs, see Marks 1998: 128.

100. Cited in von Glahn 1996: 129.

101. Montesquieu 1989: 392 (bk. 20, chap. 21).

102. Ricci 1953: 261–62.

CHAPTER 2. THE SECRETS OF PORCELAIN

1. Staehelin 1966: 70; Rowbotham 1966: 106–7; Haudrére and Le Bouëdec 1999: 5–16. Harris 1999 sketches the Jesuit information network; Adshead 2002: 211–12, 240–42, discusses the role of China Jesuits in the network.

2. See Pocock 1999: 99.

3. Cited in Rocco 2003: 99.

4. Cited in Carter 1988: 291.

5. Haudrére and Le Bouëdec 1999:11; Raffo 1982: 102; Albis and Clarke 1989; Bushell 1910: 209; Tichane 1983: 111; Dehergne 1973: 38; Thomaz de Bossierre 1982: 8–9, 33.

6. Belevitch-Stankevitch 1910: 49, 55, 71; Lach and Van Kley 1993: 3: 432; Lach 1957: 33; Rowbotham 1966: 122–23, 258; Dehergne 1973: 34; Thomaz de Bossierre 1982: xii, xv; Mungello 1977: 42.

7. Cited in Swiderski 1980–81: 138.

8. Cited in Love 1994: 67. Original emphasis.

9. Cited in Cook and Rosemount 1981: 265.

10. Cited in Lach 1957: 46–47; 52, 68–69. Leibniz's views of China are examined in Mungello 1977 and Spence 1998: 82–88.

11. Cited in Pocock 1999: 98–99; see Bien 1986: 363–64.

12. Cited in Pocock 1999: 104.

13. The phrase comes from Burton 1932: 3: 323.

14. Cited in Lach 1957: 31.

15. Leibniz 1970: 5: 591.

16. Cited in Lach 1957: 30.

17. Wiener 1951: 598.

18. Navarette 1960: 1:xlv; 154; 137; see also 2:366.
19. Cited in Lach 1957: 36.
20. Cole 1943: 11–18, 32–43, 57–58, 269–72; Lach 1957: 27.
21. Cited in Kerr and Wood 2004: 770 n. 227.
22. Smith 1976: 1:437, 439.
23. Smith 1976: 1:443.
24. Cited in Harris 2004: 168.
25. Foster 1899: 1:134.
26. Cited in Lemire 1991: 21.
27. Bernier 1968: 223.
28. Hanway 1756: 302.
29. Cited in Koerner 1999: 96, 116–17, 136.
30. Cited in Fang 2003: 819.
31. Cited in Braudel 1981: 1: 186.
32. Cited in Yonan 2004: 658.
33. Haudrère 1999: 202–3; Stein and Stein 2000: 156.
34. Boxer 1965: 111–12, 222; de Vries and Woude 1997: 433–34; Turner 2004: 183–224, 291.
35. Cited in Davies 1961: 55.
36. Smith 1976: 1:525; 2: 636.
37. Cited in Goody 1993: 210. The connection between European silver exports, Asian imports, and the rise of fashion is suggested in Pomerantz 2000: 159–61.
38. Cited in Styles 2000: 135 n. 23.
39. Cited in Porter 1999–2000.
40. Cited in Lemire 1991: 36, 41. Gilray's engraving is in a private collection.
41. Cited in Saunders 2002: 70.
42. Ricci 1953: 18.
43. Chou 1999–2000; Hayward 1972.
44. Cited in Jourdain and Jenyns 1948: 144.
45. Cited in Hayward 1972: 60; see Chou 1999–2000.
46. Miller 2001: 3; Whitehead 1993.
47. Payne 1951: 124–25, 127–28.
48. Defoe 1974: 255, 265–66.
49. Defoe 1977: 90; see also Liu 1999.
50. Defoe 1979: 205–6.
51. Cited in Kuchta 2002: 123.
52. Cited in Kerr and Wood 2004: 752 n. 177.
53. Cited in Pietsch 2004: 179.
54. Ströber 2001. The porcelain craze is recounted in Plumb 1972.
55. Cited in Cassidy-Geiger 2003: 152.
56. Postelthwayt 1774: vol. 2, note in entry on "porcelain," n.p.
57. Röntgen 1984: 31–32; Bevor 1998: 140.
58. Cited in Emerson 1991: 4.
59. Schönfeld 1998: 723–24; Röntgen 1984: 26; Pietsch 2004.

60. Watson and Whitehead 1991.
61. Cited in Patterson 1979: 28.
62. Cited in Le Corbeiller 1990: 6.
63. Schönfeld 1998 scrutinizes Böttger's claims for having created porcelain.
64. Cited in Coutts 2001: 237 n. 57.
65. Cited in Pietsch 2004: 181.
66. Postelthwayt 1774: vol. 2, entries on "manufacturers" and "mechanical arts," n.p.
67. Cited in Lemire 1991: 30. Original emphasis.
68. Smith 2002; see also Stein and Stein 2000: 109.
69. Mercier 1929: 119.
70. Cited in Stein and Stein 2000: 119.
71. Saint-Simon 1856–58: 7: 226; see also Giacomotti 1963: 30.
72. Cited in Scheurleer 1974: 111.
73. Cited in Giacomotti 1963: 32.
74. Du Halde 1738–41: 1: 338.
75. Dames 1921: 2: 213–14.
76. Ökte 1988: 1:141.
77. Cited in Lightbown 1969: 230, 231.
78. Boxer 1953: 127.
79. Cited in Liu 1999: 749.
80. Bacon 1944: 462; Browne 1964: 137.
81. Cited in Volker 1954: 21.
82. Cited in Kerr and Wood 2004: 744.
83. Cited in Divis 1983: 29.
84. Piccolpasso 1980: 2:6; Browne 1964: 136.
85. Sayer 1951: 34–35.
86. Sayer 1951: 27.
87. Marks 1999: 85, 93.
88. Ricci 1953: 14.
89. Graça 1977: 45–47.
90. Cited in Bertini 2000: 53.
91. Cited in Pinto de Matos 1999: 27.
92. Cited in Atwell 1998: 395 n. 68.
93. Carswell 1985a: 13–14.
94. Hogendorn and Johnson 1986: 15; Peng 1994: 1:9–10; Magalhães-Godinho 1969: 389–98.
95. Ibn Battuta 1929: 243, 267; Ray 1993: 100; Pires 1944: 100, 170, 181.
96. Cited in Johnson 1970: 19–20.
97. Ibn Battuta 1929: 334.
98. Wright 1854: 263, 265, 267, 283, 345–46, 363.
99. For the following, see Casteleden 1990: 104–5; Guy 1996–97: 59; Carvalho 2000: 16; Sandon 1992; Woldbye 1984; Walcha 1981: 106–7; Cort 2000: 135; Glassie 1997: 311; Mills and Ferguson 2008: 341–42.
100. Susenier's painting is in Dordrechts Museum, Dordrecht, the Netherlands.

101. Gaskell 1989: 70–80; Barnes and Rose 2002: 86. The Goednert painting is in a private German collection; the first Kalf painting is in the Thyssen-Bornemisza Collection, Lugano, Switzerland, and the second is in the Cleveland Museum of Art. The anonymous still life in the style of Kalf is in the New York Gallery of Fine Arts; the Berghe painting is in the Philadelphia Museum of Art; the de Heem painting is in the John and Mable Ringling Museum of Art, the State Art Museum of Florida, Sarasota; the Peeters painting is in the Staatliche Kunsthalle, Karlsruhe, Germany.

102. Cited in Kemp 1995: 185; see Moura Sobral 2007: 415–16.

103. Mosco 1999.

104. See Dance 1986: 143–48.

105. Harrisson 1962; Raphael 1931–32.

106. Dillon 1976: 45; Thomaz de Bossierre 1982: 114.

107. Sayer 1951: 123.

108. Cited in Plinval de Guillebon 1999: 83.

109. Cutler 2003: 10–13, 97–98, 162.

110. Cited in Lamb 2004: 16.

111. Cited in Oldroyd 1996: 51.

112. The relationship between basaltic rock and volcanoes is explained in Fortey 2004: 53–55, 76–77, 79–81.

113. Burn 1997; Vickers 1997.

114. Cited in Thackray 1996: 71.

115. Darwin 1989: 239; see Desmond and Moore 1991: 160–62; Fortey 2004: 18–21.

116. Cited in Desmond and Moore 1991: 420.

117. Desmond and Morris 1991: 420–21; Browne 1995: 390–93.

118. For the following, see Sigurdsson 1999: 112–17, 153–55; Oldroyd 1996: 50–51, 92–94, 105; Dean 1992: 13, 47, 84–85.

119. Cited in McKendrick 1973: 309.

120. Uglow 2002: 138–39, 152–53.

121. Cited in Torrens 2005: 261.

122. Cited in Dolan 2004: 180–81.

123. Jenkins and Sloan 1996: 182.

124. Uglow 2002 surveys the common interests of Hutton, Wedgwood, and Watt.

125. Reilly 1992: 29; McKendrick 1961.

126. Cited in Richards 1999: 211.

127. Farrer 1903–6: 3:89.

CHAPTER 3. THE CREATION OF PORCELAIN

1. Hillel 1991: 25.

2. Vainker 1991: 124; Rhodes 1968: 18; Pierson 1996: 9–14, 55–56; Addis 1980–81.

3. Sung 1966: 148.

4. Bushell 1910: 65.

5. Burton 1906: 249; Fortey 2004: 262–63.

6. Gaimster 1997 surveys the subject of German stoneware; see 79, 82, 106–7, 117, 124–

25; Gaimster 1999. *A Kitchen Scene* is in the National Gallery of Art, Washington, D.C.; *Young Woman at Her Toilet* is in the Museum Boijmans Van Beuningen, Rotterdam.

7. See Polyani 1958: 52.

8. Wood 1999: 167, 185, 241, 243; see also Kerr 1993: 161–62.

9. Discussion of geologic change and Chinese ceramics is developed from suggestions in Wood 1999: 27–29, 91–92; 1999–2000; see also Kerr and Wood 2004: 49–50.

10. Zhang et al. 1984; Dewey et al. 1985; Erickson 2001: 53–103, 156–80.

11. Wood and Kerr 1992: 39.

12. For the following, see Smalley 1968; Hillel 1991: 5–21; Vandiver 1990: 110; Zhou 1986; Vainker 1993: 214–15; Wood 1999: 196–97; Shelach 2001: 30; Kerr and Wood 2004: 90–96.

13. Ricci 1953: 305.

14. Golas 1999: 185.

15. For the following, see Barnard 1976, 1983; Wu 1995: 46–47; Mino and Tsiang 1986: 14–15; Kerr 1986: 301–4; Chêng 1973; Hearn 1980.

16. Rawson 1997.

17. Rawson 1993a: 808–9.

18. Cited in Rawson 1993b: 74.

19. Wu 1999: 729; Vainker 1991: 49.

20. So 1980: 326.

21. Needham 1964: 9, 21–22; Elvin 1973: 84–87; Hartwell 1967.

22. McNeill 1963: 23–24, 29–69, 217–32; Chang 1986: 242–45, 295–307, 409–13; Potts 1997: 153–56, 161.

23. For the following, see Falkenhausen 1999: 489–93, 529–30; Rhodes 1968: 18–27; Pierson 1996: 49–52; Hodges 1970: 67; Vandiver 1990: 110; Kingery and Vandiver 1986: 77; Vainker 1993: 222–23.

24. Piccolpasso 1980: 2:89.

25. Development of Chinese kilns is detailed in Kerr and Wood 2004: 283–378.

26. Medley 1981: 14, 18, 24; Vandiver 1990; Zhang 1986; Watson 1970a. For Chinese glazes from both technical and aesthetic perspectives, see Wood 1999.

27. Kingery and Vandiver 1986: 107; Rhodes 1968: 263–64.

28. He 1996: 52–53.

29. Willetts 1958: 2:410–11; J. Rawson 1984: 77–85; Melikan-Chirvani 1970.

30. Cited in Palliser 1976: 236.

31. Ricci 1953: 307.

32. Cited in Elvin 1973: 105.

33. Mote 1999: 616–17 discusses the terms "yellow China" and "blue China"; see also Schafer 1967: 14–15, 34, 263.

34. Cited in Holcombe 2004: 752.

35. Wang 2000: 3–11 makes the case for the significance of a maritime focus in western Asia and a continental one in China; see also Chaudhuri 1985: 122–23, 208; Padfield 2000: 7–19.

36. See Wong 2001.

37. See Brady 1991.

38. Cited in Borschberg 2002: 33.

39. Cited in Chaudhuri 1990: 5; see Fok 1987.

40. Ricci 1953: 128–29.

41. Ricci 1953: 311; Montesquieu 1989: 278 (bk. 17, chap. 2).

42. Hartwell 1982; Ho 1956.

43. Cited in Wolters 1986: 36.

44. Cited in Shiba 1970: 187.

45. Cited in Holcombe 2001: 89.

46. Himanshu 1994: 121–61.

47. Hodges and Whitehouse 1983:130–32; Daryaee 2003. The historical coincidence of the creation of the Tang and Muslim regimes is stressed in Hourani 1951: 61–62.

48. La Vaissière and Trombert 2004; Xiong 2000.

49. Hansen 2003.

50. Cited in Skaff 2003: 501.

51. Knauer 1998; Mahler 1959.

52. Bentley 1993: 29–66 discusses the significance of the Silk Road.

53. Xuanzang's travels are recounted in Wriggins 1996.

54. Meserve 1982: 51–61; Elisseeff 1963.

55. Cited in Schafer 1963: 58; see Liu 1996: 90, 183; Beckwith 1991; Perdue 2005: 35–36.

56. Sen 2003: 15–44; Wriggins 1996: 176–77; Jörg 1997: 154.

57. Cited in Wang 2005: 73. The gender transformation of Avalokitesvara is examined in Yü 2001: 223–62, 413–19; see Wang 2005: 219–28.

58. For the following, see Watson 1983; Whitfield 1990; Medley 1970; Rawson 1986: 34–35; Cheng 1983: 79–115.

59. Willetts 1958: 2: 479; Vainker 1991: 59; Watson 1984: 145.

60. Rougelle 1996: 161–62; Flecker 2000; Guy 2001–2.

61. Cited in Deng 1995: 6.

62. Casson 1989.

63. Peterson 1979: 474–86; La Vaissière and Trombert 2004: 961–83.

64. Pelliot 1930; Beckwith 1991: 190; Harris 2003–4.

65. Cited in Blair and Bloom 1994: 107.

66. Mote 1999: 49–71, 193–221.

CHAPTER 4. THE CULTURE OF PORCELAIN IN CHINA

1. Cited in Simkin 1968: 98.

2. Simkin 1968: 97; Kerr 1986: 313; Rockhill 1914–15: 15: 421–22; So 2000: 98–101. The Song economic revolution is surveyed in Elvin 1973: 113–99; see also Mote 1999: 164–67, 323–25.

3. Cited in Clark 1991a: 383.

4. Hirth and Rockhill 1966: 78.

5. Cited in Kerr and Wood 2004: 716.

6. Lo 1952, 1970; Kwan 1985.

7. Schottenhammer 2006: 6.

8. Cited in Deng 1997: 83.

9. Cited in Umehara 1999: 19.

10. Cited in Kwan 1985: 58.

11. Cited in Sen 2003: 142.

12. Chaffee 2001.

13. Cited in Chaffee 2006: 406.

14. For the following, see He 1996: 133–34, 137; Beurdeley and Beurdeley 1984: 116; Guy 1986: 14–16; Lam 1985; Long 1994: 14; Ho 2001; So 2006: 1270–71.

15. Clark 1991; see So 2000: 186–201.

16. Cited in Schafer 1963: 11.

17. Clavijo 1928: 288–89.

18. Cited in Hsu 1988: 151–52.

19. Cited in Wong 1978: 53. Stocking the Manila galleon is described in Schurz 1939: 182–83.

20. Chaudhuri 1985: 53, 108,184, 189, 191; Jörg 1982: 129; McEwan 1992: 103–05.

21. Cited in Guy 1997: 59.

22. Cited in Godden 1982: 59–60.

23. Mino and Tsiang 1986; Tregear 1982: 7–48; Vainker 1991: 88–133; Ts'ai 1996: 112.

24. Rawson 1989; Clunas 1992–94: 48; Whitfield 1989; Falkenhausen 1993a: 842–43.

25. Eliot 1963: 180.

26. Sung 1966: 135.

27. Cited in Gerritsen 2009: 132.

28. Elman 2000: 14, 66–124; Chaffee 1985: 3, 35–41; Little 1990: 24.

29. Jang 1999; Bai 1995.

30. See Willetts 1958: 2: 424–25.

31. The mental temper of the Song is examined in Liu 1988.

32. Cited in Lee 1996: 258.

33. Jang 1999. *Eighteen Scholars of the Tang* is in the National Palace Museum, Tapei, Taiwan.

34. Cited in Watson 1973: 2.

35. Cited in Soper 1976: 36–37.

36. Rogers 1992; Clunas 1991: 93–97, 114; Curtis 1998. Gaozong's use of Confucius for political indoctrination is discussed in Murray 1992.

37. The relationship between classical, literary, and colloquial Chinese, as well as its implications for cultural continuity, is set forth in Harbsmeier 1998: 26–27, 44–46, 417.

38. Ricci 1953: 28.

39. Lewis 1999: 337–62 examines the intimate connection between textual and political authority.

40. Cited in Huang 2007: 183 n. 8.

41. Cited in Scott 1992: 80.

42. Cited in Arnold 1999: 25. Jullien 1995 surveys the significance of *shi* in Chinese politics, art, and literature.

43. Cited in Burnett 2000: 535.

44. Cited in Cherniack 1994: 26. Jensen 1997 traces the Jesuit invention of "Confucianism" and its permutations.

45. Cahill 1982: 74–77.

46. Curtis 1996–97: 103; Curtis 1993: 135, 139; Elman 2000: 30.

47. Cited in Brook 1998: 78.

48. Bushell 1910: 132.

49. Sayer 1959: 26.

50. Murray 1999: 124; Rawson 1993b: 78–79.

51. Ho and Bronson 2004: 272; Pearce 2003; Ortiz 1999: 176–77.

52. Cited in Yu 2007: 48.

53. Cited in Curtis 1998: 11–12.

54. Cited in Pierson and Barnes 2002: 59.

55. Ricci 1953: 24.

56. Wu 1996: 53. *Ranking Ancient Works* is in the National Palace Museum, Taipei, Taiwan.

57. Cao 1979: 95–96.

58. Sayers 1951: 115.

59. Sayer 1951: 97; Sayer 1959: 34.

60. Ricci 1953: 79–80, 313.

61. Cited in Laing 1975: 224; see Sayer 1951: 46, 77, 105, 114.

62. Rawson 1989: 284.

63. Sayer 1951: 93; Ricci 1953: 15.

64. Cited in Clunas 1991a: 375; see Clunas 1992–94: 48.

65. Freeman 1977; Hartwell 1967: 131; Heng 1999: 121–23, 132–33, 160–61, 205.

66. Cited in West 1997: 93; see Adshead 1997: 32–34. *Peace Reigns over the River* is described in Hansen 2000: 282–86; the painting is in the Palace Museum, Beijing.

67. Kieschnick 2003: 222–49.

68. Ricci 1953: 25.

69. *Literary Gathering* is in the National Palace Museum, Taipei, Taiwan; see Chen 2007.

70. For the following, see McElney 1998–2000; Huang 2000: 503–70; Ukers 1935: 1:1–12; Bushell 1910: 104–5, 123.

71. Pirazzoli-t'Serstevens 2002.

72. Cited in Kieschnick 2003: 271.

73. See Kieschnick 2003: 272. Qui Ying's hand scroll is in the Cleveland Museum of Art, Cleveland, Ohio.

74. Cited in Han 1993: 43.

75. Cited in Xiong 2000: 190; see Gernet 1982: 264.

76. Boxer 1953: 140; see Cooper 2001: 279.

77. Cited in Han 1993: 44.

78. Bushell 1910: 95–96.

79. Lu 1974: 109. Original emphasis.

80. Lu 1974: 111; see Krahl 2004: 62.

81. Bushell 1910: 105.

82. Bushell 1910: 97, 124.

83. Cited in Scott 2002: 8.

84. Bushell 1910: 138.

85. For the following, see Lo 1986: 12–21, 33–37, 66, 250; Vainker 1991: 173–75; Ukers 1935: 2: 436–37; Kerr and Wood 2004: 273–77.

86. Sayer 1951: 94; on connoisseurs and Yixing teapots, see Wong 2006.

87. Cited in Ukers 1935: 2:488. Original emphasis.

88. Bushell 1910: 126, 127.

89. Coutts 2001: 71–72; Styles 2000: 146–47.

90. Cited in Coutts 2001: 234 n. 26.

91. Jörg 1995: 247.

92. *Still Life with Tea Things* is in the State Museum, Berlin, Germany; *Still Life with Silver and Ebony Casket* is in the Boymans-van Beunigen Museum, Rotterdam, the Netherlands.

93. Lu 1974: 60.

94. Cited in Macfarlane and Macfarlane 2003: 48–49.

95. Cited in Montanari 1994: 126; see Boxer 1965: 198.

96. Cited in Schama 1988: 172; see Murris 1925: 129.

97. Cited in Schama 1988: 172

98. Postelthwayt 1774: vol. 1: entry on "China," n.p.

99. Cited in Plutschow 2003: 27–28.

100. Ricci 1953: 64–65.

101. Cooper 2001: 277–78. Emphasis added.

102. The role of tea in combating disease and increasing population is highlighted in McNeill 1989: 259–61; see also Macfarlane and Macfarlane 2003: 168–77, 255–72; Ukers 1935: 1:552–59.

103. Cited in Needham 2000: 81.

104. Montaigne 1983: 159.

105. Cited in Goldwaithe 1989: 21.

106. Cited in Richards 1999: 166.

107. Richards 1999: 167. The anonymous engraving, now in the National Library of Medicine, London, is reproduced in Weeden 1984: 81.

108. Cited in Seok Chee 1993: 34.

109. Bushell 1910: 106, 125.

110. Bushell 1910: 122.

111. Kerr 1986: 311; Goody 1993: 368.

112. Bushell 1910: 111, 166.

113. Sayer 1951: 72.

114. Lindquist 1991: 340–42; Neill 1985: 244. Krahl 2004 discusses the problem of identifying and classifying the Five Great Wares.

115. Cited in Ukers 1935: 2: 485.

116. Cited in Li 1998: 31. The chemical basis for this formulation is explained in Yap and Hua 1994; Guo 1987; Kerr and Wood 2004: 122–35.

117. Cited in Beurdeley and Beurdeley 1984: 94; see Richards 1999: 202–3.

118. Wechsler 1985: 178–91; Rawson 1995.

119. Sung 1966: 300, 303.

120. Cited in Wills 1964: 85.

121. Ho and Bronson 2004: 238–39.

122. Nakamura 2005: 1017–22.

123. Cited in Yang 1996: 230.

124. Cited in Chai and Chai 1967: 2:464.

125. The portrait of Confucius, the source for all later Western illustrations, is reproduced in Jensen 1997: 82.

126. Falkenhausen 1993: 25–28, 118, 132, 202.

127. Cited in Schafer 1967: 155.

128. Bush and Shih 1985: 237.

129. Bushell 1910: 107–8; 161.

130. Evelyn 1955: 2:47.

131. Vandiver and Kingery 1984: 190, 216–18; Vainker 1991: 99–108; Pierson 1996: 21–23.

132. Bushell 1910: 46.

133. Cited in Vainker 1991: 99.

134. Cited in Mino and Tsiang 1986: 13.

135. Cited in Dillon 1976: 20.

136. Cited in Li 1998: 41.

137. Cited in He 1996: 142.

138. Changes in pottery production from the Song are examined in Dillon 1976: 20–26, 150–56.

139. Beamish 1995.

140. Cited in Nickles 2002: 234; see Teo 2002; Chen 1993.

141. Tite et al. 1984; Guo 1987: 8–9; Liu 1989: 72; Emerson, Chen, and Gates 2000: 21, 51–52; Kerr and Wood 2004: 228–39.

142. Sung 1966: 147.

CHAPTER 5. THE CREATION OF BLUE-AND-WHITE PORCELAIN

1. Bushell 1910: xxv. "Muslim blue" has been substituted for "Mohammedan cobalt blue" in this citation.

2. Israeli 1982: 86; Shangraw 1985: 40; Feng 1987: 59.

3. Ibn Battuta 1929: 236.

4. Hodges and Whitehouse 1983: 151, 156–57.

5. Chen and Lombard 1988; So 2000: 42–49.

6. Cited in So 2000: 55.

7. So 2000: 108–111, 114–15.

8. Cited in Hourani 1951: 64; see Simkin 1968: 81.

9. Rougelle 1991; Pierson 2002–3: 33; Sasaki 1994: 323.

10. Cited in Stern 1967: 10.

11. Wink 1991: 16–23; Aubin 1959; Bosworth 1968: 1–23.

12. Ashtor 1983: 270–300; Risso 1995: 20–22, 37–40; Lopez 1971.

13. Wink 1997: 43–78; Hall 2004: 234.

14. Wink 1991: 2–74, 101; Curtin 1984: 106–8; Bouchon 1988.

15. Cited in Wink 2004: 205.

16. Pires 1944: 82.

17. Hall 2004: 237; Hodgson 1974: 2:532–51; Risso 1995: 46–50; Wink 2004: 215–43.

18. Pires 1944: 182, 253.

19. Andaya and Ishii 1999: 182–83.

20. Phelan 1959: 8; Majul 1966; Bellwood and Omar 1980: 158.

21. Blair and Robertson 1915: 1:78; see 103.

22. Blair and Robertson 1915: 3:146.

23. Cited in Fernández-Armesto 1992: 18; see Headley 1995: 634.

24. Risso 1995: 71–72; Voll 1994: 219–25.

25. Ibn Battuta 1929: 78; see Pires 1944: 12–13.

26. Cited in Braudel 1982: 2: 558.

27. Ibn Battuta 1929: 292; 269–70, 288.

28. Cited in So 2000: 115.

29. Ibn Battuta 1929: 288; original emphasis. In this citation, "Abode" has been substituted for "land."

30. Ibn Battuta 1929: 283.

31. Cited in Temple 1986: 91.

32. Cited in Golombek 1996: 127–28.

33. Barry 1996: 13, 20.

34. Mason and Tite 1994.

35. Mason and Tite 1997; Caiger-Smith 1973: 45–46.

36. Cited in Chaudhuri 1985: 58.

37. Carswell 1999: 7; Mikami 1980–81; Sasaki 1994: 328.

38. Whitman 1978: 1: 25; see Barry 1996: 251, 253; Matson 1986.

39. Caiger-Smith 1985: 197–209.

40. Cited in Canby 1997: 112; see Caiger-Smith 1985: 59.

41. Cited in Blair and Bloom 1997: 113.

42. Cited in Melikan-Chirvani 1986: 103.

43. Ibn Battuta 1929: 169; see Spandounes 1997: 127; Bağci and Zeren Tanindi 2005: 448.

44. Ibn Battuta 1929: 90.

45. Caiger-Smith 1985: 36, 42, 66, 69.

46. Medley 1974: 34; Kingery and Vandiver 1986: 9, 53; Hodges 1972.

47. Weatherford 2004; Mote 1999: 425–36; Waley 1931: 93; Petrushevsky 1968: 483–91; Caiger-Smith 1973: 44.

48. Pegolotti 1936: 22; see Weatherford 2004: xviii.

49. Cited in Allsen 1997b: 2. Details on the Mongols and cross-cultural exchange are provided in Allsen 2001.

50. Cited in Adshead 2000: 136; see Smith 2000: 40–41; Weatherford 2004: 136.

51. Waley 1931: 93, 107.

52. Cited in Allsen 2001: 83.

53. Bira 1999: 241–43; Mote 1999: 690; Huang 1986; Rossabi 1981; Armijo-Hussein 1987: 197–215; Chaffee 2006: 415–16.

54. Arnold 1999; Amitai-Preiss 1999: 58–59; Schein 1979: 812.

55. Montgomery 1966: 56–57.

56. For the following, see Grabar 1968: 653–55; Yuka 2002; Barry 2000: 11, 550–52, 804–5; Gray 1972: 11.

57. Allsen 1997a explains the political and cultural significance of Central Asian textiles.

58. Cited in Jullien 1995: 155; see Shelach 2001.

59. Bailey 1996a: 59; Whitman 1978.

60. Olschki 1944; Komaroff 2002; Evelyn 1955: 1:79; Kubiski 2001. The Bedford Master's painting is in the Bibliothèque nationale de France, Paris.

61. Mack 2004; Arnold 1999: 112, 120–21; Hoeniger 1991. Martini's *Saint Louis* is in the Museo Nazionale di Capodimonte, Naples; his *Annuciation* is in the Uffizi, Florence.

62. Cited in Howard 2007: 63.

63. Pinsky 1994: 135 (Canto XVII, lines 11–15). Uccello's painting is in the National Gallery, London. Delacampagne and Delacampagne 2003 presents illustrations of dragons, many based on the Chinese version, that appear in late medieval and Renaissance art.

64. Pegolotti 1936: 138, 427–28; see Spuler 1985: 355–61.

65. Parry 1974: 69.

66. Lane 1961; Watson and Whitehead 1991: 17–19. Stripped of its mountings, the vase is now in the National Museum of Ireland, Dublin; a 1713 watercolor illustration of the vase, held by the Bibliothèque nationale de France, is reproduced in Kerr 2004: 46.

67. Harrison and Shariffuddin 1969: 34; Beamish 1995: 249.

68. Kim 1986.

69. Medley 1972: 2–3; Medley 1984; Addis 1980–81: 58–60; Joseph 1985; Garner 1970: 7–13.

70. Medley 1976: 195; Krahl 1985: 51.

71. Scott 2002: 10; Liu 1993.

72. Dillon 1976: 24.

73. Cited in Teo 2002: 246; see Chaffee 2006: 412–14. So 2000: 186–201 provides a case study of the rise in export ceramics from Quanzhou in the Yuan period.

74. Cited in Wang 2000: 18.

75. Brook 1998: 72; Metzger 1970.

76. Cited in Ng 1997: 243.

77. Allsen 1989; Endicott-West 1989.

78. Sung 1966: 155.

79. Bushell 1910: 60, 18, 69, 150; see He 1996: 211–12; Kerr and Wood 2004: 659 n. 161.

80. Wen et al. 2007 explains the chemical complexities of native and foreign cobalt oxide during the Ming period.

81. Shangraw 1985: 38; Wood 1999: 66.

82. Carswell 1985a discusses the David vases, which are in the Percival David Foundation of Chinese Art, London.

83. Dreyer 1982: 34–52.

84. Cited in Dillon 1976: 28.

85. Lau 1993; Rogers 1990: 64–65; Krahl 1991: 56; Zhang 1991.

86. Dreyer 1982: 70; Dardess 1970: 539.

87. Cited in Kerr and Wood 2004: 202. On Yongle's preference for white porcelain, see Stuart 1993a: 24; Vainker 1991: 184; Yü 1998: 913; Dreyer 1982: 1–11.

88. On the Baoen Temple and its influence, see Liu 1989: 54, 62, 73; Grigsby 1993; Conner 1979: 20.

89. Rogers 1990: 64; Cahill 1976: 129–30.

90. Krahl 2002–3: 29; Liu 1993, 1999b; Stuart 1995: 36.

91. Cited in Pierson and Barnes 2002: 24.

92. Shangraw 1985: 42; Krahl 1985; Bushell 1910: 142.

93. Medley 1976: 178–91; Medley 1990–91: 42; Macintosh 1973: 36–38; Stuart 1995.

94. Medley 1976: 180–82; Scheurleer 1974: 193–205.

95. Ricci 1953: 27.

96. For the following, see Stuart 1993a; Scott 1992; Pierson 2001: 14–18; Bartholomew 1994–97; He 1996: 253–58. The intricacies of Chinese pictorial punning are detailed in Ni 2003–4.

97. Chaffee 1985: 177–81.

98. Bai 2002: 58; Neill 1985: 206–7; Wan 2003.

99. Sugimara 1986: 18–44; J. Rawson 1984: 176, 191–92. The complex assimilation of Chinese motifs in Southwest Asia is highlighted in Whitman 1988.

100. Feng 1987; Carswell 1966.

101. Sayer 1959: 54–55; see Krahl 1986.

102. Medley 1972: 3–5; Gray 1940–41; Watson 1974: 90–91.

103. Melikan-Chirvani 1976; Schimmel 1984: 9–11, 25, 32–33, 110–14.

104. The complexity of geometric star-and-polygon *(girih)* patterns on mosque tiles is made clear in Lu and Steinhardt 2007.

105. Spuhler 1986: 712–14; Ford 1981: 118–25; Goody 1993: 112.

106. Flood 1991–92.

107. Hodges 1972: 82. Golombek 1988 argues for the dominance in Islamic societies of decoration modeled on use of textiles.

108. Jullien 1995: 75–149 spells out the philosophy and precepts of Chinese aesthetics.

109. See Ortiz 1999. *Dream Journey over Xiao Xiang* is in the Tokyo National Museum.

110. See Kerr 1999.

111. Bush and Shih 1985: 62.

112. Ricci 1953: 79.

113. Bush and Shih 1985: 149.

114. Bush and Shih 1985: 178.

115. Cited in Fong 1992: 446. Bush and Shih 1985: 145–50 provides ancient texts on the significance of pine trees in Chinese art.

116. For the same distinction between geometric regularity and natural movement in the styles, respectively, of Arabic and Chinese calligraphy, see Gaur 1994; Ledderose 1986.

CHAPTER 6. THE PRIMACY OF CHINESE PORCELAIN

1. Ricci 1953: 7–8. Ricci's mapmaking for the Ming court is recounted in Spence 1984: 64–65, 96–97, 148–149.

2. Mancall 1984: 10; Holcombe 2001: 5, 30–77, 211; Woodside 2006: 17–76. Holcombe 2001 analyzes Chinese influence on Korea, Japan, and Vietnam through the Tang period.

3. Wood and Kerr 1992: 39; Portal 1997: 100.

4. See Harrisson 1986: 4.

5. Cited in Soper 1942: 373.

6. Cited in J. Kim 1994: 1:113–14. Original emphasis.

7. The relationship between the Yaozhou kilns, southern China, and Korea is explained in Vainker 1991: 50, 112–15, 129.

8. For the following, see Best 1991: 147–48, 150, 157; Kim 1991; Nelson 1993: 233, 249; Frape 1998–2001: 52.

9. Cited in Nelson 1993: 249; see Chung 1998: 229.

10. Geographic considerations in Korean history are stressed in Nelson 1993: 12, 16, 220, 483.

11. Cited in C. Kim 1994: 109; see Itoh 1992: 50.

12. Palais 1995: 414–18; Woodside 2006: 28.

13. Umehara 1999: 22.

14. Cited in Kim and Kim 1966: 56. Original emphasis.

15. J. Kim 1994: 1:110, 116–17; McKillop 1992: 36, 38.

16. Cited in Chung 1998: 234.

17. Cited in McKillop 1992: 40.

18. Mino 1991; J. Kim 1994: 113–15; Palais 1995: 424; Rossabi 1981: 97–98.

19. Lancaster, Suh, and Yu 1996.

20. Wells 2000; Chung 2000; Lee 1999.

21. Turnbull 2002; Elisonas 1988: 264–90; Chase 2003: 186.

22. Elisonas 1991: 290–91.

23. Yun 1994: 126–27.

24. Cort 1986: 347; Day 1992–94: 56.

25. For the following, see Mellott 1990: 56; Hempel 1983: 12, 14, 22, 119; Jenyns 1971: 67; Epprecht 2007.

26. Cited in McCullough 1985: 156–57.

27. Holcombe 2001: 191; Ury 1988: 343–44.

28. Gang 1997b: 260–61; Souyri 2001: 150; Yamamura 1990a; Magalhães-Godinho 1969: 402.

29. Hall 1981; Collcutt 1988.

30. Tregear 1976: 819–20.

31. Varley 1990: 447–48, 453–54.

32. Cooper 2001: 317.

33. Murasaki 1987: 345.

34. Morris 1971: 34, 131, 168.

35. Murasaki 1987: 418, 929.

36. See Weigl 1980: 263–64.

37. Cooper 2001: 321.

38. Kawazoe 1990: 3: 409–19.

39. Bito 1991: 397–98.

40. Collcutt 1990: 584–86, 606, 644–45; Varley 1990: 489.

41. Jenyns 1971: 69, 74–75; McCullough 1988: 392–93.

42. Cited in Varley and Elison 1981: 193; see Keene 2003: 140–41; Plutschow 2003: 166–67.

43. Rousmaniere 1996; Tregear 1976.

44. Cited in Plutschow 2003: 27.

45. Tregear 1976: 820 n. 6; Jenyns 1971: 122.

46. Cited in Murai 1989: 16.

47. Kawai 2002: 36–39; Collcutt 1988: 18–19.

48. Varley 1977; Keene 2003: 103–04; Plutschow 2003: 42, 57; Murai 1989: 14.

49. Cited in Keene 2003: 5. Keene 2005 deals with the artistic interests of Yoshimasa, including his approach to the tea ceremony.

50. Cited in Varley 1997: 203.

51. Cited in Ludwig 1989: 73. Haga 1989: 221–22 points out the common appeal of monochrome painting and the Way of Tea.

52. Sheaf 1993: 176–77; Little 1982.

53. Cited in Haga 1989: 197.

54. Shimizu 1988: 350, 357; Plutschow 2003: 95, 111, 127–28.

55. Cited in Cooper 1989: 112–13.

56. Cooper 1965: 261; see Watsky 2004: 144.

57. Cooper 2001: 283.

58. Cooper 2001: 286.

59. Cooper 2001: 287.

60. Carletti 1964: 102.

61. Cooper 2001: 294; 283–94; see Varley and Elison 1981: 212.

62. Cited in Watsky 1995: 53; see Plutschow 2003: 66.

63. Cooper 2001: 291.

64. Plutschow 2003: 59, 66; Ludwig 1989: 77; Kumakura 1989: 35. The social world of merchant tea men is examined in Berry 1997: 259–79.

65. Watsky 1995; Varley and Elison 1981: 213–14; Plutschow 2003: 60–61, 82.

66. Cited in Ikegami 2005: 122. In this citation, "Way of Politics" has been substituted for "way of doing politics"; see Plutschow 2003: 83.

67. Berry 1982: 189.

68. For the following, see Bodart 1977: 55–58; Varley and Elison 1981: 215–20; Kumakura 1989: 35–37.

69. Bodart 1977: 52. A reconstruction of Hideyoshi's Golden Tea Room is in the Osaka Castle Museum, Osaka.

70. Kumakura 1989: 39–40; Ludwig 1989: 87.

71. Cited in Bodart 1977: 74; see Kumakura 1989: 47.

72. Cited in Berry 1997: 242.

73. Cited in Watsky 2004: 145.

74. Collcutt 1988: 28, 42–43; Takeuchi 2003.

75. Cited in Furukawa 2003: 100.

76. Varley 1989: 161–82.

77. Cited in Wilson 1989: 72; see 66–67.

78. Butler 2002: 249, 258, 262; Rousmaniere 2002: 150.
79. Ho 1994; Volker 1954: 50, 66, 172–73; Impey 1999; Little 1983: 1–15.
80. Volker 1954: 174.
81. Totman 1967: 64–65; J. Lee 1999: 8–9.
82. Impey 1984: 691–95; 2002: 13–18.
83. Espir 2001; Shono 1973: 9–26, 56–57.
84. For the following, see Diem 1997; Brown and Sjostrand 2000; Nguyen-Long 2001; Crick 1997–98; Flecker 2003.
85. Stevenson 1997a: 23–29.
86. See Murray 1987: 7–9, 14; Tana 2006.
87. Holcombe 2001: 145–64.
88. Cited in Schafer 1967: 159.
89. Guy 1997: 12–13; Wang 1998: 316–17.
90. Stevenson 1997b; 111–12.
91. Whitmore 1985: 89–112.
92. Diem 1999; Stevenson 1997a: 118. The impact of Chinese military technology on Vietnam is recounted in Sun 2003: 509–14.
93. Krahl 1997; Stevenson 1997a.
94. Guy 1989: 52–54; 1997b: 57; 1996–97: 44–45.
95. Dupoizat 2003; Richards 1995: 46; Guy 1997a.
96. Ho 1994: 39, 45–47; Nguyen-Long 1999; Junker 1999: 202.
97. Lombard 1990: 2:31–46 emphasizes the integration of East Asian seas from the fifteenth century.
98. Chandler 2000: 13–14; Smith 1999.
99. Mannikka 1996 deals with the nature and iconography of sacral kingship at Angkor Wat.
100. Cited by Hall 1985: 139.
101. Cited in Andaya and Ishii 1999: 203.
102. Rooney 1987: 5, 14, 24–27.
103. Richards 1995: 50; Groslier 1995.
104. Cited in Hall 1975: 330; see Higham 2001: 136, 140, 153.
105. Grave et al. 2000; Guy 1996–97; Cort 2000: 138–39.
106. Itoi 1989: 2–10; McBain 1979; Rooney 1989: 42.
107. Kasetsiri 1992; Wink 2004: 40–41; Daniels 1996: 413.
108. Viraphol 1977: 35–39.
109. Baker et al. 2005: 139.
110. Cited in Kasetsiri 1992: 75.
111. Schouten 1671: 148.
112. Schouten 1671: 126–27.
113. Chaudhuri 1990: 176; Cruysse 2002: 136.
114. Cited in Crick 1999: 52.
115. Cited in Jarry 1981: 64. Original emphasis.
116. Love 1994 provides an account of the Versailles reception and reaction to it; see Le Bonheur 1986.

117. Smith 1976a: 54.

118. Cited in Love 1994: 67.

119. Cited in Love 1994: 67.

120. Montesquieu 1989: 117 (bk. 7, chap. 6); see Hulliung 1976: 51–53; Pocock 1999: 111.

121. Cited in Andaya 1999: 37.

CHAPTER 7. THE TRIUMPH OF CHINESE PORCELAIN

1. Ricci 1953: 89; see also 7, 23, 58, 167, 314; see Spence 1984: 80.

2. Ricci 1953: 515.

3. Cited in Deng 1997a: 257.

4. Serruys 1975: 25; Wang 1953: 63–64, 95, 98; Zheng and Zheng 1980–83: 2:1196.

5. Dardess 1970; Wang 2000: 24.

6. Wills 1988: 225; see Deng 1997a: 255.

7. Cited in Wang Shixin 2000: 50.

8. Finlay 2008: 333–35.

9. Zheng and Zheng 1980–83: 2:854.

10. Chen Ching-kuang 1993; Stuart 1993a: 40.

11. Gould 2000: 93–98; see Needham 1971: 481–482. On the controversial question of the dimensions of the treasure ships, see Church 2005.

12. On Portuguese ships and personnel, see Boxer 1969: 52–53; Diffie and Winius 1977: 223.

13. On Zheng He, see Zheng and Zheng 1980–83: 1:1–38, 157; Ma 1970: 73; Aubin 2005: 58–66.

14. Dreyer 1982: 203, 212–13, 233.

15. Ma 1970: 174.

16. Cited in Chaudhuri 1989.

17. Zheng and Zheng 1980–83: 1:22–23, 34–42, 52–62.

18. Luo n.d.: 182–91; see Ptak 1986: 169–77; Finlay 1992.

19. Manguin 1986, 1991; Lombard-Salmon 1973; Graaf and Pigeaud 1984: 135–38.

20. Ma 1970: 73.

21. Cited in Souza 1986: 2; see Zheng and Zheng 1980–83: 2:1297–1353; Finlay 1991.

22. Carletti 1964: 153.

23. Cited in Braudel 1984: 3:198.

24. Finlay 2008: 336–38. A list of exports and imports is given in Ray 1993: 113–16; Chinese exports also are itemized in a sixteenth-century Spanish account in Blair and Richardson 1903–7: 16:180–83.

25. Lin 1985; Shangraw 1985: 39–40; Medley 1972: 4.

26. Deng 1995: 13.

27. Cited in Ng 1997: 245; see Ptak 2001; Reid 1992; 1996: 26:181.

28. Reid 1996: 26; Kong 1987.

29. Lombard 1990: 2:41–45; Reid 1996.

30. Ma 1970: 97; for the same evaluation, see Fei 1996: 44, 52, 55, 58, 71, 77, 97, 102, 103, 105.

31. Ptak 1986: 76–77; Luo n.d.: 92.

32. Lin 1985; Jörg and Flecker 2001: 34; Harrisson 1958; Brown 1997.

33. Moreland 1934: 108.

34. Twitchett and Stargardt 2002: 30–31, 59–60.

35. Blair and Roberston 1915: 34:188–89.

36. Cole 1912: 23.

37. Christie 1985; Chin 1977, 1977a; Shariffuddin and Omar 1978.

38. Gutman 2002; Harrison 1955, 1967; Adhyatman and Ridho 1984: 49–50.

39. Forrest 1969: 232, 105.

40. Adhyatman and Ridho 1984: 50–51; Rooney 1987: 19; Treolar 1972; Harrisson 1986: 27; Beauclair 1972.

41. Zheng and Zheng 1980–83: 2:1848; Junker 1994; Majul 1966a: 147–49; Sullivan 1960–62: 71–74; Tingley 1993: 46; Fox 1959.

42. Junker 1998: 299–300, 313; 1999: 202, 219.

43. Reid 1996: 21; Christie 1998: 355–56; Nieuwenhuis 1986; Kinney 2003; Adhyatman and Ridho 1984: 56.

44. Pigafetta 1968: 41, 94, 98.

45. Junker 1999: 183–220 focuses on Philippine chiefdoms for a case study of the cultural and political uses of porcelain in maritime Southeast Asia.

46. Helms 1993: 7, 29, 49, 91, 163–64; 1994 stresses the ideological perspective on objects made far away, especially the supernatural contexts in which indigenous peoples placed such objects.

47. Harrisson 1986: 23; Adhyatman and Ridho 1984: 53–54; Barbosa 1992: 77.

48. Cited in Volker 1954: 208.

49. Adhyatman 1990: 51; Adhyatman and Ridho 1984: 54–55; Harrisson 1986: 27; Kaboy and Moore 1967: 19–22.

50. Adhyatman and Ridho 1984: 53, 55; Kaboy and Moore 1967: 26; Adhyatman 1990: 41; Barbosa 1992: 76.

51. Adhyatman and Ridho 1984: 52; Chin 1977a; Kaboy and Moore 1967: 25–26.

52. Chin 1977a; Harrisson 1986: 28; Adhyatman and Ridho 1984: 52.

53. Solheim 1965: 261; Volker 1954: 22.

54. Zheng and Zheng 1980–83: 2:964; Ma 1970: 149.

55. Horton and Middleton 2000: 9, 16, 72–77, 101; Mathew 1963: 108; Wright 1993: 668–72; Masao and Mutoro 1981: 588–99, 600–601, 603–4.

56. Ibn Battuta 1929: 112.

57. Hirth and Rockhill 1966: 32; see Ibn Battuta 1929: 236, 260.

58. Abungu and Mutoro 1993: 702–03; Masao and Mutoro 1981: 614–15; Wright 1993: 669; see Heesterman 2003.

59. Ibn Battuta 1929: 110–11; see Middleton 2003: 516.

60. Kusimba 1999; Horton and Middleton 2000: 18, 91, 111–13; Middleton 2003: 516–19; Wright 1993: 667.

61. Abungu 1994; Garlake 1966: 36, 47, 62–63; Allen 1993: 248.

62. Middleton 1992: 39–40; Hodges and Whitehouse 1983: 142–43; Mathew 1964: 112, 123, 125–26; Chittick 1974: 1:306–9; Golombek 1996: 130–31.

63. Kirkman 1958; Abungu 1994: 155.

64. Diffie and Winius 1977: 340–47; Middleton 2003: 519–20.

65. Allen 1977: 63, 66.

66. Ma 1970: 137.

67. Cited in Ferrier 1986: 449.

68. Cited in Dias 2004: 88.

69. Foster 1899: 1: 134.

70. Chardin 1927: 267; see Mason and Golombek 2003.

71. Cited in Volker 1954: 66.

72. Guy 2004: 67.

73. Moreland 1931: 55.

74. Huyler 1996: 60–61; Miller 1985: 57, 132–40, 155–56; Douglas 1966: 41–43, 157.

75. Schimmel 2005: 290–91; Shakeb 1995: 19, 23.

76. Kramer 1997: 109–33; Greensted and Hardie 1982: 9–10.

77. Ma 1970: 165–66, 170–71.

78. See Hodgson 1974: 3:16–27, 50–51, 101.

79. Manz 1989: 2–16, 57; Grabar 2000: 56; Bailey 1996a: 12.

80. Lentz and Lowry 1989: 228–29; Clavijo 1928: 224; Golombek 1996a: 129.

81. Barry 1996: 115; Denny 1974: 77–78.

82. Clavijo 1928: 224, 269–70, 288–89; Golombek 1996a: 126–27; Grube 1993–94. Rossabi 1973 shows that the first voyage of Zheng He was not launched as a military response to the threat from Timur.

83. For the following, see Lentz and Lowry 1989: 63, 114; Bailey 1996: 11–12; Golombek 1996a: 126, 130; Sugimara 1986: 106–8; Gray 1972.

84. *Procession Scene* is in the Topkapi Library, Istanbul.

85. Sims 1992; Grabar 2000: 100. *A Royal Feast in the Garden* is in the Cleveland Museum of Art.

86. Robinson 1967: 174; Bailey 1996: 13; Golombek 1996a: 129–30; Crowe 1976: 301.

87. Bloom 2001: 14, 168–70, 186–88.

88. Eisenstein 1979: 1: 75–76.

89. Szuppe 2004.

90. Robinson 1967: 175; Hodgson 1974: 3: 31; Grabar 2000: 67–74; Pinto 1989: 170.

91. Bembo 2007: 299, 324, 335–36, 350.

92. Pope 1956: 3–18; Blair 2003: 132, 134; Savory 1980: 144–47.

93. Matthee 1999: 66–68; Crowe 1979–80.

94. The Safavid tile panel is in the Metropolitan Museum of Art, New York.

95. Michell and Zebrowski 1999: 2. Balabanlilar 2007 emphasizes the influence of Timurid institutions and traditions on Mughal rulers.

96. See Piotrovsky and Rogers 2004: 151; Balabanlilar 2007: 7–8. *The Rulers of the Mughal Dynasty* is in a private collection; *The Sons of Shah Jahan* is in the San Diego Museum of Art, California; see Schimmel 2005: 49.

97. Cited in Balabanlilar 2007: 4.

98. Babur 1995: 295; see 367–68. *Timur Handing the Imperial Crown to Babur* is in the Victoria and Albert Museum, London.

99. Pal 1983: 125; Beach 1987: 48, 60; Guy 2004: 63–64. *Akbar Receiving the Iranian Ambassador* is in the Victoria and Albert Museum.

100. Richards 1993: 34–49 discusses Akbar's political and religious projects. *Akbar Discourses with two Jesuit Priests* is in the Chester Beatty Library, Dublin, Ireland.

101. Spence 1984: 105; Bailey 2004: 151, 160; Lal 2005: 29–33, 148–49, 179–80, 216–17; see Schimmel 2005: 279, 290. *The Nativity* is in the Collection Frits Lugt, Institut Néerlandais, Paris.

102. Cited in Vassalo e Silva 2000: 128; see Gschwend 2000. *Jahangir Entertains Shah 'Abbas* is in the Freer Gallery of Art, Smithsonian Institution, Washington, D.C.; see Vassalo e Silva 2000: 128–30.

103. Cited in Irwin 1955: 113.

104. *Jahangir and Prince:* the painting is in the Freer Gallery of Art, Smithsonian Institution; see Findly 1993: 52, 235–37.

105. Cited in Rawson and Portal 1993: 274; see Gray 1964–66: 33. The painting of Mumtaz Mahal is reproduced in Preston and Preston 2007.

106. Bernier 1968: 359; see Lal 2005: 9–10, 35–36. Beach 1987: 90–110 considers Jahangir's preoccupation with connoisseurship and courtly culture.

107. Richards 1993: 100–103, 121–24, 171–75, 220–24, 290–97.

108. Begley 1979 analyzes the iconographic program of the Taj Mahal; on its construction, see Preston and Preston 2007: 178–84.

109. Balabanlilar 2007: 5–6; Henderson and Raby 1989.

110. Rogers and Ward 1988: 28–29; Bağci and Tanindi 2005: 266, 447; Blair and Bloom 1994: 97, 105; Raby and Yücel 1986: 1: 29–30.

111. Hess 1973; Scammell 1969: 408; Gibbs 1998–99: 33–35.

112. Bağci and Tanindi 2005: 265; Carnegy 1993: 33–34; Carswell 1998: 53–56, 96; Raby and Yücel 1986: 1:54; Mack 2002: 99; Necipoglu 1990: 157.

113. Raby and Yücel 1986: 1:29; Mack 2002: 99, 106–07.

114. Cited in Fontana 2007: 287.

115. Carswell 1993. *The Feast of the Gods* is in the National Gallery of Art, Washington, D.C.

116. Ibn Battuta 1929: 314; Blair and Bloom 1994: 124, 126–29; Kurtz 1975.

117. Mack 2002: 98; Glick 1995: 169; Blake 1980.

118. Wilson 1987; Spallanzani 1978: 55–69, 98–102; Caiger-Smith 1985: 134.

119. Piccolpasso 1980: 2:105; for the following, see Clunas 1997: 186; Thornton 1997: 25, 161.

120. For the following, see Bortolotto and Dumortier 1990; Pleguezuelo 2002; Gaimster 1999; Rapp 1975; Britton 1987: 18–21; Fourest 1980: 182–83; Israel 1989: 5–6.

CHAPTER 8. THE DECLINE AND FALL OF CHINESE PORCELAIN

1. Borschberg 2002.
2. On the value of the guilder, see Israel 1989: 67; Gelderbloom and Jonker 2004: 648, 650.
3. Cited in Borschberg 2002: 31.
4. Cited in Volker 1954: 23.

5. Scheurleer 1974: 46; Parmentier 1996: 121–22.

6. Giovio 1956–58: 1:171.

7. Boyajian 1993: 8, 48–49, 324; Padfield 2000: 17.

8. Cited in Boxer 1986a: 54.

9. Pires 1944: xxvii–xxx; Chang 1969: 48–53, 71, 83.

10. Higgins 1980: 43–46, 60–61; Elisonas 1991: 250–51.

11. Cullen 2003: 19–22.

12. Cited in Boxer 1968: 6. On Dutch triangular trade, see Israel 1989: 177–78.

13. Blair and Robertson 1915: 2:57; see Bjork 1998.

14. Von Glahn 1996: 140; Schurz 1939; Flynn and Giráldez 1995, 2002; Atwell 1982.

15. For the following, see Israel 1990: 199–200; 1989: 70–71, 73, 175, 184; Lach and Van Kley 1993: 3:23–27.

16. Gelderbloom and Jonker 2004.

17. Cited in Woodward 1974: 16.

18. Schama 1988: 174–88, 304–19, describes the superiority of Dutch diet, homes, and arts.

19. Cited in Jörg and Flecker 2001: 34.

20. Cited in Volker 1954: 61.

21. De Vries and Woude 1997: 309; Scholten 1995: 194–203; Caiger-Smith 1973: 129; Dam 1984: 27.

22. Leonard 1988; Mote 1999: 834–35, 849. On China developing a strong maritime focus if Zheng Chenggong and the imperial government had collaborated, see Wong 2002: 458–60.

23. Cited in Marks 1998: 167.

24. De Jonge 1964; Montias 1982: 300–306.

25. Baart 1990.

26. Hsu 1988; Medley 1990; Little 1983: 1–2, 28; Little 1996; Curtis 1995.

27. Blussé 1996; Klein 1989; Ho 1994: 45–49.

28. Pepys 1970–83: 5:49–50.

29. See Lenman 1990; Burke 1993; Wills 1993.

30. Martin 1993: 152–53.

31. See De Vries 2008: 20–37, 111–13.

32. For the following, see Paston-Williams 1993: 75–76; Dyer 1989: 173; Weatherill 1986: 90–91, 95–97; Gaimster and Nenk 1997; MacGregor 1989: 375; Johnson 1996: 182.

33. Gaba-van Dongen 2004 argues that genre paintings are fairly realistic depictions.

34. Velázquez's work is in a private collection; see Clarke 1996: 132–53. *Prayer before the Meal* is in the Philadelphia Museum of Art; see Barnes and Rose 2002: 132; *The King's Drink* is in a private collection; *Tavern of the Crescent Moon* is in the Museum of Fine Arts, Budapest; see Weller 2002: 121, 148.

35. Cited in Richards 1999: 92.

36. Martin 1989 explains the disadvantages of pewter.

37. Cited in Martin 1989: 20.

38. Cited in Gabay 1991: 100.

39. McCracken 1988: 35, 39; Clifford 1999.

40. Cited in MacGregor 1989: 374; see Winterbottom 2002: 25.
41. Cited in Clifford 1999: 151.
42. Cited in Casanovas 1999: 62.
43. Paston-Williams 1993: 250; Muldrew 2001: 88, 109; Gabay 1991: 101.
44. See Roche 1987: 141–43.
45. Ricci 1953: 64; 64–68.
46. Cooper 2001: 103.
47. Pinto 1989: 515.
48. *Banquet of the Officers* is in the Frans Halsmuseum, Haarlem, the Netherlands; see Westermann 1996: 147–48, 181. Adshead 1997: 38–41 contrasts the organization of the meal in China and Europe.
49. Petroski 1992: 8; Elias 1982: 57, 68–69, 126–29.
50. Cited in Shackel 1993: 145.
51. Frame 1983: 27; Montaigne 1993: 1230–31, 1114.
52. Cited in Camporesi 1993: 5.
53. Cited in Richards 1999: 153.
54. Smollett 1997: 44.
55. Elias 1982: 56–129 discusses the emergence of increasing personal control in the context of changes in dining etiquette.
56. Goldthwaite 1989: 21–22; Meister and Reber 1983: 102.
57. Bushell 1910: 152; see Stuart 1993a: 34; Scheurleer 1974: 112.
58. Howard and Ayers 1980: 18–19; Howard 1974: 68; Meister and Reber 1983: 101–2; J. Rawson 1984: 11.
59. Walcha 1981: 108, 121; Dauterman 1973.
60. Cited in Gabay 1991: 101.
61. Savill 1982; De Bellaigue 1986: 1–3, 8, 21.
62. For the following, see Pound 1994: 419; Jörg and Campen 1997: 306; Roche 2000: 126; Gaimster and Nenk 1997: 179; Martin 2004–5: 52; Strong 2002: 231–33.
63. See Elias 1982: 61–62, 82–83.
64. Cited in Roche 2000: 239–40.
65. Cited in Barker-Benfield 1992: 159.
66. Goldthwaite 1989: 17–18; Welch 2002; Snowdin and Howard 1996: 205.
67. Strong 2002: 149, 152–53; 166–67.
68. Cited in Volker 1954: 30; see Jörg 1993.
69. Cited in Jörg 1982: 102.
70. Mennell 1985: 62–101; Peterson 1994: 163–208; Camporesi 1993: 91; see Stols 2004.
71. Pitte 2002: 94; Robinson 1983: 96–97.
72. Montanari 1994: 59–67, 119; Halikowski Smith 2007.
73. Cited in Laurence 1991: 138.
74. Cited in Halikowski Smith 2007: 252.
75. Adshead 1997: 35.
76. Mennell 1985: 40–61; Visser 1991: 163–66, 161–62, 189–90; P. Rawson 1984: 200.
77. Belozerskaya 2002: 96–97, 139; Camporesi 1993: 68–69, 157; Strong 2002: 197–98; Witteveen 1991.

78. Cited in Stols 2004: 252.

79. The dinner for August II is described in Watanabe-O'Kelly 2002: 50; the sugar temple is described in Witteveen 1991: 213.

80. Cited in Witteveen 1991: 216.

81. Cited in Paston-Williams 1993: 250.

82. Cited in Chilton 2001: 179. On August II and his courtiers parading in costume, see Chilton 2001: 175–77.

83. Eriksen and de Bellaigue 1987: 108–9.

84. Cited in McLeod 1998: 46.

85. Cited in Brown 2004: 36.

86. Redford 1992: 3:70–71.

87. Cited in Walvin 1997: 13.

88. Cited in McCants 2008: 177. Voltaire is cited in Roche 2000: 245.

89. Cited in Hochstrasser 2007: 340 n. 178; see Scheurleer 1974: 113.

90. Cited in Schroder 2004: 4; see Foster 1934: 213; Allen 2002–3: 121.

91. Defoe 1991: 65; see Jörg and van Campen 1997: 135; Wilson 1972; Turpin 1999.

92. Cited in Emerson, Chen, and Gates 2000: 137.

93. Impey 1990: 59; Meister and Reber 1983: 18.

94. Hills 1998: 164; Kingery and Vandiver 1986: 13–14, 135–47.

95. Cited in Patterson 1979: 30. Gleeson 1998 provides an account of Böttger's career.

96. See Schönfeld 1998; Lo 1986: 250; Raffo 1982: 84; Walcha 1981: 43–45.

97. Cited in P. Rawson 1984: 64.

98. Howard 1994: 15.

99. Cited in Kerr 1989: 31.

100. Cited in Carter 1988: 440.

101. See Porter 2001: 21–33; Jarry 1981; Honour 1961.

102. Cited in Demel 1991: 49.

103. Du Halde 1738–41: 1:n.p.

104. Cited in Leites 1980: 74.

105. Cited in Ledderose 1991; see Cummins 1993: 200.

106. Vainker 1991: 200–202; Clunas 1997: 78; Pearce 1987–88: 26.

107. Curtis 1994; Kingery and Vandiver 1986; Scott 1987; Scott 1993a.

108. On changes in Western views of color in the early modern period, see Finlay 2007: 418–29.

109. Cited in Dursum 1984: 10; see Loehr 1962–63; Scott 1993a.

110. See Carnegy 1993: 31; Wilson 1987.

111. Lamb 1987: 281; see Fang 2003.

112. On the contrast between Western and Chinese pictorial space on porcelain, see Rawson 1984b: 181–86.

113. Cited in Batchelor 2003: 85; see Snowdin and Howard 1996: 206–7; Jackson-Stops 1993.

114. Watney 1973.

115. Conner 1979: 77–78, 82; Desmond 1995: 46–49.

116. Ceballos-Escalera 1968; Emerson, Chen, and Gates 2000: 144–45.

117. Cited in Ducret 1976: 102; see Honour 1967: 371; Robinson 1973: 25–87.
118. Cited in Emerson, Chen, and Gates 2000: 263.
119. Haskell 1987; Ramage 1990; Jenkins 1996; Coltman 2001.
120. Walcha 1981: 128–30; Raffo 1982; Gleeson 1998: 230–36.
121. Savill 1982; De Bellaigue 1986: 21; Eriksen and de Bellaigue 1987; Schwartz 1992.
122. See Kowaleski-Wallace 1995–96.
123. Dearing 1974: 1:292.
124. Thomas 1743: lines 44, 209–224.
125. Cited in Guest 2000: 82.
126. Cited in Richards 1999: 148.
127. Cited in Conner 1979: 59.
128. Cited in Nicholson 1994: 145.
129. Ingrassia 1998: 17–39; Brown 2001: 95–109; see Dickson 1967: 267.
130. Cited in Wahrman 2004: 204.
131. Hanway 1756: 244, 282.
132. Cited in Jacobson 1993: 123.
133. Cited in Porter 1999: 34.
134. Goldsmith 1966: 65.
135. Wilson 1995: 72–74, 192–93.
136. Smollett 1997: 58, 60.
137. Rousseau 1990: 6:340, 394.
138. The citation is from Jones 1998: 6; see Porter 2002: 398.
139. Cited in Raven 1992: 172–73; see Bindman 2003.
140. Hanway 1756: 273. Original emphasis.
141. Cited in Harris 2002: 123–24.
142. Howard 1994: 15; 2000: 34.
143. Voltaire 1947: 111.
144. Cited in Pocock 1998: 198.
145. Gwynn 1766: 91–92.
146. Cited in Boime 1987: 57.
147. Hill and Powell 1934–50: 3: 269, 339.
148. Cited in Jenkins and Sloan 1996: 183; see Robinson 1973; Mallet 1974.
149. Cited in Clark 1995: 56.
150. Cited in Roberts 1997: 173. Original emphasis.
151. Cited in McKendrick 1982: 114.
152. Farrer 1903–6: 3:356; original emphasis. See Uglow 2003; Ottomeyer 2004.
153. Uglow 2002: Vickers 1997: 270; Mankowitz 1952: 41; Raeburn 1989; Young 1995: 134, 147.
154. Cited in Coutts 2001: 185.
155. Cited in Vickers 1987: 132.
156. Reilly 1992: 31; Anderson 1986; Vainker 1991: 158.
157. Farrer 1903–6: 1:128.
158. Cited in Reilly 1994: 71. On Wedgwood's use of scientific experiment for creating ceramic formulas, see McKendrick 1973.

159. Cited in Jones 2002: 108.
160. Cited in Roberts 1997: 167.
161. McKendrick 1961.
162. Farrer 1903–6: 1:305; original emphasis. See Pollard 1965: 182–84.
163. Clark 1995: 51–53; Reilly 1992; McKendrick 1973: 308–10; Uglow 2002: 357–58.
164. Beaver 1964: 10–12; Uglow 2002: 110–17.
165. Cited in Gascoigne 1994: 232.
166. McKendrick 1960, 1970.
167. Cited in Gascoigne 1994: 214; see McKendrick 1982: 103.
168. Cited in Robinson 1973: 29.
169. Farrer 1903–6: 2:239.
170. Farrer 1903–6: 2:6; see McKendrick 1960: 427–30.
171. Cited in Dolan 2004: 325.
172. Cited in Röntgen 1984: 120; see Roberts 1997: 166.
173. Ceballos-Escalera 1968: 365; Smith 1968: 263.
174. Farrer 1903–6: 1:357; see Richards 1999: 42–43.
175. Farrer 1903–6: 1:301–2; original emphasis. See Dolan 2004: 326.
176. Israel 1995: 1009; Fourest 1980: 184–85; Hess 2002: 13.
177. Farrer 1903–6: 2: 79; see Lister and Lister 1984: 95.
178. Cited in Roberts 1990: 98.
179. Cited in Pearce 1987–88: 35.
180. Cited in Roberts 1997: 173.
181. Cited in Roberts 1997: 173. On failure of the mission to open China to trade, see Hevia 1995.
182. The Fitzwilliam Museum of Cambridge University possesses a malformed Jingdezhen copy of the Portland Vase.
183. See Wong 2002: 461–62.
184. Defoe 1974: 248–49. On the pivotal role of the opium trade, see Trocki 1999.
185. Cited in Adas 1989: 185.
186. Conner 1979: 161.
187. Leonard 1972, 1984.
188. On Chinese agriculture and textile production being caught in a "high level equilibrium trap," thereby making radical, comprehensive improvements practically unattainable, see Elvin 1973.

EPILOGUE

1. Mercier 1929: 20, 32.
2. J. Rawson 1984: 33–88; Riegel 1992: 187–206; Chêng 1969: 79–115.
3. Barry 1996: 115; Whitman 1978: 1:125.
4. For the following, see Adhyatman 1987: 4–6; Rooney 1984: 99; 1987: 11–13; Guy 1997a: 57; Portal 1997: 103; Jörg and van Campen 1997: 36, 45; Khoo 1991: 24–26; Thompson 1991: 70; Whitman 1978: 1:195–203.
5. Anderson 2004; Vikan 1991.

6. Eiland 1995–96: 114; Watson 1984: 145–47; Willetts 1958: 2:472–73.

7. Feng 1987: 67; Gray 1964–66: 30; Whitman 1978: 1: 90–92; Mosco 1999: 45.

8. Le Corbeiller 1988; Kletke 2004: 129.

9. Cited in Mudge 1985: 53; see Kuwayama 1997: 24–25, 78; 2000.

10. Irwin 1972; Snowdin and Howard 1996: 202; Neill 1985; Copland 1980: 33–35, 45; Esten, Wahlund, and Fischell 1987: 131.

REFERENCES

A	*Archaeometry*
AB	*Art Bulletin*
BM	*Burlington Magazine*
BOCSHK	*Bulletin of the Oriental Ceramic Society of Hong Kong*
ECS	*Eighteenth-Century Studies*
HJAS	*Harvard Journal of Asiatic Studies*
JESHO	*Journal of the Economic and Social History of the Orient*
JRASGBI	*Journal of the Royal Asiatic Society of Great Britain and Ireland*
JWH	*Journal of World History*
MA	*Magazine Antiques*
O	*Orientations*
OA	*Oriental Art*
SCC	*Science and Civilisation in China.* Cambridge: Cambridge University Press.
SMJ	*Sarawak Museum Journal*
TOCS	*Transactions of the Oriental Ceramics Society*
WA	*World Archaeology*

Abungo, George H. O. 1994. "Islam on the Kenyan Coast: An Overview of Kenyan Coastal Sacred Sites." In *Sacred Sites, Sacred Places,* edited by David L. Carmichael, Jane Hubert, Brian Reeves, and Audhild Schanche, 152–62. London: Routledge.

———, and Henry W. Mutoro. 1993. "Coast-Interior Settlements and Social Relations in the Kenya Coastal Hinterland." In Shaw et al. 1993.

Adas, Michael. 1989. *Machines as the Measure of Men: Science, Technology, and Ideologies of Western Dominance.* Ithaca, NY: Cornell University Press.

Addis, John. 1980–81. "Porcelain-Stone and Kaolin: Late Yuan Developments at Hutian." *TOCS* 45: 55–66.

Adhyatman, Sumarah. 1987. *Kendi: Traditional Drinking Water Containers.* Jakarta: Ceramic Society of Indonesia.

———. 1990. *Antique Ceramics found in Indonesia: Various Uses and Origins.* 2nd ed. Jakarta: Ceramic Society of Indonesia.

———, and Abu Ridho. 1984. *Tempayan di Indonesia/Martavans in Indonesia.* Exhib. cat. 2nd ed. Jakarta: Himpunan Keramik Indonesia.

Adshead, S. A.M. 1997. *Material Culture in Europe and China, 1400–1800: The Rise of Consumerism.* London: St. Martin's Press.

———. 2000. *China in World History.* 3rd ed. New York: St. Martin's Press.

———. 2004. *T'ang China: The Rise of the East in World History.* New York: Palgrave Macmillan.

Agnew, Jean-Christophe. 1993. "Consumer Culture in Historical Perspective." In *Consumption and the World of Goods,* edited by John Brewer and Roy Porter, 19–39. London: Routledge.

Albis, Antoine d', and T.M. Clarke. 1989. "Vincennes Porcelain for Orry de Fulvy." *Apollo* 129/6: 379–83.

Allen, James de Vere, ed. and trans. 1977. *Al-Inkishafi: Catechism of a Soul.* Nairobi: East African Literature Bureau.

———. 1993. *Swahili Origins: Swahili Culture and the Shungwaya Phenomenon.* London: James Currey.

Allen, Phillip. 2002–3. "The Uses of Oriental Porcelain in English Houses." *TOCS* 67: 121–28.

Allsen, Thomas T. 1989. "Mongolian Princes and Their Merchant Partners, 1200–1260." *Asia Major* 2/2: 83–126.

———. 1997a. *Commodity and Exchange in the Mongol Empire: A Cultural History of Islamic Textiles.* Cambridge: Cambridge University Press.

———. 1997b. "Ever Closer Encounters: The Appropriation of Culture and the Apportionment of Peoples in the Mongol Empire." *Journal of Early Modern History* 1/1: 2–23.

———. 2001. *Culture and Conquest in Mongol Eurasia.* Cambridge: Cambridge University Press.

Amiran, Ruth. 1965. "The Beginnings of Pottery-Making in the Near East." In Matson 1965a: 240–47.

Amitai-Preiss, Reuven. 1999. "Mongol Imperial Ideology and the Ilkhanid War against the Mamluks." In *The Mongol Empire and Its Legacy,* edited by David O. Morgan and Reuven Amitai-Preiss, 57–72. Leiden: Brill.

Andaya, Barbara Watson, and Yoneo Ishii. 1999. "Religious Development in Southeast Asia, c. 1500–1800." In *The Cambridge History of Southeast Asia,* vol. 2: *From c. 1500 to c. 1800,* edited by Nicholas Tarling, 164–227. Cambridge: Cambridge University Press.

Andaya, Leonard. 1999. "Interactions with the Outside World and Adaptation in Southeast Asian Society, 1500–1800." In *The Cambridge History of Southeast Asia,* vol. 2: *From c. 1500 to c. 1800,* edited by Nicholas Tarling, 1–57. Cambridge: Cambridge University Press.

Anderson, E.N. 1988. *The Food of China.* New Haven, CT: Yale University Press.

Anderson, William. 2004. "An Archaeology of Late Antique Pilgrim Flasks." *Anatolian Studies* 54: 79–93.

Anderson, William L. 1986. "Cherokee Clay, from Duché to Wedgwood: The Journal of Thomas Griffiths, 1767–1768." *North Carolina Historical Review* 63/4: 477–510.

Appadurai, Arjun, ed. 1986. *The Social Life of Things: Commodities in Cultural Perspective.* Cambridge: Cambridge University Press.

Arapova, Tatyana B., and Olga P. Deshpande. 1999. "Sino-Thai Ceramics in the Hermitage: Contacts between Thailand and Russia at the End of the Nineteenth Century." *Apollo* 150/11: 19–24.

Armijo-Hussein, Jacqueline Misty. 1997. "Sayyid 'Ajall Shams al-Din: A Muslim from Central Asia Serving the Mongols in China and Bringing 'Civilization' to Yunnan." Ph.D. dissertation, Harvard University.

Arnold, Lauren. 1999. *Princely Gifts and Papal Treasures: The Franciscan Mission to China and Its Influence on the Art of the West, 1250–1350.* San Francisco, CA: Desiderata Press.

Ashton, Leigh. 1933–34. "China and Egypt." *TOCS* 11: 62–72.

Ashtor, Eliyahu. 1983. *Levant Trade in the Later Middle Ages.* Princeton, NJ: Princeton University Press.

Atterbury, Paul, ed. 1982. *The History of Porcelain.* New York: William Morrow.

Atwell, William S. 1982. "International Bullion Flows and the Chinese Economy *circa* 1530–1650." *Past and Present* 95/2: 68–90:

———. 1998. "Ming China and the Emerging World Economy, c. 1470–1650." In *The Cambridge History of China,* vol. 8:2: *The Ming Dynasty, 1368–1644,* edited by Denis Twitchett and Frederick W. Mote, 376–416. Cambridge: Cambridge University Press.

Aubin, Françoise. 2005. "Zheng He, héros ethnique des Hui ou musulmans chinois." In *Zheng He: Images & Perceptions,* edited by Claudine Salmon and Roderich Ptak, 57–73. Wiesbaden: Harrassowitz Verlag.

Aubin, Jean. 1959. "La ruine de Siraf et les routes du Golfe Persique aux XIe et XIIe siècles." *Cahiers de l'Histoire Médiévale* 10–13: 187–99.

Baart, Jan M. 1990. "Ceramic Consumption and Supply in Early Modern Amsterdam: Local Production and Long-distance Trade." In *Work in Towns, 850–1850,* edited by P. J. Corfield and Derek Keene, 74–85. Leicester: Leicester University Press.

Babur, Zhahir al-Din Muhammad. 1995. *The Baburnama: Memoir of Babur, Prince and Emperor.* Translated and edited by Wheeler M. Thackston. Washington, DC: Smithsonian Institution Press.

Bacon, Francis. 1944. *Advancement of Learning and Novum Organum.* New York: Wiley.

Baer, Eva. 1998. *Islamic Ornament.* New York: New York University Press.

Bai, Qianshen. 1995. "The Irony of Copying the Elite: A Preliminary Study of the Poetry, Calligraphy and Painting on 17th-century Jingdezhen Porcelain." *OA* 41/3: 10–21.

———. 2002. "Inscriptions, Calligraphy, and Seals on Jingdezhen Porcelains from the Shunzi Era." In *Shunzi Porcelain: Treasures from an Unknown Reign, 1644–1661,* edited by Michael Butler, Julia B. Curtis, and Stephen Little, 56–67. Exhib. cat. Alexandria, VA: Arts Services International.

Bağci, Serpil, and Zeren Tanindi. 2005. "Art of the Ottoman Court." In *Turks: A Journey of a Thousand Years,* edited by David J. Roxburgh, 260–71. Exhib. cat. London: Royal Academy of the Arts.

Bailey, Gauvin Alexander. 1996. "The Stimulus: Chinese Porcelain Production and Trade with Iran." In Golombek 1996: 7–15.

———. 1996a. "The Response II: Transformation of Chinese Motifs." In Golombek 1996: 57–108.

———. 2004. "Between Religions: Christianity in a Muslim Empire." In Flores and Vassallo e Silva 2004: 148–61.

Baker, Chris, Dhiravat na Pombejra, Alfons van der Kraan, and David Wyatt, eds. 2005. *Van Vliet's Siam.* Chiang Mai, Thailand: Silkworm Books.

Balabanlilar, Lisa. 2007. "Lords of the Auspicious Conjunction: Turco-Mongol Imperial Identity on the Subcontinent." *JWH* 18/1: 1–39.

Barbosa, Atremio C. 1992. "Heirloom Jars in Philippine Rituals." In *A Thousand Years of Stoneware Jars in the Philippines,* edited by Cynthia O. Valdes, 70–94. Manila: Jars Collectors.

Barker-Benfield, G. J. 1992. *The Culture of Sensibility: Sex and Society in Eighteenth-Century Britain.* Chicago: University of Chicago Press.

Barley, Nigel. 1994. *Smashing Pots: Works of Clay from Africa.* Washington, DC: Smithsonian Institution Press.

Barnard, Noel. 1976. "The Role of the Potter in the Discovery and the Development of Metallurgy in Ancient China—With Particular Reference to Kiln and Furnace Construction." *BOCSHK* 2: 1–33.

———. 1983. "Further Evidence to Support the Hypothesis of Indigenous Origins of Metallurgy in Ancient China." In *The Origins of Chinese Civilization,* edited by David N. Keightley, 237–77. Berkeley: University of California Press.

Barnes, Donna R., and Peter G. Rose. 2002. *Matters of Taste: Food and Drink in Seventeenth-Century Dutch Art and Life.* Exhib. cat. Syracuse, NY: Albany Institute of History and Art and Syracuse University Press.

Barry, Michael. 1996. *Design and Color in Islamic Architecture: Eight Centuries of the Tile-Maker's Art.* New York: Vendome Press.

———, trans. 2000. *La Pavillon des Sept Princesses par Nezâmî de Gandjeh.* Paris: Gallimard.

Bartholomew, Terese Tse. 1994–97. "Chinese Botanical Motifs and Rebuses from the Collection of the Asian Art Museum of San Francisco." *BOCSHK* 11: 41–57.

Batchelor, Robert. 2003. "Concealing the Bounds: Imagining the British Nation through China." In *The Global Eighteenth Century,* edited by Felicity A. Nussbaum, 79–92. Baltimore, MD: Johns Hopkins University Press.

Bäumel, Jutta. 2004. "Augustus the Strong as the Sun God Apollo." In Siebel 2004: 27–29.

Beach, Milo Cleveland. 1987. *The New Cambridge History of India,* vol. 1:3: *Mughal and Rajput Painting.* Cambridge: Cambridge University Press.

Beamish, Jane. 1995. "The Significance of Yuan Blue and White Exported to South East Asia." In *South East Asia & China: Art, Interaction & Commerce,* edited by Rosemary Scott and John Guy, 225–51. London: Percival David Foundation of Chinese Art.

Beauclair, Inez de. 1972. "Jar Burial in Botel Tobago." *Asian Perspectives* 15/2: 167–76.

Beaver, Stanley H. 1964. "The Potteries: A Study in the Evolution of a Cultural Landscape." *Transactions of the Institute of British Geographers* 34/2: 1–31.

Beckwith, Christopher I. 1991. "The Impact of the Horse and Silk Trade on the Economics of T'ang China and of the Uighur Empire." *JESHO* 34/2: 183–98.

Beckwith, Martha. 1970. *Hawaiian Mythology.* Honolulu: University of Hawai'i Press.

Begley, Wayne E. 1979. "The Myth of the Taj Mahal and a New Theory of Its Symbolic Meaning." *AB* 61/1: 7–37.

Belevitch-Stankevitch, H. 1910. *Le Goût chinois en France au temps de Louis XIV.* Paris: Jean-Schemit.

Bellwood, Peter. 2005. *First Farmers: The Origins of Agricultural Societies.* Oxford: Blackwell.

———, and Matussin Omar. 1980. "Trade Patterns and Political Developments in Brunei and Adjacent Areas, A.D. 700–1500." *Brunei Museum Journal* 4/4: 155–79.

Bembo, Ambrosio. 2007. *The Travels and Journals of Ambrosio Bembo.* Translated by Clara Bargellini and edited by Anthony Welch. Berkeley: University of California Press.

Belozerskaya, Marina. 2002. *Rethinking the Renaissance: Burgundian Arts across Europe.* Cambridge: Cambridge University Press.

Bentley, Jerry H. 1993. *Old World Encounters: Cross-Cultural Contacts and Exchanges in Pre-Modern Times.* New York: Oxford University Press.

Bernier, François. 1968. *Travels in the Mogul Empire, A.D. 1656–1668.* Translated by Archibald Constable. New Delhi: S. Chand.

Berry, Mary Elizabeth. 1982. *Hideyoshi.* Cambridge, MA: Harvard University Press.

———. 1997. *The Culture of Civil War in Kyoto.* Berkeley: University of California Press.

Bertini, Giuseppe. 2000. "The Marriage of Alessandro Farnese and D. Maria of Portugal in 1565: Court Life in Lisbon and Parma." In *Cultural Links between Portugal and Italy in the Renaissance,* edited by K. J. P. Lowe, 45–74. Oxford: Oxford University Press.

Berzock, Kathleen Bickford. 2005. *For Hearth and Altar: African Ceramics from the Keith Achepohl Collection.* Exhib. cat. Chicago: Art Institute of Chicago.

Best, Jonathan W. 1991. "Tales of Three Paekche Monks Who Traveled Afar in Search of the Law." *HJAS* 51/1: 139–97.

Beurdeley, Cécile, and Michel Beurdeley. 1971. *Giuseppe Castiglione: A Jesuit Painter at the Court of the Chinese Emperors.* Translated by Michael Bullock. Rutland, VT: Charles E. Tuttle.

———. 1984. *A Connoisseur's Guide to Chinese Ceramics.* Translated by Katherine Watson. New York: Alpine Fine Arts Collection.

Beurdeley, Michel. 1962. *Porcelaine de la Compagnie des Indes.* Fribourg: Office du livre.

Bien, Gloria. 1986. "Chénier and China." *Studies on Voltaire and the Eighteenth Century* 242: 363–75.

Bindman, David. 2003. "How the French Became Frogs: English Caricature and a National Stereotype." *Apollo* 158/11: 15–20.

Bira, Sh. 1999. "Qubilai Qa'an and 'Phags-pa bLa-ma." In *The Mongol Empire and Its Legacy,* edited by David O. Morgan and Reuven Amitai-Preiss, 240–49. Leiden: Brill.

Bito Masahide. 1991. "Thought and Religion, 1550–1700." In *The Cambridge History of Japan,* vol. 4: *Early Modern Japan,* edited by John Whitney Hall, 373–424. Cambridge: Cambridge University Press.

Bjork, Katharine. 1998. "The Link That Kept the Philippines Spanish: Mexican Merchant Interests and the Manila Trade, 1571–1815." *JWH* 9/1: 25–50.

Blair, E. H., and J. A. Robertson, eds. 1915. *The Philippine Islands, 1493–1898.* 55 vols. Cleveland, OH: Arthur H. Clark.

Blair, Sheila S. 2003. "The Ardabil Carpets in Context." In *Society and Culture in the Early Modern Middle East,* edited by Andrew J. Newman, 125–43. Leiden: Brill.

———. 1997. *Islamic Art.* London: Phaidon.

———, and Jonathan M. Bloom. 1994. *The Art and Architecture of Islam, 1250–1800.* New Haven, CT: Yale University Press.

Blake, H. 1980. "The 'Bacini' of North Italy." In *La Céramique médiévale en Méditeranée occidentale: Xe–XVe siècle,* 93–111. Paris: Centre national de la recherche scientifique.

Bloom, Jonathan M. 2001. *Paper before Print: The History and Impact of Paper in the Islamic World.* New Haven, CT: Yale University Press.

Blussé, Leonard. 1996. "No Boats to China: The Dutch East India Company and the Changing Pattern of the China Sea Trade, 1635–1690." *Modern Asian Studies* 30/1: 51–76.

Bodart, Beatrice M. 1977. "Tea and Counsel: The Political Role of Sen Rikyu." *Monumenta Nipponica* 32/1: 49–74.

Boime, Albert. 1987. *Art in the Age of Revolution, 1750–1800.* Chicago: University of Chicago Press.

Bond, Donald F., ed. 1965. *The Spectator.* 5 vols. Oxford: Clarendon Press.

Bordoy, Guillermo Rosselló. 1992. "The Ceramics of al-Andalus." In *Al-Andalus: The Art of Islamic Spain,* edited by Jerrilynn D. Dodds, 191–205. New York: Metropolitan Museum of Art.

Borschberg, Peter. 2002. "The Seizure of the *Sta. Catarina* Revisited: The Portuguese Empire in Asia, VOC Politics and the Origins of the Dutch-Johor Alliance." *Journal of Southeast Asian Studies* 33/1: 31–62.

Bortolotto, Angelica Alverà, and and Claire Dumortier. 1990. "Les majoliques anversoises 'à la façon de venise' de la première moitié du XVIe siècle." *Revue belge d'archéologie et d'histoire de l'art* 59: 55–74.

Bosworth, C. E. 1968. "The Political and Dynastic History of the Iranian World (A.D. 1000–1217)." In Boyle 1968: 1–202.

Bottéro, Jean. 2001. *Religion in Ancient Mesopotamia.* Translated by Teresa Lavender Fagan. Chicago: University of Chicago Press.

Bouchon, Geneviève. 1988. "Un microcosme: Calicut au 16e siècle." In Lombard and Aubin 1988: 49–57.

Boxer, C. R., ed. 1953. *South China in the Sixteenth Century.* London: Hakluyt Society.

———. 1965. *The Dutch Seaborne Empire, 1600–1800.* London: Penguin Books.

———. 1968. *Fidalgos in the Far East, 1550–1770.* London: Oxford University Press.

———. 1969. *The Portuguese Seaborne Empire, 1415–1825.* London: Hutchinson.

———, ed. 1986. *The Tragic History of the Sea, 1589–1622.* London: Hakluyt Society.

———. 1986a. "Carreira and Cabotagen: Some Aspects of Portuguese Trade in the Indian Ocean and the China Sea, 1500–1650." *Renaissance and Modern Studies* 30: 45–59.

Boyajian, James C. 1993. *Portuguese Trade in Asia under the Habsburgs, 1580–1640.* Baltimore, MD: Johns Hopkins University Press, 1993.

Boyle, J. A., ed. 1968. *The Cambridge History of Iran,* vol. 5: *The Saljuq and Mongol Periods.* Cambridge: Cambridge University Press.

Brady, Thomas A., Jr. 1991. "The Rise of Merchant Empires, 1400–1700: A European Counterpoint." In *The Political Economy of Merchant Empires,* edited by James D. Tracy, 117–60. Cambridge: Cambridge University Press.

Braudel, Fernand. 1973. *The Mediterranean and the Mediterranean World in the Age of Philip II.* 2 vols. Translated by Siân Reynolds. New York: Harper and Row.

———. 1981–84. *Civilization and Capitalism, 15th—18th Centuries.* 3 vols. Vol. 1 (1981): *The Structures of Everyday Life.* Vol. 2 (1982): *The Wheels of Commerce.* Vol. 3 (1984): *The Perspective of the World.* Translated by Siân Reynolds. New York: Harper and Row.

Britton, Frank. 1987. *London Delftware.* London: Jonathan Horne.

Brook, Timothy. 1981. "The Merchant Network in Sixteenth-Century Ming China: A Discussion and Translation of Chang Han's 'On Merchants.'" *JESHO* 24/2: 165–214.

———. 1998. *The Confusions of Pleasure: Commerce and Culture in Ming China.* Berkeley: University of California Press.

Brown, Jonathan. 1995. *Kings & Connoisseurs: Collecting Art in Seventeenth-Century Europe.* Princeton, NJ: Princeton University Press.

Brown, Laura. 2001. *Fables of Modernity: Literature and Culture in the English Eighteenth Century.* Ithaca, NY: Cornell University Press.

Brown, Peter. 2004. "Time for Tea." *Apollo* 159/4: 36–44.

Brown, Roxanna M. 1977. *The Ceramics of Southeast Asia: Their Dating and Identification.* Kuala Lumpur: Oxford University Press.

———. 1997. "Xuande-Marked Trade Wares and the 'Ming Gap.'" *OA* 43/2: 2–6.

———, and Sten Sjostrand. 2000. *The Turiang: A Fourteenth-Century Wreck in Southeast Asian Waters.* Pasadena, CA: Pacific Asia Museum.

Browne, Janet. 1995. *Charles Darwin: A Biography.* Princeton, NJ: Princeton University Press.

Browne, Thomas. 1964. *The Works of Sir Thomas Browne,* vol. 2: *Pseudodoxia Epidemica.* Edited by Geoffrey Keynes. Chicago: University of Chicago Press.

Burke, Peter. 1993. "*Res et Verba:* Conspicuous Consumption in the Early Modern World." In *Consumption and the World of Goods,* edited by John Brewer and Roy Porter, 148–61. London: Routledge.

Burn, Lucilla. 1997. "Sir William Hamilton and the Greekness of Greek Vases." *Journal of the History of Collections* 9/2: 241–52.

Burnett, Katherine P. 2000. "A Discourse of Originality in Late Ming Chinese Painting Criticism." *Art History* 23/4: 522–58.

Burton, Robert. 1932. *The Anatomy of Melancholy (1621).* 3 vols. in 1. Edited by Holbrook Jackson. Reprint, New York: Vintage Books.

Burton, William. 1906. *Porcelain: A Sketch of Its Nature and Manufacture.* London: Cassell.

Bush, Susan, and Hsio-yen Shih, eds. 1985. *Early Chinese Texts on Painting.* Cambridge, MA: Harvard University Press.

Bushell, Stephen W. 1910. *Description of Chinese Pottery and Porcelain: Being a Translation of the T'ao Shuo.* Oxford: Clarendon Press.

Butler, Lee. 2002. *Emperor and Aristocracy in Japan, 1467–1680: Resilience and Renewal.* Cambridge, MA: Harvard University Asia Center.

Cahill, James. 1976. *Hills beyond a River: Chinese Painting of the Yüan Dynasty, 1279–1368.* New York: Weatherhill.

——. 1982. *The Compelling Image: Nature and Style in Seventeenth-Century Chinese Painting.* Cambridge, MA: Harvard University Press.

Caiger-Smith, Alan. 1973. *Tin-Glaze Pottery in Europe and the Islamic World: The Tradition of 1000 Years in Maiolica, Faience and Delftware.* London: Faber and Faber.

——. 1985. *Lustre Pottery: Technique, Tradition and Innovation in Islam and the Western World.* London: Faber and Faber.

——. 1993–94. "Continuity and Innovation in Ceramics." *TOCS* 58: 51–61.

Camporesi, Piero. 1993. *The Magic Harvest: Food, Folklore and Society.* Translated by Joan Krakover Hall. Cambridge, MA: Polity Press.

Canby, Sheila B. 1997. "Islamic Lustreware." In Freestone and Gaimster 1997: 110–15.

Cao Xueqin. 1979. *The Story of the Stone,* vol 1: *The Golden Days.* Translated by David Hawkes. Bloomington: Indiana University Press.

Carboni, Stefano, ed. 2007. *Venice and the Islamic World, 828–1797.* Exhib. cat. New York and New Haven: Metropolitan Museum of Art and Yale University Press.

Carletti, Francesco. 1964. *My Voyage around the World.* Translated by Herbert Weinstock. New York: Pantheon Books.

Carnegy, Daphne. 1993. *Tin-Glazed Earthenware from Maiolica, Faience and Delftware to the Contemporary.* London: Adam and Charles Black.

Carswell, John. 1966. "An Early Ming Porcelain Stand from Damascus." *OA* 12/3: 176–82.

——, ed. 1985. *Blue and White: Chinese Porcelain and Its Impact on the Western World.* Exhib. cat. Chicago: University of Chicago and the David and Alfred Smart Gallery.

——. 1985a. "Blue-and-White Porcelain in China." In Carswell 1985: 13–35.

——. 1985b. "Blue-and-White in China, Asia and the Islamic World." In Carswell 1985: 27–35.

——. 1993. "*The Feast of the Gods:* The Porcelain Trade between China, Istanbul and Venice." *Journal of Asian Affairs* 24/2: 180–85.

——. 1998. *Iznik Pottery.* London: British Museum Press.

——. 1999. "China and the Middle East." *OA* 45/1: 2–14.

Carter, H. B. 1988. *Sir Joseph Banks, 1743–1820.* London: British Museum.

Carvalho, Pedro Moura. 2000. "Macao as a Source for Works of Art of Far Eastern Origin." *OA* 46/3: 13–21.

Casanovas, María Antonia. 2003. "Ceramics in Domestic Life in Spain." In Gavin et al. 2003: 49–75.

Cassidy-Geiger, Maureen. 2003. "Fabled Beasts: Augustus the Strong's Meissen Menagerie." *MA* 164/4: 152–61.

Casson, Lionel, ed. and trans. 1989. *The Periplus Maris Erythraei.* Princeton, NJ: Princeton University Press.

Castleden, Rodney. 1990. *Minoans: Life in Bronze Age Crete.* New York: Routledge.

Cauvin, Jacques. 2000. *The Birth of the Gods and the Origins of Agriculture.* Translated by Trevor Watkins. Cambridge: Cambridge University Press.

Ceballos-Escalera, I. 1968. "Porcelain from a Garden Factory." *Apollo* 87/5: 363–69.

Chaffee, John W. 1985. *The Thorny Gates of Learning in Sung China: A Social History of Examinations.* Cambridge: Cambridge University Press.

——. 2001. "The Impact of the Song Imperial Clan on the Overseas Trade of Quanzhou." In Schottenhammer 2001: 13–46.

———. 2006. "Diasporic Identities in the Historical Development of the Maritime Muslim Communities of Song-Yuan China." *JESHO* 49/4: 395–420.

Chai, Ch'u, and Winberg Chai, eds. 1967. *Li Chi: Book of Rites: An Encyclopaedia of Ancient Ceremonial Usages, Religious Creeds, and Social Institutions.* Translated by James Legge. 2 vols. New Hyde Park, NY: University Books.

Chandler, David P. 2000. *A History of Cambodia.* Boulder, CO: Westview Press.

Chang, Kwang-chih. 1986. *The Archaeology of Ancient China.* 4th ed. New Haven, CT: Yale University Press.

———. 1999. "China on the Eve of the Historical Period." In Loewe and Shaughnessy 1999: 37–73.

Chang, T'ien-tsê. 1969. *Sino-Portuguese Trade from 1514 to 1644.* Leiden: Brill.

Chardin, John. 1927. *Sir John Chardin's Travels in Persia.* London: Argonaut Press

Chaudhuri, K. N. 1985. *Trade and Civilization in the Indian Ocean: An Economic History from the Rise of Islam to 1750.* Cambridge: Cambridge University Press.

———. 1989. "A Note on Ibn Taghri Birdi's Description of Chinese Ships in Aden and Jedda." *JRASGBI* 1: 113.

———. 1990. *Asia before Europe: Economy and Civilisation of the Indian Ocean from the Rise of Islam.* Cambridge: Cambridge University Press.

Chen Baiquan. 1993. "The Development of Song Dynasty Qingbai Wares from Jingdezhen." In Scott 1993: 13–32.

Chen Ching-kuang. 1993. "Sea Creatures on Ming Imperial Porcelains." In Scott 1993: 101–22.

Chen Dasheng and Denys Lombard. 1988. "Le rôle des étrangers dans le commerce maritime de Quanzhou ('Zaitun') aux 13e et 14e siècles." In Lombard and Aubin 1988: 23–29.

Chen Yaocheng, Guo Yanyi, and Chen Hong. 1993–94. "Sources of Cobalt Pigment Used on Yuan Blue and White Porcelain Wares." *OA* 40/1: 14–19.

Chen Yunru. 2007. "At the Emperor's Invitation: Literary Gathering and the Emergence of Imperial Garden Space in Northern Song Painting." *O* 38/1: 56–61.

Chêng, Tê-k'un. 1969. *Archaeology in Sarawak.* Cambridge: W. Heffer and Sons.

———. 1973. "The Beginning of Chinese Civilization." *Antiquity* 47/187: 197–209.

———. 1983. *Studies in Chinese Art.* Hong Kong: Chinese University Press.

Cherniack, Susan. 1994. "Book Culture and Textual Transmission in Sung China." *HJAS* 54/1: 5–125.

Chilton, Meredith. 2001. *Harlequin Unmasked: The Commedia dell'Arte and Porcelain Sculpture.* New Haven, CT: George R. Gardiner Museum of Ceramic Art with Yale University Press.

Chin, Lucas. 1977. "Trade Pottery Discovered in Sarawak from 1948 to 1976." *SMJ* 25/46: 1–7.

———. 1977a. "Impact of Trade Ceramic Objects on Some Aspects of Local Culture." *SMJ* 25/46: 67–69.

———. 1980. "Chinese Ceramics in Sarawak." *O* 11/8: 30–36.

Chittick, H. N. 1974. *Kilwa: An Islamic Trading City on the East African Coast.* 2 vols. Nairobi: British Institute in Eastern Africa.

Chou Kung-shin. 1999–2000. "French Jesuits and Chinese Lacquer in the Late 17th Century." *OA* 45/4: 33–37.

Christian, David. 2004. *Maps of Time: An Introduction to Big History.* Berkeley: University of California Press.

Christie, Jan Wisseman. 1985. "On Po-no: The Santubong Sites of Sarawak." *SMJ* 24/55: 77–89.

———. 1998. "Javanese Markets and the Asian Sea Trade Boom of the Tenth to Thirteenth Centuries A.D." *JESHO* 41/3: 344–381.

Chung Yang-mo. 1998. *The Art of the Korean Pottery: From the Neolithic Period to the Choson Dynasty.* Edited by Judith G. Smith. Exhib. cat. New York: Metropolitan Museum of Art.

———. 2000. "Korean Ceramics of the Choson Dynasty." In *Earth, Spirit, Fire: Korean Masterpieces of the Choson Dynasty (1392–1910),* edited by Claire Roberts andMichael Brand, 20–25. Exhib. cat. Sydney: Powerhouse Publishing.

Church, Sally K. 2005. "Zheng He: An Investigation into the Plausibility of 450-ft. Treasure Ships." *Monumenta Serica* 53: 1–43.

Clark, Garth. 1995. *The Potter's Art.* London: Phaidon Press.

Clark, Hugh R. 1991. *Community, Trade, and Networks: Southern Fujian Province from the Third to the Thirteenth Century.* Cambridge: Cambridge University Press.

———. 1991a. "The Politics of Trade and the Establishment of the Quanzhou Trade Superintendency." In *China and the Maritime Silk Route: UNESCO Quanzhou International Seminar on China and the Maritime Routes of the Silk Roads,* edited by Quanzhou International Seminar on China and the Maritime Routes of the Silk Roads Organization, 376–94. Fuzhou: Fujian People's Publishing House.

———. 1995. "Muslims and Hindus in the Culture and Morphology of Quanzhou from the Tenth to the Thirteenth Century." *JWH* 6/1: 49–74.

Clarke, Michael. 1996. *Velázquez in Seville.* Exhib. cat. Edinburgh: Trustees of the National Gallery of Scotland.

Clavijo, Ruy Gonzalez de. 1928. *Clavejo: Embassy to Tamerlane, 1403–1406.* Translated by Guy Le Strange. London: Routledge.

Clifford, Helen. 1999. "A Commerce with Things: The Value of Precious Metalwork in Early Modern England." In *Consumers and Luxury: Consumer Culture in Europe, 1650–1850,* edited by Maxine Berg and Helen Clifford, 147–68. Manchester: Manchester University Press.

Clunas, Craig, ed. 1987. *Chinese Export Art and Design.* Exhib. cat. London: Victoria and Albert Museum.

———. 1991. *Superfluous Things: Material Culture and Social Status in Early Modern China.* Urbana: University of Illinois Press.

———. 1991a. "The Art of Social Climbing in Sixteenth-Century China." *BM* 133/5: 368–75.

———. 1992–94. "The Cost of Ceramics and the Cost of Collecting Ceramics in the Ming Period." *BOCSHK* 10: 47–53.

———. 1997. *Art in China.* Oxford: Oxford University Press.

Cole, Charles Woolsey. 1943. *French Mercantilism, 1683–1700.* New York: Columbia University Press.

Cole, Fay-Cooper. 1912. *Chinese Pottery in the Philippines.* Anthropological Series, vol. 12. Chicago: Field Museum of Natural History.

Collcutt, Martin. 1988. "Daimyo and daimyo culture." In *Japan: The Shaping of Daimyo Cul-*

ture, 1185–1868, edited by Yoshiaki Shimizu, 1–46. Exhib. cat. Washington, DC: National Gallery of Art.
———. 1990. "Zen and the *Gozan.*" In Yamamura 1990: 583–652.
Collett, D. P. 1993. "Metaphors and Representations Associated with Precolonial Iron-Smelting in Eastern and Southern Africa." In Shaw et al. 1993: 499–511.
Coltman, V. 2001. "Sir William Hamilton's Vase Publications (1766–1776): A Case Study in Reproduction and Dissemination of Antiquity." *Journal of Design History* 14/1: 1–16.
Conner, Patrick. 1979. *Oriental Architecture in the West.* London: Thames and Hudson.
Cook, Daniel J., and Henry Rosemount Jr. 1981. "The Pre-established Harmony between Leibniz and Chinese Thought." *Journal of the History of Ideas* 42/2: 253–68.
Coomaraswamy, A. K., and F. S. Kershaw. 1928–29. "A Chinese Buddhist Water Vessel and Its Indian Prototype." *Artibus Asiae* 3/2–3: 122–41.
Cooper, Michael, ed. 1965. *They Came to Japan: An Anthology of European Reports on Japan, 1543–1640.* Berkeley: University of California Press.
———. 1989. "The Early Europeans and Tea." In Varley and Kumakura 1989: 101–33.
———, ed. and trans. 2001. *João Rodrigues's Account of Sixteenth-Century Japan.* London: Hakluyt Society.
Copland, Robert. 1980. *Spode's Willow Pattern and Other Designs after the Chinese.* London: Rizzoli, 1980.
Cort, Louise Allison. 1986. "Korean Influences in Japanese Ceramics: The Impact of the Teabowl Wars of 1592–1598." In Kingery 1986: 331–62.
———. 2000. "Khmer Stoneware Ceramics." In *Asian Traditions in Clay: The Hague Gifts,* edited by Louise Allison Cort, Massumeh Farhad, and Ann C. Gunter, 91–145. Exhib. cat. Washington, DC: Smithsonian Institution.
Coutts, Howard. 2001. *The Art of Ceramics: European Ceramic Design, 1500–1830.* New Haven, CT: Yale University Press.
Cowen, Pamela. 1993. "The Trianon de Porcelaine at Versailles." *MA* 143/1: 136–43.
Crick, Monique. 1997–98. "Hongzhi (1488–1505) and Zhengde (1506–21) Ceramics Found on Shipwrecks off the Coast of the Philippines." *TOCS* 62: 69–81.
———. 1999. "Trade with France." *OA* 45/1: 52–59.
Crowe, Yolanda. 1976. "The Islamic Potter and China." *Apollo* 103/4: 298–301.
———. 1979–80. "Aspects of Persian Blue and White and China in the Seventeenth Century." *TOCS* 44: 15–30.
Cruysse, Dirk van der. 2002. *Le noble désir de courir le monde: Voyager en Asie au XVIIe siècle.* Paris: Fayard.
Cullen, L. M. 2003. *A History of Japan, 1582–1941: Internal and External Worlds.* Cambridge: Cambridge University Press.
Cummins, J. S. 1993. *A Question of Rites: Friar Domingo Navarette and the Jesuits in China.* Aldershot: Ashgate.
Curtin, Philip D. 1984. *Cross-Cultural Trade in World History.* Cambridge: Cambridge University Press.
Curtis, Emily Byrne. 1994. "Vitreous Art: Colour Materials for Qing Dynasty Enamels." *Arts of Asia* 24/6: 96–100

Curtis, Julia B. 1993. "Markets, Motifs and Seventeenth-Century Porcelains from Jingdezhen." In Scott 1993: 123–50.

———. 1995. "Chinese Porcelains of the Seventeenth Century: Landscapes, Scholars' Motifs and Narratives." *O* 26/4: 58–62.

———. 1996–97. "If Pots Could Speak: The Old Art History and the New." *TOCS* 61: 101–19.

———. 1998. "'Glorious Dynasty Transmitting Antiquity': Chinese Porcelain Decoration and Politics, 1670–1700." *OA* 44/2: 11–15.

Cutler, Alan. 2003. *The Seashell on the Mountaintop.* New York: Dutton.

Dam, Jan Daniel van. 1984. "A Survey of Dutch Tiles." In *Dutch Tiles in the Philadelphia Museum of Art,* edited by Ella Schapp, 25–39. Philadelphia: University of Pennsylvania Press.

Dames, Mansel Longworth, ed. and trans. 1921. *The Book of Duarte Barbosa.* 2 vols. London: Hakluyt Society.

Dance, Peter. 1986. *A History of Shell Collecting.* Leiden: Brill.

Daniels, Christian. 1996. *SCC,* vol. 6: *Biology and Biological Technology: Agro- Industries: Sugarcane Technology,* edited by Joseph Needham.

Dardess, John W. 1970. "The Transformation of Messianic Revolt and the Founding of the Ming Dynasty." *Journal of Asian Studies* 29/3: 539–58.

Darwin, Charles. 1989. *Voyage of the Beagle.* Edited by Janet Browne and Michael Neve. London: Penguin Books.

Daryaee, Touraj. 2003. "The Persian Gulf Trade in Late Antiquity." *JWH* 14/1: 1–16.

Dauterman, Carl Christian. 1973. "The Mastery of Meissen." *Apollo* 98/8: 92–99.

David, Nicholas, Judy Sterner, and Kodzo Gavua. 1988. "Why Pots Are Decorated." *Current Anthropology* 29/3: 365–89.

Davies, D. W. 1961. *A Primer of Dutch Seventeenth Century Overseas Trade.* The Hague: Martinus Nijhoff.

Day, Judith H. 1992–94. "Influence and Innovation: Transplanted Korean Potters and Japanese Ceramics." *BOCSHK* 10: 54–58.

Dean, Dennis R. 1992. *James Hutton and the Origins of Geology.* Ithaca, NY: Cornell University Press.

Dearing, Vinton A. 1974. *John Gay: Poetry and Prose.* 2 vols. Oxford: Clarendon Press.

De Bellaigue, Geoffrey. 1986. *The Louis XVI Service.* Cambridge: Cambridge University Press.

Defoe, Daniel. 1974. *The Farther Adventures of Robinson Crusoe, Being the Second and Last Part of His Life.* Edited by George A. Aitken. London: J. M. Dent.

——— 1977. *The Life and Strange Adventures of Robinson Crusoe, of York, Mariner.* London: Everyman's Library.

———. 1979. "A Brief Deduction of the Original *[sic],* Progress, and Immense Greatness of the British Woolen Manufacture." In *The Versatile Defoe: An Anthology of Uncollected Writings by Daniel Defoe,* edited by Laura Ann Curtis, 171–207. Totawa, NJ: Rowman and Littlefield.

———. 1991. *A Tour through the Whole Island of Great Britain.* Edited by P. N. Furbank and W. R. Owens. New Haven, CT: Yale University Press.

Dehergne, Joseph. 1973. *Répertoire des Jésuites de Chine de 1552 à 1800.* Rome: Institutum Historicum Societatis Iesu.

de Jonge, C. H. 1964. "Delftware at Vught: The Fentener van Vlissingen Collection." *Apollo* 80/11: 384–89.

Delacampagne, Arianne, and Christian Delacampagne. 2003. *Here Be Dragons: A Fantastic Bestiary.* Princeton, NJ: Princeton University Press.

Demel, Walter. 1991. "China in the Political Thought of Western and Central Europe, 1570–1750." In *China and Europe: Images and Influences in the Sixteenth to Eighteenth Centuries,* edited by Thomas H. C. Lee, 45–64. Hong Kong: Chinese University Press.

Deng, Gang. 1995. "An Evaluation of the Role of Admiral Zheng He's Voyages in Chinese Maritime History." *International Journal of Maritime History* 7/2: 1–19.

———. 1997. *Chinese Maritime Activities and Socioeconomic Development, c. 2100 B.C.–1900 A.D.* Westport, CT: Greenwood Press.

———. 1997a. "The Foreign Staple Trade of China in the Pre-Modern Era." *International History Review* 19/2: 253–83.

———. 1999. *Maritime Sector, Institutions, and Sea Power of Premodern China.* Westport, CT: Greenwood Press.

Denny, Walter B. 1974. "Blue-and-White Islamic Pottery on Chinese Themes." *Boston Museum Bulletin* 72/368: 76–99.

Desmond, Adrian, and James Moore. 1991. *Darwin.* London: Penguin Books.

Desmond, Ray. 1995. *Kew: The History of the Royal Botanic Gardens.* London: Harvill Press.

de Vries, Jan. 2008. *The Industrious Revolution: Consumer Behavior and the Household Economy, 1650 to the Present.* Cambridge: Cambridge University Press.

———, and Ad van der Woude. 1997. *The First Modern Economy: Success, Failure, and Perseverance of the Dutch Economy, 1500–1815.* Cambridge: Cambridge University Press.

Dewey, J. F., S. Cande, and W. C. Pitman III. 1985. "Tectonic Evolution of the India/Eurasia Collision Zone." *Ecologae Geologiche Helvetae* 82: 717–34.

Dias, Pedro. 2004. "The Palace of the Viceroys in Goa." In Flores and Vassallo e Silva 2004: 68–97.

Dickson, P. G. M. 1967. *The Financial Revolution in England: A Study in the Development of Public Credit, 1688–1756.* New York: St. Martin's Press.

Diem, Allison I. 1997. "The Pandanan Wreck, 1414: Centuries of Regional Interchange." *OA* 43/2: 7–9.

———. 1999. "Ceramics from Vijaya, Central Vietnam: Internal Motivations and External Influences (14th–Late 15th Century). *OA* 45/3: 55–64.

Diffie, Bailey W., and George Winius. 1977. *Foundations of the Portuguese Empire, 1415–1580.* Minneapolis: University of Minnesota Press.

Digby, Simon. 1982. "The Maritime Trade of India." In *The Cambridge Economic History of India,* vol. 1: *c. 1200–1750,* edited by Tapan Ray Chaudhuri and Irfan Habib, 125–59. Cambridge: Cambridge University Press.

Dillon, Michael. 1976. "A History of the Porcelain Industry in Jingdezhen." Ph.D. dissertation, University of Leeds.

———. 1992. "Transport and Marketing in the Development of the Jingdezhen Porcelain Industry during the Ming and Qing Dynasties." *JESHO* 25/3: 278–90.

Divis, Jan. 1983. *European Porcelain.* Translated by Iris Urwin. London: Peerage Books.

Dolan, Brendan. 2004. *Wedgwood: The First Tycoon.* New York: Viking.

Douglas, Mary. 1966. *Purity and Danger: An Analysis of Concepts of Pollution and Taboo.* New York: Routledge.

Dreyer, Edward. 1982. *Early Ming China: A Political History, 1355–1435*. Stanford, CA: Stanford University Press.

Dryden, John. 1958. *The Poems of John Dryden*. 4 vols. Edited by James Kinsley. Oxford: Clarendon Press.

Dupoizat, Marie-France. 2003. "Mojopahit et la couleur: Le cas des carreaux de revêtement mural." *Archipel* 66: 47–61.

Du Halde, J.-B. 1735. *Description géographique, historique, chronologique, politique, et physique de l'empire de la Chine et de la Tartarie Chinoise*. 4 vols. Paris: P. G. Le Mercier.

———. 1738–41. *A Description of the Empire of China and Chinese-Tartar containing the Geography and History (Natural as Well as Civil) of those Countries*. 3 vols. Translated by T. Gardner. London: E. Cave.

Dursum, Brian A. 1984. "Famille-Rose Enamelled Dish of the Yongzheng Period." *O* 15/7: 10–15.

Dyer, Christopher. 1989. *Standards of Living in the Later Middle Ages: Social Change in England, c. 1200–1520*. Cambridge: Cambridge University Press.

Eiland, Murray Lee. 1995–96. "Ceramics of the Silk Road: Parthia and China." *TOCS* 60: 105–20.

Eisenstein, Elizabeth. 1983. *The Printing Press as an Agent of Change*. 2 vols. Cambridge: Cambridge University Press.

Elias, Norbert. 1982. *The Civilizing Process*, vol. 1: *The History of Manners*. Translated by Edmund Jephcott. New York: Pantheon Books.

Eliot, T. S. 1963. *Collected Poems, 1909–1962*. New York: Harcourt Brace.

Elisonas, Jurgis. 1991. "The Inseparable Trinity: Japan's relations with China and Korea." In *The Cambridge History of Japan*, vol. 4: *Early Modern Japan*, edited by John Whitney Hall, 235–300. Cambridge: Cambridge University Press.

Elisseeff, Vadime. 1963. "The Middle Empire, a Distant Empire, an Empire without Neighbors." *Diogenes* 42: 60–64.

Elman, Benjamin A. 2000. *A Cultural History of Civil Examinations in Late Imperial China*. Berkeley: University of California Press.

Elvin, Mark. 1973. *The Pattern of the Chinese Past: A Social and Economic Interpretation*. Stanford, CA: Stanford University Press.

Emerson, Julie, Jennifer Chen, and Mimi Gardner Gates, eds. 2000. *Porcelain Stories: From China to Europe*. Exhib. cat. Seattle, WA: Seattle Art Museum.

Endicott-West, Elizabeth. 1989. "Merchant Associations in Yüan China: The Ortoy." *Asia Major* 2/2: 127–54.

Epprecht, Katharina. 2007. "Kannon–Divine Compassion: Early Buddhist Art from Japan." *O* 38/2: 122–27.

Eriksen, Svend, and Geoffrey de Bellaigue. 1987. *Sèvres porcelain: Vincennes and Sèvres, 1740–1800*. Translated by R. J. Charleston. London: Faber and Faber.

Erickson, Jon. 2001. *Plate Tectonics*. New York: Checkmark Books.

Espir, Helen. 2001. "Jingdezhen to Arita via Delft: The Transformation of Chinese Porcelain into Japanese by Dutch Enamellers in the 18th Century." *OA* 47/2: 25–30.

Esten, John, Olof Wahlund, and Rosalind Fischell. 1987. *Blue and White China*. Boston: Little, Brown.

Ettinghausen, Richard. 1960. *Medieval Near Eastern Ceramics in the Freer Gallery of Art.* Washington, DC: Smithsonian Institution.

Evelyn, John. 1955. *The Diary of John Evelyn.* 6 vols. Edited by E. S. de Beer. Oxford: Clarendon Press.

Falkenhausen, Lothar von. 1993. *Suspended Music: Chime-Bells in the Culture of Bronze Age China.* Berkeley: University of California Press.

———. 1993a. "On the Historiographical Orientation of Chinese Archaeology." *Antiquity* 67/257: 839–49.

———. 1999. "The Waning of the Bronze Age: Material Culture and Social Developments, 770–481 B.C." In Loewe and Shaughnessy 1999: 450–544.

Fang, Karen. 2003. "Empire, Coleridge, and Charles Lamb's Consumer Imagination." *Studies in English Literature 1500–1900* 43/4: 815–43.

Farrer, Katherine Eufemia, ed. 1903–6. *Letters of Josiah Wedgwood.* 3 vols. London: Women's Printing Society.

Farrington, Anthony. 2002. *Trading Places: The East India Company and Asia, 1600–1834.* Exhib. cat. London: British Library.

Fehervari, Geza. 1970. "Near Eastern Wares under Chinese Influence." In Watson 1970: 27–34.

Fei Xin. 1996. *Hsing-ch'a-sheng lan: The Overall Survey of the Star Raft.* Translated by J. V.. Mills. Revised and edited by Roderich Ptak. Wiesbaden: Harrossowitz Verlag.

Feng Xianming. 1987. "Yongle and Xuande Blue-and-white Porcelain in the Palace Museum." *O* 18/11: 56–71.

Fernández-Armesto, Felipe. 1992. *Columbus.* Oxford: Oxford University Press.

Ferrer, María Paz Soler. 2003. "The Use of Spanish Ceramics in Architecture." In Gavin et al. 2003: 77–101.

Ferrier, Ronald. 1986. "Trade from the Mid-Fourteenth Century to the End of the Safavid Period." In Jackson and Lockhart 1986: 412–90.

Findly, Ellison Banks. 1993. *Nur Jahan, Empress of India.* New York: Oxford University Press.

Finlay, Robert. 1991. "The Treasure-Ships of Zheng He: Chinese Maritime Imperialism in the Age of Discovery." *Terrae Incognitae: The Journal for the History of Discoveries* 23: 1–12.

———. 1992. "Portuguese and Chinese Maritime Imperialism: Camões's *Lusiads* and Luo Maodeng's *Voyage of the San Bao Eunuch.*" *Comparative Studies in Society and History* 34/2: 225–41.

———. 1998. "The Pilgrim Art: The Culture of Porcelain in World History." *JWH* 9/2: 141–87.

———. 2007. "Weaving the Rainbow: Visions of Color in World History." *JWH* 18/4: 383–431.

———. 2008. "The Voyages of Zheng He: Ideology, State Power, and Maritime Trade in Ming China." *Journal of the Historical Society* 8/3: 327–47.

Flecker, Michael. 2000. "A Ninth-Century A.D. Arab or Indian Shipwreck in Indonesian Waters." *International Journal of Nautical Archaeology* 29/2: 199–217.

———. *2003*. "The Thirteenth-Century Java Sea Wreck." Mariner's Mirror *89/4: 388–404.*

Flood, F. B. 1991–92. "The Tree of Life as a Decorative Device in Islamic Window-Fillings: The Mobility of a Motif." *OA* 37/4: 209–22.

Flores, Jorge, and Nuno Vassallo e Silva, eds. 2004. *Goa and the Great Mughal.* Exhib. cat. Lisbon: Calouste Gulbenkian Foundation.

Flynn, Dennis O., and Arturo Giráldez. 1995. "Born with a Silver Spoon: The Origin of World Trade in 1571." *JWH* 6/2: 201–21.

———, and Arturo Giráldez. 2002. "Cycles of Silver: Global Economic Unity through the Mid-Eighteenth Century." *JWH* 13/2: 391–428.

Fok, K. C. 1987. "Early Ming Images of the Portuguese." In *Portuguese Asia: Aspects in History and Economic History (Sixteenth and Seventeenth Centuries),* edited by Roderich Ptak, 143–55. Stuttgart: Steiner Verlag Wiesbaden.

Fong, Wen, ed. 1980. *The Great Bronze Age of China.* Exhib. cat. New York: Metropolitan Museum of Art.

———. 1992. *Beyond Representation: Chinese Painting and Calligraphy, 8th to 14th Century.* New York and New Haven, CT: Metropolitan Museum of Art and Yale University Press.

Fontana, Maria Vittoria. 2007. "Islamic Influence on the Production of Ceramics in Venice and Padua." In Carboni 2007: 280–93.

Ford, P. R. J. 1981. *The Oriental Carpet: A History and Guide to Traditional Motifs, Patterns, and Symbols.* New York: Harry N. Abrams.

Fortey, Richard. 2004. *Earth: An Intimate History.* New York: Alfred A. Knopf.

Forrest, Thomas. 1969. *A Voyage to New Guinea and the Moluccas, 1774–1776.* Kuala Lumpur: Oxford University Press.

Foster, George M. 1965. "The Sociology of Pottery: Questions and Hypotheses Arising from Contemporary Mexican Work." In Matson 1965: 43–61.

Foster, William, ed. 1899. *The Embassy of Sir Thomas Roe to the Court of the Great Mogul, 1615–1619, as Narrated in His Journal and Correspondence.* 2 vols. London: Hakluyt Society.

———, ed. 1934. *The Voyage of Thomas Best to the East Indies, 1612–14.* London: Hackluyt Society, 1934

Fourest, H.-P. 1980. *Delftware: Faience Production at Delft.* Translated by Katherine Watson. New York: Rizzoli.

Foust, Clifford M. 1992. *Rhubarb: The Wondrous Drug.* Princeton, NJ: Princeton University Press.

Fox, R. B. 1959. "The Calatagan Excavations: Two 15th-Century Burial Sites in Batangas, Philippines." *Philippines Studies* 7: 325–90.

Frame, Donald M., ed. and trans. 1983. *Montaigne's Travel Journal.* San Francisco: North Point Press.

Franits, Wayne, 2004. *Dutch Seventeenth-Century Genre Painting: Its Stylistic and Thematic Evolution.* New Haven, CT: Yale University Press.

Frank, Andre Gunder, and Barry K. Gills, eds. 1993. *The World System: Five Hundred Years or Five Thousand?* London: Routledge.

Frape, Chris. 1998–2001. "Some Thoughts on the Origin, Development and Influence of Koryo Celadons." *BOCSHK* 12: 50–54.

Freeman, Michael. 1977. "Sung." In *Food in Chinese Culture: Anthropological and Historical Perspectives,* edited by K. C. Chang, 141–92. New Haven, CT: Yale University Press.

Freestone, Ian, and David Gaimster, eds. 1997. *Pottery in the Making: Ceramic Traditions.* Washington, DC: Smithsonian Institution Press.

Furukawa, Hideaki. 2003. "The Tea Master Oribe." In *Turning Point: Oribe and the Arts of*

Sixteenth-Century Japan, edited by Miyeko Murase, 98–101. Exhib. cat. New York and New Haven, CT: Metropolitan Museum of Art and Yale University Press.

Gaba-van Dongen, Alexandra. 2004. "Between Fantasy and Reality: Utensils in Seventeenth-Century Dutch Art." In *Senses and Sin: Dutch Painters of Daily Life in the Seventeenth Century,* edited by Jeroen Giltaij, 30–38. Exhib. cat. Ostfildern-Ruit, Germany: Hatje Cantz Verlag.

Gabay, Elizabeth. 1991. "The Political and Social Implications of Tableware for Feasting." In *Proceedings of the Oxford Symposium on Food and Cookery 1990: Feasting and Fasting,* edited by Hiram Walker, 99–103. London: Prospect Books.

Gaimster, David. 1997. *German Stoneware, 1200–1900: Archaeology and Cultural History.* London: British Museum.

———. 1999. "Maiolica in the North: the Shock of the New." In *Maiolica in the North: The Archaeology of Tin-Glazed Earthenware in North-West Europe, c. 1500–1600,* 1–3. British Museum Occasional Papers, no. 22. London: British Museum.

———, and Beverley Nenk. 1997. "English Households in Transition c. 1450–1550: The Ceramic Evidence." In *The Age of Transition: The Archaeology of English Culture, 1400–1600,* edited by David Gaimster and Paul Stamper, 171–95. Oxford: Society for Medieval Archaeology.

Garlake, Peter S. 1966. *The Early Islamic Architecture of the East African Coast.* Nairobi: Oxford University Press.

Garner, Harry. 1970. *Oriental Blue and White.* London: Faber and Faber.

Gascoigne, John. 1994. *Joseph Banks and the English Enlightenment: Useful Knowledge and Polite Culture.* Cambridge: Cambridge University Press.

Gaskell, Ivan. 1989. *Seventeenth-Century Dutch and Flemish Painting: The Thyssen-Bornemisza Collection.* London: Sotheby's Publications.

Gaur, Albertine. 1994. *A History of Calligraphy.* New York: Cross River Press.

Gavin, Robin Farwell. 2003. "Introduction." In Gavin et al. 2003: 1–23.

———, Donna Pierce, and Alfonso Pleguezuelo, eds. 2003. *Cerámica y Cultura: The Story of Spanish and Mexican Mayólica.* Exhib. cat. Alburquerque: University of New Mexico Press.

Gelderblom, Oscar, and Joost Jonker. 2004. "Completing a Financial Revolution: The Finance of the Dutch East India Trade and the Rise of the Amsterdam Capital Market, 1595–1612." *Journal of Economic History* 64/3: 641–72.

Gernet, Jacques. 1982. *A History of Chinese Civilization.* Translated by J. R. Foster. Cambridge: Cambridge University Press.

———. 1985. *China and the Christian Impact: A Conflict of Cultures.* Translated by Janet Lloyd. Cambridge: Cambridge University Press.

Gerritsen, Anne. 2009. "Fragments of a Global Past: Ceramic Manufacture in Song-Yuan-Ming." *JESHO* 52/1: 117–52.

Giacomotti, Jeanne. 1963. *French Faience.* New York: Universe Books.

Gibbs, Edward. 1998–99. "Mamluk Ceramics, 648–923 A.H./A.D. 1250–1517." *TOCS* 63: 19–44.

Gilbert, Michelle. 1989. "The Cracked Pot and the Missing Sheep." *American Ethnologist* 16/2: 213–29.

Giovio, Paolo. 1956–58. *Lettere.* Edited by Giuseppe Guido Ferrero. 2 vols. Rome: Istituto Poligrafico dello Stato.

Glassie, Henry. 1997. *Art and Life in Bangladesh.* Bloomington: Indiana University Press.

———. 1999. *Material Culture.* Bloomington: Indiana University Press.

Gleeson, Janet. 1998. *The Arcanum: The Extraordinary True Story of the Invention of European Porcelain.* London: Bantam Press.

Glick, Thomas F. 1995. *From Muslim Fortress to Christian Castle: Social and Cultural Change in Medieval Spain.* Manchester: Manchester University Press.

Godden, Geoffrey. 1979. *Oriental Export Market Porcelain and Its Influence on European Wares.* London: Granada.

———. 1982. "China for the West." In Atterbury 1982: 55–77.

Golas, Peter J. 1999. *SCC,* vol. 5, pt. 13: *Chemistry and Chemical Technology: Mining,* edited by Joseph Needham.

Goldsmith, Oliver. 1966. *Collected Works of Oliver Goldsmith,* vol. 2: *The Citizen of the World,* edited by Arthur Friedman. Oxford: Clarendon Press.

Goldwaithe, Richard. 1989. "The Economic and Social World of Renaissance Maiolica." *Renaissance Quarterly* 42/1: 1–32.

Golombek, Lisa. 1988. "The Draped Universe of Islam." In *Content and Context of Visual Arts in the Islamic World,* edited by Priscilla P. Soucek, 25–50. University Park: Pennsylvania State University Press.

———, ed. 1996. *Tamerlane's Tableware: A New Approach to the Chinoiserie Ceramics of Fifteenth- and Sixteenth-Century Iran.* Exhib. cat. Ontario: Mazda Publishers and the Royal Ontario Museum.

———. 1996a. "The Ceramic Industry in Fifteenth-Century Iran: An Interpretation." In Golombek 1996: 124–39.

Goody, Jack. 1993. *The Culture of Flowers.* Cambridge: Cambridge University Press.

Gould, Richard A. 2000. *Archaeology and the Social History of the Ship.* Cambridge: Cambridge University Press.

Graaf, H. J. de, and Theodore G. Th. Pigeaud. 1984. *Chinese Muslims in Java in the 15th and 16th Centuries: The Malay Annals of Semerang and Cerbon.* Melbourne: Monash Papers on Southeast Asia, Monash University.

Grabar, Oleg. 1968. "The Visual Arts, 1050–1350." In Boyle 1968: 626–58.

———. 1992. *The Mediation of Ornament.* Princeton, NJ: Princeton University Press.

———. 2000. *Mostly Miniatures: An Introduction to Persian Painting.* Princeton, NJ: Princeton University Press.

Graça, Jorge. 1977. "The Portuguese Porcelain Trade with China." *Arts in Asia* 7: 45–51.

Grave, Peter, Mike Barbetti, Mike Hotchkis, and Roger Bird. 2000. "The Stoneware Kilns of Sisatchanalai and Early Modern Thailand." *Journal of Field Archaeology* 27/2: 169–182.

Gray, Basil. 1940–41. "The Influence of Near Eastern Metalwork on Chinese Ceramics." *TOCS* 18: 47–60.

———. 1964–66. "The Export of Chinese Porcelain to India." *TOCS* 36: 21–36.

———. 1972. "Chinese Influence in Persian Painting: 14th and 15th Centuries." In Watson 1972: 11–19.

———. 1984. *Sung Porcelain and Stoneware.* London: Faber and Faber.

——. 1986. "The Arts in the Safavid Period." In Jackson and Lockhart 1986: 877–912.

Greensted, Mary, and Peter Hardie. 1982. *Chinese Ceramics: The Indian Connection.* Exhib. cat. Bristol: City of Bristol Museum and Art Gallery.

Grigsby, Leslie B. 1993. "Johan Nieuhoff's *Embassy:* An Inspiration for Relief Decoration on English Stoneware and Earthenware." *MA* 143/1: 172–83.

Groeneveldt, W. P. 1880. *Notes on the Malay Archipelago and Malacca, Compiled from Chinese Sources.* Batavia, Indonesia: Bruining.

Groslier, Bernard-Philippe. 1995. "La céramique angkorienne (6e–14e siècles)." *Péninsule: Étude interdisciplinaires sur l'Asie du Sud-Est Péninsulaire.* 31/2: 5–60.

Grube, Ernest J. 1993–94. "Timurid Ceramics: Filling a Gap in the Ceramic History of the Islamic World." *TOCS* 58: 77–86.

Gschwend, Annemarie Jordan. 2000. "A Masterpiece of Indo-Portuguese Art: The Mounted Rhinoceros Cup of Maria of Portugal, Princess of Parma." *OA* 46/3: 48–57.

Guest, Harriet. 2000. *Small Change: Women, Learning, Patriotism, 1750–1810.* Chicago: University of Chicago Press.

Guo Yanyi. 1987. "Raw Materials for Making Porcelain and the Characteristics of Porcelain Wares in North and South China in Ancient Times." *A* 29/1: 3–19.

Gutman, Pamela. 2002. "The Martaban Trade: An Examination of the Literature from the Seventh until the Eighteenth Century." *Asian Perspectives* 40/1: 108–18.

Guy, Basil. 1963. "The French Image of China before and after Voltaire." *Studies on Voltaire and the Eighteenth Century,* vol. 21. Geneva: Librairie E. Droz.

Guy, John S. 1982. "Vietnamese Trade Ceramics." In *Vietnamese Ceramics,* edited by Carol M. Young, Marie-France Dupoizat, and Elizabeth W. Lane, 28–35. Exhib. cat. Singapore: Oxford University Press.

——. 1986. *Oriental Trade Ceramics in South-East Asia, Ninth to Sixteenth Centuries.* Singapore: Oxford University Press.

——. 1989. *Ceramic Traditions of South-East Asia.* Oxford: Oxford University Press.

——. 1996–97. "A Reassessment of Khmer Ceramics." *TOCS* 61: 39–63.

——. 1997. "Vietnamese Ceramics and Cultural Identity." In Stevenson and Guy 1997: 11–22.

——. 1997a. "Vietnamese Ceramics in International Trade." In Stevenson and Guy 1997: 47–60.

——. 2001–2. "Early Asian Ceramic Trade and the Belitung ('Tang') Cargo." *TOCS* 66–13–27.

——. 2004. "Asian Trade and Exchange before 1600." In Jackson and Jaffer 2004: 52–67.

Gwynn, John. 1766. *London and Westminster Improved.* Reprint, Farnborough, U.K.: Gregg.

Haga Koshiro. 1989. "The *Wabi* Aesthetic through the Ages." In Varley and Kumakura 1989: 195–230.

Halikowski Smith, Stefan. 2007. "Demystifying a Change in Taste: Spices, Space, and Social Hierarchy in Europe, 1380–1750." *International History Review* 39/2: 237–57.

Hall, John Whitney. 1981. "Japan's Sixteenth-Century Revolution." In *Warlords, Artists, and Commoners: Japan in the Sixteenth Century,* edited by George Elison and Bardwell L. Smith, 7–21. Honolulu: University Press of Hawai'i.

Hall, Kenneth R. 1975. "Khmer Commercial Development and Foreign Contacts under Suryavarman I." *JESHO* 18/3: 318–36.

———. 1985. *Maritime Trade and State Development in Early Southeast Asia.* Honolulu: University of Hawai'i Press.

———. 2004. "Local and International Trade and Traders in the Straits of Melaka Region: 600–1500." *JESHO* 47/2: 213–60.

Hallberg, Paul, and Christian Koninckx, eds. 1996. *A Passage to China: Colin Campbell's Diary of the First Swedish East India Company Expedition to Canton.* Göteborg, Sweden: Royal Society of Arts and Sciences in Göteborg.

Haller, William B. 1967. *Foxe's Book of Martys and the Elect Nation.* London: Jonathan Cape.

Han Wei. 1993. "Tang Dynasty Tea Utensils and Tea Culture: Recent Discoveries at Famen Temple." *Chanoyu Quarterly* 74: 38–58.

Hansen, Valerie. 1990. *Changing Gods in Medieval China, 1127–1276.* Princeton, NJ: Princeton University Press.

———. 2000. *The Open Empire: A History of China to 1600.* New York: Norton.

———. 2003. "The Hejia Village Hoard: A Snapshot of China's Silk Road Trade." *O* 34/2: 14–19.

Hanway, Jonas. 1756. *A Journal of Eight Days' Journey, to which is added An Essay on Tea.* London: H. Woodfall.

Harbsmeier, Christoph. 1998. *SCC,* vol. 7, pt. 1: *Language and Logic,* edited by Joseph Needham.

Harris, Bob. 2002. *Politics and the Nation: Britain in the Mid-Eighteenth Century.* Oxford: Oxford University Press.

Harris, Conrad. 2003–04. "Chinese Ceramic Horses and How They Changed." *TOCS* 68: 75–78

Harris, Jonathan Gil. 2004. *Sick Economics: Drama, Mercantilism, and Disease in Shakespeare's England.* Philadelphia: University of Pennsylvania Press.

Harris, Steven J. 1999. "Mapping Jesuit Science: The Role of Travel in the Geography of Knowledge." In *The Jesuits: Cultures, Sciences, and the Arts, 1540–1773,* edited by Gauvin Alexander Bailey, Steven J. Harris, and T. Frank Kennedy, 212–40. Toronto: University of Toronto Press.

Harrison, Tom. 1955. "Ceramics Penetrating Central Borneo." *SMJ* 6: 549–60.

———. 1958. "'The Ming Gap' and Kota Batu, Brunei." *SMJ* 8: 273–77.

———. 1967. "Ceramic Crayfish and Related Vessels in Central Borneo, the Philippines and Sweden." *SMJ* 15/30–31: 1–9.

Harrison-Hall, Jessica. 1997. "Chinese Porcelain from Jingdezhen." In Freestone and Gaimster 1997: 194–99.

Harrisson, Barbara. 1962. "Stonewares: 'Marco Polo Ware' in South-East Asia." *SMJ* 11/19–20: 412–16.

———. 1986. *Pusaka: Heirloom Jars of Borneo.* Singapore: Oxford University Press.

———, and P. M. Shariffuddin. 1969. "Sungai Lumut: A 15th-Century Burial Ground." *Brunei Museum Journal* 1/1: 24–56.

Hartwell, Robert. 1967. "A Cycle of Economic Change in Imperial China: Coal and Iron in Northeast China, 750–1350." *JESHO* 10/1: 102–59.

———. 1982. "Demographic, Political, and Social Transformations of China, 750–1550." *HJAS* 42/2: 365–442.

Haskell, Francis. 1987. "The Baron d'Hancarville: An Adventurer and Art Historian in Eighteenth-Century Europe." In *Past and Present in Art and Taste*, 30–45. New Haven, CT: Yale University Press.

Haudrére, Philippe. 1999. "The French India Company and Its Trade in the Eighteenth Century." In *Merchants, Companies and Trade: Europe and Asia in the Early Modern Era*, edited by Sushil Chaudhury and Michel Morineau, 202–11. Cambridge: Cambridge University Press.

———, and Gérard Le Bouëdec. 1999. *Les Compagnies des Indes*. Rennes: Editions Ouest-France.

Hay, John. 1986. *Kernels of Energy, Bones of Earth: The Rock in Chinese Art*. Exhib. cat. New York: China Institute in America.

Hayden, Brian. 2003. *Shamans, Sorcerers, and Saints: A Prehistory of Religion*. Washington, DC: Smithsonian Institution Press.

Hayward, Helena. 1972. "The Chinese Influence on English Furniture from the 16th to the 18th Century." In Watson 1972: 57–61.

He Li. 1996. *Chinese Ceramics*. New York: Rizzoli.

Headley, John M. 1995. "Spain's Asian Presence, 1565–1590: Structures and Aspirations." *Hispanic American Historical Review* 75/4: 623–46.

Hearn, Maxwell K. 1980. "The Terracotta Army of the First Emperor of Qin (221–206 B.C.)." In Fong 1980: 353–68.

Heesterman, Jan. 2003. "The Tides of the Indian Ocean, Islamization and the Dialectic of Coast and Inland." In *Circumambulations in South Asian History: Essays in Honour of Dirk H. A. Kolff*, edited by Jos Gommans and Om Prakash, 29–46. Leiden: Brill.

Hevia, James L. 1995. *Cherishing Men from Afar: Qing Guest Ritual and the Macartney Embassy of 1793*. Durham, NC: Duke University Press.

Helms, Mary W. 1993. *Craft and the Kingly Ideal: Art, Trade, and Power*. Austin: University of Texas Press.

———. 1994. "Essay on Objects: Interpretations of Distance Made Tangible." In *Implicit Understandings: Observing, Reporting, and Reflecting on the Encounters between Europeans and Other Peoples in the Early Modern Era*, edited by Stuart B. Schwartz, 355–77. Cambridge: Cambridge University Press.

Hempel, Rose. 1990. *The Golden Age of Japan, 794–1192*. Translated by Katherine Wilson. New York: Rizzoli.

Henderson, J., and J. Raby. 1989. "The Technology of Fifteenth-Century Turkish Tiles: An Interim Statement on the Origins of the Iznik Industry." *WA* 21/1: 115–32.

Heng Chye Kiang. 1999. *Cities of Aristocrats and Bureaucrats: The Development of Medieval Cityscapes*. Singapore: Singapore University Press.

Hess, Andrew. 1973. "The Ottoman Conquest of Egypt (1517) and the Beginning of the Sixteenth-Century World War." *International Journal of Middle Eastern Studies* 4/1: 55–76.

Hess, Catherine. 2002. *Italian Ceramics: Catalogue of the J. Paul Getty Collection*. Los Angeles, CA: J. Paul Getty Museum.

Higgins, Roland, Oliver. 1981. "Piracy and Coastal Defense in the Ming Period: Government Responses to Coastal Disturbances, 1523–1549." Ph.D. dissertation, University of Minnesota.

Higham, Charles. 2001. *The Civilization of Angkor.* Berkeley: University of California Press.

Hill, George Birkbeck, and L. F. Powell, eds. 1934–50. *Boswell's Life of Johnson.* 6 vols. Oxford: Clarendon Press.

Hillel, Daniel J. 1991. *Out of the Earth: Civilization and the Life of the Soil.* New York: Free Press.

Hills, Paul. 1998. "Venetian Glass and Renaissance Self-Fashioning." In *Concepts of Beauty in Renaissance Art,* edited by Francis Ames-Lewis and Mary Rogers, 163–77. Aldershot: Ashgate.

Himanshu, P. Ray. 1994. *The Winds of Change: Buddhism and the Maritime Links of Early South Asia.* New Delhi: Oxford University Press.

Hirth, Friedrich, and W. W. Rockhill, trans. 1966. *Chau Ju-kua: On the Chinese and Arab Trade in the Twelfth and Thirteenth Centuries, entitled "Chu-Fan-Chi."* New York: Paragon Book Reprint Corp.

Ho, Chuimei. 1994. "The Ceramic Trade in Asia, 1602–82." In *Japanese Industrialization and the Asian Economy,* edited by A. J. H. Latham and Heita Kawakasu, 35–70. London: Routledge.

———. 2001. "The Ceramic Boom in Minnan during Song and Yuan Times." In Schottenhammer 2001: 237–81.

———, and Bennet Bronson. 2004. *Splendors of China's Forbidden City: The Glorious Reign of Emperor Qianlong.* Exhib. cat. London: Merrell.

Ho, Ping-ti. 1956. "Early-Ripening Rice in Chinese History." *Economic History Review* 9/2: 200–18.

———. *1969.* "The Loess and the Origins of Chinese Agriculture." *American Historical Review* 75/1: 1–36.

Hobson, R. L. 1922–23. "The Significance of Samarra." *TOCS* 5: 29–32.

Hodges, Henry. 1970. "Interaction between Metalworking and Ceramic Technologies in the T'ang Period." In Watson 1970: 64–67.

———. 1972. "The Technical Problems of Copying Chinese Porcelains in Tin Glaze." In Watson 1972: 79–87.

Hochstrasser, Julie Berger. 2007. *Still Life and Trade in the Dutch Golden Age.* New Haven, CT: Yale University Press.

Hodges, Richard, and David Whitehouse. 1983. *Mohammed, Charlemagne and the Origins of Europe.* Ithaca, NY: Cornell University Press.

Hodgson, Marshall G. S. 1974. *The Venture of Islam: Conscience and History in a World Civilization.* 3 vols. Chicago: University of Chicago Press.

Hoeniger, Cathleen S. 1991. "Cloth of Gold and Silver: Simone Martini's Techniques for Representing Luxury Textiles." *Gesta: International Center of Medieval Art* 30/2: 154–62.

Hogendorn, Jan, and Marion Johnson. 1986. *The Shell Money of the Slave Trade.* Cambridge: Cambridge University Press.

Holcombe, Charles. 2001. *The Genesis of East Asia, 221 B.C.–A.D. 907.* Honolulu: Association for Asian Studies and University of Hawai'i Press.

———. 2004. "Southern Integration: The Sui-Tang (581–907) Reach South." *Historian* 66/4: 749–71.

Honour, Hugh. 1961. *Chinoiserie: The Vision of Cathay.* London: John Murray.

———. 1967. "Statuettes after the Antique: Volpato's Roman Porcelain Factory." *Apollo* 85/5: 371–73.
Horton, Mark, and John Middleton. 2000. *The Swahili: The Social Landscape of a Mercantile Society.* Oxford: Blackwell.
Hourani, George F. 1951. *Arab Seafaring in the Indian Ocean in Ancient and Early Medieval Times.* Princeton, NJ: Princeton University Press.
Howard, David S. 1974. *Chinese Armorial Porcelain.* London: Faber and Faber.
———. 1994. *The Choice of the Private Trader: The Private Market in Chinese Export Porcelain Illustrated from the Hodroff Collection.* London: Zwemmer.
———. 1997. *A Tale of Three Cities: Canton, Shanghai and Hong Kong.* Exhib. cat. London: Sotheby's.
———. 2000. "The East India Company and the Pearl River." *OA* 46/3: 32–36.
———, and John Ayers. 1980. *Masterpieces of Chinese Export Porcelain from the Mottahedeh Collection in the Virginia Museum.* London: Sotheby Parke Bernet.
Howard, Deborah. 2007. "Venice as an 'Eastern City.'" In Carboni 2007: 58–71.
Hsu Wen-chin. 1988. "Social and Economic Factors in the Chinese Porcelain Industry in Jingdezhen during the Late Ming and Early Qing Period." *JRASGBI* 1: 135–59.
Huang, Chun-Chieh. 2007. "The Defining Character of Chinese Historical Thinking." *History and Theory* 46/2: 180–88.
Huang, H. T. 2000. *SCC,* vol. 6, pt. 5: *Biology and Biological Technology: Fermentation and Food Science,* edited by Joseph Needham.
Huang Shijian. 1986. "The Persian Kanguage in China during the Yuan Dynasty." *Papers on Far Eastern History* 34: 83–95.
Hulliung, Mark. 1976. *Montesquieu and the Old Regime.* Berkeley: University of California Press.
Huyler, Stephen P. 1996. *Gifts of Earth: Terracottas and Clay Sculptures of India.* New Delhi: Indira Gandhi National Centre for the Arts.
Ibn Battuta. 1929. *Travels in Asia and Africa, 1325–1354.* Translated by H. A. R. Gibb. London: Routledge & Kegan Paul.
Ikegami, Eiko. 2005. *Bounds of Civility: Aesthetic Networks and the Political Origins of Japanese Culture.* Cambridge: Cambridge University Press.
Impey, Oliver. 1982. "Japanese Porcelain." In Atterbury 1982: 43–53.
———. 1984. "Japanese Export Art of the Edo Period and Its Influence on European Art." *Modern Asian Studies* 18/4: 685–97.
———. 1990. "Porcelain for Palaces." In *Porcelain for Palaces: The Fashion for Japan in Europe, 1650–1750,* 56–69. Exhib. cat. London: Oriental Ceramic Society.
———. 1999. "Chinese Porcelain Exported to Japan." *OA* 45/1: 15–21.
———. *2002. Japanese Export Porcelain: Catalogue of the Collection of the Ashmolean Museum, Oxford.* Amsterdam: Hotei Publishing.
Ingrassia, Catherine. *1998. Authorship, Commerce, and Gender in Early Eighteenth-Century England.* Cambridge: Cambridge University Press.
Irwin, John. 1955. "Origins of the 'Oriental Style' in English Decorative Art." BM *625/4: 106–14.*
———. 1972. "The Chinese Element in Indian Chintz." In Watson 1972: 26–32.

Israel, Jonathan A. 1989. *Dutch Primacy in World Trade, 1585–1740.* Oxford: Clarendon Press.

———. 1990. "Spain, the Spanish Embargoes, and the Struggle for the Mastery of World Trade, 1585–1660." In *Empires and Entrepots: The Dutch, the Spanish Monarchy and the Jews, 1585–1713,* 189–212. London: Hambledon Press.

———. 1995. *The Dutch Republic: Its Rise, Greatness, and Fall, 1477–1806.* Oxford: Clarendon Press.

Israeli, Raphael. 1982. "Islam in the Chinese Environment." *Contributions to Asian Studies* 17: 79–94.

Itoh, Ikutaro. 1992. "Koreanization in Koryo Celadon." *O* 23/12: 46–50.

Itoi, Kenji. 1989. *Thai Ceramics from the Sosai Collection.* Translated by Kaworu Uno and Carolyn Nakamura. Exhib. cat. Singapore: Oxford University Press.

Jackson, Anna, and Amin Jaffer, eds. 2004. *Encounters: The Meeting of Asia and Europe, 1500–1800.* Exhib. cat. London: V & A Publications.

Jackson, Peter, and Laurence Lockhart, eds. 1986. *The Cambridge History of Islam,* vol. 6: *The Timurid and Safavid Periods.* Cambridge: Cambridge University Press.

Jackson-Stops, Gervase. 1993. "Sharawadgi Rediscovered: The Chinese House at Stowe." *Apollo* 137/4: 217–22.

Jacobson, Dawn. 1993. *Chinoiserie.* London: Phaidon Press.

Jang, Scarlett. 1999. "Representations of Exemplary Scholar-Officials, Past and Present." In *Arts of the Sung and Yüan: Ritual, Ethnicity, and Style in Painting,* edited by Cary Y. Liu and Dora C. Y. Ching, 19–67. Princeton, NJ: Art Museum, Princeton University.

Jarry, Madeleine. 1981. *Chinoiserie: Chinese Influence on European Decorative Art, 17th and 18th Centuries.* New York: Vendome Press.

Jenkins, Ian. 1996. "'Contemporary Minds': Sir William Hamilton's Affair with Antiquity." In Jenkins and Sloan: 40–64.

———, and Kim Sloan, eds., 1996 *Vases and Volcanoes: Sir William Hamilton and His Collection.* Exhib. cat. London: British Museum.

Jensen, Lionel M. 1997. *Manufacturing Confucianism: Chinese Traditions and Universal Civilization.* Durham, NC: Duke University Press.

Jenyns, Soame. 1951. *Later Chinese Porcelain: The Ch'ing Dynasty (1644–1912).* London: Faber and Faber.

———. 1971. *Japanese Pottery.* London: Faber and Faber.

Johnson, Marion. 1970. "The Cowrie Currencies of West Africa. Part I." *Journal of African History* 11/1: 17–49.

Johnson, Matthew. 1996. *An Archaeology of Capitalism.* Oxford: Blackwell.

Jones, Colin. 2002. *Madame de Pompadour: Images of a Mistress.* London: National Gallery Company.

Jones, Robert W. 1998. *Gender and the Formation of Taste in Eighteenth-Century Britain.* Cambridge: Cambridge University Press.

Jörg, C. J. A. 1982. *Porcelain and the Dutch China Trade.* The Hague: Martinus Nijhoff.

———. 1986. *The Geldermalsen: History and Porcelain.* Groningen: Kemper.

———. 1993. "Porcelain for the Dutch in the Seventeenth Century: Trading Networks and Private Enterprise." In Scott 1993: 183–205.

———. 1995. *Oosters Porselein / Oriental Porcelain.* Rotterdam: Museum Boymans-van Beuningen.
———. 2002–3. "Treasures of the Dutch Trade in Chinese Porcelain." *OA* 48/5: 20–26.
———, and Jan van Campen. 1997. *Chinese Ceramics in the Collection of the Rijksmuseum, Amsterdam.* Amsterdam: Rijksmuseum.
———, and Michael Flecker. 2001. *Porcelain from the Vung Tau Wreck: The Hallstrom Excavation.* Singapore: Sun Tree Publishing.
Joseph, Adrian Malcolm. 1985. "The Mongol Influence on Blue-and-white Porcelain." In *Jingdezhen Wares: The Yuan Evolution,* 44–49. Hong Kong: Oriental Ceramic Society of Hong Kong.
Jourdain, Margaret, and Soame Jenyns. 1948. "Chinese Export Lacquer of the 17th and 18th Century." *OA* 1/3: 143–48.
Jullien, François. 1995. *The Propensity of Things: Toward a History of Efficacy in China.* Translated by Janet Lloyd. New York: Zone Books.
Junker, Laura Lee. 1994. "Trade Competition, Conflict, and Political Transformations in Sixth- to Sixteenth-Century Philippine Chiefdoms." *Asian Perspectives* 33/2: 229–60.
———. 1998. "Integrating History and Archaeology in the Study of Contact Period Philippine Chiefdoms." *International Journal of Historical Archaeology* 2/4: 291–320.
———. 1999. *Raiding, Trading, and Feasting: The Political Economy of Philippine Chiefdoms.* Honolulu: University of Hawai'i Press.
Kaboy, Tuton, and Eine Moore. 1967. "Ceramics and Their Uses among the Coastal Melanaus." *SMJ* 15/30–31: 10–29.
Kasetsiri, Charnvit. 1992. "Ayudhya: Capital-Port of Siam and Its 'Chinese Connection' in the Fourteenth and Fifteenth Centuries." *Journal of the Siam Society* 80/1: 75–79.
Kawai, Masatomo. 2002. "Reception Room Display in Medieval Japan." In *Kazari: Decoration and Display in Japan, 15th–19th Centuries,* edited by Nicole Coolidge Rousmaniere, 32–41. Exhib. cat. London: British Museum.
Kawazoe, Shoji. 1990. "Japan and East Asia." In Yamamura 1990: 396–446.
Kee Il Choi Jr. 1999. "Chinese Export Porcelain for America." *OA* 45/1: 90–97.
Keene, Donald. 2003. *Yoshimasa and the Silver Pavilion: The Creation of the Soul of Japan.* New York: Columbia University Press.
Kelly, Jack. 2004. *Gunpowder: Alchemy, Bombards, and Pyrotechnics.* New York: Basic Books.
Kemp, Martin. 1995. "'Wrought by No Artist's Hand': The Natural, the Artificial, the Exotic, and the Scientific in Some Artifacts from the Renaissance." In *Reframing the Renaissance: Visual Culture in Europe and Latin America 1450–1650,* edited by Claire Farago, 176–96. New Haven, CT: Yale University Press.
Kerr, Rose. 1986. "The Interaction between Bronzes and Ceramics in China: Song to Qing Periods, ca. A.D. 1100 to 1900." In Vickers 1986: 301–13.
———. 1989. "The Chinese Porcelain at Spring Grove Dairy: Sir Joseph Bank's Manuscript. *Apollo* 129/1: 30–34.
———. 1993. "Jun Wares and Their Qing Dynasty Imitation at Jingdezhen." In Scott 1993: 151–64.
———. 1999. "Celestial Creatures: Chinese Tiles in the Victoria and Albert Museum. *Apollo* 149/3: 3–14.

——. 2004. "Chinese Porcelain in Early European Collections." In Jackson and Jaffer 2004: 44–51.

——, and Nigel Wood. 2004. *SCC*, vol. 5, pt. 12: *Chemistry and Chemical Technology: Ceramic Technology.* Edited by Rose Kerr.

Khoo, Joo Ee. 1991. *Kendi: Pouring Vessels in the University of Malaya Collection.* Singapore: Oxford University Press.

Kieschnick, John. 2003. *The Impact of Buddhism on Chinese Material Culture.* Princeton, NJ: Princeton University Press.

Kim, Chewon, and Won-Yong Kim. 1966. *Treasures of Korean Art: 2000 Years of Ceramics, Sculpture, and Jeweled Arts.* New York: Harry N. Abrams.

Kim Chong-tae. 1994. "A History of Korean Handicrafts." In Son 1994: 104–09.

Kim Hongnam. 1991. "China's Earliest Datable White Stonewares from the Tomb of King Muryong (d. A.D. 523), Paekche, Korea." *OA* 37/1: 17–34.

Kim Jae-yol. 1994. "Jade Green: Koryo Celadon." In Son 1994: 110–117.

Kim, Wondong. 1986. "Chinese Ceramics from the Wreck of a Yuan Ship in Sinan, Korea—With Particular Reference to Celadon Wares." 2 vols. Ph.D. dissertation, University of Kansas.

Kingery, W. D., ed. 1986. *Technology and Style.* Columbus, OH: American Ceramic Society.

——, and Pamela B. Vandiver. 1986. "The Eighteenth-Century Change in Technology and Style from the Famille-Verte Palette to the Famille-Rose Palette." In Kingery 1986: 363–81.

——, and Pamela B. Vandiver. 1986a. *Ceramic Masterpieces: Art, Structure, and Technology.* New York: Free Press.

Kinney, Ann R., with Marijke J. Klokke and Lydia Kieven. 2003. *Worshiping Siva and Buddha: The Temple Art of East Java.* Honolulu: University of Hawai'i Press.

Kirkman, James. 1958. "The Great Pillars of Malindi and Mambrui." *OA* 4/2: 55–67.

Klein, Peter W. 1989. "The China Seas and the World Economy between the Sixteenth and Nineteenth Centuries: The Changing Structures of Trade." In *Interactions in the World Economy: Perspectives on International Economic History,* edited by Carl-Ludwig Holtfrerich, 61–89. New York: New York University Press.

Kletke, Daniel. 2004. "Dresden's Princely Splendor at Home and Abroad." *MA* 166/4: 124–31.

Knauer, Elfriede Regina. 1998. *The Camel's Load in Life and Death: Iconography and Ideology of Chinese Pottery Figurines from Han to Tang and Their Relevance to Trade along the Silk Routes.* Zurich: Haeberlin and Partner.

Koerner, Lisbet. 1999. *Linnaeus: Nature and Nation.* Cambridge, MA: Harvard University Press.

Komaroff, Linda. 2002. "The Transmission and Dissemination of a New Visual Language." In *The Legacy of Genghis Khan: Courtly Art and Culture in Western Asia,* edited by Linda Komaroff and Stefano Carboni, 169–95. Exhib. cat. New York: Metropolitan Museum of Art.

Kong Yuan Zhi. 1987. "A Study of Chinese Loanwords (from South Fujian Dialects) in the Malay and Indonesian Languages." *Bijdragen tot de Taal-, Land- en Volkenkunde* 143/4: 452–67.

Kowaleski-Wallace, Beth. 1995–96. "Women, China, and Consumer Culture in Eighteenth-Century England." *ECS* 29/2: 153–67.

Krahl, Regina. 1985. "Longquan Celadon of the Yuan and Ming Dynasties in the Topkapi Saray, Istanbul." *TOCS* 49: 41–57.

———. 1986. "Export Porcelain Fit for the Chinese Emperor: Early Chinese Blue-and-White in the Topkapi Saray, Istanbul." *JRASGBI* 1: 68–92.

———. 1997. "Vietnamese Blue-and-White and Related Wares." In Stevenson and Guy 1997: 147–57.

———. 2002–3. "By Appointment to the Emperor: Imperial Porcelains from Jingdezhen and Their Various Destinations." *OA* 48/5: 27–32.

———. 2004. "Famous Brands and Counterfeits: Problems of Terminology and Classification in Song Ceramics." In *Song Ceramics: Art History, Archaeology and Technology*, edited by Stacey Pierson, 61–79. London: Percival David Foundation of Chinese Art and the School of Oriental and African Studies, University of London.

Kramer, Carol. 1997. *Pottery in Rajasthan: Ethnoarcheology in Two Indian Cities.* Washington, DC : Smithsonian Institution Press.

Kubiski, Joyce. 2001. "Orientalizing Costume in Early Fifteenth-Century French Manuscript Painting (*Cité des Dames*, Limbourg Brothers, Boucicaut Master, and Bedford Master)." *Gesta: International Center of Medieval Art* 40/2: 161–80.

Kuchta, David. 2002. *The Three-Piece Suit and Modern Masculinity: England, 1550– 1850.* Berkeley: University of California Press.

Kumakura Isao. 1989. "Sen no Rikyu: Inquiries into His Life and Tea." In Varley and Kumakura 1989: 33–69.

Kurtz, O. 1975. "The Strange History of an Alhambra Vase." *Al-Andalus* 40: 205–12.

Kusimba, Chapurukha M. 1999. "Material Symbols among the Precolonial Swahili of the East African Coast." In *Material Symbols: Culture and Economy in Prehistory*, edited by John E. Robb, 318–41. Carbondale: Southern Illinois University.

Kuwayama, George. 1997. *Chinese Ceramics in Colonial Mexico.* Los Angeles: Los Angeles County Museum of Art, 1997.

———. 2000. "Chinese Ceramics in Colonial Peru." *OA* 46/1: 2–15.

Kwan, K. K. 1985. "Canton, Pulau Tioman and Southeast Asian Maritime Trade." In *A Ceramic Legacy of Asia's Maritime Trade*, edited by Peter Y. K. Lam, 49–63. Singapore: Oxford University Press with the Southeast Asian Ceramic Society.

Lach, Donald F. 1957. *The Preface to Leibniz' Novissima Sinica: Commentary, Translation, Text.* Honolulu: University of Hawai'i Press.

———, and Edwin J. Van Kley. 1993. *Asia in the Making of Europe*, vol. 3: *A Century of Advance.* Chicago: University of Chicago Press.

Laing, Ellen Johnston. 1975. "Chou Tan-chüan Is Chou Shih-ch'en." *OA* 21/3: 224–28.

Lal, Ruby. 2005. *Domesticity and Power in the Early Mughal World.* Cambridge: Cambridge University Press.

Lam, Peter Y. K. 1985. "Northern Song Guangdong Wares." In *A Ceramic Legacy of Asia's Maritime Trade*, edited by Peter Y. K. Lam, 1–30. Singapore: Oxford University Press with the Southeast Asian Ceramic Society.

———. 1998–99. "Tang Ying (1682–1756): The Imperial Superintendent at Jingdezhen." *TOCS* 63: 65–82.

Lamb, Charles. 1987. *Elia and the Last Essays of Elia.* Edited by Jonathan Bate. Oxford: Oxford University Press.

Lamb, Simon. 2004. *Devil in the Mountain: A Search for the Origins of the Andes.* Princeton, NJ: Princeton University Press.

Lancaster, Lewis R., Kikun Suh, and Chai-shin Yu, eds. 1996. *Buddhism in Koryo: A Royal Religion.* Berkeley: Institute of East Asian Studies, University of California.

Lane, Arthur. 1961. "The Gaignières-Fonthill Vase: A Chinese Porcelain of about 1300." *BM* 103/8: 124–32.

Lane, Frederic C. 1973. *Venice, a Maritime Republic.* Baltimore, MD: Johns Hopkins University Press.

Lau, Christine. 1993. "Ceremonial Monochrome Wares of the Ming Dynasty." In Scott 1993: 83–100.

Laurence, Janet. 1991. "Royal Feasts." In *Proceedings of the Oxford Symposium on Food and Cookery 1990: Feasting and Fasting,* edited by Hiram Walker, 138–51. London: Prospect Books.

La Vaissière, Étienne de, and Éric Trombert. 2004. "Des Chinois et des Hu: Migrations et intégration des Iraniens orientaux en milieu chinois durant le haut Moyen Êge." *Annales* 59/5–6: 931–69.

Le Bonheur, Albert, ed. 1986. *Phra Narai, roi de Siam et Louis XIV.* Exhib. cat. Paris: Musée de l'Orangerie.

Le Corbeiller, Clare. 1988. "A Medici Porcelain Pilgrim Flask." *J. Paul Getty Museum Journal* 16: 119–26.

———. 1990. "German Porcelain of the Eighteenth Century." *Metropolitan Museum of Art Bulletin* 47: 4–56.

Ledderose, Lothar. 1986. "Chinese Calligraphy: Its Aesthetic Dimension and Social Function." *O* 17/8: 35–50.

———. 1991. "Chinese Influence on Europe Art, Sixteenth to Eighteenth Centuries." In *China and Europe: Images and Influences in Sixteenth to Eighteenth Centuries,* edited by Thomas H. C. Lee, 221–37. Hong Kong: Chinese University Press.

———. 2000. *Ten Thousand Things: Module and Mass Production in Chinese Art.* Princeton, NJ: Princeton University Press.

Lee, Heekyung. 1999. "The Introduction of Blue-and-White Wares from Ming China to Choson Korea." *OA* 45/2: 26–37.

Lee, Hui-shu. 1996. "Art and Imperial Images at the Late Southern Sung Court." In *Arts of the Sung and Yüan Court,* edited by Maxwell K. Hearn and Judith G. Smith, 249–69. New York: Metropolitan Museum of Art.

Lee, John. 1999. "Trade and Economy in Preindustrial East Asia, c. 1500–c. 1800: East Asia in the Age of Global Integration." *Journal of Asian Studies* 58/1: 2–26.

Lee, Sherman E. 1982. *A History of Far Eastern Art.* New York: Harry N. Abrams.

Leibniz, Gottfried Wilhelm, Freiherr von. 1970. *Sämtliche Schriften und Briefe. Hrsg. von der Deutschen Akademie der Wissenschaften zu Berlin.* 6 vols. Reprint, Hildesheim, Germany: G. Olms.

Leites, Edmund. 1980. "Confucianism in Eighteenth-Century England: Natural Morality and Social Reform." In *Les Rapports entre la Chine et l'Europe au temps des lumières,* 173–84. Paris: Les Belles Lettres.

Lemire, Beverly. 1991. *Fashion's Favorite: The Cotton Trade and the Consumer in Britain, 1660–1800.* Oxford: Oxford University Press.

Lenman, Bruce P. 1990. "The English and Dutch East India Companies and the Birth of Consumerism in the Augustan World." *Eighteenth-Century Life* 14/1: 47–65.

Lentz, Thomas W. Lentz, and Glenn D. Lowry. 1989. *Timur and the Princely Vision: Persian Art and Culture in the Fifteenth Century.* Los Angeles: Los Angeles County Museum of Art.

Leonard, Jane Kate. 1972. "Chinese Overlordship and Western Penetration in Maritime Asia: A Late Ch'ing Reappraisal of Chinese Maritime Relations." *Modern Asian Studies* 6/2: 151–74.

———. 1984. *Wei Yuan and China's Rediscovery of the Maritime World.* Cambridge, MA: Harvard University, Council on East Asian Studies.

———. 1988. "Geopolitical Reality and the Disappearance of the Maritime Frontier in Qing Times." *American Neptune* 48/4: 230–36.

Lévi-Strauss, Claude. 1988. *The Jealous Potter.* Translated by Bénédicte Chorier. Chicago: University of Chicago Press.

Lewis, Mark Edward. 1999. *Writing and Authority in Early China.* Albany: State University of New York Press.

Li Zhiyan. 1998. "The Development and Background of White and *Qingbai* Wares from Tang to Song." In *Bright as Silver, White as Snow: Chinese White Ceramics from Late Tang to Yuan Dynasty,* edited by Kai-Yin Lo, 31–45. Hong Kong: Yungmingtang.

Lightbown, R. W. 1969. "Oriental Art and the Orient in Late Renaissance and Baroque Italy." *Journal of the Warburg and Courtauld Institutes* 32: 228–79.

Lin Shimin. 1985. "Zheng He xia xiyang yu ciqi waixiao" (Zheng He's Expeditions to the Western Ocean and the Export of Chinaware). In *Zhenghe xia xiyan lunwenji* (Essays on Zheng He's Voyages), edited by Research Association of Chinese Navigational History, 42–49. Beijing: Peoples' Communication Publishing House.

Lindquist, Cecilia. 1991. *China: Empire of Living Symbols.* Translated by Joan Tate. New York: Addison-Wesley.

Lister, Florence, and Robert H. Lister. 1984. "The Potter's Quarters of Colonial Puebla, Mexico." *Historical Archaeology* 18/1: 87–115.

Little, Stephen. 1982. "Ko-sometsuke in the Asian Art Museum of San Franciso." *O* 13/4: 12–23.

———. 1983. *Chinese Ceramics of the Transitional Period: 1620–1683.* Exhib. cat. New York: China House Gallery, China Institute in America.

———. 1990. "Narrative Themes and Woodblock Prints in the Decoration of Seventeenth-Century Chinese Porcelain." In *Seventeenth-Century Chinese Porcelain from the Butler Family Collection,* edited by Michael Butler, Margaret Medley, and Stephen Little, 21–33. Exhib. cat. Alexandria, VA: Art Services International.

———. 1996. "Economic Change in Seventeenth-Century China and Innovations at the Jingdezhen Kilns." *Ars Orientalis* 26: 47–54.

Liu, James T. C. 1988. *China Turning Inward: Intellectual-Political Changes in the Early Twelfth Century.* Cambridge, MA: Harvard University Press.

Liu, Lydia. 1999. "Robinson Crusoe's Earthenware Pot." *Critical Inquiry* 25/4: 729–57.

Liu Xinru. 1996. *Silk and Religion: An Exploration of Material Life and the Thought of People, A.D. 600–1200.* New Delhi: Oxford University Press.

Liu Xinyuan. 1989. *Imperial Porcelain of the Yongle and Xuande Periods Excavated from the Site of the Ming Imperial Factory at Jingdezhen.* Hong Kong: Urban Council of Hong Kong and the Jingdezhen Museum of Ceramic History.

———. 1993. "Yuan Dynasty Official Wares from Jingdezhen." In Scott 1993: 33–46.

———. 1995. "Amusing the Emperor: The Discovery of Xuande Period Cricket Jars from the Ming Imperial Kilns." *O* 26/8: 62–77.

Lo, Jung-pang. 1955. "The Emergence of China as a Sea Power during the Late Sung and Early Yuan Periods." *Far Eastern Quarterly* 14/4: 489–503.

———. 1970. "Chinese Shipping and East-West Trade from the Tenth to the Fourteenth Century." In *Sociétés et compagnies de commerce en orient et dans l'Océan Indien,* 167–76. Paris: SEVPEN.

Lo, K. S. 1986. *The Stoneware of Yixing.* London: Sotheby's Publications.

Loehr, George. 1962–63. "Missionary Artists at the Manchu Court." *TOCS* 34: 51–67.

Loewe, Michael, and Edward L. Shaughnessy, eds. 1999. *The Cambridge History of Ancient China: From the Origins of Civilization to 221 B.C.* Cambridge: Cambridge University Press.

Lombard, Denys. 1990. *Le Carrefour javanais: Essai d'histoire globale, les réseaux asiatiques.* 3 vols. Paris: Éditions de l'École des hautes études en sciences sociales.

———, and Jean Aubin, eds. 1988. *Marchands et hommes d'affaires asiatiques dan l'Océan Indien et la Mer de Chine 13e–20e siècles.* Paris: Editions de l'Ecole des hautes études en sciences sociales.

———, and Claudine Salmon. 1985. "Islam et sinité." *Archipel* 30: 73–94.

Lombard-Salmon, Claudine. 1973. "A propos de quelques cultes chinois particuliers a Java." *Arts Asiatiques* 26: 243–64.

Long, So Kee. 1994. "The Trade Ceramics Industry in Southern Fukian during the Song." *Journal of Sung-Yuan Studies* 24: 1–19.

Lopez, Robert S. 1971. *The Commercial Revolution of the Middle Ages.* New Haven, CT: Yale University Press.

Loureiro, Rui. 1999. "Portugal em busca da China: Imagens e miragens (1498–1514)." *Ler História* 19: 31–43.

Love, Ronald S. 1994. "The Making of an Oriental Despot: Louis XIV and the Siamese Embassy of 1686." *Journal of the Siam Society* 82/1: 57–78.

Lu, Peter J., and Paul J. Steinhardt. 2007. "Decagonal and Quasi-Crystalline Tilings in Medieval Islamic Architecture." *Science* 315/5815: 1106–10.

Lu Yü. 1974. *The Classic of Tea.* Translated by Francis Ross Carpenter. Boston: Little, Brown.

Ludwig, Theodore M. 1989. "*Chanoyu* and Momoyama: Conflict and Transformation in Rikyu's Art." In Varley and Kumakura 1989: 71–100.

Luo Maodeng. n.d. *San Bao taijian xia xiyang.* Edited by Shen Yunjia. Shanghai: n.p.

Ma Chengyuan. 1980. "The Splendor of Ancient Chinese Bronzes." In Fong 1980: 1–19.

Ma Huan. 1970. *Ying-Yai Sheng-Lan: "The Overall Survey of the Ocean's Shores" [1433]*. Edited by J. V. G. Mills and translated by Feng Ch'eng-chün. Cambridge: Cambridge University Press.

Macfarlane, Alan, and Iris Macfarlane. 2003. *Green Gold: The Empire of Tea*. London: Ebury.

MacGregor, Arthur. 1989. *The Late King's Goods: Collections, Possessions and Patronage of Charles I in the Light of the Commonwealth Sales Inventories*. Oxford: Oxford University Press.

Macintosh, Duncan. 1973. "Beloved Blue and White: An Introduction to the Porcelains of the Yuan and Ming Dynasties." *O* 4/2: 27–41.

———. 2001. "Chinese Blue-and-White and Europe, 1500–1800." *OA* 47/2: 43–48.

Mack, Rosamond E. 2002. *Bazaar to Piazza: Islamic Trade and Italian Art, 1300–1600*. Berkeley: University of California Press.

———. 2004. "Oriental Carpets in Italian Renaissance Paintings: Art Objects and Status Symbols." *MA* 166/6: 82–89.

Magalhães-Godinho, Vitorino. 1969. *L'Économie de l'Empire portugais aux XVe et XVIe siècles*. Paris: SEVPEN.

Mahler, Jane Gaston. 1959. *The Westerners among the Figurines of the T'ang Dynasty of China*. Rome: Instituto Italiano per il Medio ed Estremo Oriente.

Majul, Cesar Adib. 1966. "Islamic and Arab Cultural Influences in the South of the Philippines." *Journal of Southeast Asian History* 7/2: 61–73.

———. 1966a. "Chinese Relationships with the Sultanate of Sulu." In *The Chinese in the Philippines*, 1:143–66. 2 vols. Manila: Solidaridad Publishing House.

Mallet, J. V. G. 1974. "Wedgwood and the Rococo." *Apollo* 99/5: 320–31.

Mancall, Mark. 1984. *China at the Center: 300 Years of Foreign Policy*. New York: Free Press.

Manguin, Pierre-Yves. 1986. "Shipshape Societies: Boat Symbolism and Political Systems in Insular Southeast Asia." In *Southeast Asia in the 9th to 14th Centuries*, edited by D. G. Marr and A. C. Milner, 187–213. Singapore: Institute of Southeast Asian Studies.

———. 1991. "The Merchant and the King: Political Myths of Southeast Asian Coastal Polities." *Indonesia* 52: 41–54.

Mankowitz, Wolf. 1952. *The Portland Vase and the Wedgwood Copies*. London: Andre Deutsch.

Mannikka, Eleanor. 1996. *Angkor Wat: Time, Space, and Kingship*. Honolulu: University of Hawai'i Press.

Manz, Beatrice Forbes. 1989. *The Rise and Rule of Tamerlane*. Cambridge: Cambridge University Press.

Marks, Robert B. 1998. *Tigers, Rice, Silk and Silt: Environment and Economy in Late Imperial South China*. Cambridge: Cambridge University Press.

———. 1999. "Maritime Trade and the Agro-Ecology of South China, 1685–1850." In *Pacific Centuries: Pacific and Pacific Rim History since the Sixteenth Century*, edited by Dennis O. Flynn, Lionel Frost, and A. J. H. Latham, 85–109. London: Routledge.

Martin, Ann Smart. 1989. "The Role of Pewter as Missing Artifact: Consumer Attitudes Toward Tablewares in Late 18th-Century Virginia." *Historical Archaeology* 23/2: 1–27.

———. 1993. "Markets, Buyers, and Users: Consumerism as a Material Culture Framework." *Winterthur Portfolio* 28/2–3: 141–57.

Martin, Jean. 2004–05. "Chinese 16th and 17th Century Porcelain—*Kraak*, 'Swatow,' and 'Transitional' Styles." *TOCS* 69: 49–61.

Masao, F. T., and H. W. Mutoro. 1981. "The East African Coast and the Comoro Islands." In *General History of Africa*, vol. 3: *Africa from the Seventh to the Eleventh Century*, edited by M. Elfasi and I. Hrbek, 586–615. Berkeley: University of California Press.

Mason, R. B., and L. Golombek. 2003. "The Petrography of Iranian Safavid Ceramics." *Journal of Archaeological Science* 30/2: 251–61.

———, and M. S. Tite. 1994. "The Beginnings of Islamic Stonepaste Technology." *A* 36/1: 77–91.

———, and M. S. Tite. 1997. "The Beginning of Tin-Opacification of Pottery Glazes." *A* 39/1: 41–58.

Mathew, Gervase. 1956. "Chinese Porcelain in East Africa and on the Coast of South Arabia." *OA* 2/2: 50–55.

———. 1963. "The East African Coast until the Coming of the Portuguese." In *History of East Africa*, edited by Roland Oliver and Gervase Mathew, 94–127. Oxford: Clarendon Press.

Matson, Frederick R., ed. 1965. *Ceramics and Man*. Chicago: Aldine.

———. 1986. "Glazed Brick from Babylon—Historical Setting and Microprobe Analysis." In Kingery 1986: 133–56.

———. 1989. "Ceramics: The Hub of Ancient Craft Interplay." In *Cross-Craft and Cross-Cultural Interaction in Ceramics*, edited by Patrick E. McGovern and M. D. Notis, 13–28. Westerville, OH: American Ceramic Society.

Matthee, Rudolph P. 1999. *The Politics of Trade in Safavid Iran: Silk for Silver, 1600–1730*. Cambridge: Cambridge University Press.

McBain, Audrey Y. 1979. "Sukhothai and Swankhalok Ceramics." *Brunei Museum Journal* 4/3: 78–99.

McCants, Anne E. 2008. "Poor Consumers as Global Consumers: The Diffusion of Tea and Coffee Drinking in the Eighteenth Century." *Economic History Review* 61/supplement 1: 172–200.

McCracken, Grant. 1988. *Culture and Consumption: New Approaches to the Symbolic Character of Consumer Goods and Activities*. Bloomington: Indiana University Press.

McCullough, Helen. 1985. *Brocade by Night: 'Kokin Wakashu' and the Court Style in Japanese Classical Poetry*. Stanford, CA: Stanford University Press.

———. 1990. "Aristocratic Culture." In *The Cambridge History of Japan*, vol. 2: *Heian Japan*, edited by Donald H. Shively and William H. McCullough, 390–448. Cambridge: Cambridge University Press.

McElney, Brian. 1998–2000. "Chinese Tea Drinking Customs and Changes over the Centuries." *BOCSHK* 12: 37–42.

McEwan, Bonnie G. 1992. "The Role of Ceramics in Spain and Spanish America during the 16th Century." *Historical Archaeology* 26/1: 92–108.

McKendrick, Neil. 1960. "Josiah Wedgwood: An Eighteenth-Century Entrepreneur in Salesmanship and Marketing Techniques." *Economic History Review* 12/3: 408–33.

———. 1961. "Josiah Wedgwood and Factory Discipline." *Historical Journal* 4/1: 30–55.

———. 1970. "Josiah Wedgwood and Cost Accounting in the Industrial Revolution." *Economic History Review* 23/1: 45–67.

———. 1973. "The Role of Science in the Industrial Revolution: A Study of Josiah Wedgwood as a Scientist and Industrial Chemist." In *Changing Perspectives in the History of Science: Essays in Honour of Joseph Needham,* edited by Mikulás Teich and Robert Young, 274–319. Boston: D. Reidel.

———. 1982. "Josiah Wedgwood and the Commercialization of the Potteries." In McKendrick et al. 1982: 100–145.

———, John Brewer, and J. H. Plumb. 1982. *The Birth of a Consumer Society: The Commercialization of Eighteenth-Century England.* Bloomington: Indiana University Press.

McKillop, Beth. 1992. *Korean Art and Design.* Cambridge: Victoria and Albert Museum.

McLeod, Bet. 1998. "Horace Walpole and Sèvres Porcelain: The Collection at Strawberry Hill." *Apollo* 140/1: 42–47.

McNeill, J. R., and William H. McNeill. 2003. *The Human Web: A Bird's-Eye View of World History.* New York: Norton.

McNeill, William H. 1963. *The Rise of the West: A History of the Human Community.* Chicago: University of Chicago Press.

———. 1982. *The Pursuit of Power: Technology, Armed Force, and Society since* A.D. *1000.* Chicago: University of Chicago Press.

———. 1989. "The Historical Significance of the Way of Tea." In Varley and Kamakura 1989: 255–63.

McQuade, Margaret E. Connors. 2003. "The Emergence of a Mexican Tile Tradition." In Gavin et al. 2003: 204–25.

Medley, Margaret. 1966. "Ching-tê Chên and the Problem of the 'Imperial Kilns.'" *Bulletin of the School of Oriental Studies* 29: 326–38.

———. 1972. "Chinese Ceramics and Islamic Design." In Watson 1972: 79–87.

———. 1974. *Yüan Porcelain and Stoneware.* London: Pitman.

———. 1976. *The Chinese Potter: A Practical History of Chinese Ceramics.* Oxford: Phaidon Press.

———. 1981. *T'ang Pottery and Porcelain.* London: Faber and Faber.

———. 1984. "Islam and Chinese Porcelain in the Fourteenth and Early Fifteenth Centuries." *BOCSHK* 6: 36–47.

———. 1990. "Trade, Craftsmanship, and Decoration." In *Seventeenth-Century Chinese Porcelain from the Butler Family Collection,* edited by Michael Butler, Margaret Medley, and Stephen Little, 11–20. Exhib. cat. Alexandria, VA: Art Services International.

———. 1990–91. "Imperial Patronage and Early Ming Porcelain." *TOCS* 55: 29–42.

Meister, Peter Wilhelm, and Horst Reber. 1983. *European Porcelain of the 18th Century.* Translated by Ewald Osers. Ithaca, NY: Cornell University Press.

Melikan-Chirvani, Assadullah Souren. 1970. "Iranian Silver and Its Influence in T'ang China." In Watson 1970: 12–18.

———. 1976. "Iranian Metal-Work and the Written Word." *Apollo* 103/4: 286–91.

———. 1986. "Silver in Islamic Iran: The Evidence from Literature and Epigraphy." In Vickers 1986: 89–106.

Mellott, Richard L. 1990. "Ceramics of the Asuka, Nara, and Heian Periods (A.D. 552–1185)." In *The Rise of a Great Tradition: Japanese Archaeological Ceramics from the Jomon through Heian Periods (10,500* B.C.*–*A.D. *1185),* 56–66. New York: Japan Society.

Mennell, Stephen. 1985. *All Manners of Food: Eating and Taste in England and France from the Middle Ages to the Present.* Oxford: Blackwell.

Mercier, Louis-Sébastien. 1929. *The Picture of Paris.* Translated by Wilfred Jackson and Emilie Jackson. London: Routledge and Sons.

———. 1999. *Panorama of Paris: Selections from Tableau de Paris.* Translated by Helen Simpson and Jeremy D. Popkin and edited by Jeremy D. Popkin. University Park: Pennsylvania State University Press.

Meserve, Ruth I. 1982. "The Inhospitable Land of the Barbarian." *Journal of Asian History* 16/1: 51–89.

Metzger, Thomas. 1970. "The State and Commerce in Imperial China." *Asian and African Studies* 6: 23–46.

Michell, George, and Mark Zebrowski. 1999. *The New Cambridge History of India,* vol. 1:7: *Architecture and Art of the Deccan Sultanate.* Cambridge: Cambridge University Press.

Middleton, John. 1992. *The World of the Swahili: An African Mercantile Civilization.* New Haven, CT: Yale University Press.

———. 2003. "Merchants: An Essay in Historical Ethnography." *Journal of the Royal Anthropological Institute* 9/3: 509–26.

Mikami, Tsugio. 1980–81. "China and Egypt: Fustat." *TOCS* 45: 67–89

Miller, Daniel. 1985. *Artefacts as Categories: A Study of Ceramic Variability in Central India.* Cambridge: Cambridge University Press.

Miller, Susan. 2001. "Images of Asia in French Luxury Goods: Jean-Antoine Fraisse at Chantilly, c. 1729–36." *Apollo* 154/11: 3–12.

Mills, Barbara J., and T. J. Ferguson. 2008. "Animate Objects: Shell Trumpets and Ritual Networks in the Greater Southwest." *Journal of Archaeological Method and Theory* 15/4: 338–61.

Mino, Yutaka. 1991. "Koryo and Choson Dynasty Ceramics." In *The Radiance of Jade and the Clarity of Water: Korean Ceramics from the Ataka Collection,* 27–34. Exhib. cat. New York: Hudson Hills Press.

———, and Katherine R. Tsiang. 1986. *Ice and Green Clouds: Traditions of Chinese Celadon.* Exhib. cat. Indianapolis: Indiana University Press.

Mitchell, Steven. 2004. *Gilgamesh: A New English Version.* New York: Free Press.

Montaigne, Michel de. 1993. *The Complete Essays.* Translated and edited by M. A. Screech. London: Penguin.

Montanari, Massimo. 1994. *The Culture of Food.* Translated by Carl Ipsen. Oxford: Blackwell.

Montesquieu, Baron de la Bréde et de. 1989. *The Spirit of the Laws.* Translated and edited by Anne M. Cohler, Basia Carolyn Miller, and Harold Samuel Stone. Cambridge: Cambridge University Press.

Montgomery, James M., trans. 1966. *The History of Yaballaha III, Nestorian Patriarch, and of His Vicar, Bar Sauma, Mongol Ambassador to the Frankish Courts at the end of the Thirteenth Century.* New York: Octogon Books.

Montias, John Michael. 1982. *Artists and Artisans in Delft: A Socio-Economic Study of the Seventeenth Century.* Princeton, NJ: Princeton University Press.

Moore, A. M. T. 1995. "The Inception of Potting in Western Asia and Its Impact on Economy and Society." In *The Emergence of Pottery: Technology and Innovation in Ancient Societies,* edited by William K. Barnett and John W. Hoopes, 39–53. Washington, DC: Smithsonian Institution Press.

Moreland, W. H., ed. 1931. *Relations of Golconda in the Early Seventeenth Century.* London: Hakluyt Society.

———, ed. 1934. *Peter Floris: His Voyage to the East Indies in the Globe, 1611–1615.* London: Hakluyt Society.

Morga, Antonio de. 1971. *Sucesos de las Islas filipinas.* Translated and edited by J. S. Cummins. Cambridge: Hakluyt Society.

Morris, Ivan, trans. and ed. 1971. *The Pillow Book of Sei Shonagon.* New York: Penguin Books.

Mosco, Marilena. 1999. "The Shells of the Medici: The Nautiluses of the Pitti Palace in Florence." *FMR* 98 (1999): 37–68.

Mote, Frederick W. 1999. *Imperial China, 900–1800.* Cambridge, MA: Harvard University Press.

———, and Denis Twitchett, eds. 1988. *The Cambridge History of China,* vol. 7:1: *The Ming Dynasty, 1368–1644.* Cambridge: Cambridge University Press.

Moura Sobral, Luís de. 2007. "The Expansion and the Arts: Transfers, Contaminations, Innovations." In *Portuguese Oceanic Expansion, 1400–1800,* edited by Francisco Bethencourt and Diogo Ramada Curto, 390–459. Cambridge: Cambridge University Press.

Mudge, Jean McClure. 1985. "Hispanic Blue-and-White Faience in the Chinese Style." In Carswell 1985a: 43–54.

Muldrew, Craig. 2001. "'Hard Food for Midas': Cash and Its Social Value in Early Modern England." *Past and Present* 170/1: 78–120.

Mungello, David E. 1977. *Leibniz and Confucianism: The Search for Accord.* Honolulu: University of Hawai'i Press.

———, ed. 1994. *The Chinese Rites Controversy: Its History and Meaning.* Aldershot: Ashgate.

Murai, Yasuhiko. 1989. "The Development of *Chanoyu:* Before Rikyu." In Varley and Kamakura 1989: 3–32.

Murasaki, Shikibu. 1987. *The Tale of Genji.* Translated by Edward G. Seidensticker. New York: Alfred A. Knopf.

Murray, Dian H. 1987. *Pirates of the South China Coast, 1790–1810.* Stanford, CA: Stanford University Press.

Murray, Julia K. 1992. "The Hangzhou *Portraits of Confucius and Seventy-Two Disciples* (*Sheng xian tu*): Art in the Service of Politics." *AB* 74/1: 7–18.

———. 1999. "Patterns of Evolution in Chinese Illustration: Expansion or Epitomization?" In *Arts of the Sung and Yüan: Ritual, Ethnicity, and Style in Painting,* edited by Cary Y. Liu and Dora C. Y. Ching, 121–51. Princeton, NJ: Art Museum, Princeton University.

Murris, R. 1925. *La Hollande et les Hollandais au XVIIe et au XVIIIe siècles vus par les Français.* Paris: Honoré Champion.

Myers, Ramon H., and Yeh-chien Wang. 2002. "Economic Developments, 1644–1800." In *The Cambridge History of China,* vol. 9:1: *The Ch'ing Empire to 1800,* edited by Denis Twitchett and John K. Fairbank, 563–645. Cambridge: Cambridge University Press.

Myrtle, J. H. 1960. "Notes on Certain Technical Defects in 14th and 15th Century Blue-and-White Chinese Porcelains." *Far Eastern Ceramic Bulletin* 12: 39–42.

Nakamura, Shin'ichi. 2005. "Le riz, le jade et la ville: Évolution des sociétiés néolithique du Yangzi." *Annales* 60/5: 1009–34.

Navarette, Domingo. 1960. *The Travels and Controversies of Friar Domingo Navarette.* 2 vols. Translated and edited by J. S. Cummins. London: Hakluyt Society.

Necipoglu, Gülru. 1990. "From International Timurid to Ottoman: A Change of Taste in Sixteenth-Century Ottoman Tiles." *Muqarnas* 7: 136–70.

Needham, Joseph. 1964. *The Development of Iron and Steel Technology in China.* Cambridge: W. Heffer and Sons.

———. 1971. *SCC,* vol. 4, pt. 3: *Civil Engineering and Nautics.*

———. 1973. "The Role of Science in the Industrial Revolution: A Study of Josiah Wedgwood as a Scientist and Industrial Chemist." In *Changing Perspectives in the History of Science: Essays in Honour of Joseph Needham,* edited by Mikulás Teich and Robert Young, 274–319. Boston: D. Reidel.

———. 2000. *SCC,* vol. 6, pt. 6: *Biology and Biological Technology: Medicine.* Edited by Nathan Sivin.

Neill, Mary Gardner. 1985. "The Flowering Plum in the Decorative Arts." In *Bones of Jade, Soul of Ice: The Flowering Plum in the Decorative Arts,* edited by Maggie Bickford, 193–244. New Haven, CT: Yale University Art Gallery.

Nelson, Sarah Milledge. 1993. *The Archaeology of Korea.* Cambridge: Cambridge University Press.

Newitt, Malyn. 2005. *A History of Portuguese Overseas Expansion, 1400–1688.* New York: Routledge.

Ng Chin-keong. 1997. "Maritime Frontiers, Territorial Expansion, and *Hai-fang* during the Late Ming and High Ch'ing." In *China and Her Neighbors: Borders, Visions of the Other, Foreign Policy, 10th to 19th Century,* edited by Sabine Dabringhaus and Roderich Ptak, 211–57. Wiesbaden: Harrassowitz.

Nguyen-Long, Kerry. 1999. "Vietnamese Ceramic Trade to the Philippines in the Seventeenth Century." *Journal of Southeast Asian Studies* 30/1: 1–21.

———. 2001. "Treasures from the Hoi An Hoard. Vietnamese Ceramics and the Cham Island Site: Significance and Implications." *Arts of Asia* 31/1: 90–97.

Nicholson, Colin. 1994. *Writing and the Rise of Finance: Capital Satires of the Early Eighteenth Century.* Cambridge: Cambridge University Press.

Nickles, Estelle. 2002. "Further Reading on *Qingbai* Ware: Some Bibliographical Notes." In Pierson 2002: 234–40.

Nieuwenhuis, F. J. Domela. 1986. "Terracotta Art of Majapahit." *OA* 32/1: 67–78.

Ni Yibin. 2003–4. "The Anatomy of Rebus in Chinese Decorative Art." *OA* 49/3: 12–23.

Ökte, Ertugrul Zekâi, ed. 1988. *Kitabi-I Bayriye Pirî Reis.* Translated by Robert Bragner. Istanbul: Historical Research Foundation, Istanbul Research Center.

Oldroyd, David R. 1996. *Thinking about the Earth: A History of Ideas in Geology.* Cambridge, MA: Harvard University Press.

Olschki, Leonardo. 1944. "Asiatic Exoticism in Italian Art of the Early Renaissance." *AB* 26/2: 95–106.

Origen. 1998. *Homilies on Jeremiah.* Translated by John Clark Smith. Washington, DC: Catholic University of America Press.

Ortiz, Valérie Malenfer. 1999. *Dreaming the Southern Song Landscape: The Power of Illusion in Chinese Painting.* Leiden: Brill.

Ottomeyer, Hans. 2004. "The Metamorphosis of the Neoclassical Vase." In *Vasemania—Neoclassical Form and Ornament in Europe: Selections from the Metropolitan Museum of Art,* edited by Stefanie Walker, 15–29. Exhib. cat. New Haven, CT: Yale University Press.

Padfield, Peter. 2000. *Maritime Supremacy & the Opening of the Western Mind.* New York: Overlook Press.

Pagani, Catherine. 1995. "Clockmaking in China under the Kangxi and Qianlong Emperors." *Arts Asiatiques* 50: 76–84.

Pal, Pratapaditya. 1983. *Court Paintings of India, 16th–19th Centuries.* New York: Navin Kumar.

Palais, James B. 1995. "A Search for Korea's Uniqueness." *HJAS* 55/2: 409–25.

Palliser, D. M. 1976. *The Staffordshire Landscape.* London: Hodder and Stoughton.

Parker, Geoffrey. 1998. *The Grand Strategy of Philip II.* New Haven, CT: Yale University Press.

Parmentier, Jan. 1996. *Tea Time in Flanders: The Maritime Trade between the Southern Netherlands and China in the 18th Century.* Exhib. cat. Bruges: Ludion Press.

Parry, J. H. 1974. *The Discovery of the Sea.* Berkeley: University of California Press.

Paston-Williams, Sara. 1993. *The Art of Dining: A History of Cooking and Eating.* London: National Trust.

Patterson, Jerry. 1979. *Porcelain.* Washington, DC: Smithsonian Institution.

Payne, William L., ed. 1951. *The Best of Defoe's Review.* New York: Columbia University Press.

Pearce, Nick. 2003. "Images of Guanxiu's Sixteen Luohan in Eighteenth-Century China." *Apollo* 158/8: 25–31.

Pearce, N. J. 1987–88. "Chinese Export Porcelain for the European Market: The Years of Decline, 1770–1820." *TOCS* 52: 21–38.

———. 2001. *Designs as Signs: Decoration and Chinese Ceramics.* Pearson, Richard, and Li Min. 2002. "Quanzhou Archaeology: A Brief Review." *International Journal of Historical Archaeology* 6/1: 23–59.

Pegolotti, Francesco Balducci. 1936. *La Practica della Mercatura.* Edited by Allan Evans. Cambridge, MA: Medieval Academy of America, no. 24.

Pelliot, Paul. 1930. "Des artisans Chinois à la capitale Abbasid en 751–762." *T'oung Pao* 26: 110–12.

Peng Xinwei. 1994. *A Monetary History of China.* 2 vols. Translated by Edward H. Kaplan. Bellingham: Center for East Asian Studies, Western Washington University.

Pepys, Samuel. 1970–83. *The Diary of Samuel Pepys.* Edited by Robert Latham and William Matthews. 11 vols. Berkeley: University of California Press.

Perdue, Peter C. 2005. *China Marches West: The Qing Conquest of Central Asia.* Cambridge, MA: Harvard University Press.

Peterson, C. A. 1979. "Court and Province in Mid and Late T'ang." In *The Cambridge History of China,* vol. 3:1: *Sui and T'ang China, 589–906,* edited by Denis Twitchett, 464–560. Cambridge: Cambridge University Press.

Peterson, T. Sarah. 1994. *Acquired Taste: The French Origins of Modern Cooking.* Ithaca, NY: Cornell University Press, 1994.

Petroski, Henry. 1992. *The Evolution of Useful Things.* New York: Random House.

Petrushevsky, I. P. 1968. "The Socio-Economic Conditions of Iran under the Il-Khans." In Boyle 1968: 483–537.

Phelan, John Leddy. 1959. *The Hispanization of the Philippines: Spanish Aims and Filipino Responses, 1565–1700.* Madison: University of Wisconsin Press.

Piccolpasso, Cipriano. 1980. *The Three Books of the Potter's Art.* 2 vols. Translated and edited by Ronald Lightbown and Alan Caiger-Smith. London: Scolar Press.

Pierce, Donna. 2003. "Mayólica in the Daily Life of Colonial Mexico." In Gavin et al. 2003: 245–69.

Pierson, Stacey. 1996. *Earth, Fire and Water: Chinese Ceramic Technology.* London: Percival David Foundation of Chinese Art.

———. 2001 *Designs as Signs: Decoration and Chinese Ceramics.* London: Percival David Foundation of Chinese Art and School of Oriental and African Studies, University of London.

———, ed. 2002. *Qingbai Ware: Chinese Porcelain of the Song and Yuan Dynasties.* London: Percival David Foundation of Chinese Art.

———. 2002–3. "Archaeology and Chinese Ceramics: Export Wares." *OA* 48/5: 33–35.

———, and Amy Barnes. 2002. *A Collector's Vision: Ceramics of the Qianlong Emperor.* London: Percival David Foundation of Chinese Art and School of Oriental and African Studies, University of London.

Pietsch, Ulrich. 2004. "The Royal Porcelain Collection." In Siebel 2004: 178–181.

Pigafetta, Antonio. 1969. *The Voyage of Magellan: The Journal of Antonio Pigafetta.* Translated by Paula Spurlin Paige. Englewood Cliffs, NJ: Prentice-Hall.

Pinksy, Robert. 1994. *The Inferno of Dante: A New Verse Translation.* New York: Farrar, Straus and Giroux.

Pinto de Matos, Maria Antónia. 1999. "The Portuguese Trade." *OA* 45/1: 22–29.

———. 2002–3. "Chinese Porcelain in Portuguese Written Sources." *OA* 48/5: 36–40.

Pinto, Fernão Mendes. 1989. *The Travels of Mendes Pinto.* Translated and edited by Rebecca D. Catz. Chicago: University of Chicago Press.

Piotrovsky, M. B., and J. M. Rogers. 2004. *Heaven on Earth: Art from Islamic Lands.* Munich: Prestal Verlag.

Pirazzoli-t'Serstevens, Michèle. 2002. "From the Ear-Cup to the Round-Cup: Changes in Chinese Drinking Vessels (2nd to 6th Century A.D.)." *OA* 48/3: 17–27.

Pires, Tomé. 1944. *The Suma Oriental of Tomé Pires.* Translated and edited by Armando Cortesão. London: Hakluyt Society.

Pitte, Jean-Robert. 2002. *French Gastronomy: The History and Geography of a Passion.* Translated by Jody Gladding. New York: Columbia University Press.

Pleguezuelo, Alfonso. 2002. "Jan Floris (ca. 1520–1567), a Flemish Tile Maker in Spain." In *Majolica and Glass, From Italy to Antwerp and Beyond: The Transfer of Technology in the 16th–Early 17th century,* edited by John Veekman and Sarah Jennings, 123–44. Antwerp: Stad Antwerpen.

Plinval de Guillebon, Régine de. 1999. "The Manufacture and Sale of Soft-Paste Porcelain in Paris in the Eighteenth Century." In *Discovering the Secrets of Soft-Paste Porcelain at the Saint-Cloud Manufactory, ca. 1690–1766,* edited by Bertrand Rondot, 83–96. Exhib. cat. New Haven, CT: Yale University Press and Bard Graduate Center for the Studies in the Decorative Arts, New York.

Plumb, J. H. 1972. "The Royal Porcelain Craze." In *In the Light of History,* 57–68. London: Allen Lane.

Plutschow, Herbert. 2003. *Rediscovering Rikyu and the Beginnings of the Japanese Tea Ceremony.* Kent: Global Oriental.

Pocock, J. G. A. 1999. *Barbarism and Religion,* vol. 2: *Narratives of Civil Government.* Cambridge: Cambridge University Press.

Pocock, Tom. 1998. *Battle for Empire: The Very First World War, 1756–63.* London: Michael O'Mara.

Pollard, Sidney. 1965. *The Genesis of Modern Management: A Study of the Industrial Revolution in Great Britain.* Cambridge, MA: Harvard University Press.

Polyani, Michael. 1958. *Personal Knowledge: Towards a Post-Critical Philosophy.* New York: Harper and Row.

Pomerantz, Kenneth. 2000. *The Great Divergence: China, Europe, and the Making of the Modern World Economy.* Princeton, NJ: Princeton University Press.

Pope, John A. 1956. *Chinese Porcelains from the Ardebil Shrine.* Washington, DC: Freer Gallery of Art, Smithsonian Institution.

Portal, Jane. 1997. "Korean Celadons of the Koryo Dynasty." In Freestone and Gaimster 1997: 98–103.

Porter, David. 1999. "Chinoiserie and the Aesthetics of Illegitimacy." *Studies in Eighteenth-Century Culture* 28: 27–54.

———. 1999–2000. "A Peculiar But Uninteresting Nation: China and the Discourse of Commerce in Eighteenth-Century England." *ECS* 33/2: 181–99.

———. 2001. *Ideographia: The Chinese Cipher in Early Modern Europe.* Stanford, CA: Stanford University Press.

———. 2002. "Monstrous Beauty: Eighteenth-Century Fashion and the Aesthetics of Chinese Taste." *ECS* 35/3: 395–411.

Postelthwayt, Malachy. 1774. *The Universal Dictionary of Trade and Commerce.* 2 vols. 4th ed. Reprint, New York: Augustus M. Kelley.

Potts, D. T. 1997. *Mesopotamian Civilization: The Material Foundations.* Ithaca, NY: Cornell University Press.

Preston, Diana, and Michael Preston. 2007. *Taj Mahal: Passion and Genius at the Heart of the Mughal Empire.* New York: Wallace & Company.

Ptak, Roderich. 1986. *Cheng Hos Abenteuer im Drama und Roman der Ming-Zeit.* Stuttgart: Franz Verlag Wiesbaden.

———. 2001. "Quanzhou: At the Northern Edge of a Southeast Asian Mediterranean." In Schottenhammer 2001: 395–427.

Raby, Julian, and Ünsal Yücel. 1986. "Chinese Porcelain at the Ottoman Court." In *Chinese Ceramics in the Topkapi Saray Museum, Istanbul,* edited by John Ayers, catalogue by Regina Krahl, 1:27–54. 3 vols. London: Sotheby's Publications.

Raeburn, Michael. 1989. *The Green Frog Service: A Portrait of 18th-Century England on Wedgwood's Imperial Service.* London: Barrie and Jenkins.

Raffo, Pietro. 1982. "The Development of European Porcelain." In Atterbury 1982: 79–125.

Ramage, Nancy H. 1990. "Sir William Hamilton as Collector, Exporter, and Dealer: The Acquisition and Dispersal of His Collections." *American Journal of Archaeology* 96/3: 469–80.

Raphael, Oscar C. 1931–32. "Chinese Porcelain Jar in the Treasury of San Marco, Venice." *TOCS* 10: 13–15.

Rapp, Richard T. 1975. "The Unmaking of the Mediterranean Trade Hegemony: International Trade Rivalry and the Commercial Revolution." *Journal of Economic History* 35/3: 499–525.

Raven, James. 1992. *Judging New Wealth: Popular Publishing and Responses to Commerce in England, 1750–1800.* Oxford: Clarendon Press.

Rawson, Jessica. 1984. *Chinese Ornament: The Lotus and the Dragon.* London: British Museum.

———. 1986. "Tombs or Hoards: The Survival of Chinese Silver of the Tang and Song Periods, Seventh to Thirteenth Centuries A.D." In Vickers 1986: 31–56.

———. 1989. "Chinese Silver and Its Influence on Porcelain Development." In *Cross- Craft and Cross-Cultural Interaction in Ceramics,* edited by Patrick E. McGovern and M. D. Notis, 275–99. Westerville, OH: American Ceramic Society.

———. 1993. "Ancient Chinese ritual bronzes: The Evidence from Tombs and Hoards of the Shang (*c.* 1500–1050 B.C.) and Western Zhou (*c.* 1050–771 B.C.) Periods." *Antiquity* 67/257: 805–23.

———, ed. 1993a. *The British Museum Book of Chinese Art.* New York: Thames and Hudson.

———. 1993b. "Jades and Bronzes for Ritual." In Rawson 1993a: 44–83.

———. 1995. *Chinese Jade from the Neolithic to the Qing.* London: British Museum.

———. 1997. "Overturning Assumptions: Art and Culture in Ancient China." *Apollo* 145/3: 3–9.

———, and Jane Portal. 1993. "Luxuries for Trade." In Rawson 1993a: 256–91.

———, M. Tite, and M. J. Hughes. 1987–88. "The Export of Tang *Sancai* Wares: Some Recent Research." *TOCS* 52: 39–58.

Rawson, Philip. 1984. *Ceramics.* Philadelphia: University of Pennsylvania Press.

Ray, Anthony. 1991. "Sixteenth-Century Pottery in Castile—A Documentary Study." *BM* 134/4: 298–305.

Ray, Haraprasad. 1993. *Trade and Diplomacy in India-China Relations: A Study of Bengal during the Fifteenth Century.* New Delhi: Radiant Publishers.

Redford, Bruce, ed. 1992. *The Letters of Samuel Johnson.* 5 vols. Princeton, NJ: Princeton University Press.

Reid, Anthony. 1992. "The Rise and Fall of Sino-Javanese Shipping." In *Looking in Odd Mirrors: The Java Sea,* edited by V. J. H. Houben, H. M. J. Maier, and W. van der Molen, 177–211. Leiden: Vakgroep Talen en Culturen van Zuidoost-Azië en Oceanië.

———. 1996. "Flows and Seepages in the Long-Term Chinese Interaction with Southeast Asia." In *Soujourners and Settlers: Histories of Southeast Asia and the Chinese in Honour of Jennifer Cushman,* edited by Anthony Reid, 15–49. Sydney: Allen and Unwin.

Reilly, Robin. 1992. *Josiah Wedgwood, 1730–1795.* London: Macmillan.

———. 1994. *Wedgwood Jasper.* New York: Thames and Hudson.

Rhodes, Daniel. 1968. *Kilns: Design, Construction, and Operation.* Radnor, PA: Chilton Book Company.

Ricci, Matteo. 1953. *China in the Sixteenth Century: The Journals of Matthew Ricci, 1583–1610.* Translated and edited by Louis J. Gallagher. New York: Random House.

Rice, Prudence M. 1999. "On the Origins of Pottery." *Journal of Archaeological Method and Theory* 6/1: 1–54.

Richards, Dick. 1995. *South-East Asian Ceramics: Thai, Vietnamese, and Khmer: From the Collection of the Arts Gallery of South Australia.* Kuala Lumpur: Oxford University Press.

Richards, John. 1993. *The New Cambridge History of India,* vol. 1:5: *The Mughal Empire.* Cambridge: Cambridge University Press.

Richards, Sarah. 1999. *Eighteenth-Century Ceramics: Products for a Civilised Society.* Manchester: Manchester University Press.

Riegel, Alois. 1992. *Problems of Style: Foundations for a History of Ornament.* Translated by Evelyn Kain. Princeton, NJ: Princeton University Press.

Risso, Patricia. 1995. *Merchants and Faith: Muslim Commerce and Culture in the Indian Ocean.* Boulder, CO: Westview Press.

Ritter, Hellmut. 2003. *The Ocean of the Soul: Man, the World and God in the Stories of Farid al-Din 'Attar.* Translated by John O'Kane. Leiden: Brill.

Roberts, Gaye Blake. 1990. "Patterns of Trade in the 18th Century." *English Ceramic Circle* 14/1: 93–105.

———. 1997. "'To Astonish the World with Wonders': Josiah Wedgwood I 1730–1795." *English Ceramic Circle* 16/2: 156–74.

Robertson, Maureen. 1983. "Periodization in the Arts and Patterns of Change in Traditional Chinese Literary History." In *Theories of the Arts in China,* edited by Susan Bush and Christian Murck, 3–26. Princeton, NJ: Princeton University Press.

Robinson, B. W. 1967. "Six Centuries of Persian Painting." *OA* 13/3: 173–79.

Robinson, Dwight E. 1973. "The Styling and Transmission of Fashions Historically Considered: Winckelmann, Hamilton and Wedgwood in the 'Greek Revival.'" In *Fashion Marketing,* edited by Gordon Wills and David Midgley, 25–87. London: George Allen & Unwin.

Robinson, John Martin. 1983. *Georgian Model Farms: A Study of the Decorative and Model Farm Buildings in the Age of Improvement, 1700–1846.* Oxford: Clarendon Press.

Rocco, Fiammetta. 2003. *The Miraculous Fever Tree: Malaria and the Quest for a Cure That Changed the World.* New York: HarperCollins.

Roche, Daniel. 1987. *The People of Paris: An Essay in Popular Culture in the 18th* Century. Translated by Marie Evans. Berkeley: University of California Press.

———. 2000. *A History of Everyday Things: The Birth of Consumption in France, 1600–1800.* Translated by Brian Pearce. Cambridge: Cambridge University Press.

Rockhill, W. W. 1914–15. "Notes on the Relations and Trade of China with the Eastern Archipelago and the Coasts of the Indian Ocean during the Fourteenth Century." *T'oung Pao* 15: 419–47; 16: 61–159, 236–71, 374–92, 435–67, 604–26.

Rogers, J. M., and R. M. Ward. 1988. *Süleyman the Magnificent.* Exhib. cat. London: British Museum.

Rogers, Mary Ann. 1990. "In Praise of Errors." *O* 21/9: 62–78.

———. 1992. "The Mechanics of Change: The Creation of a Song Imperial Ceramic Style." In *New Perspectives on the Art of Ceramics in China,* edited by George Kuwayama, 64–79. Honolulu: University of Hawai'i Press.

Röntgen, Robert E. 1984. *The Book of Meissen.* Exton, PA: Schiffer Publishing.

Rooney, Dawn F. 1984. *Khmer Ceramics.* Singapore: Oxford University Press.

———. 1987. *Folk Pottery in South-East Asia.* Singapore: Oxford University Press.

———. 1989. "The Fish Motif in Sukhothai Ceramics." *OA* 35/1: 35–43.

Rosenthal, Franz, trans. 1989. *The History of al-Tabari (Ta'rikh al-rusul wa'l-muluk),* vol. 1: *From the Creation to the Great Flood.* Albany: State University of New York Press.

Rossabi, Morris. 1973. "Cheng Ho and Timur: Any Relation?" *Oriens Extremus* 20: 129–36.

———. 1988. *Khubilai Khan: His Life and Times.* Berkeley: University of California Press.

———. 1992. *Voyager from Xanadu: Rabban Sauma and the First Journey from China to the West.* New York: Kodansha International.

Rougelle, Axelle. 1991. "Les importations de céramiques chinoises dans le Golfe arabo-persique (8–11èmes siècles)." *Archélogie Islamique* 2: 5–46.

———. 1996. "Medieval Trade Networks in the Western Indian Ocean (8–14th Centuries): Some reflections from the Distribution Pattern of Chinese Imports in the Islamic World." In *Tradition and Archaeology: Early Maritime Contacts in the Indian Ocean,* edited by Himanshu Prabha Ray and Jean-François Salles, 159–80. New Delhi: Manohar.

Rousmaniere, Nicole Coolidge. 1996. "Defining Temmoku: Jian Ware Tea Bowls Imported into Japan." In *Hare's Fur, Tortoiseshell, and Patridge Feathers: Chinese Brown- and Black-Glazed Ceramics, 400–1400,* edited by Robert D. Mowry, 43–58. Exhib. cat. Cambridge, MA: Harvard University Art Museums.

Rousseau, Jean-Jacques. 1990. *Julie, or the New Héloïse.* In *The Collected Writings of Rousseau,* vol. 6, translated and edited by Philip Stewart and Jean Vaché. Hanover, NH: University Press of New England.

Rowbotham, Arnold W. 1966. *Missionary and Mandarin: The Jesuits at the Court of China.* New York: Russell and Russell.

Saint-Simon, Louis de Rouvroy, duc de. 1856–58. *Mémoires complets et authentiques du duc de Saint-Simon sur le siècle de Louis XIV et la régence.* 14 vols. Edited by M. Chéruel. Paris: Hachette.

Sáiz, María Concepción García. 2003. "Mexican Ceramics in Spain." In Gavin et al. 2003: 187–203.

Salles-Reese, Verónica. 1997. *From Viracocha to the Virgin of Copacabana: Representation of the Sacred at Lake Titicaca.* Austin: University of Texas Press.

Sandon, John. 1992. "Shell-Decorated Worcester Porcelain." *MA* 141/6: 946–55.

Sasaki, Tatsuo. 1994. "Trade Patterns of Zhejiang Ware Found in West Asia." In *New Light on Chinese Yue and Longquan Wares: Archaeological Ceramics Found in Eastern and Southern Asia,* A.D. *800–1400,* edited by Chuimei Ho, 322–32. Hong Kong: Centre of Asian Studies.

Saunders, Gill. 2002. *Wallpaper in Interior Design.* London: Victoria and Albert Museum.

Savill, Rosalind. 1982. " 'Cameo Fever': Six Pieces from the Sèvres Porcelain Dinner Service made for Catherine II of Russia." *Apollo* 116/11: 304–11.

Savory, Roger. 1980. *Iran under the Safavids.* Cambridge: Cambridge University Press.

Sayer, Geoffrey R., trans. and ed. 1951. *Ching-tê-chên t'ao-lu, or The Potteries of China.* London: Routledge & Kegan Paul.

———, trans. and ed. 1959. *T'ao Ya or Pottery Refinements.* London: Routledge & Kegan Paul.

Scammell, G. V. 1969. "The New Worlds and Europe in the Sixteenth Century." *Historical Journal* 12/3: 389–412.
Schafer, Edward H. 1963. *The Golden Peaches of Samarkand: A Study of T'ang Exotics.* Berkeley: University of California Press.
———. 1967. *The Vermilion Bird: T'ang Images of the South.* Berkeley: University of California Press.
Schama, Simon. 1988. *An Embarrassment of Riches: An Interpretation of Dutch Culture in the Golden Age.* Berkeley: University of California Press.
Schein, Sylvia. 1979. "*Gesta Dei per Mongolos* 1300: The Genesis of a Non-Event." *English Historical Review* 94/373: 805–19.
Scheurleer, D. F. Lunsingh. 1974. *Chinese Export Porcelain / Chine de Commande.* London: Pitman.
Schimmel, Annemarie. 1984. *Calligraphy and Islamic Culture.* New York: New York University Press.
———. 2005. *The Empire of the Great Mughals: History, Art and Culture.* Translated by Corinne Attwood and edited by Burzine K. Waghmar. London: Reaktion Books.
Schneider, Laura T. 1973. "The Freer Canteen." *Ars Orientalis* 9: 137–56.
Scholten, Frits T. 1995. "The Variety of Decoration on Dutch Delft, 1625–1675." *MA* 147/1: 194–203.
Schönfeld, Martin. 1998. "Was There a Western Inventor of Porcelain?" *Technology and Culture* 39/4: 716–27.
Schottenhammer, Angela, ed. 2001. *The Emporium of the World: Maritime Quanzhou, 1000–1400.* Leiden: Brill.
———. 2006. "The Sea as Barrier and Contact Zone: Maritime Space and Sea Routes in Traditional Chinese Books and Maps." In *The Perception of Maritime Space in Traditional Chinese Sources,* edited by Angela Schottenhammer and Roderich Ptak, 3–13. Wiesbaden: Harrassowitz Verlag.
Schouten, Joost. 1671. *A Description of the Mighty Kingdoms of Japan and Siam.* Translated by Roger Manley. London: Robert Boulter.
Schroder, Timothy. 2004. "Lord Burghley's Silver Spice Dishes." *Apollo* 159/2: 3–12.
Schurz, William Lytle. 1939. *The Manila Galleon.* New York: E. P. Dutton.
Schwartz, Selma. 1992. *Sèvres Porcelain Service for Marie Antoinette's Dairy at Rambouillet: An Exercise in Archaeological Neo-Classicism.* Paris: French Porcelain Society.
Scott, Rosemary E. 1987. "18th-Century Overglaze Enamels: the Influence of Technological Development on Painting Style." In *Style in the East Asian Tradition,* edited by Rosemary E. Scott and Graham Hutt, 149–68. London: Percival David Foundation of Chinese Art.
———. 1992. "Archaism and Invention: Sources of Ceramic Design in the Ming and Qing Dynasties." In *New Perspectives on the Art of Ceramics in China,* edited by George Kuwayama, 80–96. Honolulu: University of Hawai'i Press.
———, ed. 1993. *The Porcelains of Jingdezhen.* London: Percival David Foundation of Chinese Art.
———. 1993a. "Jesuit Missionaries and the Porcelains of Jingdezhen." In Scott 1993: 233–56.

———. 2002. "*Qingbai* Porcelain and Its Place in Chinese Ceramic History." In Pierson 2002: 6–12.

Sen, Tansen. 2003. *Buddhism, Diplomacy, and Trade: The Realignment of Sino-Indian Relations, 600–1400*. Honolulu: Association for Asian Studies and University of Hawai'i Press.

Seok Chee, Eng-Lee. 1993. "Passions and Pursuits of the Chinese Scholar." In *Ceramics in Scholarly Taste*, edited by Maura Rinaldi, 30–40. Exhib. cat. Singapore: Southeast Asian Ceramic Society.

Serruys, Henry. 1975. *Trade Relations: The Horse Fairs (1400–1600)*. Brussels: Institut Belge des Hautes Études Chinoises.

Shackel, Paul A. 1993. *Personal Discipline and Material Culture: An Archaeology of Annapolis, Maryland, 1695–1870*. Knoxville: University of Tennessee Press.

Shakeb, M. Z. A. 1995. "Aspects of Golconda–Iran Commercial Contacts." *Islamic Culture* 69/2: 1–39.

Shangraw, Clarence F. 1985. "Fifteenth-Century Blue-and-White Porcelain in the Asian Art Museum of San Francisco." *O* 16/5: 34–46.

Shaw, Thurstan, Paul Sinclair, Bassey Andah, and Alex Okpoko, eds. 1993. *The Archaeology of Africa: Food, Metals and Towns*. London: Routledge.

Sheaf, Colin. 1993. "Chinese Ceramics and Japanese Tea Taste in the Late Ming Period." In Scott 1993: 165–82.

———, and Richard Kilburn. 1988. *The Hatcher Porcelain Cargoes: The Complete Record*. Oxford: Phaidon.

Shelach, Gideon. 2001. "The Dragon Ascends to Heaven, the Dragon Dives into the Abyss: Creation of the Chinese Dragon Symbol." *OA* 47/3: 29–40.

Shiba, Yoshinobu. 1970. *Commerce and Society in Sung China*. Translated by Mark Elvin. Ann Arbor: Center for Chinese Studies, University of Michigan.

Shoemaker, Nancy. 1997. "How Indians Got to Be Red." *American Historical Review* 102/3: 625–44.

Shono, Masako. 1973. *Japanisches Aritaporzellan im sogenannten 'Kakiemonstil' als Vorbild für die Meissener Porzellanmanufaktur*. Translated by Richard Rasch and Rainer Rückert. Munich: Editions Schneider.

Shulsky, Linda Rosenfeld. 1998. "Philip II of Spain as Porcelain Collector." *OA* 44/2: 51–54.

Siebel, Sabine, ed. 2004. *The Glory of Baroque Dresden: The State Art Collections Dresden*. Exhib. cat. Jackson: Mississippi Commission for International Exchange.

Sigurdsson, Haraldur. 1999. *Melting the Earth: The History of Ideas on Volcanic Eruptions*. Oxford: Oxford University Press.

Simkin, C. G. F. 1968. *The Traditional Trade of Asia*. London: Oxford University Press.

Simpson, St. John. 1997. "Early Urban Ceramic Industries in Mesopotamia." In Freestone and Gaimster 1997: 50–55.

Sims, Eleanor. 1992. "The Illustrated Manuscripts of Firdausi's *Shahnama* Commissioned by Princes of the House of Timur." *Ars Orientalis* 22: 43–68.

Skaff, Jonathan Karam. 2003. "The Sogdian Trade Diaspora in East Turkestan during the Seventh and Eighth Centuries." *JESHO* 46/4: 475–524.

Smalley, Ian James. 1968. "The Loess Deposits and Neolithic Culture of Northern China." *Man* 3: 22–41.

Smith, Adam. 1976. *An Inquiry into the Nature and Causes of the Wealth of Nations.* 2 vols. Edited by R. H. Campbell and A. K. Skinner. Oxford: Clarendon Press.

———. 1976a. *Theory of Moral Sentiments.* Oxford: Clarendon Press.

Smith, David Kammerling. 2002. "Structuring Politics in Early Eighteenth-Century France: The Political Innovations of the French Council of Commerce." *Journal of Modern History* 74/3: 490–537.

Smith, John Masson, Jr. 2000. "Dietary Decadence and Dynastic Decline in the Mongol Empire." *Journal of Asian History* 34/1: 35–52.

Smith, Lawrence, Victor Harris, and Timothy Clark. 1990. *Japanese Art: Masterpieces in the British Museum.* New York: Oxford University Press.

Smith, Monica. 1999. "'Indianization' from the Indian Point of View: Trade and Cultural Contacts with Southeast Asia in the Early First Millennium C.E." *JESHO* 42/1: 1–26.

Smith, Robert C. 1968. *The Art of Portugal, 1600–1800.* New York: Meredith Press.

Smollett, Tobias. 1997. *Travels through France and Italy.* Evanston, IL: Marlboro Press and Northwestern University Press.

Snowdin, Michael, and Maurice Howard. 1996. *Ornament: A Social History since 1450.* New Haven, CT: Yale University Press.

So, Billy K. L. 2000. *Prosperity, Region, and Institutions in Maritime China: The South Fukien Pattern, 946–1368.* Cambridge, MA: Harvard University Asia Center.

———. 2006. "Logiques de marché dans le Chine maritime: Espace et institutions dans deux régions préindustrielles." *Annales* 61/6: 1261–88.

So, Jenny F. 1980. "The Waning of the Bronze Age: The Western Han Period (206 B.C.–A.D. 8)." In Fong 1980: 323–27.

Solheim II, Wilhelm G. 1965. "The Functions of Pottery in Southeast Asia: From the Present to the Past." In Matson 1965: 254–73.

Son Chu-hwan, ed. 1994. *Koreana: Korean Cultural Heritage,* vol. 1: *Fine Arts.* Seoul: Korea Foundation.

Soper, Alexander C. 1942. "The Rise of Yamato-e." *AB* 24/4: 351–79.

———. 1976. "The Relationship of Early Chinese Painting to Its Own Past." In *Artists and Traditions: Uses of the Past in Chinese Culture,* edited by Christian F. Murck, 21–47. Princeton, NJ: Princeton University Press.

Souyri, Pierre François. 2001. *The World Turned Upside Down: Medieval Japanese Society.* New York: Columbia University Press.

Souza, George Bryan. 1986. *The Survival of Empire: Portuguese Trade and Society in China and the South China Sea, 1630–1754.* Cambridge: Cambridge University Press.

Spallanzani, Marco. 1978. *Ceramiche orientali a Firenze nel rinascimento.* Florence: Cassa di Risparmio.

Spandounes, Theodore. 1997. *On the Origin of the Ottoman Emperors.* Translated and edited by Donald M. Nicol. Cambridge: Cambridge University Press.

Spence, Jonathan D. 1984. *The Memory Palace of Matteo Ricci.* New York: Penguin Books.

———. 1998. *The Chan's Great Continent: China in Western Minds.* New York: Norton.

Spuhler, F. 1986. "Carpets and Textiles." In Jackson and Lockart 1986: 698–727.

Spuler, Bertold. 1985. *Die Mongolen in Iran: Politik, Verwaltung und Kultur der Ilchanzeit 1220–1350.* Leiden: Brill.

Staehelin, Walter A. 1966. *The Book of Porcelain: The Manufacture, Transport and Sale of Export Porcelain during the Eighteenth Century, Illustrated by a Contemporary Series of Chinese Watercolors.* Translated by Michael Bullock. New York: Macmillan.

Stein, Stanley J., and Barbara H. Stein. 2000. *Silver, Trade, and War: Spain and America in the Making of Early Modern Europe.* Baltimore, MD: Johns Hopkins University Press.

Stern, S. M. 1967. "Ramisht of Siraf, a Merchant Millionaire of the Twelfth Century." *JRASGBI* 1: 10–14.

Stevens, Wallace. 1982. *The Collected Poems of Wallace Stevens.* New York: Vintage Books.

Stevenson, John. 1997a. "The Evolution of Vietnamese Ceramics." In Stevenson and Guy 1997c: 23–45.

———. 1997b. "Ivory-Glazed Wares of Ly and Tran." In Stevenson and Guy 1997c: 111–27.

———, and John Guy, eds. 1997c. *Vietnamese Ceramics: A Separate Tradition.* Chicago: Art Media Resources.

Stols, Eddy. 2004. "The Expansion of the Sugar Market in Western Europe." In *Tropical Babylons: Sugar and the Making of the Atlantic World,* edited by Stuart B. Schwartz, 237–88. Chapel Hill: University of North Carolina Press.

Ströber, Eva. 2001. *'La maladie de porcelaine': Ostasiatisches Porzellan aus der Sammlung Augusts des Starken.* Exhib. cat. Leipzig: Edition Leipzig.

Strong, Roy. 2002. *Feast: A History of Grand Eating.* New York: Hartcourt.

Stuart, Jan. 1993. "Imperial Porcelain and Court Values." *O* 24/8: 24–30.

———. 1993a. "Layers of Meaning." In *Joined Colors: Decoration and Meaning in Chinese Porcelain,* edited by Louise Allison Cort and Jan Stuart, 33–61. Washington, DC: Arthur M. Sackler Gallery and the Smithsonian Institution.

———. 1995. "Unified Style in Chinese Painting and Porcelain in the 18th Century." *OA* 41/2: 32–46.

Styles, John. 2000. "Product Innovation in Early Modern London." *Past and Present* 168/3: 124–69.

Sugimara, Toh. 1986. *The Encounter of Persia with China: Research into Cultural Contacts Based on Fifteenth-Century Persian Pictorial Materials.* Senri Ethnological Studies, no. 18. Osaka: National Museum of Ethnology.

———. 1992. "Chinese Influence on Persian Paintings of the Fourteenth and Fifteenth Centuries." In *Significance of Silk Roads in the History of Human Civilizations,* edited by Umesao Tadao and Sugimara Toh, 135–43. Senri Ethnological Studies, no. 32. Osaka: National Museum of Ethnology.

Sullivan, Michael. 1960–62. "Notes on Chinese Export Wares in Southeast Asia." *TOCS* 33: 61–77.

Sun Laichen. 2003. "Military Technology Transfers from Ming China and the Emergence of Northern Mainland Southeast Asia (*c.* 1390–1527)." *Journal of Southeast Asian Studies* 34/3: 495–517.

Sung Ying-Hsing. 1966. *T'ien-kung K'ai-wu: Chinese Technology in the Seventeenth Century.* Translated by E-tu Zen Sun and Shiou-Chuan Sun. University Park: Pennsylvania State University Press.

Swiderski, Richard M. 1980–81. "Bouvet and Leibniz: A Scholarly Correspondence." *ECS* 14/2:135–50.

Szuppe, Maria. 2004. “Circulation des lettrés et cercles littéraires: Entre Asie centrale, Iran et Inde du Nord (xve[e]–xviiie siècle).” *Annales* 59/5–6: 997–1018.

Takeuchi, Jun'ichi. 2003. “Furata Oribe and the Tea Ceremony.” In *Turning Point: Oribe and the Arts of Sixteenth-Century Japan,* edited by Miyeko Murase, 17–29. Exhib. cat. New York and New Haven, CT: Metropolitan Museum of Art and Yale University Press.

Tana, Li. 2006. “A View from the Sea: Perspectives on the Northern and Central Vietnamese Coast.” *Journal of Southeast Asian Studies* 37/1: 83–102.

Tedlock, Dennis, trans. 1985. *Popol Vuh: The Mayan Book of the Dawn of Life.* New York: Simon and Schuster.

Temple, Robert. 1986. *The Genius of China: 3,000 Years of Science, Discovery, and Invention.* New York: Simon and Schuster.

Teo, Catherine. 2002. “*Qingbai* Ware for Export.” In Pierson 2002: 244–51.

Thackray, John. 1996. “ ‘The Modern Pliny’: William Hamilton and Vesuvius.” In Jenkins and Sloan 1996: 65–74.

Thomas, Elizabeth. 1743. *The Metamorphoses of the Town; or A View of the Present Fashions.* 4th ed. London: J. Wilford.

Thomaz de Bossierre, Yves de. 1982. *François Xavier Dentrecolles et l'apport de la Chine a l'Europe du XVIIIe siécle.* Paris: Les Belles Lettres.

Thornton, Dora. 1997. *The Scholar in His Study: Ownership and Experience in Renaissance Italy.* New Haven, CT: Yale University Press.

Tichane, Robert. 1983. *Ching-te-chen: Views of a Porcelain City.* New York: New York State Institute for Glaze Research.

Tingley, Nancy. 1993. “Thai Ceramics: The James and Elaine Connell Collection.” *O* 24/12: 38–44.

Tite, M. S., I. C. Freestone, and M. Bimson. 1984. “A Technological Study of Chinese Porcelain of the Yuan Dynasty.” *A* 26/2: 139–54.

Torrens, H. S. 2005. “Erasmus Darwin's Contributions to the Geological Sciences.” In *The Genius of Erasmus Darwin,* edited by C. U. M. Smith and Robert Arnott, 259–72. Aldershot: Ashgate.

Totman, Conrad D. 1967. *Politics in the Tokugawa Bakufu, 1600–1843.* Cambridge, MA: Harvard University Press.

Tregear, Mary. 1976. “Chinese Ceramic Imports to Japan between the Ninth and Fourteenth Centuries.” *BM* 118/4: 816–24.

———. 1982. *Song Ceramics.* London: Thames and Hudson.

Treolar, F. E. 1972. “Stoneware Bottles in the Sarawak Museum: Vessels for the Mercury Trade.” *SMJ* 20/40–41: 377–84.

Trocki, Carl A. 1999. *Opium, Empire and the Global Political Economy: A Study of the Asian Opium Trade, 1750–1950.* London: Routledge.

Ts'ai Mei-fen. 1996. “A Discussion of Ting Ware with Unglazed Rims and Related Twelfth-Century Official Porcelain.” In *Arts of the Sung and Yüan Court,* edited by Maxwell K. Hearn and Judith G. Smith, 109–31. New York: Metropolitan Museum of Art.

Tsiang, Katherine R. 1981. “Chinese Jade.” *Arts in Asia* 11/5: 77–83.

Turnbull, Stephen. 2002. *Samurai Invasion: Japan's Korean War, 1592–1598.* London: Cassell.

Turner, Jack. 2004. *Spice: The History of a Temptation.* New York: Alfred A. Knopf.

Turpin, Adriana. 1999. "A Table for Queen Mary's Water Galley at Hampton Court." *Apollo* 149/1: 3–14.

Twitchett, Dennis, and Janice Stargardt. 2002. "Chinese Silver Bullion in a Tenth-Century Indonesian Wreck." *Asia Major* 15/1: 23–72.

Uglow, Jenny, 2002. *The Lunar Men: Five Friends Whose Curiosity Changed the World.* New York: Farrar, Straus and Giroux.

———. 2003. "Vase Mania." In *Luxury in the Eighteenth Century: Debates, Desires and Delectable Goods.* Edited by Maxime Berg and Elizabeth Eger, 151–62. London: Palgrave.

Ukers, William H. 1935. *All About Tea.* 2 vols. New York: Tea and Coffee Trade Journal Company.

Umehara, Kaoru. 1999. "The Context and Spirit of Song Ceramics." In *Song Ceramics,* edited by the Museum of Oriental Ceramics, 16–23. Osaka: Asahi Shimbun.

Ury, Marian. 1988. "Chinese Learning and Intellectual Life." In *The Cambridge History of Japan,* vol. 2: *Heian Japan,* edited by Donald H. Shively and William H. McCullough, 341–89. Cambridge: Cambridge University Press.

Vainker, S. J. 1991. *Chinese Pottery and Porcelain.* London: British Museum.

———. 1993. "Ceramics for Use." In Rawson 1993a: 212–54.

Vandiver, Pamela B. 1990. "Ancient Glazes." *Scientific American* 262/4: 106–13.

———, and W. D. Kingery. 1984. "Variations in the Microstructure and Microcomposition of Pre-Song, Song, and Yuan Dynasty Ceramics." In *Ceramics and Civilization: Ancient Technology to Modern Sciences,* edited by W. D. Kingery, 181–233. Columbus, OH: American Ceramic Society.

Varley, H. Paul. 1977. "Ashikaga Yoshimitsu and the World of Kitayama: Social Change and Shogunal Patronage in Early Muromachi Japan." In *Japan in the Muromachi Age,* edited by John W. Hall and Toyoda Takeshi, 183–204. Berkeley: University of California Press.

———. 1989. "*Chanoyu* from the Genroku Epoch to Modern Times." In Varley and Kumakura 1989: 161–194.

———. 1990. "Cultural Life in Medieval Japan." In Yamamura 1990: 447–99.

———. 1997. "Cultural Life of the Warrior Elite in the Fourteenth Century." In *The Origins of Japan's Medieval World: Courtiers, Clerics, Warriors, and Peasants in the Fourteenth Century,* edited by Jeffrey P. Maas, 192–208. Stanford, CA: Stanford University Press.

———, and George Elison. 1981. "The Culture of Tea: From Its Origins to Sen no Rikyu." In *Warlords, Artists, and Commoners: Japan in the Sixteenth Century,* edited By George Elison and Bardwell L. Smith, 187–222. Honolulu: University Press of Hawai'i.

———, and Kumakura Isao, eds. 1989. *Tea in Japan: Essays on the History of Chanoyu.* Honolulu: University of Hawai'i Press.

Vassalo e Silva, Nuno. 2000. "Precious Stones, Jewels and Cameos: Jacques de Coutre's Journey to Goa and Agra." In Flores and Vassallo e Silva 2004: 116–33.

Vickers, Michael, ed. 1986. *Pots and Pans: Proceedings of the Colloquium on Precious Metal and Ceramics in the Muslim, Chinese and Graeco-Roman Worlds.* Oxford: Oxford University Press.

———. 1987. "Value and Simplicity: Eighteenth-Century Taste and the Study of Greek Vases." *Past and Present* 116/4: 98–137.

———. 1997. "Hamilton, Geology, Stone Vases, and Taste." *Journal of the History of Collections* 9/2: 263–74.

———, and David Gill. 1994. *Artful Crafts: Ancient Greek Silverware and Pottery.* Oxford: Clarendon Press.

Vikan, Gary. 1991. "'Guided by Land and Sea': Pilgrim Art and Pilgrim Travel in Early Byzantium." In *Tesserae: Festshrift für Josef Engemann*, 74–92. Jahrbbuch für Antike und Christentum, Ergänzungsband 18. Münster: Aschendorffsche Verlagsbuchhandlung.

Viraphol, Sarasin. 1977. *Tribute and Profit: Sino-Siamese Trade, 1562–1853.* Cambridge, MA: Harvard University Press.

Visser, Margaret. 1991. *The Rituals of Dinner: The Origins, Evolution, Eccentricities, and Meaning of Table Manners.* New York: Grove Press.

Volker, T. 1954. *Porcelain and the Dutch East India Company as Recorded in the Dagh-Registers of Batavia Castle, Those of Hirada and Deshima and Other Contemporary Papers, 1602–1682.* Leiden: Rijksmuseum voor Volkenkunde.

———. 1959. *The Japanese Porcelain Trade of the Dutch East India Company after 1683.* Leiden: Brill.

Voll, John Obert. 1994. "Islam as a Special World-System." *JWH* 5/2: 213–26.

Voltaire. 1947. *Candide.* Translated by Philip Littell. New York: Modern Library.

Von der Porten, Edward P. 1972. "Drake and Cermeno in California: Sixteenth-Century Chinese Ceramics." *Historical Archaeology* 6/1: 1–22.

Von Glahn, Richard. 1996. *Fountain of Fortune: Money and Monetary Policy in China, 1000–1700.* Berkeley: University of California Press.

Wade, Geoffrey. 2000. "The *Ming shi-lu* as a Source for Thai History: Fourteenth to Seventeenth Centuries." *Journal of Southeast Asian Studies* 31/2: 249–94.

Wahrman, Dror. 2004. *The Making of the Modern Self: Identity and Culture in Eighteenth-Century England.* New Haven, CT: Yale University Press.

Walcha, Otto. 1981. *Meissen Porcelain.* New York: G. P. Putnam.

Waley, Arthur, trans. 1931. *The Travels of an Alchemist: The Journey of the Taoist Ch'ang-ch'un from China to the Hindukush at the Summons of Chingiz Khan.* London: Routledge & Kegan Paul.

Walvin, James. 1997. *Fruits of Empire: Exotic Produce and British Taste, 1660–1800.* New York: New York University Press.

Wan, Maggie C. K. 2003. "Motifs with Intentions: Reading the Eight Trigrams on Official Porcelain of the Jiajing Period (1522–1566)." *Artibus Asiae* 63/2: 191–221.

Wang, Eugene Y. 2005. *Shaping the Lotus Sutra: Buddhist Visual Culture in Medieval China.* Seattle: University of Washington Press.

Wang Gungwu. 1970. "China and South-East Asia, 1402–1424." In *Studies in the Social History of China and South-East Asia: Essays in Memory of Victor Purcell*, edited by Jerome Ch'en and Nicholas Tarling, 375–401. Cambridge: Cambridge University Press.

———. 1998. "Ming Foreign Relations: Southeast Asia." In *The Cambridge History of China*, vol. 8:2: *The Ming Dynasty, 1368–1644*, edited by Denis Twitchett and Frederick W. Mote, 301–32. Cambridge: Cambridge University Press.

———. 2000. *The Chinese Overseas: From Earthbound China to the Quest for Autonomy.* Cambridge, MA: Harvard University Press.

Wang Shixin. 2000. "Capitalism in the Late Ming Dynasty." In *Chinese Capitalism, 1522–1840*, edited by Xu Dixin and Wu Chengming and translated by Li Zhengde, Liang Miaoru, and Li Siping, 23–110. New York: St. Martin's Press.

Wang Yi-t'ung. 1953. *Official Relations between China and Japan, 1368–1549*. Cambridge, MA: Harvard University Press.

Ward, Cheryl. 2001. "The Sadana Island Shipwreck: An Eighteenth-Century A.D. Merchantman off the Red Sea Coast of Egypt." *WA* 32/3: 368–82.

Wästfelt, Berit, Bo Gyllensvärd, and Jörgen Weibull. 1990. *Porcelain from the East Indiaman "Götheburg."* Translated by Jeanne Rosen. Höganäs, Denmark: Wiken.

Watanabe-O'Kelly, Helen. 2002. *Court Culture in Dresden: From Renaissance to Baroque*. New York: Palgrave.

Watney, Bernard. 1973. "Chinoiserie and English Porcelain of the Eighteenth Century: A Fairy-Tale World." In Watson 1972: 94–97.

Watsky, Andrew M. 1995. "Commerce, Politics, and Tea: The Career of Imai Sokyu." *Monumenta Nipponica* 50/1: 47–65.

———. 2004. *Chikubushima: Deploying the Sacred Arts in Momoyama Japan*. Seattle: University of Washington Press.

Watson, Francis, and John Whitehead. 1991. "An Inventory Dated 1689 of the Chinese Porcelain in the Collection of the Grand Dauphin, Son of Louis XIV, at Versailles." *Journal of the History of Collections* 3/1:13–51.

Watson, James L. 1985. "Standardizing the Gods: The Promotion of T'ien Hou ('Empress of Heaven') along the South China Coast, 960–1960." In *Popular Culture in Late Imperial China*, edited by David Johnson, Andrew J. Nathan, and Evelyn S. Rawski, 292–324. Berkeley: University of California Press.

Watson, William, ed. 1970. *Pottery and Metalwork in T'ang China*. London: Percival David Foundation of Chinese Art.

———. 1970a. "On T'ang Soft-Glazed Pottery." In Watson 1970: 41–49.

———, ed. 1972. *The Westward Influence of the Chinese Arts from the 14th to the 18th Century*. London: Percival David Foundation of Chinese Art.

———. 1973. "On Some Categories of Archaism in Chinese Bronze." *Ars Orientalis* 9: 1–14.

———. 1974. *Style in the Arts of China*. Harmondsworth: Penguin Books.

———. 1983. "Iran and China." In *The Cambridge History of Iran*, vol. 3:1: *The Seleucid, Parthian and Sasanian Periods*, edited by Ehsan Yarshater, 537–67. Cambridge: Cambridge University Press.

———. 1984. *Tang and Liao Ceramics*. New York: Rizzoli.

Watt, James C. Y. 2002. "A Note on Artistic Exchange in the Mongol Empire." In *The Legacy of Genghis Khan: Courtly Art and Culture in Western Asia*, edited by Linda Komaroff and Stefano Carboni, 63–73. Exhib. cat. New York: Metropolitan Museum of Art.

Weatherford, Jack. 2004. *Genghis Khan and the Making of the Modern World*. New York: Three Rivers Press.

Weatherill, Lorna. 1986. *The Growth of the Pottery Industry in England, 1660–1815*. New York: Garland.

Wechsler, Howard J. 1985. *Offerings of Jade and Silk: Ritual and Symbol in the Legitimation of the T'ang Dynasty*. New Haven, CT: Yale University Press.

Weeden, Richard P. 1984. *Poison in the Pot: The Legacy of Lead.* Carbondale: Southern Illinois University Press.

Weigl, Gail Capitol. 1980. "The Reception of Chinese Painting Models in Muromachi Japan." *Monumenta Nipponica* 35/3: 257–72.

Weinberg, Saul W. 1965. "Ceramics and the Supernatural: Cult and Burial Evidence in the Aegean World." In Matson 1965: 187–201.

Welch, E. 2002. "Public Magnificence and Private Display: Giovanni Pontano's *De Splendore* (1498) and the Domestic Arts." *Journal of Design History* 15/4: 211–27.

Weller, Dennis P. 2002. *Jan Miense Molenaer: Painter of the Dutch Golden Age.* Exhib. cat. Raleigh: North Carolina Museum of Art.

Wells, Kenneth M. 2000. "Korean Cultural Values in the Choson Dynasty." In *Earth, Spirit, Fire: Korean Masterpieces of the Choson Dynasty (1392–1910),* edited by Claire Roberts and Michael Brand, 16–19. Exhib. cat. Sydney: Powerhouse Publishing.

Wen, R., C. S. Wang, Z. W. Mao, Y. Y. Huang, and A. M. Pollard. 2007. "The Chemical Composition of Blue Pigment on Chinese Blue-and-White Porcelain of the Yuan And Ming Dynasties (A.D. 1271–1644)." *A* 49/1: 101–15.

West, Stephen H. 1997. "Playing with Food: Performance, Food, and the Aesthetics of Artificiality in the Sung and Yuan." *HJAS* 57/1: 67–106.

Westermann, Mariët. 1996. *A Wordly Art: The Dutch Republic, 1585–1718.* New York: Harry N. Abrams.

Whitehead, John. 1993. "George IV: Furnishing in the French Taste: The Place of Dominique Daguerre in the Royal Collection." *Apollo* 138/7: 153–59.

Whitehouse, David. 1983. "Maritime Trade in the Gulf: The 11th and 12th Centuries." *WA* 14/3: 328–34.

Whitfield, Roderich, and Anne Farrer. 1990. *Caves of the Thousand Buddhas: Chinese Art from the Silk Route.* New York: George Braziller.

Whitman, Marina D. 1978. "Persian Blue-and-White Ceramics: Cycles of Chinoiserie." 2 vols. Ph.D. dissertation, New York University.

———. 1988. "The Scholar, the Drinker, and the Ceramic Pot-Painter." In *Content and Context of Visual Arts in the Islamic World,* edited by Priscilla P. Soucek, 255–61. University Park: Pennsylvania State University Press.

Whitmore, John K. 1985. *Vietnam, Hô Quy Ly, and the Ming (1371–1421).* New Haven, CT: Yale Southeast Asia Studies.

Wiener, Philip P., ed. 1951. *Leibniz: Selections.* New York: Charles Scribner's Sons.

Willetts, William. 1958. *Chinese Art.* 2 vols. New York: George Braziller.

Wills, Geofffrey. 1964. *Jade.* London: Acro Publications.

Wills, John E., Jr. 1988. "Tribute, Defensiveness, and Dependency: Uses and Limits of Some Basic Ideas about Mid-Qing Dynasty Foreign Relations." *American Neptune* 48/4: 225–29.

———. 1993. "European Consumption and Asian Production in the Seventeenth and Eighteenth Centuries." In *Consumption and the World of Goods,* edited by John Brewer and Roy Porter, 133–47. London: Routledge.

Wilson, Joan. 1972. "A Phenomenon of Taste: The China Ware of Queen Mary II." *Apollo* 96/8: 116–23.

Wilson, Kathleen. 1995. *The Sense of the People: Politics, Culture and Imperialism in England, 1715–1785.* Cambridge: Cambridge University Press, 1995.

Wilson, Richard L. 1989. "The Tea Ceremony: Art and Etiquette for the Tokugawa Era." In *The Japan of the Shoguns: The Tokugawa Collection,* edited by Denise L. Bissonnette, 63–73. Translated by Judith Terry and Shiro Nada. Exhib. cat. Montreal: Montreal Museum of Fine Arts.

Wilson, Timothy. 1987. *Ceramic Art of the Renaissance.* Exhib. cat. Austin: University of Texas Press.

Wink, André. 1991–2004. *Al-Hind: The Making of the Indo-Islamic World.* 3 vols. Vol. 1 (1991): *Early Medieval India and the Expansion of Islam.* Vol. 2 (1997): *The Slave Kings and the Islamic Conquest.* Vol. 3 (2004): *Indo-Islamic Society, 14th–15th* Centuries. Leiden: Brill.

Winterbottom, Matthew. 2002. " 'Such Massy Pieces of Plate': Silver Furnishings in the English Royal Palaces, 1660–1702." *Apollo* 156/8: 19–26.

Witteveen, Joop. 1991. "Of Sugar and Porcelain: Table Decoration in the Netherlands in the 18th Century." In *Feasting and Fasting: Proceedings of the Oxford Symposium on Food & Cookery,* 212–20. London: Prospect Books.

Woldbye, Vibek. 1984. "Shells and the Decorative Arts." *Apollo* 120/1: 56–61.

Wolters, O. W. 1986. "Restudying Some Chinese Writings on Sriwijaya." *Indonesia* 42: 1–41.

Wong, Anita, ed. 2006. *Tea, Wine, and Poetry: Qing Dynasty Literati and Their Drinking Vessels.* Exhib. cat. Hong Kong: University Art Museum and Art Gallery, University of Hong Kong.

Wong, Grace. 1978. "Chinese Blue-and-White Porcelain and Its Place in the Maritime Trade of China." In *Chinese Blue and White Ceramics,* edited by S. T. Yeo and Jean Martin, 51–92. Singapore: Oxford University Press.

Wong, R. Bin. 2001. "Entre monde et nation: Les regions Braudéliennes en Asie." *Annales* 66/1: 9–16.

———. 2002. "The Search for European Differences and Domination in the Early Modern World: A View from Asia." *AHR* 107/2: 447–69.

Wood, Nigel. 1999. *Chinese Glazes: Their Origins, Chemistry, and Recreation.* London and Philadelphia: A. & C. Black and University of Pennsylvania Press.

———. 1999–2000. "Chinese Glazes." *TOCS* 64: 93–94.

———, and Rose Kerr. 1992. "Graciousness to Wild Austerity: Aesthetic Dimensions of Korean Ceramics Explored through Technology." *O* 23/12: 39–42.

Woodside, Alexander. 2006. *Lost Modernities: China, Vietnam, Korea, and the Hazards of World History.* Cambridge, MA: Harvard University Press.

Woodward, C. S. 1974. *Oriental Ceramics at the Cape of Good Hope, 1652–1795.* Cape Town: A. A. Balkema.

Wriggins, Sally Hovey. 1996. *Xuanzang: A Buddhist Pilgrim on the Silk Road.* Boulder, CO: Westview Press.

Wright, H. T. 1993. "Trade and Politics on the Eastern Littoral of Africa, A.D. 800–1300." In Shaw 1993: 658–72.

Wright, Thomas, ed. 1854. *The Travels of Marco Polo.* Translated by William Marsden. London: Henry G. Bohn.

Wu Hung. 1995. *Monumentality in Early Chinese Art and Architecture.* Stanford, CA: Stanford University Press.

———. 1996. "The Painted Screen." *Critical Inquiry* 23/1: 37–79.

———. 1999. "The Art and Architecture of the Warring States Period." In Loewe and Shaughnessy 1999: 651–744.

Xiong, Victor Cunrui. 2000. *Sui-Tang Chang'an: A Study in the Urban History of Medieval China.* Ann Arbor: University of Michigan, Center for Chinese Studies.

Yamamura, Kozo, ed. 1990. *The Cambridge History of Japan,* vol. 3: *Medieval Japan.* Cambridge: Cambridge University Press.

———. 1990a. "The Growth of Commerce in Medieval Japan." In Yamamura 1990: 344–95.

Yang Yang. 1996. "The Chinese Jade Culture." In *Mysteries of Ancient China: New Discoveries from the Early Dynasties,* edited by Jessica Rawson, 225–31. Exhib. cat. New York: George Braziller.

Yap, C. T., and Younan Hua. 1994. "A Study of Chinese Porcelain Raw Materials from Ding, Xing, Gongxian and Dehua Wares." *A* 36/1: 63–76.

Yonan, Michael E. 2004. "Veneers of Authority: Chinese Lacquers in Maria Theresa's Vienna." *ECS* 37/4: 652–72.

Young, Hilary. 1995. "From the Potteries to St Petersburg: Wedgwood and the Making and Selling of Ceramics." In *The Genius of Wedgwood,* edited by Hilary Young, 9–20. Exhib. cat. London: Victoria and Albert Museum.

———. 1999. *English Porcelain, 1745–95: Its Makers, Design, Marketing and Consumption.* London: Victoria and Albert Museum.

Yü, Chün-fang. 1998. "Ming Buddhism." In Mote and Twitchett 1998: 893–952.

———. 2001. *Kuan-yin: The Chinese Transformation of Avalokitesvara.* New York: Columbia University Press.

Yu Peichin. 2007. "The Imprint of Collecting: The Circulation of Northern Song Ru Ware in the 18th Century Qing Palace." *O* 38/1: 48–55.

Yuan, Tsing. 1978. "The Porcelain Industry at Ching-te-chen, 1500–1700." *Ming Studies* 6: 45–54.

Yuka Kadoi. 2002. "Cloud Patterns: The Exchange of Ideas between China and Iran under the Mongols." *OA* 48/2: 25–36.

Yun Yong-i. 1994. "Satsuma and Arita Pottery." In Son 1994: 126–31.

Zacks, Richard. 2002. *The Pirate Hunter: The True Story of Captain Kidd.* New York: Hyperion.

Zainie, Carla M. 1979. "The Sinan Shipwreck and Early Muromachi Art Collections." *OA* 25/1: 103–14.

Zhang, Z. H., J. G. Liou, and R. G. Coleman. 1984. "An Outline of the Plate Tectonics of China." *Bulletin of the Geological Society of America* 95: 295–312.

Zhang Fukang. 1986. "The Origin of High-Fired Glazes in China." In *Scientific and Technological Insights on Ancient Pottery and Porcelain,* edited by Shanghai Institute of Ceramics, 40–45. Beijing: Science Press.

Zhang Pusheng. 1991. "Notes on Some Recently Discovered Glazed Tiles from the Former Ming Palace in Nanjing." *O* 22/3: 62–63.

Zheng Hesheng and Zheng Yijun, eds. 1980–83. *Zheng He xia xiyang yangzi liao huibian* (Collected Sources on Zheng He's Voyages). 3 vols. Jinan: Qilu Publishing House.

INDEX

TEXT
10/12.5 Minion Pro
DISPLAY
Minion Pro
COMPOSITOR
Integrated Composition Systems

The publisher gratefully acknowledges the generous support
of the Simpson Humanities Endowment Fund of the
University of California Press Foundation.

SIMPSON

IMPRINT IN HUMANITIES

The humanities endowment
by Sharon Hanley Simpson and
Barclay Simpson honors
MURIEL CARTER HANLEY
whose intellect and sensitivity
have enriched the many lives
that she has touched.